Handbook of Research on Advanced Mechatronic Systems and Intelligent Robotics

Maki K. Habib
The American University in Cairo, Egypt

IGI Global
DISSEMINATOR OF KNOWLEDGE

A volume in the Advances in Computational Intelligence and Robotics (ACIR) Book Series

Published in the United States of America by
IGI Global
Engineering Science Reference (an imprint of IGI Global)
701 E. Chocolate Avenue
Hershey PA, USA 17033
Tel: 717-533-8845
Fax: 717-533-8661
E-mail: cust@igi-global.com
Web site: http://www.igi-global.com

Library of Congress Cataloging-in-Publication Data

Names: Habib, Maki K., 1955- editor.
Title: Handbook of research on advanced mechatronic systems and intelligent robotics / Maki K. Habib, editor.
Description: Hershey, PA : Engineering Science Reference, [2020] | Includes bibliographic references.
Identifiers: LCCN 2019016790 | ISBN 9781799801375 (hardcover) | ISBN 9781799801382 (ebook)
Subjects: LCSH: Mechatronics--Research. | Robotics--Research. | Autonomous robots--Research. | Artificial intelligence--Research.
Classification: LCC TJ163.12 .H36 2020 | DDC 621--dc23
LC record available at https://lccn.loc.gov/2019016790

This book is published in the IGI Global book series Advances in Computational Intelligence and Robotics (ACIR) (ISSN: 2327-0411; eISSN: 2327-042X)

British Cataloguing in Publication Data
A Cataloguing in Publication record for this book is available from the British Library.

For electronic access to this publication, please contact: eresources@igi-global.com.

Advances in Computational Intelligence and Robotics (ACIR) Book Series

Ivan Giannoccaro
University of Salento, Italy

ISSN:2327-0411
EISSN:2327-042X

Mission

While intelligence is traditionally a term applied to humans and human cognition, technology has progressed in such a way to allow for the development of intelligent systems able to simulate many human traits. With this new era of simulated and artificial intelligence, much research is needed in order to continue to advance the field and also to evaluate the ethical and societal concerns of the existence of artificial life and machine learning.

The **Advances in Computational Intelligence and Robotics (ACIR) Book Series** encourages scholarly discourse on all topics pertaining to evolutionary computing, artificial life, computational intelligence, machine learning, and robotics. ACIR presents the latest research being conducted on diverse topics in intelligence technologies with the goal of advancing knowledge and applications in this rapidly evolving field.

Coverage

- Computer Vision
- Computational Logic
- Artificial Life
- Fuzzy Systems
- Heuristics
- Intelligent control
- Robotics
- Cognitive Informatics
- Artificial Intelligence
- Cyborgs

IGI Global is currently accepting manuscripts for publication within this series. To submit a proposal for a volume in this series, please contact our Acquisition Editors at Acquisitions@igi-global.com or visit: http://www.igi-global.com/publish/.

The Advances in Computational Intelligence and Robotics (ACIR) Book Series (ISSN 2327-0411) is published by IGI Global, 701 E. Chocolate Avenue, Hershey, PA 17033-1240, USA, www.igi-global.com. This series is composed of titles available for purchase individually; each title is edited to be contextually exclusive from any other title within the series. For pricing and ordering information please visit http://www.igi-global.com/book-series/advances-computational-intelligence-robotics/73674. Postmaster: Send all address changes to above address.

Titles in this Series

For a list of additional titles in this series, please visit: www.igi-global.com/book-series

Edge Computing and Computational Intelligence Paradigms for the IoT
G. Nagarajan (Sathyabama Institute of Science and Technology, India) and R.I. Minu (SRM Institute of Science and Technology, India)
Engineering Science Reference • ©2019 • 347pp • H/C (ISBN: 9781522585558) • US $285.00

Semiotic Perspectives in Evolutionary Psychology, Artificial Intelligence, and the Study of Mind Emerging Research and Opportunities
Marcel Danesi (University of Toronto, Canada)
Information Science Reference • ©2019 • 205pp • H/C (ISBN: 9781522589242) • US $175.00

Handbook of Research on Human-Computer Interfaces and New Modes of Interactivity
Katherine Blashki (Victorian Institute of Technology, Australia) and Pedro Isaías (The University of Queensland, Australia)
Engineering Science Reference • ©2019 • 488pp • H/C (ISBN: 9781522590699) • US $275.00

Machine Learning and Cognitive Science Applications in Cyber Security
Muhammad Salman Khan (University of Manitoba, Canada)
Information Science Reference • ©2019 • 321pp • H/C (ISBN: 9781522581000) • US $235.00

Multi-Criteria Decision-Making Models for Website Evaluation
Kemal Vatansever (Alanya Alaaddin Keykubat University, Turkey) and Yakup Akgül (Alanya Alaaddin Keykubat University, Turkey)
Engineering Science Reference • ©2019 • 254pp • H/C (ISBN: 9781522582380) • US $185.00

Handbook of Research on Deep Learning Innovations and Trends
Aboul Ella Hassanien (Cairo University, Egypt) Ashraf Darwish (Helwan University, Egypt) and Chiranji Lal Chowdhary (VIT University, India)
Engineering Science Reference • ©2019 • 355pp • H/C (ISBN: 9781522578628) • US $295.00

Computational Intelligence in the Internet of Things
Hindriyanto Dwi Purnomo (Satya Wacana Christian University, Indonesia)
Engineering Science Reference • ©2019 • 342pp • H/C (ISBN: 9781522579557) • US $225.00

IGI Global
DISSEMINATOR OF KNOWLEDGE

701 East Chocolate Avenue, Hershey, PA 17033, USA
Tel: 717-533-8845 x100 • Fax: 717-533-8661
E-Mail: cust@igi-global.com • www.igi-global.com

List of Contributors

Table of Contents

Detailed Table of Contents

to help the physician to capture and manage their data during CGA procedures. The robotic platform was tested in some healthcare centers and results show elderly participants accepted the technology. Clarc is a tool that facilitates the work of health professionals given the ease of access to patient data and their evaluations, freeing these professionals to focus their activity on purely clinical tasks that require their knowledge.

This chapter deals with the development of a new active body weight support, based on a parallel robotic structure with 5 degrees of freedom. Initially, a revision of the main structures applied in body weight support for gait rehabilitation is presented, along with a conceptual model of the proposed structure. Therefore, the inverse kinematics and singularity analysis are performed. A dynamic model is then built based on the obtained kinematics, and all parts of the structure are optimized using an evolutional algorithm. Finally, a CAD model of the structure is built based on the optimized dimensions and the model is applied to computational simulations of the gait rehabilitation, being thus ready for prototype construction.

In this chapter, authors focus on mechatronic solutions for low-cost control devices for the homecare of elderly and people with reduced mobility, having the goal of assisting people in daily-life activities. The development of such systems can be exploited in the form of a toolkit to be flexible and applied either to rehabilitation or assistive systems, in order to aid movements with controlled position, force/torque, and acceleration, possibly introducing reliability, repetitiveness, and low cost in the related sector. In particular, the development of a low-cost control system design allows the applicability of the solution to a broad range of devices for the home-care. The authors aim to develop low-cost technologies for the homecare of elderly and people with motor impairments trying to reduce significantly the overall costs to facilitate wider use of assisting and rehabilitation systems. In particular, a Sit-To-Stand (STS) assisting device is considered as a paradigmatic example to illustrate design considerations.

Infrared thermography has been applied in different research projects involving leakages, micro-climate monitoring, energy control, security, and predictive maintenance. The chapter focuses on image and data process techniques improving intelligent inspection systems able to extract hidden information. Image

loads are applied to the end of the robotic arm until significant deformation occurs. The same procedure is done for modified designs in the analysis software. The results considered include displacement and stress. Based on the results, the critical stress areas are near to the rotating joints of the robotic arm, the back of the gripper and the sharp edge of the second arm. Proposed modifications include increasing the servo motor shaft radius and edge filleting the affected area, increasing the thickness and reducing the length of the gripper base plate, and implementing a new design for the second arm.

In this chapter, a two-degree-of-freedom controller that exploits the flat properties of a three degree-of-freedom wing type flat-plate for an Ornithopter Autonomous robot is proposed. A set of kinematical patterns inspired by nature is used to simulate the wing's movement around two wingtip trajectories; also, the effects of the aerodynamical forces as a function of the wind velocity and the wing's angle of attack are considered. In order to control the system, the effects of these forces are viewed as disturbs that affect the wing's dynamics. The proposed control scheme drives the device through the desired path by generating a set of desired inputs which are compensated by a feedback loop when the aerodynamical forces actuate upon the system.

This chapter presents a comparative study of the topographical structure of three common biological robotic inspirations: human, canine, and feline feet. It is shown that the metrological roughness of each of the examined feet is customized for the specific locomotion demands of the species. The textural parameters manifest close correlation to the pressure distribution experienced in movement and gait. This correlation enhances the durability and structural integrity of the bio-analogue. It is also shown that the metrological function of the human (plantigrade) feet pads combine that of the back and the front feet pads of the digitigrade mammals examined. It is argued that integrating the targeted engineering of roughness within the design process of robotic feet can enhance the function of walking robots. Further, it offers elegant solutions to some of the current problems encountered in design of humanoids and other bio-inspired walking robots.

Recently, robot architectures with various structures have been developed to improve the human quality of life. Such a robot needs various capabilities such as learning, inference, and prediction for human interaction, and such capabilities are interconnected with each other as a whole system. In the development of a socially-embedded robot partner, human-robot interaction plays an important role. Therefore, in order to develop a socially embedded robot partner, we must consider human communication system. Human Cognition, Emotion, and Behavior should be considered in the development process of the robot partner, and if these factors are fully reflected in the robot partner, then the robot can be used as a socially-friendly robot partner. This book chapter is organized as follows: First, we describe the hardware and software structures. Next, we discuss the cognitive model of the robot partners. Third, we discuss interaction content design for various services. Finally, we discuss the contents of society implementation, and discuss the applicability of robots for social utilization.

An amusement robot that can recognize and respond to human facial expressions is proposed. The integration of human facial recognition into the control systems of robots is necessary to improve their entertainment value. We are proposing a new technology that uses a camera to read and respond to human facial expressions. As the preliminary research, a portable bubble ejection robot, which produces soap bubbles like a fountain, has been fabricated. The ejected soap bubbles are illuminated by LEDs and the direction of the ejection nozzle and the amount of the ejected bubbles are controlled by using an Android tablet. Subsequently, human facial expressions are read by a facial expression detection system. This paper will explain the design of the bubble ejection robot, its control system, which reads human facial expressions, and the experimental results of human reaction to the bubble ejection. Consequently, the relationship between the soap bubble ejection and the human emotions is revealed.

Robotics has developed into a solid discipline of study that incorporates the background, knowledge, and creativity of mechanical, electronics, electrical, computer, industrial, and manufacturing engineering. The versatility of a robot can be translated into increased productivity, improved product quality, and decreased production costs in numerous ways. The robot can be wired or wireless, and can have a controller device. Radio frequency control system is one of the most efficient controls for a robot. Other than control, image processing is one of the important parts of robotics. Moreover, it also requires hardware for interfacing with the system for radio frequency control. It is one of the key technologies which can be used in defense, safety, intelligent transport system, and efficient management traffic. In recent years, there has been an increased scope for image processing in robotics. The main goal of this work is to show a system that solves the practical problem of identification of real scenes. All the simulations have used MAT LAB software.

The topic of this chapter is the implementation and analysis of vision algorithms for the detection of static and dynamic objects in videos. These algorithms are typically components of the visual perception module of modern driver-assistance systems or even autonomous cars. Whereas the vast majority of today's papers from the vision community use convolutional deep neural networks (CNNs), this chapter explores the more traditional approaches, namely HOG (Histogram of Oriented Gradients) as well as SVM (Support Vector Machine). The static and dynamic objects to be recognized are traffic signs and motorcycles, respectively. These two object classes have been chosen since traffic signs are relatively easy to detect and motorcycles state a much more complex task. Thus, this chapter tackles differently difficult tasks with a single set of algorithms.

This chapter proposes a new trajectory planning approach by improving A* algorithm, which is a widely-used, path-planning algorithm. This algorithm is a heuristic method used in maps such as the occupancy grid map. As the resolution increases in these maps, obstacles can be defined more precisely. However, the cell/grid size must be larger than the size of the mobile robot to prevent the robot from crashing into the borders of the working environment or obstacles. The second constraint of the algorithm is that it does not provide continuous headings. In this study, an avoidance area is calculated on the map for the mobile robot to avoid collisions. Then curve-fitting methods, general polynomial and b-spline, are applied to the path calculated by traditional A* algorithm to obtain smooth rotations and continuous headings by staying faithful to the original path calculated. Performance of the proposed trajectory planning method is compared to others for different target points on the grid map by using a software developed in Labview Environment.

This chapter presents the mechanical design, dynamic model, and walking control law of an insect-like, asymmetric hexapod robot. The proposed model is an original walking mechanism designed with three actuators to provide quasi-omnidirectionality. One of the motivational aims is to reduce the number of actuators preserving similar holonomy as compared to popular 18-servo redundant hexapods with three servos per leg. This work includes the Klann mechanism as limb, two-drive differential robot's

Preface

Mechatronics systems and intelligent robotics represent the successfully evolving interdisciplinary and intelligent engineering science paradigm that brings out innovative and creative possibilities of synergies by fusing different knowledge, technologies and disciplines into a unified framework that enhances design sustainability of products and manufacturing processes. It is the current and future horizons in all fields due to the stimulation of critical and creative thinking, fusing new way of work with effective design process supported by effective interdisciplinary contributions. The adoption of such synergized inter- or trans-disciplinary approach to engineering design implies a greater understanding of the design process.

The synergy in engineering creative design and development enables a higher level of research and development that leads to high quality performance, smart and high functionality, precision, robustness, power efficiency, application flexibility and modularity, improved quality and reliability, enhance adaptability, intelligence, maintainability, better spatial integration of subsystems (embodied systems), miniaturization, embedded lifecycle design, sustainability, and cost effective approaches.

While the technologies are advancing in different direction and the continuous evolution of interdisciplinary development in terms of research, education and product development, there is continuous and growing interest in the field of robotics and mechatronics. This new handbook aims to capture the state-of-art of research developments in the subject area of advanced mechatronic systems and intelligent robotics and to provide the latest practical and theoretical knowledge in the field. This handbook includes 17 chapters, being divided into four groups.

The first group covers Chapters 1-4, which presents biomedical robotics focusing mainly on biomechanics and rehabilitation as an interdisciplinary engineering science. This group of chapters covers topics on: Design and Implementation of a Smart Mechatronic Elbow Brace: Design and Implementation of a Smart Mechatronic Elbow Brace; CLARC, Robotic Platform to Assist the CGA Evaluation in Elderly Patients; Development of a novel parallel structure for gait rehabilitation; and Mechatronic Design of Low-Cost Control Systems for Rehabilitation and Assisting Devices.

The second group includes Chapters 5-8. It introduces research topics related to advancement in robotics and industrial technologies, applications and safety. This group covers topics on: Infrared Thermography for Intelligent Robotic Systems in Research Industry Inspections: Thermography in Industry Processes; Safe Development Environments for Radiation Tracing Robots; Membrane Micro Electro-Mechanical Systems for Industrial Applications; and Analyses on Engineering Mechanics of Robotic Arm for Sorting Multi-Materials.

The third group introduces Chapters 9-10. It focuses on Bioinspiration and innovative development. The topics of this group include: Control of a Wing Type Flat-Plate for an Ornithopter Autonomous Robot with Differential Flatness; and Locomotion Interfaces for Legged Robots- Design Inspiration from Natural Locomotion Interfaces.

The fourth group consists of Chapters 11-17. These chapters discuss different topics from service robots to advanced Mechatronics systems supporting different research development and applications. This group of chapters covers topics on: A Modular Structured Architecture Using Smart Devices for Socially Embedded Robot Partners; Incorporation of Human Facial Expression into Robot Control; Design and Implementation of a Wireless Robot for Image Processing; Detection of Traffic Signs and Road Users from a Moving Vehicle; A Proposed Trajectory Planning Algorithm for Mobile Robot Navigation Based on A* Algorithm; Dynamic Modelling and Control of an Underactuated Quasi-Omnidirectional Hexapod; and Robust and Intelligent Control Techniques for Twin Rotor System

The development and advancement of the field of mechatronics systems and intelligent robotics helps to motivate graduate students, engineers and scientists who are looking to broaden their knowledge and expertise to develop innovative, intelligent, and sustainable ideas, and techniques for autonomous and smart interdisciplinary products and systems to meet today's and future's most pressing challenges.

Maki K. Habib
The American University in Cairo, Egypt

Chapter 1

Design and Implementation of a Smart Mechatronic Elbow Brace

Vu Trieu Minh
Tallinn University of Technology, Estonia

Mart Tamre
Tallinn University of Technology, Estonia

Aleksei Safonov
Tallinn University of Technology, Estonia

Victor Musalimov
ITMO University, Russia

Pavel Kovalenko
ITMO University, Russia

Iurii Monakhov
ITMO University, Russia

ABSTRACT

This project develops a design and implementation of a smart mechatronic brace used as a rehabilitation device for elbow injury patients training their bone ligaments and muscles, and recovering at home. The proposed brace is designed to be low cost and to fit large different users since it is able to be regulated and combined with reliable mechanical and electronic parts, and to correspond to high safety requirements. Electromyography (EMG) sensors are used above the skin to measure the biopotential signals from the muscles to detect the human motion intention, and then, to recognize the flexion/extension movement. Data of the human motion intention and direction are processed and converted into the pulse-width modulation signals (PWM) to run DC motors. A prototype for this brace is fabricated and tested with real human motion intentions and directions. The DC motors can follow well the human motions with the error of less than 7^0 over a moving angle of 120^0.

DOI: 10.4018/978-1-7998-0137-5.ch001

INTRODUCTION

Elbow injuries are prevelant. According to recent statistics, about 80% of injuries are related to the damages of human activity system and half of them are on the upper limbs or the elbow injuries (Smania, Berto, Melotti, & Miditi, 2012). Due to the complexity of the human neuromuscular system, the postoperative recovery may take 12-24 months. And it is important to maintain the stiffness of the joints and avoid motion limitations, muscular contractures, and the development of secondary deformities. Therefore, it is needed mechatronic devices integrated with updated technologies for upper limb rehabilitations.

The main purpose of rehabilitation period is to return normal mobility for the injured joints via physical exercises with trained therapists. In bio-mechanics, many smart devices are emerging to replace expensive training sections with therapists as they allow the patients training at home. This project develops a home-based rehabilitation orthosis for the elbow injured patients with low cost, high accuracy and integration with modern technologies.

The main movements occurring at the elbow joint are flexion, extension, pronation and supination. The range of elbow motion is from 0° to approximately 145° in the normal person. Flexion here is due to the action of biceps, brachialis and brachioradialis muscles. Extension is achieved from the action of the triceps muscle located in the posterior aspect of the arm. The biomechanics of the elbow joint are affected by the bones, muscles and ligaments. Weakness in muscle or injury to ligaments can result in abnormal forces in the elbow, which can ultimately over time cause degeneration of the articular cartilage of bone (Shahid, Fletcher, Robati, & Pemmaraju, 2015).

The passive devices or the mechanical braces containing with meatal and rubber bands are simple, safe and low cost. But after some periods of time, they lose the rehabilitation properties. And the patients must come to the clinic to replace them. Therefore nowadays, there are many active devices for longtime home-based treatment.

Active devices provide active motion assistance via their mechatronic system. Such assistance is supporting patients, who are too weak to perform physical exercises. There is a wide range of such active devices on market and namely mechatronic wearable orthoses.

A review of modern active elbow orthoses with short description and disadvantages is shown in table 1.

The mechatronic systems from these active elbow orthoses are used to deliver power to joint and manipulate movements of the elbow. So, they must be high precious and have enough power to provide adequate torques. There are many types of actuators for these devices, such as electric, hydraulic, pneumatic, pneumatic artificial muscles, series elastic actuator and inverse pneumatic artificial muscles. The review of modern actuators and their transmission system is presented in table 2.

In order to obtain the appropriate human intention signal for the actuation system, sensors and control system are installed based on six sources of (1) Brain activity – EEG, MEG, (2) Muscle activation – EMG, (3) Muscle contraction – MK, MT, (4) Body segment movement – IMU, (5) Joint rotation – Goniometer, and (6) Force/Pressure deformation. The review of sensors and control system is shown in table 3.

According to tables from 1 to 3, all orthoses have disadvantages in mechanical design. Most of the devices have motor placed only on one side which leads to undesired torsion. Further, most of orthoses are built with quite heavy and expensive components. Their weight exceeds 2.5 kg and the cost exceeds $1000.

In order to follow the joint movements, the human intention signal must be measured and translated to PWM to run the motor. EMG sensors and subsequent signal processing subsystems are always the best solution. Based on the EMG signals, the user intention motion can be identified from the estima-

Table 1. Review of recent smart elbow orthoses

Reference	General Information	Disadvantages
("Myomo", 2016)	Successfully commercial powered elbow – wrist – hand orthosis designed to support weak or deformed arm.	Motor is placed only on one side so undesired torsion is possible also I can't find any information about mechanical stoppers
(Pedrocchi, Ferrante, Ambrosini, & Gandolla, 2013)	Powered device for upper limb support provides 2 DOF for shoulder and 1 for elbow. Weight is 0.51 kg and 230V powered.	Not wearable, only with wheelchair
(Ripel, Krejsa, Hrbacek, & Cizmar, 2014)	Powered device with 1 DOF for shoulder with -5 to 150 degree range of motion (ROM) and 94Nm torque	It's not wearable and quite heavy
(Budaker, 2014)	Active elbow orthosis for elbow injury rehabilitation	Motor is placed only on one side so undesired torsion, no mechanical stoppers. No information about length adjusting
(Durandau & Suleiman, 2016)	1 DOF orthosis based on an active compliant actuator. ROM is between 50 to 150 degree	It's not wearable and quite heavy
(Herrnstadt & Menon, 2016)	1 DOF powered tremor suppression orthosis with 0-120 degree ROM and weight is about 2kg	Heavy construction for wearable devices, no mechanical stoppers and may slide off the arm
(Looned, Webb, Xiao, & Menon, 2014)	1 DOF assisting drinking elbow orthosis with ROM limited at 110 degree	Heavy construction, motor is placed only on one side so undesired torsion
(Mohammadi, Zohoor, & Khadem, 2014)	Active Assistive Exoskeletal Robot for Rehabilitation of Elbow and Wrist with 3 DOF and 0 – 150 ROM	Motor is placed only on one side so undesired torsion
(Desplenter, Lobo-Prat, Stienen, & Trejor, 2016)	Wearable mechatronic systems with 5 DOF including elbow flexion/extension	Motor is placed only on one side so undesired torsion No information about length adjusting
(Hekman et al., 2009)	Exoskeleton combines functional exercises resembling activities of daily living with impairment-targeted force-coordination training and has 4 DOF including elbow flexion/ extension with 0-135 degree ROM	Not wearable and heavy construction
(Hawkes, Christensen, & Okamura, 2016)	Proposed powered elbow orthosis and maximum flexion up to 100 degrees	Don't have mechanical stoppers

tion of joint torque, force and angle. There are some different methods for such estimation in (Ding, Xiong, Zhao, & Han, 2011; Mountjoy, 2008; Kiguchi & Hayashi, 2013; Aung & Al-Jumaily, 2013; Ehrampoosh, Yousefi-koma, & Ayati, 2015; Wilson Antuvan, 2016; Youn & Kim, 2010; Chen, Zhao, & Han, 2013; Zhang, Hosoda, & Venture, 2013). Some of estimation algorithms are quite complicated but only four of them (Mountjoy, 2008), (Kiguchi & Hayashi, 2013), (Youn & Kim, 2010), and (Zhang, Hosoda,, & Venture, 2013) can control the motor tracking accuracy of higher than 92% (see Table 3). Latest development of mathematical modeling and control of human mechatronic devices are referred to in references (Ullah & Kim, 2009), (Minh & Rani, 2009), (Minh, Hashim, & Awang, 2012), (Minh & Hashim, 2010), (Habib, 2008), and (Habib, 2007).

Table 2. Review of actuators and actuation transmissions

Reference	Type of Actuator	Actuation Transmission
("Myomo", 2016)	DC motor	Motor located on opposite sides of the joint is connected to the gear head. The actuation mechanism generates up to 14 Nm.
(Pedrocchi, Ferrante, Ambrosini, & Gandolla, 2013)	DC motor	DC motor with a planetary gearhead (A-max 22 and GP 22, Maxon motor, Switzerland) with torque 1Nm
(Ripel, Krejsa, Hrbacek, & Cizmar, 2014)	DC motor	The worm gear transmission is driven by a direct-current (DC) motor Maxon RE36 with nominal power of 70 W. Furthermore, the motor is equipped with the planetary pre-gearbox GP32A. Maximum torque is 94Nm
(Budaker, 2014)	DC motor	Electrical 40W motor with planetary gearbox are able to produce torque about of 13Nm
(Durandau & Suleiman, 2016)	DC motor	The actuator uses a servomotor as velocity source, two magnetorheological brakes, a differential mechanism and an electronic drive
(Herrnstadt & Menon, 2016)	DC motor	Actuation system consist of suppression and driving motor Maxon EC 45 connected with 26:1 Maxon Spur Gearhead and can produce torque about 3Nm
(Looned, Webb, Xiao, & Menon, 2014)	DC motor	System was designed to generate 10 Nm of output torque by a brushless DC motor with customized gearbox
(Mohammadi, Zohoor, & Khadem, 2014)	DC motor	13 watt DC gear motor (1.64.068.501, Buhler Co, Germany) and worm gearbox with ratio of 1:200 generates a torque 6.7 Nm
(Desplenter, Lobo-Prat, Stienen, & Trejor, 2016)	DC motor	Brushless DC motor (Maxon EC20 Flat 5W 12V, maxon motor ag, Switzer- land) in combination with a planetary gearbox with 1:128 reduction ratio (Maxon GP22 C, Maxon Motor ag, Switzerland) and a bevel gear of 1:3 reduction ratio (SDP/SI, NY, USA)
(Hekman et al., 2009)	Hydraulic	The Dampace uses hydraulic disk brakes, which can resist rotations with up to 50 N m and have a torque bandwidth of 10 Hz for multisine torques of 20 N m
(Hawkes, Christensen, & Okamura, 2016)	Inverse pneumatic artificial muscle	One artificial muscle with maximum force 83 N
(Andrikopoulos, Nikolakopoulos, & Manesis, 2011)	Pneumatic artificial muscle	There are presented some rehabilitation devices with artificial muscle, but no detailed information
(Parivash & Bamdad, 2015)	Series elastic actuator	The cable actuation mechanism consists of a motor which provides tension force in the cable, and a disk that changes the direction of exerted tension force to the robot link. Maximum torque is up to 6Nm
(Ates, Sluiter, Lammertse, & Stienen, 2014)	Series elastic actuator	In the ServoSEA, a small rotational spring is attached to the output shaft. System can produce up to 2Nm

Thus, this project develops relatively simple and safe mechanical design and provides good controlling algorithms which should be able to control the motor tracking as high precise as possible but not too complicated and reliable for long calibration intervals. The purpose of this project is providing a low-cost device (less than $500), light weight (less than 2 kg), and high motion tracking accuracy (higher than 94%).

Table 3. Sensory and control methods

Reference	Method	Accuracy	Limitation
(Ding, Xiong, Zhao, & Han, 2011)	Modified Hill – based model and extended Kalman filter. This method use only BB	Root mean square error 0.142 rad	Low accuracy and long calibration process
(Mountjoy, 2008)	Hill – based model use BB and TB	92%	Long calibration process
(Kiguchi & Hayashi, 2013)	Neuro-fuzzy modifier use EEG and EMG signal from BB and TB	93%	Additional EEG sensor
(Aung & Al-Jumaily, 2013)	Back propagation neural network (BPNN), electrodes were placed on AD, PD, BB, TB	MSE for flexion/ extension of elbow 0.0176	This method require 4 electrodes and calibration period
(Ehrampoosh, Yousefi-koma, & Ayati, 2015)	The nonlinear autoregressive network with exogenous input (NARX) was used. Electrodes were placed on BB and TB	Probability of prediction 95%	Calibration period, no information about real error of joint's angle
(Wilson Antuvan, 2016)	Extreme learning machine for feed – forward neural network which provide faster learning speed. Such muscles as BB, TB, AD, PD, PM, I, T were used in synergies.	91.79%	Requires 7 muscles and calibration period. Quite complicate algorithm.
(Youn & Kim, 2010)	Artificial neural network that uses combination of EMG and mechanomyography. Electrodes were placed on BB and BRD.	93%	The model was tested only in isometric contraction
(Chen, Zhao, & Han, 2013)	Hierarchically projected regression algorithm uses only BB	91.6%	Low accuracy
	Autoregressive structure with exogenous input (ARX) mapped model with Kalman filter/ Electrodes were placed on BB and TB	91.7%	Low accuracy
(Zhang, Hosoda, & Venture, 2013)	Mathematical mapping model which use signal from BB and TB	98.4%	Isometric contraction and ROM 0 – 90 degree

MECHANICAL DESIGN

The hardware design covers three considerations: 1. Adjustability in size for different limb parameters and ability to use the device on both arms without any reconstruction; 2. Ability to track joint motions in all motion ranges; 3. High safe, reliable and comfortable.

People have different anthropometric parameters. Therefore, the hardware design must be fitted to 95% of population according to the data in (Zhang, Hosoda, & Venture, 2013) and shown in Table 4.

To support the forearm and naturally manipulate it, the lower arm cuff must be placed on the lower arm + hand center of mass, which is at 31.8% from distal distance. So, the lower part length can be calculated as:

$$l_d = l_{ah} \times COM, \quad (1)$$

where l_d is the length of lower part of device (cm), and l_{ah} is the lower arm + hand length (cm), and COM is the distance for the lower arm + hand center of mass from distal (%).

Table 4. 95% human anthropometric data

Parameter	Male	Female
Upper arm length (m)	0.389	0.358
Lower arm + hand length (m)	0.517	0.458
Lower arm + hand mass (kg)	2.29	1.74
Distance for the lower arm + hand center of mass from distal (%)	0.318	
Elbow breadth (cm)	8.2	7.4

Hence, the length of lower cuff may vary from 14 to 17 cm. It is reasonable to place upper arm cuff in the middle of upper arm, so the length of the upper cuff will vary from 17 to 20 cm. And the breadth of elbow, according to table 4, will vary from 7.4 to 8.2 cm.

The elbow movement can be modeled as a hinge joint. The mechanical elbow brace is designed with a hinge model rotating upper hand and lower hand via a hinge as shown in Figure 1.

During the rehabilitation period, it is essential to make flexion/extension exercises to increase the elbow range of motion and to improve the muscular strength. Some load, like a glass of water, may be used in such exercises. Thus, orthosis should be able to manipulate with such load mass together with lower arm + hand mass and its own mass. The necessary torque can be calculated as:

$$T = M \times g \times L = \left(m_l + m_{ah} + m_d\right) \times g \times l_d, \tag{2}$$

where, M – total manipulated mass, which consists of m_l – load mass, m_{ah} - lower arm + hand mass m_d – device mass (0.07 kg in case of ABS plastic), $g = 9.81\ \text{m}/s^2$ – earth gravity and l_d - length of lower part of device.

Substituting data from table 4 into equation (2), the motor has to produce at least torque of 5.6 Nm for males and 3.7 Nm for females. But during every-day activity, people can make complex movements, so this torque should be increased up to 6 Nm. This value is considered as a the minimum load for the motor.

The orthosis must be able to hold and fix in each position. Thus, mechanical stoppers are needed to include in the device construction. Moreover, elbow joint has certain range of motion (ROM) from 0°

Figure 1. Hinge and elbow brace design

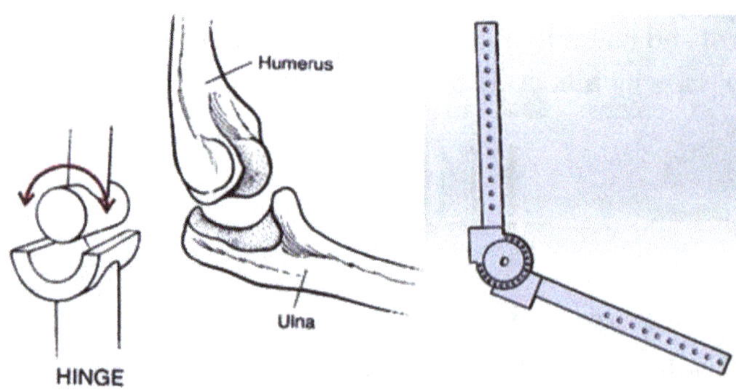

to approximately 150°. It is needed to limit motions of the device not go beyond this range. Also, during treatment, patient must not use the whole ROM but only some parts of it. So the device must have some function to divide the range into some steps and allow the movements only within these steps. Therefore, the hinge is designed to have 17 holes and mechanical stoppers are represented as a pin placed into these holes. The possible rotation with each step is equal to 9°. The mechanical stoppers and the rotating hinge design are shown in Figure 2.

For relying and convenient fixation on the limb, an arm cuff with soft material is developed. This cuff mounted on the lower and upper part of the orthosis and presses on limb. The diameter of the cuff is a little bit more than arm diameter for comfortable seating.

The orthosis mechanical design is consisted of two single – axis hinges, adjustable arm cuffs, mechanical stoppers, and two motors with the transmission system. The position of the arm cuffs may be shifted along the arm cuffs. The overall CAD model assembly is shown in Figure 3 including: Two rolling bears (1) mounted on the lower arm (2) symmetrically on both sides; The special cap (3) and upper arm part (4) covers this bearing; There are two big holes in the center of (3) for the potentiometer and in the center of the (4) for installing the DC motors.

Finite Element Analysis (FEA) for this device is conducted in ANSYS in order to estimate the total deformation, equivalent stress and safety factor under proposed loads. Figure 4 shows the total deformation under the general load of 50N and the maximum load of 70N.

According to Figure 4, the maximum deformation and the maximum stress during every day activity, load of 50N, are 0.00147 mm and 89.433 MPa and under the maximum load of 70N are 0.00206 mm and 125.18 MPa, respectively. The maximum deformation occurs in the middle of the stopper, whereas the maximum deformation occurs at the end of the stopper.

To satisfy the higher safety requirements, it is needed to widen the safety margin by increasing the factor of safety. The factor of safety means how much the system being stronger than it usually needs. The factor of safety for this device can be calculated as:

$$Factor\ of\ Safety = \frac{yield\ stress}{working\ stress} \tag{3}$$

Figure 2. Mechanical stoppers and hinge design.

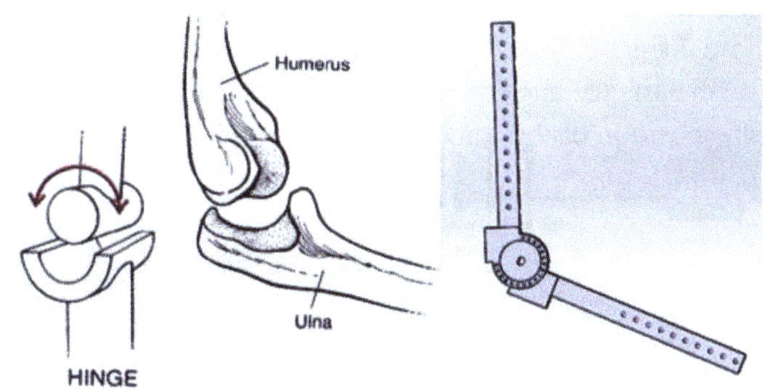

Figure 3. CAD model assembly

Figure 4. Total deformation for load of 50N (L) and 70N (R)

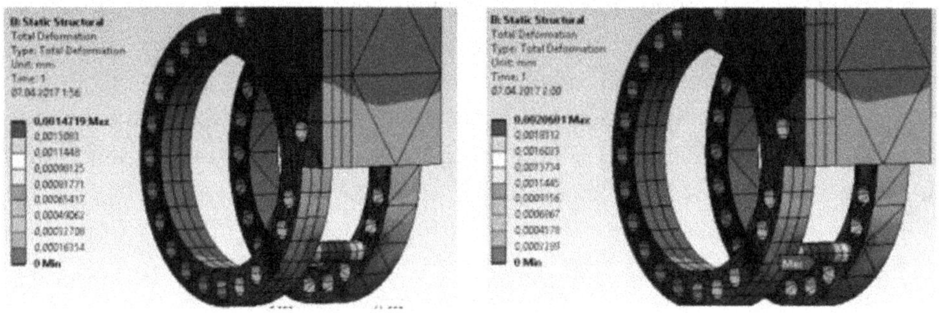

According to this simulation for the maximum load (70N), the safety factor reaches of 1.997, which means that the structure will fail almost at twice the design load.

Figure 5 shows the final system CAD design: Two DC worm geared motors (1) placed in the motor holder (2) provides the system with 11.6 Nm torque and manipulate the joint (3). This joint consists of upper arm and lower arm parts and bearing assembly as shown on Figure 3. Also, the orthosis has mechanical stoppers (4) (more details in Figure 1), which prevent harmful range of motion. For attaching the orthosis to a person, arm cuffs (5) and (6) are used. The (5) cuff also used as housing for electronic components fastening. The position of the arm cuffs may be slided along to the arrows in (6).

Figure 5. Final CAD design

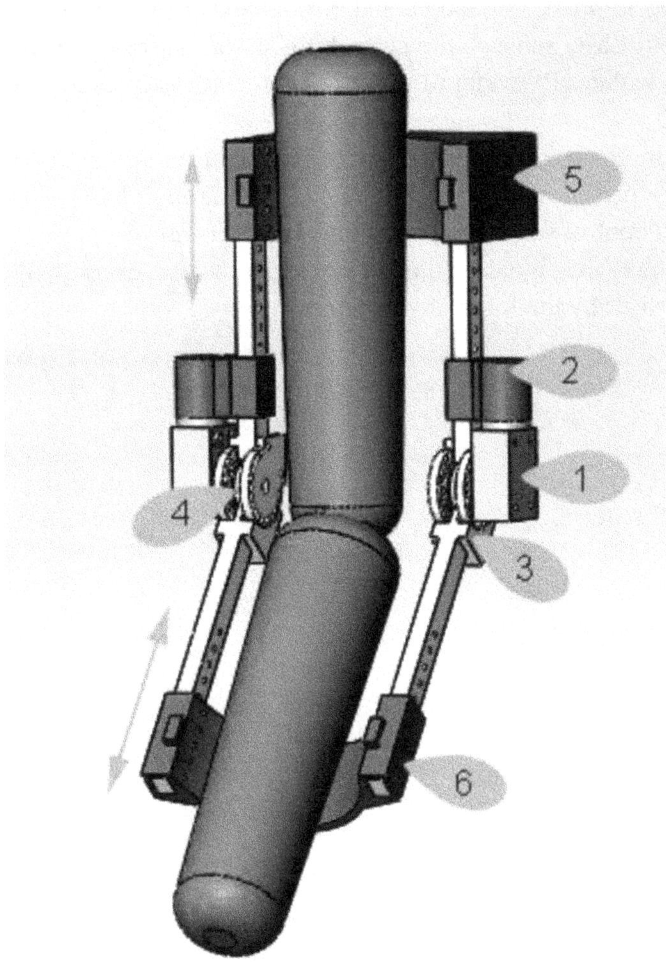

ELECTRONICS DESIGN

MyoWare muscle sensors are chosen for this project because of its non – invasive type, convenience using and adequate price. Another advantage of this sensor is that it is specially designed for microcontrollers and requires only single supply.

Particularly, sensors subsystem is presented by EMG sensors and potentiometer. EMG sensors capture biopotential from biceps and triceps and then transmit it to the Arduino UNO for subsequent data processing. Potentiometer placed in joint measures current angle of the elbow and transmits it to the microcontroller.

Data processing subsystem consists only of Arduino UNO and process incoming data for motor driver. EMG signals are calibrated, smoothed and normalized in this subsystem. Further, according to the normalized data from EMG sensors, the central processor distinguishes movement intention from muscle inactivity and calculates PWM for motor. Two motors provide torques of 5.8Nm for the orthosis from each side.

The power subsystem consists of a rechargeable LiPo battery which distributes power to each of subsystems. This battery provides 12 V for motors and 5 V through DC-DC converter to the Arduino and sensors. The overall control scheme is shown in Figure 6.

In order to smooth the EMG signals and reduce noise, a moving average filter is implemented. The moving average is described by the following equation:

Figure 6. System control scheme

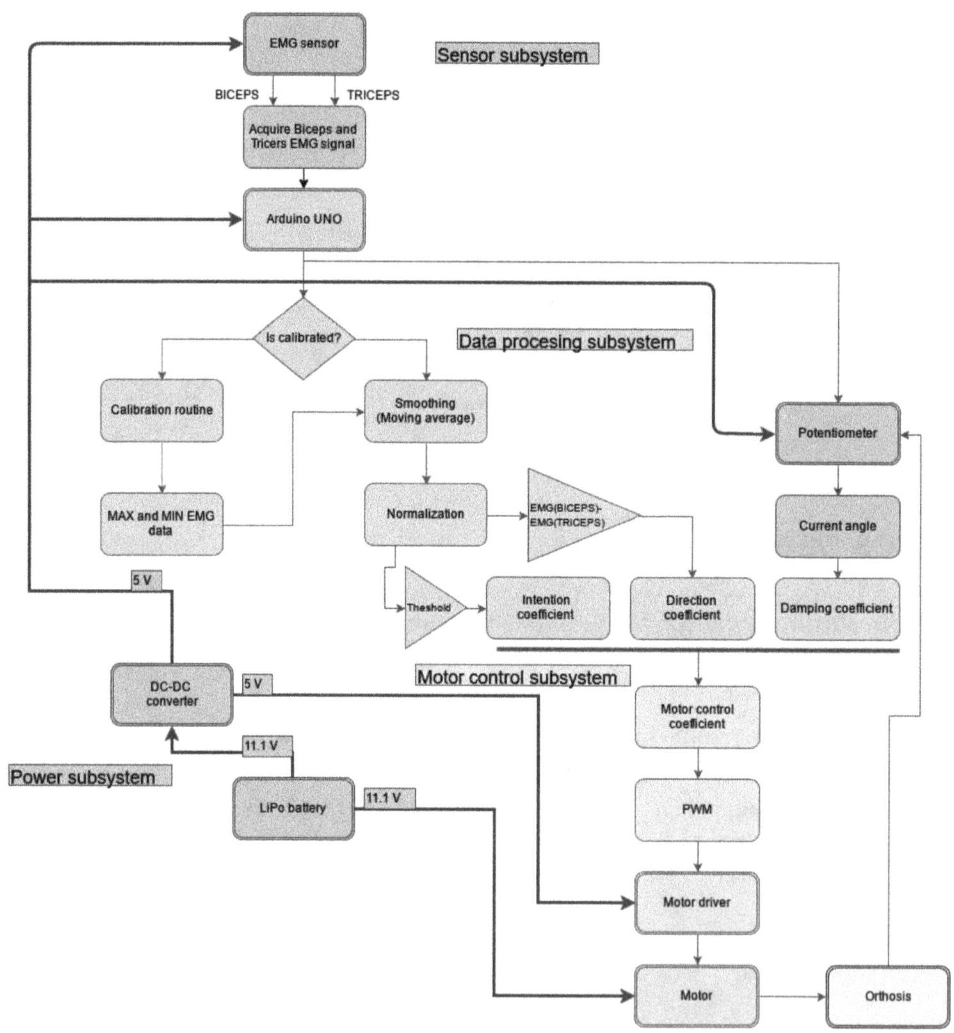

$$Mooving\,average = (EMG(N))_i = \frac{\sum_{i=1}^{N} EMG_i}{N} \tag{4}$$

where $(EMG(N))_i$ – value after moving average, EMG_i – EMG signal, N – number of samples.

The higher the number of samples, the more values will be smoothed, but the slower feedback will be responded. The result of filter smoothing with N = 10 is presented.

After smoothing and filtering of noises, the EMG signals will be normalized in such manner that it would be appropriate for each individual. Therefore, it is possible to directly map EMG activity to torque value which will be used to run the DC motor:

$$Force = \frac{(EMG\,Activity - MIN\left(EMG\,Activity\right)}{\left(MAX\left(EMG\,Activity\right) - MIN\left(EMG\,Activity\right)\right)}, \tag{5}$$

where $EMG\,Activity$ – current EMG signal, $MAX\left(EMG\,Activity\right)$ and $MIN\left(EMG\,Activity\right)$ – values obtained in calibration, $Force$ – proportion of individuals maximum contraction.

For example, if one has maximal contraction equal to EMG magnitude of 1500 and minimal of 200 and during exercise reaches 950. According to the above equation, force will be 0.58 or 58% from the maximum.

Another person with EMG maximum of 1000 and minimum of 100 reaches only 500 which seems twice less than 950 from first person. But normalized force will be 0.55 or 55% from maximum voluntary contraction (MVC). As result, both persons made an equal effort from the maximum. Thus, such mapping erases difference among physical conditions and the proposed device will be universal.

After normalization, all EMG signals are limited in the range from 0 to 1, where 0 is no contraction (muscle is flexed) and 1 is the maximum contraction. Both in static and dynamic movements lead to muscle co-contraction (the simultaneous activation of antagonist muscles around a joint). Thus, there should be some thresholds to distinguish muscle activity from inactivity. First of all, the absolute value of difference between normalized signal from biceps and triceps is obtained. Then, a 20% of MVC threshold is set. Hence:

$$abs\left(EMG\left(BICEPS\right) - EMG\left(TRICEPS\right)\right) > 0.2\,muscle\,active \tag{6}$$

$$abs\left(EMG\left(BICEPS\right) - EMG\left(TRICEPS\right)\right) < 0.2\,muscle\,inactive \tag{7}$$

Now, it is possible to determine in which moment the person intents for arm moving. Motion intention and direction are determined from the normalization graph. The direction coefficient is just the comparison between normalized biceps and triceps signal. If biceps signal is higher than triceps signal than it tends to intention and vice versa, but it should be considered that difference between normalized signal at that moment is set at threat hold of 0.2 in order to avoid false movement.

It is needed to distinguish when person tends to flex or extend its arm and when it is just a noise. Then, signals are converted to PWM to control the DC motors. The power system transmits power from a rechargeable battery to the motor and converted power to the logic supply is shown in Figure 5. General 12V power is taken from Turnigy nano-tech 3S Lipo Pack with a capacity equal to 3300 mAh. Two motors consume 800mA, thus the whole system should be working for about 4h without recharging. For Arduino, sensors and motor driver supplying 12V pass through Pololu Step-Down Voltage Regulator D24V10F5 which has an input voltage of up to 36 V and efficiently reduces it to 5 V while allowing for a maximum output current of 1 A. A physical switch allows the user to shut the system off in case of an emergency.

In Figure 7, Cal button is a calibration button for new user and after switching it on calibration routine will start. LED1 is indicating calibration period and after calibration, it will turn off.

Experiments

A prototype of the above design is installed and tested for a real 2 kg load as shown in Figure 8.

The total cost of the device consists of the components price including delivery is only $234 or far below the proposal price of $500. The prototype cost list is presented in Table 5.

Figure 7. Power electronics schematic design

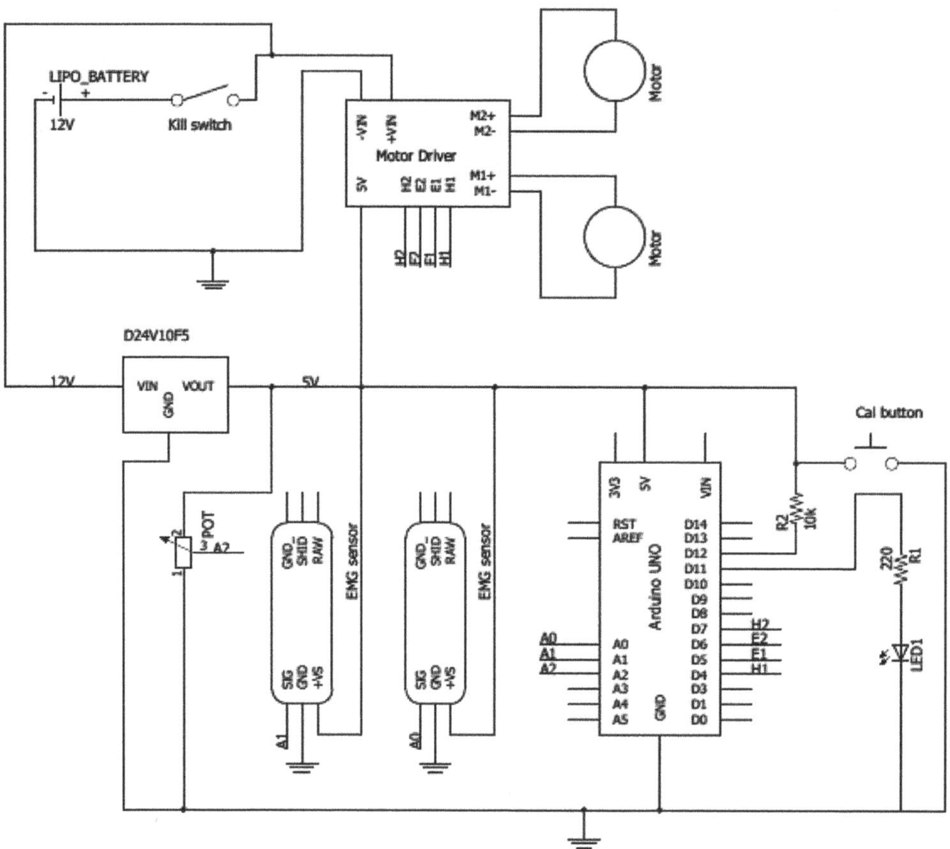

Figure 8. Prototype device and test with work load of 2 kg

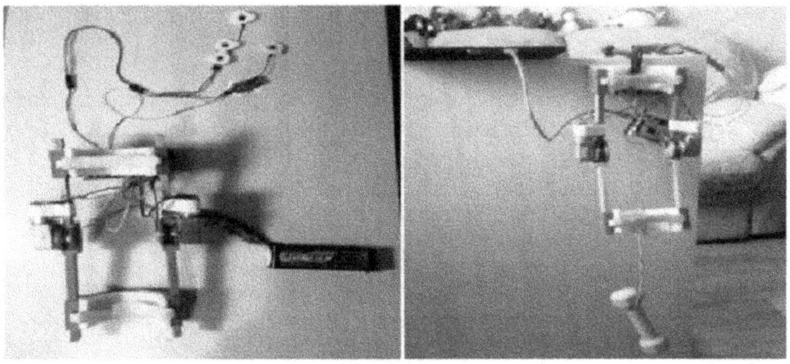

Table 5. The prototype cost.

Item	Cost (Including Delivery Service)
Myoware muscle sensor	$106.3
EMG electrodes	$8
Arduino UNO	$8
Motor shield	$15
Worm geared DC motor	$40
LiPo battery	$35
LiPo battery monitor	$3
Wires, resistors, buttons and LED	$5
Potentiometer	$1
DC-DC converter	$4.5
Standard components (bearing, screws and so on)	$4.5
Velcro straps and foam rubber	$4
Total	$234.3

The system is now tested on a real person motions as shown in figure 9.

After passing through calibration routine the following raw signals noises from muscles will go through a moving average for signals smothering and through some noised filters for filtering the disturbances as shown in Figure 10 with raw signals and smothered signals.

Signals now are normalized. After normalization, the signals are limited in range from 0 to 1 for everyone. 0 is no contraction, the muscle is flexed. And 1 is the maximum contraction of that person corresponding to his MVC obtained during calibration. Motion intention and direction of this person is detected. Horizontal lines present thresholds and if the signal is above it then the intention is detected and orthosis will start to move. Finally, the motion intention and direction are converted into PWM to run DC motors. Figure 11 shows the whole processes with normalized EMG signals; Detection of motion intention and direction; Converted PWM signals transmitting to DC motors.

Figure 9. Test on real person motions

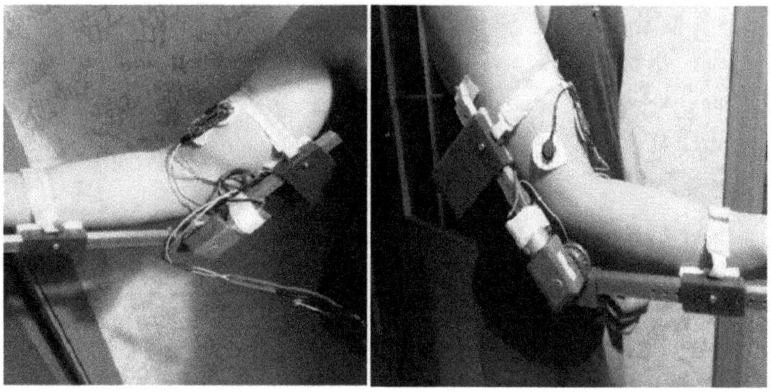

Figure 10. Raw and smothered EMG signals from biceps and triceps

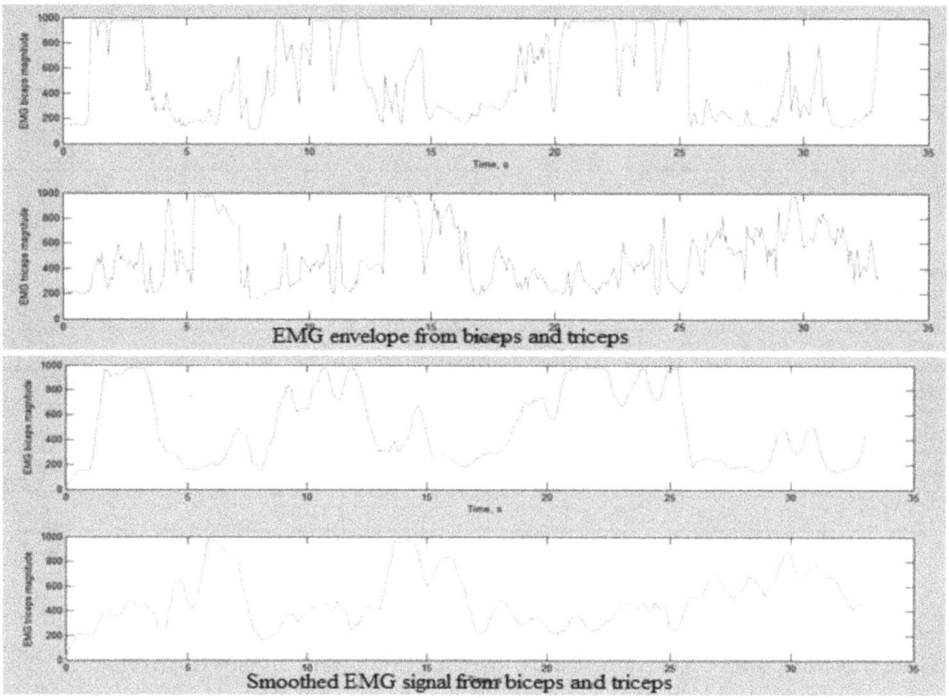

The motors start tracking the human motion intention. Figure 12 shows the real human movement and the motors tracking recoded in potentiometers. It's obvious that there three different flexion/extension movements are performed on the same time. The motor can track well the human movement with the maximum error of 7^0 over the maximum rotation angle of 120^0 or the system accuracy of more than 95%.

Figure 11. Normalized EMG, motion detection and PWM conversion

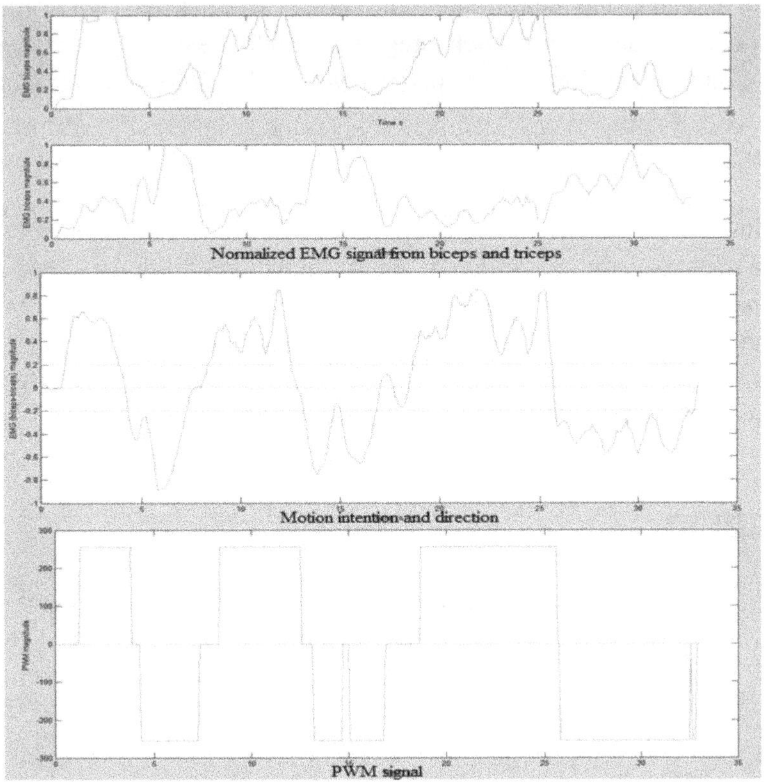

Figure 12. Human movement and motor tracking

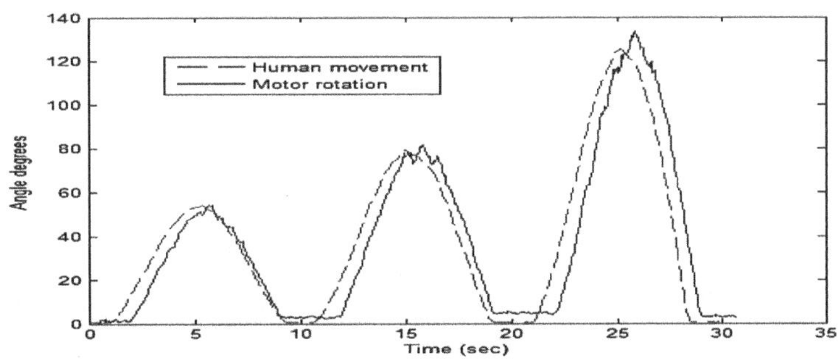

CONCLUSION

This project has provided the design and test of a mechatronic elbow orthosis. The test shows that the motor can track well the human movement with errors of less than 6%. Some contributions are highlighted from this project:

- Relatively simple and reliable mechanical construction, which provides smooth bending movement, limitation of harmful motions and convenience for different arm parameters;

- Universal actuation consists of worm geared motors. Tt is easy to increase the output torque by changing motor and choosing appropriate motor holder without redesign or modifications;
- Improvement of control algorithm which integrates with four electrical subsystems. The outcomes of integration of such algorithm provide patients with stable, easy-to-control and reliable.

Some limitations and future works:

- Control algorithms with modern mapping system can be found and implemented.
- The maximum tracking error of 6.9^0 can be improved by modifying the signal processing system. Moreover, potentiometer is not very precise for the angle detection. Some better accelerometer sensors may be used to get more precise feedbacks to control the DC motors.

YouTube for the Test on: https://www.youtube.com/watch?v=rsxIwoitJhc

ACKNOWLEDGMENT

The authors would like to thank Tallinn University of Technology (TTU) and Saint Petersburg State University of Information Technologies, Mechanics and Optics (ITMO) for supporting this research project.

REFERENCES

Andrikopoulos, G., Nikolakopoulos, G., & Manesis, S. (2011). A Survey on Applications of Pneumatic Artificial Muscles, *19th Mediterranean Conference on Control and Automation*, Corfu, Greece. 10.1109/MED.2011.5982983

Antuvan, C. W., Bisio, F., Marini, F., Yen, S.-C., Cambria, E., & Masia, L. (2016). Role of Muscle Synergies in Real-Time Classification of Upper Limb Motions using Extreme Learning Machines. *Journal of Neuroengineering and Rehabilitation*, *13*(76). PMID:27527511

Ates, S., Sluiter, V. I., Lammertse, P., & Stienen, A. H. A. (2014). ServoSEA Concept: Cheap, Miniature Series-Elastic Actuators for Orthotic, Prosthetic and Robotic Hands, *2014 5th IEEE RAS & EMBS International Conference on Biomedical Robotics and Biomechatronics (BioRob),* Sao Paulo, Brazil. 10.1109/BIOROB.2014.6913868

Aung, Y. M., & Al-Jumaily, A. (2013). Estimation of Upper Limb Joint Angle Using Surface EMG Signal. *International Journal of Advanced Robotic Systems*, 1–8.

Budaker, B. (2014). Biomechatronic System Engineering, *5th International Conference on Biomedical Robotics and Biomechatronics (BioRob)*, Sao Paulo, Brazil, pp. 768-773.

Chen, Y., Zhao, X., & Han, J. (2013). Hierarchical projection regression for online estimation of elbow joint angle using EMG signals. *Neural Computing & Applications*, *23*(3-4), 1129–1138. doi:10.100700521-012-1045-8

Desplenter, T., Lobo-Prat, J., Stienen, A. H. A., & Trejor, A. L. (2016). Extension of the WearME Framework for EMG-driven Control of a Wearable Assistive Exoskeleton, *IEEE International Conference on Advanced Intelligent Mechatronics (AIM)*, Alberta, Canada. 10.1109/AIM.2016.7576781

Ding, Q. C., Xiong, A. B., Zhao, X. G., & Han, J. D. (2011). A Novel EMG-driven State Space Model for the Estimation of Continuous Joint Movements. In *IEEE conference on Systems*. Alaska, USA: Man, and Cybernetics. doi:10.1109/ICSMC.2011.6084104

Durandau, G., & Suleiman, W. (2016). User-safe Orthosis Based on Compliant Actuators: Mechanical Design and Control Framework. In *Proceedings of the SICE Annual Conference*. 10.1109/SICE.2016.7749223

Ehrampoosh, A., Yousefi-koma, A., & Ayati, M. (2015). Estimation of elbow joint angle by NARX model using EMG data, *RSI International Conference on Robotics and Mechatronics*, Tehran, Iran.

Habib, M. K. (2008, August). Human Adaptive and Friendly Mechatronics (HAFM), *IEEE International Conference on Mechatronics and Automation*, Takamatsu-Kagawa, Japan, pp. 61-65.

Habib, M. K. (2007). Mechatronics: A Unifying Interdisciplinary and Intelligent Engineering Paradigm, IEEE Industrial Electronics Magazine, 1(2), Summer, pp. 12-24.

Hawkes, E. W., Christensen, D. L., & Okamura, A. M. (2016). Design and Implementation of a 300% Strain Soft Artificial Muscle, *IEEE International Conference on Robotics and Automation (ICRA)*, Stockholm, Sweden, pp. 1-10. 10.1109/ICRA.2016.7487592

Hekman, E. E. G., Prange, G., Aalsma, A. M., van der Helm, F. C., & van der Kooij, H. (2009). Dampace: Design of an Exoskeleton for Force-Coordination Training in Upper-Extremity Rehabilitation. *Journal of Medical Devices*, 1–10.

Herrnstadt, G., & Menon, C. (2016). Voluntary-Driven elbow Orthosis with speed-controlled Tremor suppression. *Frontiers in Bioengineering and Biotechnology*, *4*(29). PMID:27066477

Kiguchi, K., & Hayashi, Y. (2013). Motion Estimation based on EMG and EEG Signals to Control Wearable Robots, *IEEE International Conference on Systems, Man, and Cybernetics*, Washington, USA. 10.1109/SMC.2013.718

Looned, R., Webb, J., Zheng, G. X., & Menon, C. (2014). Assisting drinking with an affordable BCI-controlled wearable robot and electrical stimulation: A preliminary investigation. *Journal of Neuroengineering and Rehabilitation*, 1–13. PMID:24708603

Minh, V. T., & Hashim, M. (2010). Adaptive teleoperation system with neural network-based multiple model control, *Mathematical Problems in Engineering*, Open Access, 592054.

Minh, V.T., & Rani A. M. (2009). Modeling and control of distillation column in a petroleum process, *Mathematical Problems in Engineering*, Open Access, 404702.

Mohammadi, E., Zohoor, H., & Khadem, S. M. (2014). Control System Design of an Active Assistive Exoskeletal Robot for Rehabilitation of Elbow and Wrist. In *Proceedings of the 2nd RSI/ISM International Conference on Robotics and Mechatronics*, Tehran, Iran. 10.1109/ICRoM.2014.6991008

Mountjoy, K. (2008). Use of A Hill-based Muscle Model in the Fast Orthogonal Search Method to Estimate Wrist Force and Upper Arm Physiological Parameters, IEEE conference on Medicine and Biology, Vancouver, Canada.

"Myomo," [Online]. (2016). Available at http://www.myomo.com/myopro-upper-limb-orthosis-technologymyomo-

Parivash, F., & Bamdad, M. (2015). Independent Position-Stiffness Control for Elbow Rehabilitation Robot with Cable-Based Series Elastic Actuator, *22nd Iranian Conference on Biomedical Engineering(ICBME 2015)*, Tehran, Iran. 10.1109/ICBME.2015.7404168

Pedrocchi, A., Ferrante, S., Ambrosini, E., Gandolla, M., Casellato, C., Schauer, T. ... Reichenfelser, W. (2013). MUNDUS project: MUltimodal Neuroprosthesis for daily Upper limb Support. Journal of neuroengineering and rehabilitation, 10(1).

Ripel, T., Krejsa, J., Hrbacek, J., & Cizmar, I. (2014). Active Elbow Orthosis. *International Journal of Advanced Robotic Systems*, 1–10.

Shahid, M. K. (2015). Fletcher, Robati S and Pemmaraju G, "The Biomechanical Forces that Act on the Elbow Joint. *EC Orthopaedics*, *1*(1), 1–11.

Smania, N., Berto, G., Melotti, C., & Miditi, A. (2012). Rehabilitation of brachial plexus injuries in adults and children. *European Journal of Physical and Rehabilitation Medicine*, *43*(3), 483–506. PMID:23075907

Trieu Minh, Vu., Hashim, M., & Awang, F. (2012). Development of a real-time clutch transition strategy for a parallel hybrid electric vehicle. *Journal of Systems and Control Engineering*, *226*(2), 188–203.

Ullah, K., & Kim, J.-H. (2009). A Mathematical Model for Mapping EMG Signal to Joint Torque for the Human Elbow Joint using Nonlinear Regression, *Proceedings of the 4th International Conference on Autonomus Robots and Agents*, pp. 103-108, Wellington, New Zealand.

Youn, W., & Kim, J. (2010). Estimation of elbow flexion force during isometric muscle contraction from mechanomyography and electromyography. *Medical & Biological Engineering*, *48*(11), 1149–1157. PMID:20524072

Zhang, Q., Hosoda, R., & Venture, G. (2013). Human Joint Motion Estimation for Electromyography (EMG)-Based Dynamic Motion Control, *35th IEEE conference on EMBS*, Osaka, Japan.

Chapter 2
CLARC:
Robotic Platform to Assist the CGA Evaluation in Elderly Patients

Ana Belén Naranjo-Saucedo
Virgen del Rocio University Hospital, Spain

Alvaro Dueñas-Ruiz
Virgen del Rocío University Hospital, Spain

Cristina Suarez-Mejias
Virgen del Rocio University Hospital, Spain

Ana Iglesias
Universidad Carlos III de Madrid, Spain

Rebeca Marfil
University of Malaga, Spain

Antonio Bandera
University of Malaga, Spain

ABSTRACT

The increasing ageing of the European population is an unstoppable process causing more and more demand for health and social assistance resources. The Comprehensive Geriatric Assessment (CGA) is an indispensable tool to identify the psychosocial and functional limitations of geriatric patients. The problem is that the CGA process is very long and demands a lot of time from health professionals. It is necessary to develop a robotic platform that helps in CGA. In this sense, Clarc is a mobile robot developed to help the physician to capture and manage their data during CGA procedures. The robotic platform was tested in some healthcare centers and results show elderly participants accepted the technology. Clarc is a tool that facilitates the work of health professionals given the ease of access to patient data and their evaluations, freeing these professionals to focus their activity on purely clinical tasks that require their knowledge.

DOI: 10.4018/978-1-7998-0137-5.ch002

INTRODUCTION

The ageing of the world's population is already a reality and a worrying legacy for upcoming generations. The percentage of people over the age of 65 is increasing year by year and it seems that this trend will not cease in the coming decades (Bengtson, 2018). Nowadays and since the past 6 decades, there is a remarkable increase in life expectancy, especially in industrialized societies. European countries have the most aged populations. Moreover, Bengtson (2018) indicates that "during 2000, the world's elderly population (aged 65 and over) grew by more than 795,000 people a month". Furthermore, he indicates that "the current growth rate of the elderly population in the developing countries is expected to rise to above 3.5 percent annually from 2015 through 2030".

Health resources such as health care are becoming more and more necessary but at the same time limited. In the last few decades, life expectancy has risen, which has led to an increase of the appearance of multiple comorbidities, while there are fewer active people who can care for them. This is caused by the coexistence of the older population and the ageing of the working population, as well as by the decline in the working-age population. In addition to all these factors, there is an important decrease of the infant mortality, an increase in the birth rate and a significant improvement in health systems (Rodriguez, 2014).

Older people make greater use of social and health services and therefore generate greater expenditure both financially and in terms of resources themselves. This is the reason why, for this age group, a change in the model of health care is needed, in which medical conduct is established that employs great efforts in the prevention and early detection of diseases and functional impairments. Therefore, an early assessment of the overall health status of older people as early as possible is crucial to prevent as much as possible age-related deterioration.

Want et al. (2014) pointed out that certain protective factors may act reducing the negative consequences old age and improving their quality of life, such as physical and cognitive activities practices, a healthy social and relative environment, etc. To consider all these aspects, besides the traditional medical evaluation, a different multidisciplinary and more holistic approach has been developed: The Comprehensive Geriatric Assessment (CGA).

The CGA is a multidimensional tool used to assess and identify the needs of frail elderly patients on the basis of which a personalized care plan can be developed. CGA is a widely used method because, in addition to providing detailed clinical, functional and cognitive data on older patients, it provides information on non-medical aspects such as economic and socio-environmental conditions, which allows the creation of a holistic view of the reality of the older person (Rubenstein, 2004).

In this sense, according to the findings of Stuck and Iliffe (2011), patients over 75 years of age who received a comprehensive geriatric evaluation in a home visit could reduce functional status impairment and avoid an early admission to nursing homes. In addition, the article points out that hospitalized patients who received an CGA were significantly less likely to die or experience functional impairment. From an economic perspective, therefore, in the long term, CGA, in addition to significantly improving patient status, offers cost savings by reducing hospital readmissions and reducing the need for long-term care in these nursing homes.

On the other hand, Ellis et al. (2017) compared CGA with routine care for patients over 65 who were admitted to hospital. According to their findings, "older people who receive CGA rather than routine medical care after admission to hospital are more likely to be living at home and are less likely to be admitted to a nursing home at up to a year after hospital admission". Moreover, no evidences were found that CGA reduces risk of death or in dependence during the follow-up to a year after.

BACKGROUND

Comprehensive geriatric assessment (CGA) is defined as a multidisciplinary diagnostic and treatment process that identifies limitations in several aspects: medical, psychosocial, and functional limitations of the elderly in order to develop a coordinated plan to optimize their overall health status (Ward & Reuben, 2016). The CGA is widely used and during the last 20 years, it has become a standard clinical tool for healthcare professionals of around the world to identify medical, psychosocial, and functional limitations of frail elderly people. Given that in older subjects the occurrence of negative outcomes, such as capacity degradation, hospitalization, institutionalization, or mortality, is the result of a combination of biological, functional, psychological, pathological, and environmental factors, tools that identify high-risk older patients should adopt a multidimensional approach.

The CGA process is conducted by a number of specialists from many disciplines in the health of older people; it involves a holistic, multidimensional and interdisciplinary assessment (with inputs from physicians, nurses, doctors and other health professionals) that has been shown to be associated with better outcomes in a variety of environments. A typical schedule of CGA includes the administration to the elderly a some of evaluation instruments that focus on relevant clinical and functional areas to allow establish individual impairments or risk factors that may improve with specific interventions. The result of the CGA assessment is to have a list of needs and issues to be addressed that allows the development of an individualized, goal-oriented care and support plan that is tailored to the patient's needs, deficits, and priorities and ultimately facilitates the establishment of an integrated approach to treatment, rehabilitation, support and long-term care.

The British Geriatric Society (2019) identifies several situations in where it is recommended to implement a CGA:

- When an older person consults their primary care physician with one or more obvious frailty syndromes: such as falls, confusion, reduced mobility, increasing incontinence, repeated infections, or if they have ever had to call the ambulance service.
- When a general practitioner learns of an accident which involves fragility.
- When an older person has been discharged from the hospital after presenting a fragility syndrome (falling, reduced mobility, delirium, etc.). Although the CGA process could have been initiated in the hospital, it will be necessary to perform it once the elderly person returns to his or her environment and is stabilized.
- In nursing homes where most of their residents (both those with and without nursing care) suffer impairment of their capacities. The CGA process will help identify future treatment goals and support advance planning for needed care.

THE CGA PROCESS

Typically, the CGA is divided into three phases:

- **Phase 1:** Clinical Interview: It is the first step in the process where patients and their families meet with health professionals. This interview includes a review of his personal history, assessing past and present diagnoses, hospital admissions or in socio-sanitary centers, surgical interventions, etc. All these factors determine the functional and mental environment of the patient. In addition, the complete pharmacological history and personal health problems such as allergies and medications will also be addressed in the clinical interview.
- **Phase 2:** Multidimensional Assessments: this is the phase of the evaluation, where several multidimensional assessment tests are administrated to assess the functional, mental and social status of the elderly person. The usual scenarios where the CGA assessments are performed are hospital settings: Hospitalization Units for income patients, Day Care Hospital, or Ambulatory Care Units for ambulatory patients. This is the main functionality of the envisaged technical solution.
- **Phase 3:** Individualized care plan. This is the most important phase of the CGA process where healthcare professionals evaluate patient's information gathered during the previous phases and identify therapeutic goals important to the senior. The individualized care plan includes: additional diagnostic tests, therapeutic recommendations (medications, rehabilitation treatment, cognitive stimulation, etc.) and the more suitable setting for the patient to execute the care plan (ambulatory care unit, day care hospital, or hospitalization units). Depending of the needs detected the plan may consist of one or more actions to be carried out by a clinician or other relevant members of the multidisciplinary team. Hospital-based programs with extended ambulatory follow-up and preventive home visits based on CGA are more and more accepted by evidence from randomized controlled trials or systematic reviews on CGA (Mann et al., 2004).

The problem is that the CGA process takes in general a lot of time of Health professionals, and moreover health professionals spend their time with tasks which can be automated today: on average, between 2 and 3 hours per patient are needed to complete the assessment. The Multidimensional Assessments (Phase 2) takes over 50% of the total time of the process while the individualized care (Phase 3) plan phase only lasts 11% of the time. When CGA is performed in Ambulatory Care units the process lasts only 60 minutes. In this settings time is a handicap and the health professional needs to hurry in Phase 1 and Phase 2 in order to complete the process; but many times, the CGA process is not completed in one session and has to be continued in further sessions also in other hospital setting (usually Day Care Hospital Unit). All in all, in ambulatory care units the health professional has a lack of time to perform the process; especially for the final and most important phase, where the personalized care plan is organized.

On the other hand, the elderly patients are evaluated in the hospital, ambulatory care units, or in residences at least once a year. For this purpose, functional, motor and cognitive tests such as Barthel, Mini Mental and Get up and Go are used. Usually, these evaluations are done manually on paper. Due to it, healthcare professionals after evaluation have to register the result of the tests on the electronic health record. However, the EHR (Electronic Health Record) is not prepared to record this information in a structured format. In this sense, they have to write the results in free text without following a template or structure format. This causes difficulties for the comparison and exploitation of data, especially for the evaluation process of the patients' evolution. In this context, having a system that allows the objective patient assessment would be a great achievement. Clinicians will be able to see patients' performances in order to compare them to previous years, also with the help of visualization videos from the same system. This will also save time and will optimize the evaluation process, decrease health professionals' tiredness or fatigue perception as consequence of doing tests.

ROBOTIC TOOLS

Population is currently in the Age of Technology. Robotic is a potent tool in several areas to the human. It exists robots who are in charge of the construction of orthopedic elements as legs, or arms (Sasaki et al., 2018) even biomechatronic rehabilitation system (Jatsun & Yatsun, 2018). However, the area of the robotic related to the seniors is frequently focused in contribute of social needs as social support and companion or in improve their independence like Nao, AIBO, Paro, or iCat (Korchut, 2017). In addition, robotic solutions are proposed to reduce the percentages of the human-based care due to higher needs to attend the need to cater to the ever-increasing elderly population with many health care requirements. Vercelli et al. (2017) refer a list of applications that can be supported by technology to address various medical areas of the aforementioned population, including those that highlight: Robots can help them directly in certain activities of daily living and safety or, even, help their caregivers in daily tasks; at the same time they can assist them in controlling their behavior and health; or provide entertainment and hobbies and social contact. However, authors emphasize that population take the risks of reducing human interactions and losing the privacy and personal freedom of the individual, which is one of the major handicaps of current technology.

MAIN FOCUS OF THE CHAPTER

In this section, the CLARC, *"smart CLinic Assistant Robot for CGA"* project is exposed in depth. In this project, part of the CGA is trusting to a robot in the goal of automatizing a part of the process, the evaluate, what allow releasing healthcare professionals in their task of carrying out some evaluations, performing functional tests, as well as, evaluating and monitoring patients. This is one of the Public end-user Driven Technological Innovation (PDTI) challenges proposed by the ECHORD++ project. Firstly, the technical solution will be also exposed followed by a brief explanation of the tests implemented in the robot. Then, the pilot study is detailed where the case use is presented according to the design. Finally, results and conclusions are presented.

TECHNICAL SOLUTION

The CLARC project was addressed by a balanced European multidisciplinary team of experts coming from technical, clinical and social disciplines. Concretely, the Consortium was composed by researchers of one robotics enterprise (Metralabs GmbH), three universities (Troyes, Carlos III Madrid, and Malaga) and one hospital (Virgen del Rocío University Hospital, Seville). For achieving the goal of automatizing one part of the CGA process, the robotic platform CLARC was designed, developed and validated in real scenarios. The initial hypothesis is that a robot able to autonomously conduct geriatric tests, could reduce the time medical professionals have to spend with purely mechanistic tasks, allowing them to have more time to develop individualized care plans for their patients.

The design of the CLARC robot was driven within a participatory process, where real end-users (elderlies, caregivers and clinicians, but also engineers) were involved (Ting et al., 2017). CLARC was built over the robust SCITOS G5 mobile base from Metralabs GbmH, and will use for navigation and localization the CogniDrive software[1]. But, for managing the use cases implementing the CGA tests, CLARC will be endowed with a specific instantiation of the CORTEX architecture (Bustos et al., 2019), including the PELEA framework for Automated Planning and Monitoring (Pulido et al., 2017). Being the major emphasis on engaging and helping the patients to perform the tests, several interaction channels were provided. Thus, CLARC will be equipped with a touchscreen panel on its torso, one shotgun microphone, speakers, and one RGB-D camera (Figure 1). Using this hardware, CLARC can automatically collect data from patients through questionnaire-based tests as the Barthel test, and also through direct observation (body pose and motion for addressing the Get Up and go test).

Figure 3 (Left) shows the CLARC robot driving one CGA test at Seville. Originally, it prioritized the verbal channel, and used the touchscreen as an alternative channel for capturing the responses on the questionnaire-based tests. The uncomfortability of the touchscreen, which forces the user to maintain the arm in an unpleasant position, suggested us, after closing a first short-scale pilot, to add an additional external device to augment the interaction channels of CLARC. Figure 3 shows the remote control.

This Remote control was entirely designed in the CLARC project. The core of the device is a Raspberri Pi Zero. It incorporates a charging port, which allowed external battery charging. The buttons could be lighted, and this was used for marking the available questions to the patient in the Barthel test. The size of external chassis eases the elderly to take the device and employ it during the test.

Apart of this human-robot interaction, the CLARC framework must provide an initial scoring from the tests (see Bandera et al. (2019) for the Get up and go test) and also to allow the physicians to edit these scores. For this end, CLARC will use two cameras. One IP-camera allows the online monitoring, and a webcam is employed for recording the session. All data is collected and analyzed at runtime, while the patient was filling the questionnaires. Recorded files and scores should be transferred from the robot to an external server (the CGAmed one), so they were encrypted for guaranteeing the confidentiality of the patients. Data (video/audio files and scores, and patients' profiles) is then stored in an external server, being accessible to the physician through dedicated web interfaces. During the tests, CGAmed interface provided the user with the tools for visualizing the agenda of the robot, manual launching or stopping of a CGA test, and also, online supervising the session. One of the interfaces is shown in Figure 5. This interface also provides the tools for helping the physician on reporting the sessions. Finally, the CGAmed server must include the tools for including the acquired information within the clinical data management system (CDMS).

Figure 1.

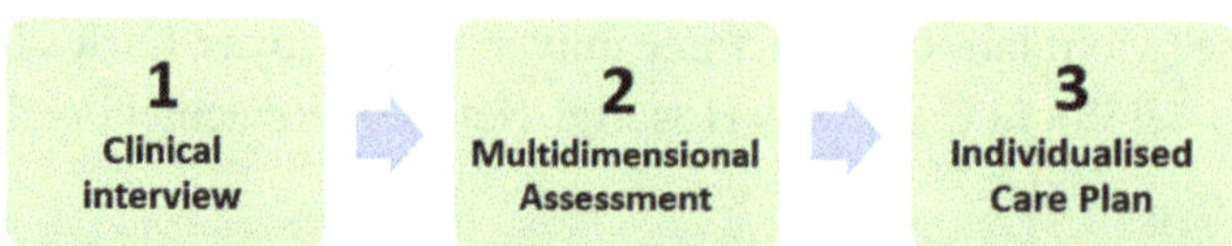

Figure 2. Inernal structure of the CLARC robot

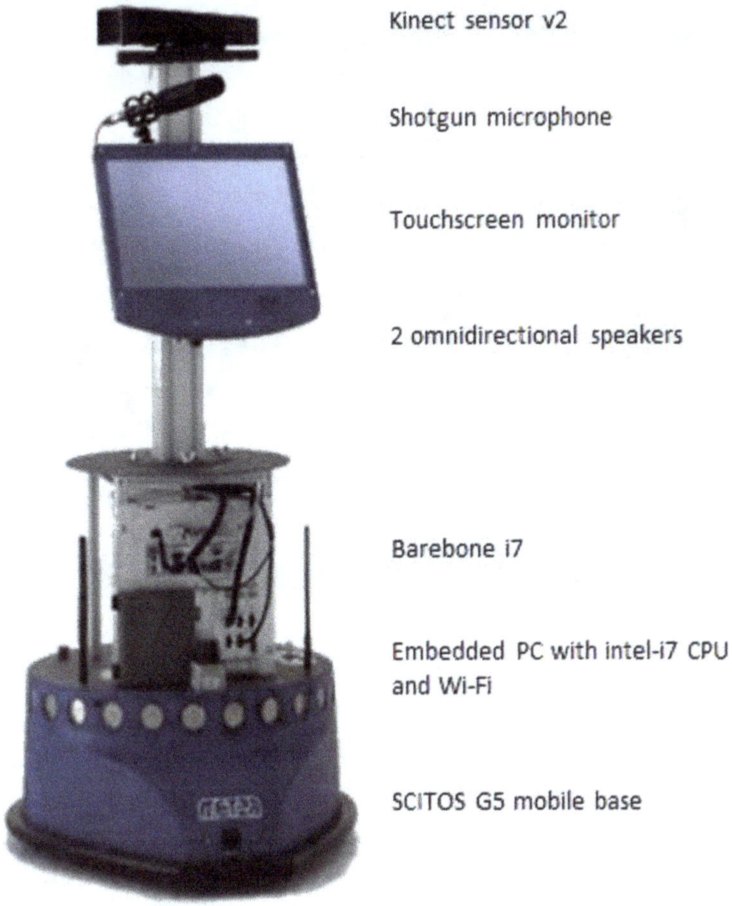

In summary, the proposed solution provides the following functionalities:

- Facilities to configuration and development of new tests with minimal need for assistance by computer science experts or robotics
- Selection of tests to include in an individual CGA by professionals directly
- Ability to interact by speaking and natural language processing (even in case of slightly slurred speech) to limited extend, interpreting a set of standard predefined answers and with multi-language support (French, English, and Spanish)
- Ability to ask patients items of selected tests
- Ability to interpret and codify patients answers in spoken language and by touch screen input of selected tests
- Ability to calculate tests scores based on codified information recorded from the tests. Health professional has the possibility of modifying and correcting tests scores
- Professionals would have a global vision of total scores and deepen when needed.

Figure 3. (Left) Evaluation of the first version of the CLARC robot at Seville (Spain), and (Right) final version of CLARC, performing a Barthel test at Viso del Alcor Primary Care Center (Seville, Spain)

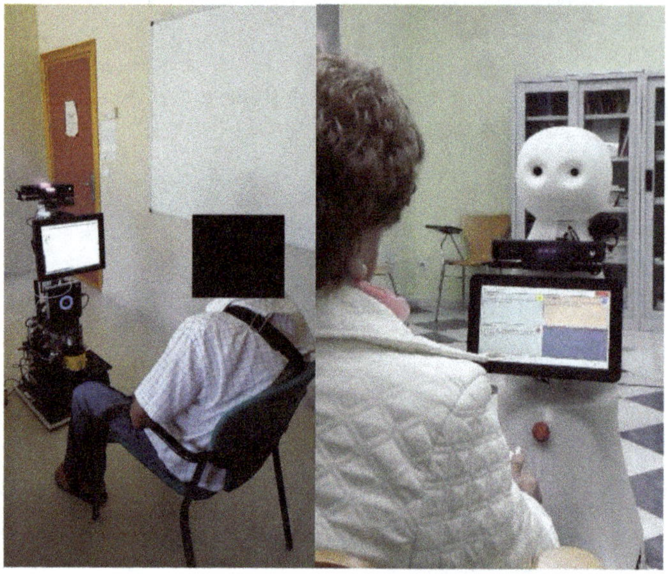

INTEROPERABILITY

CLARC robot was designed for integration with different Health Information System. For this purpose, CLARC uses IHE profiles. Concretely, XDS (Cross-Enterprise Document Sharing) for document management. This profile is focused on providing a standards-based specification for managing the sharing of documents between any healthcare enterprise, ranging from a private physician office to a clinic to an acute care in patient facility and personal health record systems. This is managed through federated document repositories and a document registry to create a longitudinal record of information about a patient within a given clinical affinity domain. In our case, Health Information System (HIS) will request CLARC System a list of document entries related to a particular patient. It will return registry metadata containing the list found to meet the specified criteria including the locations and identifier of each corresponding document in one or more Document Repositories. Then, HIS will request a particular document set of that list and, finally, CLARC System shall return this specified document.

On the other hand, for patient management, CLARC uses PIX (Patient Identifier Cross Referencing) profile. This profile describes cross-referencing patient identifiers between the various systems that register and store healthcare information. This allows domains to correlate information about a patient from multiple sources that use different identifiers. Finally, for different clinical Tests HL7 CDA (Clinical Document Architecture) were designed. The HL7 CDA is a document markup standard that specifies the structure and semantics of clinical documents for the purpose of exchange. This document will transport the information related to clinical tests from the robot system to other external systems.

Figure 4. (Top) The Remote-Control device: functionalities; and (Bottom) external aspect of the Remote Control and CLARC interacting with a patient. The patient easily holds the Remote Control over their knees

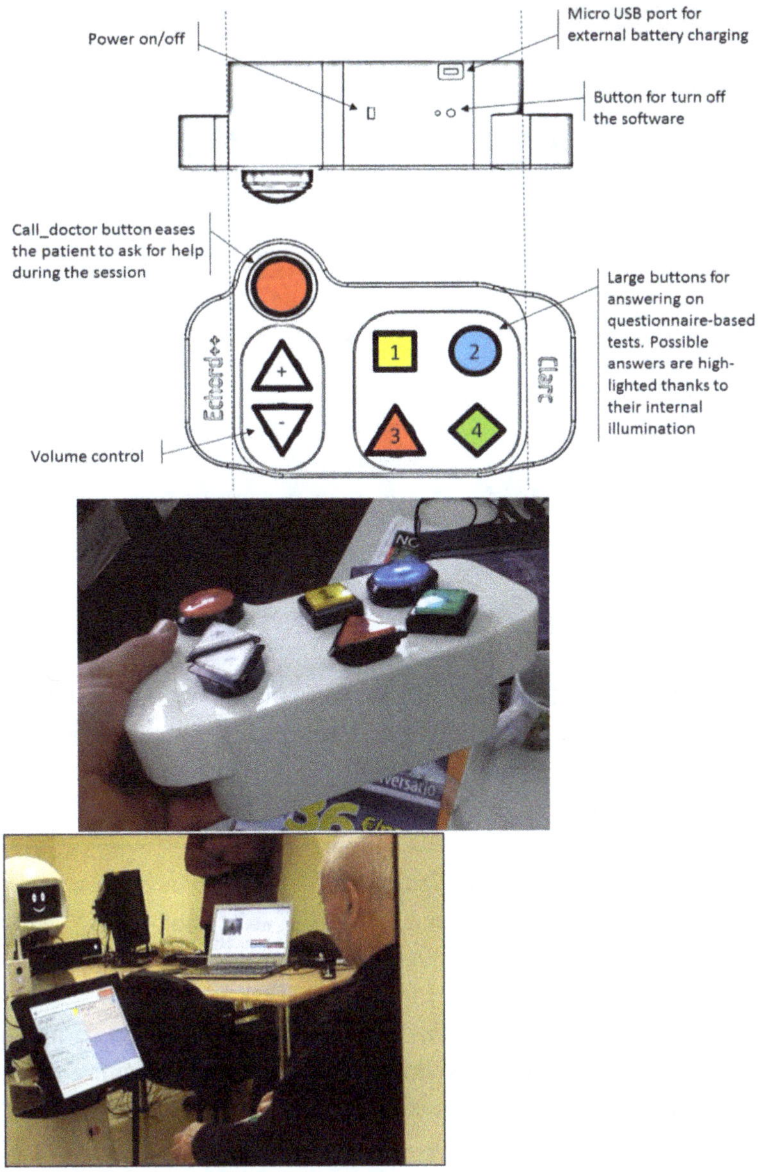

IMPLEMENTED TESTS

Health professionals defined two tests to be implemented in the technological solution, due it was tests frequently administered by them in their clinical practice in elderly. One of them was Barthel Index, who measures the person's capacity to carry out ten basic activities of daily life, obtaining a quantitative to estimate the dependence degree of the patient (Cid-Ruzafa & Damián-Moreno, 1997). The activities analyzed over 10 questions can be grouped according to personal care (feeding, grooming, bathing,

Figure 5. The Robot Control service on the CGAmed

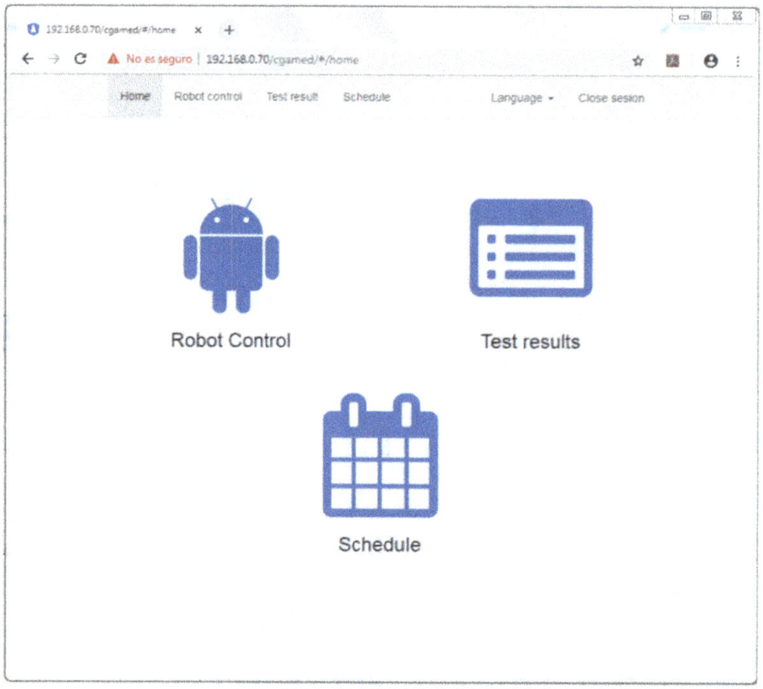

dressing, bowel and bladder care and toilet use) and mobility (ambulation, transfers, and climbing stairs). The robot was then able to ask questions using three developed interaction channels (i.e. voice, screen, or remote control) in order to adapt to patient's abilities and/or preferences. Thus, par example, in the case of elderly patients with stroke or other pathologies that make speech difficult, the system will NOT work to respond by voice.

Likewise, for each question, two, three or four possible answers are offered. The maximum score of 100 indicates that the patient is fully independent in physical functioning, while if the score is 0, or close to it, it represents a state totally dependent on the bed. Figure 6 shows the eighth question of Barther test and his options. The second is Get up and go Test and evaluates balance and gait, so the CGA is used to determine the risk of falls (Gálvez-Cano et al., 2010). The Get up and go test evaluates the patient's position when sitting, standing, walking and changing direction. Concretely, it requires the patient to stand up from a chair, walk three meters, turn around, return, and sit down again. The results are expressed on a scale from 1 to 5, where 1 is reduced risk of falling and 5 high risk of falling. The robotic platform has to give instructions to the patient, position itself in a proper location to observe the complete motion, and provide a signal to start the test. For successful automation of the test, the robot needs to perceive the gait and to analyze balance and timing issues.

To reply both tests, the elderly patients of the care centers participants in the study must fill the inclusion and exclusion criteria described in next section.

Figure 6. Eighth question of Barthel test.

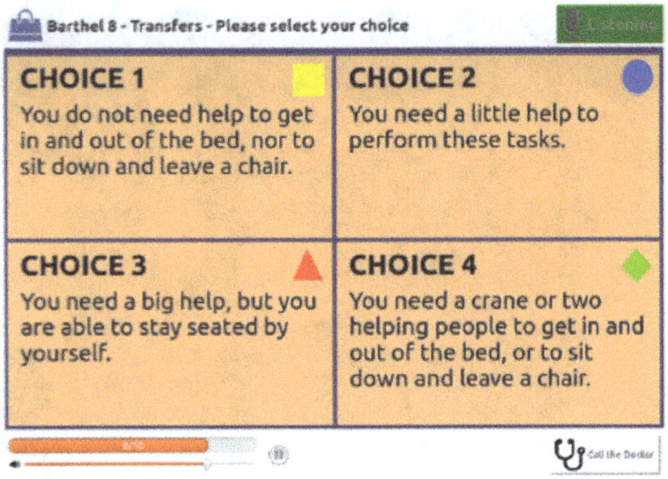

INCLUSION AND EXCLUSION CRITERIA

The inclusion criteria to participate in the study were:

- Present ability to collaborate with the study.
- Give written consent for inclusion in the study.
- Over 65 years of age.

The exclusion criteria defined to the study were:

- History of psychiatric illness that makes collaboration in the study impossible.
- No written informed consent to participate in the study.
- Minimental test less than 23 points.
- Severe hearing and visual impairment.

In fact, patients in wheelchairs or with severe motor impairments were rejected for the Get up and go test but participated in the pilot through the Barthel's test.

PILOT STUDY

The use of the robot in real scenarios had been carried out to validate the technological solution designed and developed. In order to do this, two robotic platforms were transferred to two health care centers in Seville (Andalusian, Spain) during three weeks: a primary health center and a retirement home. Figure 7 shows both robotic platforms.

Figure 7. Two CLARC robots

The two robotic platforms were identical at IT point of view and in internal components and interacted with patients in the same way. Only, they had difference in some physical appearance details that means whereas one robot had a screen with a face inside, the other one had eyes included inside the same casing, and in the material that is made of.

Participants in the research had to fill some inclusion criteria of age and minimal cognitive abilities to be able to understand and follow robot's instructions. In addition, to sign the informed consent was compulsory.

The authors would like to point out that the difference between the patients at the primary health center and those at the retirement home is that the latter weren't patients (they slept there). It was decided to carry out the evaluation in two different scenarios in order to obtain a wider range of age, since, as the results will show, the patients in the residence were older.

The design and develop of the technology found respond a real clinical need: reducing the time of evaluation employed in the CGA by clinician. Thanks to the pilot experience, it was possible to checking the usability, acceptability and interaction of the user with the robot, and in addition, to analyze the reliability of the evaluation's robot-human: those carried out by the robot in comparison with those carried out by the clinician responsible for the project.

Robotic platforms administered two of several tests most employed in the CGA, according to the request of clinician participants in CLARC project: Barthel test and Get up and go test.

In order to avoid the bias during the evaluation of a prior knowledge of the Barthel and Get up and go tests, it was decided to control this factor, altering the order of administration of the tests. It means, half of the sessions with patients were carried out following the procedure 1 (where the clinician first carried out the Test with the patient and then the patient interacted with the robot) and the other half following the procedure 2 (where the patient interacted with the robot first, being assessed by the robot, and then with the clinician), evaluating the first patient with the procedure 1 and the next one with procedure 2 and so on.

Usually, those participants who were given both tests began the evaluation session first with the Barthel test and secondly with the Get up and go test. However, it is worth mentioning that no variable has been controlled with respect to whether the order of administration of the different tests influences their results as a consequence of possible fatigue.

In patients with motor disabilities who could not walk without a technical device such as a cane, only Barthel's test was administered due to the fact that they could not walk three meters without help, as required by the Get up and go test.

On the other hand, two questionnaires related to the interaction between the robot and the patient were carried out. The *pretest* was performed once the validation of the inclusion/exclusion criteria for the patient was checked, and the *posttest* was performed after Get up and go and Barthel tests. These questionnaires include questions related to usability, accessibility, social acceptance, user experience and societal impact taking into account the familiarity of the patients with the tests and technology, age, sex or studies among other user characteristics, as well as functional and cognitive diversity.

PATIENT-ROBOT INTERACTION

Aiming at being as adaptable as possible to patients' abilities and preferences, the patient-robot interaction proposes a multimodal framework: voice technologies for both input and output (Automatic Speech Recognition -ASR- and Text-to-Speech -TTS- technologies), touch screen and a physical remote device. The patient could choose in each moment of the interaction the best mode for him/her to interact with the robot.

Initially, during the development of the robotic platform, only the inclusion of a touch screen and voice as a human-robot interaction was proposed. However, the direct contact with clinical experts in geriatrics and users in various usability tests carried out in the Virgen del Rocío University Hospital of Seville and in a retirement home in the same city, changed this initial approach. Both agreed that the touch screen was too modern an interaction for patients over the age of 65, accustomed to the conventional touch of buttons.

To make easier the learning phase of the interaction to the patients, before performing Barthel test, the robot carried out a training phase to train the patients in the use of the system and its different interaction modes. The training phase consisted in four questions which could be answered by the patient by choosing one of the four options proposed for each question.

The training phase of the Get up and go test consisted of a visualization of a demonstration video of a person performing the Get up and go test in front of the robot. Then, the patients should repeat the test watched in the video.

ROLE OF THE CLINICIAN IN CHARGE OF THE PROJECT

Two clinicians were responsible of managing the robot, completing the patient data on the interface through which the tests are launched and the robot collect the data and in administrate clinical tests in the established random order (form 1 or 2) and oversee the robot's execution.

The clinician, previous to the robot test, did not teach or anticipate how the robot would perform the test, but only performed the following actions:

- Told the patient that he was going to perform a clinical test with a robotic solution.
- Placed the patient in the chair in front of the robot.
- Gave to the patient the remote control with and said that he could interact with the robot with it.
- Inserted the patient's data into the platform and launched the tests so that the robot could begin to perform them.

Therefore, during the tests performed by the robot, the clinician did not interact with the patient unless the system/robot was turned off or the patient pressed the button to call the doctor (the called oz-wizard technique). The patient had to solve the doubts and perform the interaction and test only with the help of the robot.

While the patient was performing the test with the robot, the clinician filled in a table for the Barthel and another one for the Get up and go, several observations about the patient/robot interaction.

PILOT RESULTS

Barthel Test Results

The sample consisted in 111 patients. 66 of them were excluded because they didn't accomplish the inclusion criteria. Therefore, 45 patients were definitely enrolled in the evaluation.

There are three different reasons that explain why some participants were excluded. In particular, seven patients from the primary center and 59 in the retirement home. Between them, 57 patients did not fulfil inclusion criteria defined, as to have a Minimental punctuation less than 23 point (86.36% and reason 1). On the other hand, six of them had not enough time or lack of interest to perform the clinical tests (9.09% and reason 2). Lastly, there were three patients who due to severe hearing impairment could not be included in the study (4.54% and reason 3).

The patients from the primary health center were 74, 75 years old in mean, while the mean age in patients from the retirement home was 83.3.

Between participants enrolled in the evaluation, 29 of them belonged to the nursing home while 16 of them were evaluated in the primary health center. In reference to gender distribution there were 26 men (57,77%) and 19 women (42,22%) who had an average age of 79.67 years and a standard deviation was ± 7.33.

As figure 8 shows, remote control was the favorite mode to interact with the robot chosen by the patients, due to 41 of 45 participants chose answer to the robot with the remote control, which represents 91.11%. Concretely, 41 patients chose to answer 352 questions with it, it means, 78.22% of questions answered. In fact, only two patients used the touch screen to answer some questions (4.44% of the total number of patients). In particular, the touch screen was used 11 times (2.44% of the total questions answered). Three patients used their voice to interact with the robot (6.66%). They answered 20 questions in total, so, 4.44% of the total of questions were answered by this method. However, eighty questions were not answered, as they did not attempt to answer them. That represents the 17.77% of questions answered in the test.

Figure 8. Interaction mode with the robot.

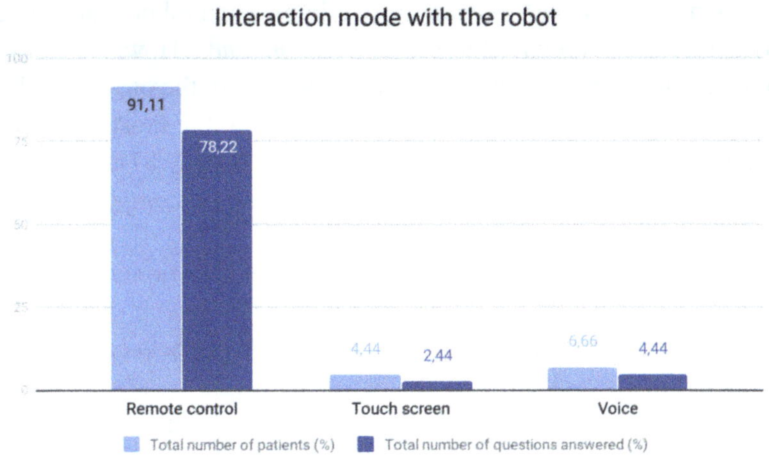

It is important to point out the percentage of the number of responses does not add up to 100% owing to some patients tried to use their voice to respond, but sometimes, the robot did not understand them and these patients finally used the remote control right afterward. This occurred because the robot is only programmed to understand a series of commands for voice interaction and these patients were trying to maintain a fluid conversation with the robot. Although this series of allowed commands were taught during the training, the patients were forgotten during the development of the test.

In the next table, table 1, the most important results of the Barthel tests are exposed, which has been analyzed by SPSS. It was an analysis made by category, according to the questions. It shows the percentage of answering to each question and their coincidence score of 100%. The table presents also the times where the Barthel test administered by clinicians had higher punctuations than robot punctuations and conversely.

Table 1. Analysis done by questions or categories.

Question	Total of Questions Answered	100% Coincident Score	Clinician Score > Robot Score	Clinician Score < Robot Score
1-Feeding	78%	51%	24%	3%
2-Bathing	87%	56%	15%	16%
3-Dressing	84%	44%	27%	13%
4-Grooming	82%	58%	17%	7%
5-Fecal incontinence	78%	38%	29%	11%
6-Urinary incontinence	80%	60%	4%	16%
7- Toilet use	84%	49%	24%	11%
8-Transfers	78%	47%	18%	13%
9-Walking	82%	42%	20%	20%
10-Climbing stairs	89%	56%	17%	16%
MEAN	82.2%	50.10%	19.5%	12.6%

The percentage of answering was high and satisfactory, 82.2%. Approximately, in 50% of the times of the evaluations the punctuations, both punctuations, clinician and robot, were coincident. The answer with higher coincidence score was referred to *urinary incontinence*. However, some patients gave different responses to the clinician than to the robot, as can be seen in the previous table.

Upon the results of the evaluations of Barthel test conducted, the most answered question was the number 10 (89%), refers to *climbing stairs*, followed by the number 2 (87%), the *bathing*. Attending to the least answered questions (78%), these referred to *feeding, fecal incontinence* and *transfers* from the bed to chair and back.

The mean of coincidence score was 50% approximately. The question with the highest coincidence was the sixth, *urinary incontinence* (60%). Then, it highlights *grooming* (58%). Though, the question about *walking* was the one in which there was the least agreement between the answers given to the robot and the clinician.

On the other hand, it occurred that, in questions were not had coincidence in the answers gave to clinician and robot, there was more frequent about than scores registered by clinicians were higher than these that patients gave to the robot: 19.5% vs 12.6% respectively.

The researchers of the study are awake that the Barthel test results show not very stimulating, but it was the first experience with the platform. Mindful of this, the research group raise to enlarge the sample and improve the technical solution to achieve stronger data and get possible the actual use in clinics.

Next, the analysis of the coincidence index of the scores given by the patients to the clinician with respect to those given to the robot was deepened. It means, a finely comparing analysis answer to answer was made for each one of the questions to determine 100% coincidence, 80% coincidence and 50% coincidence for each question. The results show 18% of patients with 100% of coincidence (question by question). Nevertheless, if the degree of coincidence decreases, the results improve gradually. If the coincidence scores admitted is 80%, the result is 35.5%, while, its increase if the degree is 50% the result shows a coincidence of 46% (figure 9).

The clinician in charge of the pilot registered some observations in regard with the interaction of the patient to the robotic platform, so, according to some comments and remarks expressed by the patients, he could see how the evaluation sessions were elapsed. Five patients manifested not to understand any

Figure 9. Coincidence Barthel Index test scores

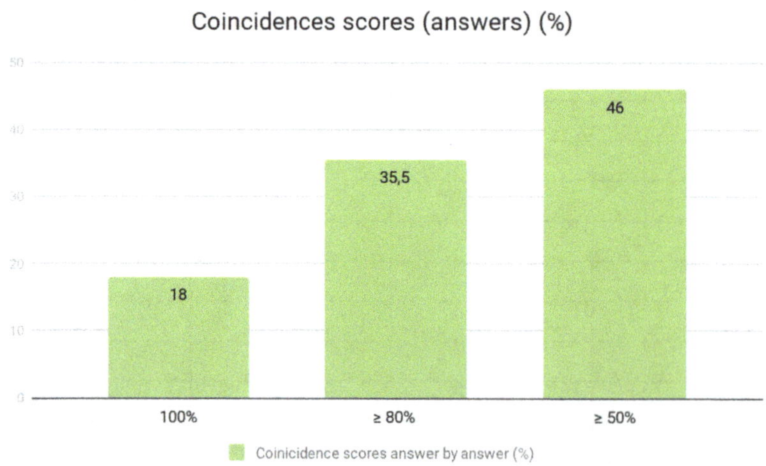

question, which happened in 20 of them. On the other hand, one patient, despite understanding questions, had doubts in three questions about the options because he didn't be identified with any option. Finally, severe difficulties as in reading robot questions were detected in eleven patients, and, besides, seven patients had severe problems with what the robot tells.

On the other hand, a few of patients tried to interact with the clinician at the same time that they answered the robot's tests, even when this person had advised the patient that this interaction during the test was not allow to avoid this risk in the bias. Concretely, 13 patients did it, (28.88% of the total number of patients), in 42 different occasions. However, only three of 45 patients evaluated tried to talk to the robot (6.66% of the total).

Relative to the interaction patient-robot with the remote control, it's must be mentioned that 11 patients showed that they didn't understand the relation between the option mentioned by the robot and his correspondence with the button of control remote. This fact represented 24.44% of the total patients, who had this problem in 85 questions.

Get Up and Go Test Results

Nineteen patients were evaluated, where, eight of them were evaluated in the primary health center and eleven of them in the retirement home. From these tests, only one of them, carried out in the health center, was correctly finished due to some difficulties happened. The robot established score unavailable if it was not sure of the development that the patient had done of the test, which happened in three occasions. However, the mean difficulty was that in certain cases, the system was not able to detect that the patient had completed the test, so it did not give a score to the patient's performance, which happened in 15 of them.

The clinician in charge of the pilot observed that the main reason why the technological system did not always score patients was because many times, patients did not respect all the indications that the robot gave them during the explanations about how it was necessary to perform the test. The system required that the patient did some specifically steps, which not always were done. In the primary health center, they were perfectly respected for seven patients, but not respected only by one patient. Nonetheless, in the retirement home the steps were correctly respected by four patients, while seven of them didn't do it.

This person detected that the robot terminated the evaluation session on 4 occasions despite the fact that such data were not dumped or reflected in the robot interface, while on 13 occasions, the sessions were not concluded.

RESULTS DEALING WITH THE INTERACTION BETWEEN THE PATIENT AND THE ROBOT

Related to previous use of the technology, none of the users interacted before with robots, but most of them (all except one) wanted to interact with robots. Related to the use of other technology, all the users know what a smartphone or personal computer is. 8.33% of the users never used a smartphone and 10.42% of the user use the smartphone continuously, using it for phone calls and instant messages (WhatsApp app) among other services. Related to personal computers or tablets, 77.08% never used a tablet or personal computer and 6.22% of the users use the tablet or personal computer continuously.

The reason that the preferred option to answer was the remote control may reside in that for the elderly, the tactile option is still a very modern concept and they prefer to feel the analog touch of the buttons. In addition, the fact that the voice option has not been used, it is because when having the screen and / or remote control, they focused on one of those two options forgetting almost all that they could also interact through the voice.

Related to social acceptability (Venkatesh et al., 2003), the geriatric patients had a very positive attitude towards the robot: they declared feeling safe, comfortable, not being intimidated, and interestingly, that they had the impression of "really interacting" with the robot (not just responding to a machine).

Usability (Nielsen, 1994) was carefully evaluated during the robot's pilots (during the design process) and the evaluations, and more importantly has been the object of careful design, so as to make the interaction as "easy" as possible. Characteristics as the users' specific needs, abilities and previous use of technology, as well as the demographic data collected in the *pre-test* questionnaire, help us to determine and improve the usability of the system. On the one hace, the authors observed either an absence or very limited use of technology. Yet, despite this lack of familiarity with technology, the usability results obtained were very positive. Most of the users were able to understand (hear or read) what the robot was asking. Moreover, most of them though that the robot is easy to use for the tests and the explanations given by the robot on each task to be performed are clear and easy to understand. Moreover, they that the robot behaves in a flexible way in terms of interaction (voice/touch screen/remote control) and they were able to answer without any problem by using one of this interaction modes.

Related to accessibility (Petrie & Bevan, 2009), the user tests allowed the observation that users were having when interacting with CLARC despite the usability evaluation and improvements done in the robot. The problem was therefore not usability, but the extent to which most of elderly patients had sensory impairments (visual and hearing) as well as cognitive impairments, due mainly to the age usual degeneration, which prevented them from interacting efficiently with the robot (or any other technology). A deep analysis of the accessibility barriers has been carried out by an accessibility expert, then an accessible design of the robot was proposed, including multimodal interaction: the patients could choose in any moment the mode of interaction with the robot: by voice, touch screen, written text – captioning -, physical buttons in a remote control, visual cues (colors, forms, etc.), among others. In this way, the main aim of this part of the evaluation is to adapt the robot to be able to interact with the patients in an inclusive way, nevertheless of the patients' characteristics, abilities/disabilities, needs and preferences, facilitating the interaction.

FUTURE RESEARCH DIRECTIONS

Nowadays we are working on defining a complete methodological framework to be able to evaluate in an orthogonal mode the performance of the robots in different scenarios from a Human-Robot Interaction (HRI) point of view. In the last years, different metrics and methodologies have been defined, evaluating factors as usability, user experience, social acceptance or societal impact among others. However, none of the current methodologies include one of the most important factors to have into account in interaction: whether all the users are able to interact with the system with the same opportunities and easiness. Therefore, Accessibility is defined as an evaluation factor to be taken into account in HRI, describing the metrics and methodology to evaluate this factor.

Moreover, the interface of CLARC robot is going to be improved to avoid the specific problems that we have detected during the evaluation tests.

Finally, the robot should be included in a more ambitious physical ecosystem integrating external and wearable sensors, virtual assistants and/or other actuators. In this scenario, the robot will be one of the interfaces for interacting with this smart environment, designed to provide additional functionalities to the inhabitants of residence homes or care units.

CONCLUSION

It is well known that robotic platforms for use in the medical field is a challenge that has been underway for several decades. The most common robotic devices are related to surgery because they have high precision and miniaturization. A good example of this are the technological advances that lead to the improvement in the surgical technique of the minimally invasive medical procedures.

At present, it exists physically assistive robots as robotic articulations, for example, who are very beneficial to his users but do not combat the increasing decline typical in the elderly population. More aligned with this need are the Socially Assistive Robots (SAR). These robots are focused on socially interactive technology that provide assistance physical and social. Abdi et al. (2018) reviewed different roles fulfilled by Socially Assistive Robots destined as tool with seniors. Among their main functions, the authors highlight they use as social facilitator, companionship, cognitive training and affective therapy and physiological therapy. Another function in health field is provide feedback on exercise quality and needed improvements and they courage and motivate to continue in these executions (Mataric, 2017). However, to provide assistance to health staff in their daily tasks such as the evaluation and treatment of elderly people is faced with major challenges as it attempts to close the huge gap between those who need services. In this line, the research just exposed presents a challenge in trying to respond to this need, not previously addressed to the extent of our knowledge. It is important to note that this type of initiative is not proposed as a replacement for health professionals. In contrast, robots can serve as clinical tools and assistants in a wide range of settings, enable them to have more time to be spent for care planning decisions itself (the analytic and comprehensive final step of CGA) instead to spend very valuable time for just doing tests. In fact, many of the older people who participated in the pilot said that when they saw the robot for the first time, they felt reluctant to participate as they thought the robot was meant to demonstrate its effectiveness and therefore replace the professionals who regularly carried out evaluations. After carrying out the tests they understood that it was really a tool for the doctors, which would facilitate their work.

In the Get up and go test is necessary to point out that most of the errors recorded during the execution of the risk fall test with the robotic platform were due to it was necessary, and essential, for the good development of the test that patients respected each step it required. For that purpose, a video with an example of a person performing the exercise correctly was shown in the platform and the detection system of the robot and the Kinect camera need the patient to be located at the beginning of the test next to the chair. In fact, the robot gave express indications at various times to explain this. If patients do not respect this premise, and started sit down, the detection of the exercise was erroneous. That's why there were some non-scored executions: the failure in the starting patient's position of the exercise, produced failures in the beginning of the detection, reason why the final punctuation was not calculated by the robot.

Our conclusion after the pilot of the Get up and go test, when performed with the robot, is that our results are poor and the tool is not yet valid to implement it in a real scenario, so it is necessary to continue with the development of the technology implemented in the robot. The authors accept that the robotic platform is so sensitive to the characteristics of the pilot's room (such as natural light, the specificities of the chairs, etc.), the degree of inclination of the Kinect camera, etc., that to obtain better results the researchers conclude that changing the detection camera is necessary. In fact, currently, part of the team that has participated in the research is implementing another patient detection system that is more robust and facilitates the punctuation of patients who perform the test of gait and balance. The reasons could be a natural light source directly in the Kinect camera, chair with armrest., etc. This experience and pilot have shown us certain limitation that will be more controlled in next opportunities to administrate Get up and go test.

Initiatives such as the one proposed in this research advocates an approach to technology. For most of the elderly people who have accessed and participated with the technology own designed and developed, has been its first occasion to interact with technological devices and, even more, with a robot. Currently, older people who use technological advances, do it to stay in touch with their relatives or to assist them in housekeeping. According with Akbar et al. (2018), technological develops are focused in provide devices to monitor medical records of health status in real time, but their physical and mental increasing deterioration need to benefit from the technology to improve the daily life. Tools as CLARC robot facilize healthcare professional quotidian work and, involve, to elderly people an opportunity to integrate technology in their environment.

REFERENCES

Abdi, J., Al-Hindawi, A., Ng, T., & Vizcaychipi, M. P. (2018). Scoping review on the use of socially assistive robot technology in elderly care. *BMJ Open*, *8*(2). doi:10.1136/bmjopen-2017-018815 PMID:29440212

Akbar, M. F., Ramadhani, N. A. Y., & Putri, R. A. (2018). Assistive and wearable technology for elderly. *Bulletin of Social Informatics Theory and Application*, *2*(1), 8–14. doi:10.31763/businta.v2i1.101

Bandera, J. P., Marfil, R., Romero-Garcés, A., & Voilmy, D. (2019). A new paradigm for autonomous human motion description and evaluation: Application to the Get Up & Go test use case. *Pattern Recognition Letters*, *118*, 51–60. doi:10.1016/j.patrec.2018.02.003

Bengtson, V. (2018). *Global aging and challenges to families*. Routledge. doi:10.4324/9781351328166

British Geriatric Society. (2019). Comprehensive Geriatric Assessment Toolkit for Primary Care Practitioners. British Geriatric Society. Retrieved from https://www.bgs.org.uk/sites/default/files/content/resources/files/2019-02-08/BGS%20Toolkit%20-%20FINAL%20FOR%20WEB_0.pdf

Bustos, P., Manso, L. J., Bandera, A., Bandera, J. P., García-Varea, I., & Martínez-Gómez, J. (2019). The CORTEX cognitive robotics architecture: Use cases. *Cognitive Systems Research*, *55*, 107–123. doi:10.1016/j.cogsys.2019.01.003

Cid-Ruzafa, J., & Damián-Moreno, J. (1997). Valoración de la discapacidad física: El índice de Barthel. *Revista Espanola de Salud Publica, 71*(2), 127–137. doi:10.1590/S1135-57271997000200004 PMID:9546856

Ellis, G., Gardner, M., Tsiachristas, A., Langhorne, P., Burke, O., Harwood, R. H., ... Wald, H. (2017). Comprehensive geriatric assessment for older adults admitted to hospital. *Cochrane Database of Systematic Reviews*, 9. PMID:28898390

Gálvez Cano, M., Varela Pinedo, L. F., Helver Chávez, J., Cieza Zevallos, J., & Méndez Silva, F. (2010). Correlación del Test" Get Up and Go. *Acta Médica Peruana, 27*(1).

Jatsun, S. F., & Yatsun, A. S. (2018). Criterion of the rehabilitation process effectiveness on the basis of biomehatronic system ExoLite Rehab. In *2018 IEEE Workshop on Advanced Robotics and its Social Impacts (ARSO)* (pp. 105-110). IEEE. 10.1109/ARSO.2018.8625840

Korchut, A., Szklener, S., Abdelnour, C., Tantinya, N., Hernández-Farigola, J., Ribes, J. C., ... Rejdak, K. (2017). Challenges for service robots—Requirements of elderly adults with cognitive impairments. *Frontiers in Neurology, 8*, 228. doi:10.3389/fneur.2017.00228 PMID:28620342

Mann, E., Koller, M., Mann, C., van der Cammen, T., & Steurer, J. (2004). Comprehensive Geriatric Assessment (CGA) in general practice: Results from a pilot study in Vorarlberg, Austria. *BMC Geriatrics, 4*(1), 4. doi:10.1186/1471-2318-4-4 PMID:15151704

Matarić, M. J. (2017). Socially assistive robotics: Human augmentation versus automation. *Science Robotics, 2*(4).

Nielsen, J. (1994). Usability engineering. Amsterdam, The Netherlands: Elsevier.

Petrie, H., & Bevan, N. (2009). The Evaluation of Accessibility, Usability, and User Experience. *The universal access handbook, 1,* 1-16.

Pulido, J. C., González, J. C., Suárez-Mejías, C., Bandera, A., Bustos, P., & Fernandez, F. (2017). Evaluating the Child–Robot Interaction of the NAOTherapist Platform in Pediatric Rehabilitation, *Int J of Soc Robotics*, 9, 343. doi:10.100712369-017-0402-2

Rubenstein, L. Z. (2004). Comprehensive geriatric assessment: From miracle to reality (JT Freeman Award Lecture). *J Gerontol A Biol. Medical Science, 59,* M473–M477.

Sasaki, T., Saraiji, M. H. D., Minamizawa, K., & Inami, M. (2018). MetaArmS: Body remapping using feet-controlled artificial arms. In *The 31st Annual ACM Symposium on User Interface Software and Technology Adjunct Proceedings* (pp. 140-142). ACM.

Stuck, A. E., & Iliffe, S. (2011). Comprehensive geriatric assessment for older adults.

Ting, K. L. H., Voilmy, D., Iglesias, A., Pulido, J. C., García, J., Romero-Garcés, A., . . . Dueñas, Á. (2017, August). Integrating the users in the design of a robot for making Comprehensive Geriatric Assessments (CGA) to elderly people in care centers. In *2017 26th IEEE International Symposium on Robot and Human Interactive Communication (RO-MAN)* (pp. 483-488). IEEE.

Venkatesh, V., Morris, M. G., Davis, G. B., & Davis, F. D. (2003). User acceptance of information technology: Toward a unified view. *Management Information Systems Quarterly*, *27*(3), 425–478. doi:10.2307/30036540

Vercelli, A., Rainero, I., Ciferri, L., Boido, M., & Pirri, F. (2018). Robots in Elderly Care. *DigitCult-Scientific Journal on Digital Cultures*, *2*(2), 37–50.

Ward, K. T., & Reuben, D. B. (2016). Comprehensive geriatric assessment. *UpToDate, Waltham, MA. Accessed, 4,* 03-19. Retrieved from http://www. uptodate. com/contents/comprehensive-geriatric-assessment

ADDITIONAL READINGS

Bengtson, V. (2018). *Global aging and challenges to families.* Routledge. doi:10.4324/9781351328166

Korchut, A., Szklener, S., Abdelnour, C., Tantinya, N., Hernández-Farigola, J., Ribes, J. C., ... Rejdak, K. (2017). challenges for service robots—Requirements of elderly adults with cognitive impairments. *Frontiers in Neurology*, *8*, 228. doi:10.3389/fneur.2017.00228 PMID:28620342

Mucchiani, C., Sharma, S., Johnson, M., Sefcik, J., Vivio, N., Huang, J., ... Lau, T. (2017, September). Evaluating older adults' interaction with a mobile assistive robot. In *2017 IEEE/RSJ International Conference on Intelligent Robots and Systems (IROS)* (pp. 840-847). IEEE. 10.1109/IROS.2017.8202246

Portet, F., Vacher, M., Golanski, C., Roux, C., & Meillon, B. (2013). Design and evaluation of a smart home voice interface for the elderly: Acceptability and objection aspects. *Personal and Ubiquitous Computing*, *17*(1), 127–144. doi:10.100700779-011-0470-5

KEY TERMS AND DEFINITIONS:

Accessibility: The extent to which a product (or robot) can be used without discrimination and equivalent user experience by the users in a specific context of use.

Barthel Index: Extensive measuring instrument used with older people for the evaluation of activities of daily living.

CGAmed: A local server able to store a database with all captured data and to provide the physicians with the tools for online monitoring and offline editing and supervision.

Comprehensive Geriatric Assessment: is a very useful and instituted multidimensional and multidisciplinary diagnostic instrument that provides personalized healthcare to the elderly, by means of evaluating their physical, functional and mental state.

ECHORD++: European Clearing House for Open Robotics Development Plus Plus (E++ for short).

Get up and go Test: Measuring instrument employed to evaluate the degree of risk of falling of patients in order to prevent and evaluate their treatment.

Technology Acceptation: The degree to which a person believes that using a product (or robot) would be useful for him/her and s/he has usage intention.

Usability: The extent to which a product (or robot) can be used with effectiveness, efficiency and satisfaction by the users in a specific context of use.

ENDNOTE

1 http://www.mira-project.org/joomla-mira/

Chapter 3
Development of a Novel Parallel Structure for Gait Rehabilitation

Rogério Sales Gonçalves
Federal University of Uberlândia, Brazil

Lucas Antônio Oliveira Rodrigues
Federal University of Uberlandia, Brazil

ABSTRACT

This chapter deals with the development of a new active body weight support, based on a parallel robotic structure with 5 degrees of freedom. Initially, a revision of the main structures applied in body weight support for gait rehabilitation is presented, along with a conceptual model of the proposed structure. Therefore, the inverse kinematics and singularity analysis are performed. A dynamic model is then built based on the obtained kinematics, and all parts of the structure are optimized using an evolutional algorithm. Finally, a CAD model of the structure is built based on the optimized dimensions and the model is applied to computational simulations of the gait rehabilitation, being thus ready for prototype construction.

INTRODUCTION

Rehabilitation is an area of medicine and physiotherapy that deals with the treatment of chronical neurological diseases and injuries, caused during birth, or by work or transit accidents. The main objectives of this science is to manage a dynamic process over patients to achieve physical recover in order to regain global functional skills (Rogério S. Gonçalves, 2010).

The rehabilitation process begins with pathology identification and its disabilities, followed by improvements projection based on the rehabilitation potential. Hereafter an appropriate treatment is prescribed, applying medical-surgical action if needed, and finally proceeding with physical stimulations to maximize residual capacities by natural recovery and regeneration processes (Barbosa, 2013).

DOI: 10.4018/978-1-7998-0137-5.ch003

Recovery of physical functions that promote functional independency for patients contributes for self esteem increase, joy for life and sensation of overcoming difficulties, implying in improving life quality (Camargos et al., 2012).

Robotics has a wide area of application which includes physical rehabilitation, acting in coadjustment and support to medical-surgical actions; facilitation and stimulation of the processes of natural recovery and regeneration; and stimulation, maximization and compensation of residual capacities (Nunes, 2012).

Treatments applying robotic structures are common to victims of Stroke (Babaiasl, Mahdioun, Jaryani, & Yazdani, 2015; Louie & Eng, 2016; Lum, Reinkensmeyer, Mahoney, Rymer, & Burgar, 2002); Cerebral Palsy (CP) (Wu, Hwang, Ren, Gaebler-spira, & Zhang, 2011); and Spinal-Cord Injury (SCI) (Jezernik, Colombo, & Morari, 2004; Mekki, Delgado, Fry, Putrino, & Huang, 2018).

The rehabilitation robotic systems can be defined as an example of a new unifying interdisciplinary engineering science, as can be classified as a Human Adaptive and Friend Mechatronic (HAFM). The main goals of this new era of Mechatronics are to provide smart environments, reducing repetitive and stressful routines, enhancing productivity and contributing to a better life. HAFM are changing the concept of the machines, where their new smart features allow them to behave as a partner rather than a simple tool (M. Habib, 2007; M. K. Habib, 2008).

The application of robotic structures in treatments mentioned above has successfully benefitted medicine and physiotherapy with active labor cost reduction for movement-based procedures, and with the expansion of the range of therapy exercise, helping chronical patients to maintain mobility by continuous therapy (Lum et al., 2002).

In addition, Stroke, in most cases, causes impairment to the victims, while being the third most common cause of death in United States and Europe (Lloyd-Jones et al., 2010), and the most frequent in Brazil, causing a large social and economic impact (Brasil, 2013). These fact highlights the importance of Stroke treatment, therefore encouraging robotic research to this application.

It is estimated that for every 100,000 Brazilians, there will be 108 annual stroke cases, with fatalities up to 30 days to 18.5% of cases and up to 12 mouths to 31% of cases (Brasil, 2017). In average, a person in United States will have a stroke every 40 seconds (Lackland, 2017).

In this chapter, a novel robotic structure applied in human gait rehabilitation of Stroke, CP and SCI patients is proposed. The conceptual model of this structure is shown in Fig. 1. Along the next sections, a literature review of the robotic structures applied in gait rehabilitation and human gait kinematics provides the background of this research.

Next, the mathematical, kinematic and dynamic models of the structure are presented, followed by a singularity analysis of the obtained kinematic chain. Then, based on the obtained models, the dimensions of each part of the structure are optimized using an evolutionary algorithm. After, CAD models of the structure are built and used to generate computational simulations.

Finally, the first actuator prototype of the structure and its control system are built. Load-less experimental tests are used to tune the control parameters to velocity and position control. The obtained results are then analyzed in order to present the next steps and perspectives of the development process.

Figure 1. Conceptual model of the novel parallel structure for gait rehabilitation.

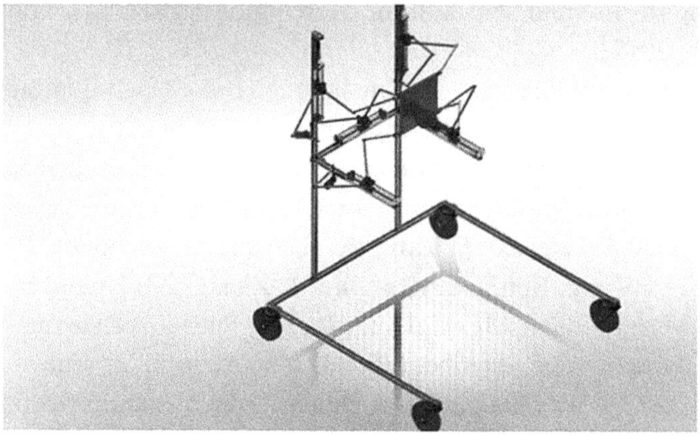

BACKGROUND

In order to provide broad definitions of rehabilitation robotics and build the conception of the proposed project, a literature review of existing rehabilitation robotic structures, applied for human lower-limb rehabilitation, is presented in the next subtopic, followed by motivations and goals to propose an active bodyweight support able to assist pelvic motion during exercises sessions.

Review of Rehabilitation Robotic Structures

Analyzing the studies and papers regarding robotic structures applied to human lower-limb rehabilitation processes, four main different types of structures can be defined (Rodrigues, 2017). These four types can be described as:

1. Treadmill based rehabilitation structures;
2. Programmable feet base rehabilitation structures;
3. Mobile base rehabilitation structures;
4. Stationary base rehabilitation structures.

A simple illustrative scheme of these four types of structures is presented in Fig. 2. In this chapter, types (1), (2) and (3) are reviewed due its dependence on bodyweight support devices, while type (4) defines only structures connected to chairs, hospital stretchers, etc.

Treadmill Based Rehabilitation Structures

One example of a treadmill-based rehabilitation structure is called LOPES (Veneman et al., 2007). In this work, an exoskeleton designed to be attached to patients' lower-limbs over a treadmill is presented. The structure has a 2 degree-of-freedom active bodyweight support that releases the passive weight

Figure 2. Types of Rehabilitation robotic Structures: (a) treadmill based; (b) Programmable feet base; (c) Mobile base; and (d) Stationary base.

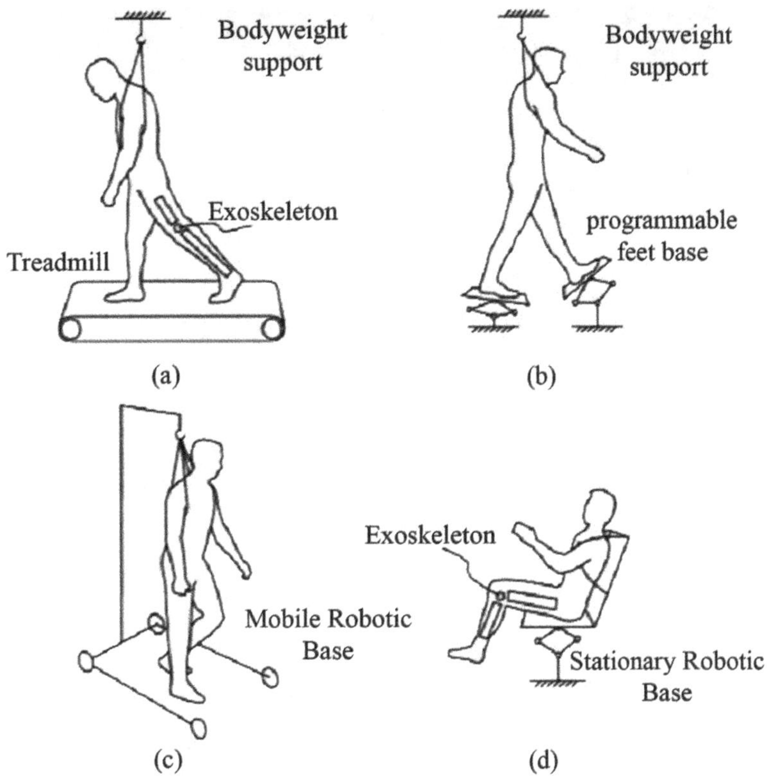

from the lower-limbs and act on the pelvic joint. The main actuators are composed by two serial robotic arms directly connected on patients' legs, using two actuators to assist hip movements and one to knee movements.

LOPES structure uses mechanical impedance control on all its actuators, allowing bidirectional interactions between user and equipment, i. e., the desired movement trajectory for both limbs can be adjusted and followed by robotic or human influence. Images of LOPES structure during operation are presented in Fig. 3.

A module proposed by Roberts (2004) also explores bodyweight supporting while active assisting pelvic movements. This structure is proposed to integrate a rehabilitation treadmill system, using a flat parallel mechanism with 2 degree-of-freedom and a prismatic joint on vertical position.

This module allows the complete rehabilitation system to release the passive weight from the lower-limbs, while automatically adjusting its position and assisting the dynamic stability and propulsion during gait. The main goal of this structure is to prevent falls and balance loss during the exercises, while keeping movements comparable to natural human gait. The Pelvic actuation of this robot is restricted to movements on the transverse plane of the joint, keeping patients' vertical movements free and non-actuated. The prototype of the structure is presented in Fig. 4.

Figure 3. Images of a patient performing rehabilitation exercises assisted by LOPES exoskeleton (Font: http://balance-fp7.eu).

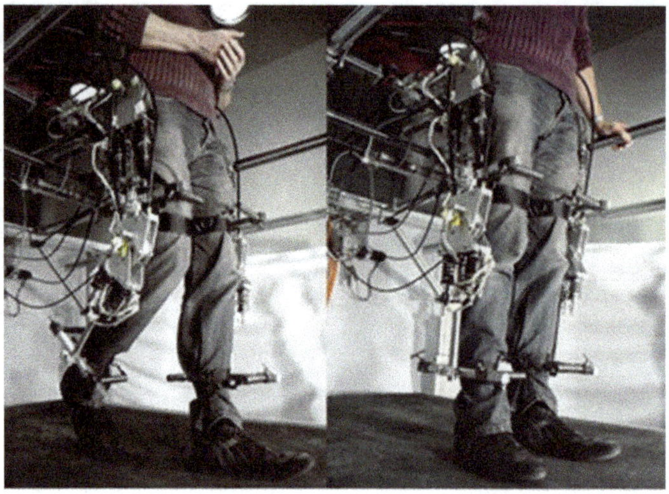

Another example of treadmill-based structures is called "Robotic Gait Rehabilitation" (RGR) (Pietrusinski et al., 2010). It consists on a rehabilitation treadmill system designed specifically to assist treatment of gait secondary deviations in post-stroke patients.

The RGR robot is connected on the pelvic joint of patient using an orthosis equipped with electromagnetic actuators that induces, whenever needed, postural corrections along the gait exercises. The structure also contains an active bodyweight support that assists the corrections. An image of the RGR during operation is presented in Fig. 5.

The "Pelvic Assist Manipulator" (PAM) (Aoyagi, Ichinose, Reinkensmeyer, & Bobrow, 2004) is another rehabilitation structure focused on assisting treadmill exercises by actuating on the pelvic joint. The movements are generated by 2 pneumatic cylinders organized on a tripod structure, measuring and controlling the pelvis position during all human gait phases over the treadmill. An illustrative scheme and an image of the PAM robot are shown in Fig. 6.

A cartesian robot was also designed to assist rehabilitation exercises actuating on pelvic joint (Salguero-Beltran et al., 2012). It is constructed in order to allow the ambulation of the user, while applying a serial structure as a bodyweight support, and using its 4 degree-of-freedom robotic arm to assist on balance and prevent falls, as shown in Fig.7. First tests of this structure demonstrate that the proposed electrical and software architecture is appropriated to stabilize the actuating mechanism.

One of the most classical rehabilitation devices that apply a bodyweight support and a treadmill is the LOKOMAT (Jezernik et al., 2004). This robotic orthosis features two 4-DOF serial structures that actuates directly on the lower limbs during rehabilitation training sessions. The system is controlled applying adaptative algorithms able to track and assist patients gait patterns on a treadmill. An image of this equipment is shown in Fig. 8.

Figure 4. Assembly of the active bodyweight support structure
(Roberts, 2004).

Programmable Feet Base Rehabilitation Structures

Hitherto, all presented structures are from type (1) (Rodrigues, 2017), designed to work associated with active treadmills. Next, the literature review shows the type (2) of rehabilitation robots which main characteristic is the actuators positioned below the patients' feet, called programmable bases.

The first example of programmable feet base structure is the "Gangtrainer GT I" robot (Hesse & Uhlenbrock, 2000), distributed by Reha-Stim®, shown in Fig. 9. This robot is equipped with a simple cable-based bodyweight support, two actuated plates, whose movements simulate stance and swing on patients' feet, and ropes attached to the patient to perform corrections on vertical and lateral positions during the exercises.

Figure 5. Image of a volunteer operating the RGR robot
(Pietrusinski et al., 2010).

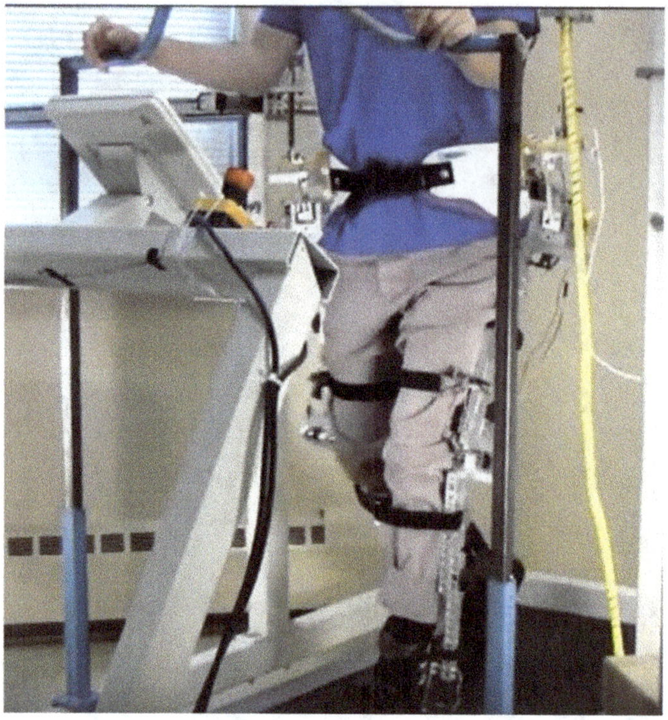

Figure 6. (a) Scheme of the PAM robot; and (b) an image of a volunteer using its prototype
(Aoyagi et al., 2004).

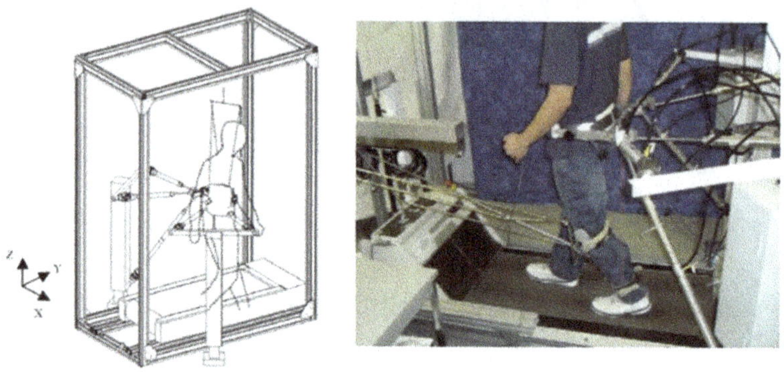

(a) (b)

Clinical studies has shown that the Gangtrainer GT I has, at least, an equivalent efficiency of treatment when compared to conventional methods, with the benefit of promoting independency to the health professionals during the exercises (Hesse, Schmidt, Werner, & Bardeleben, 2003; Peurala et al., 2009).

Figure 7. (a) Image of a volunteer operating the serial structure; and (b) its assembly scheme (Salguero-Beltran et al., 2012).

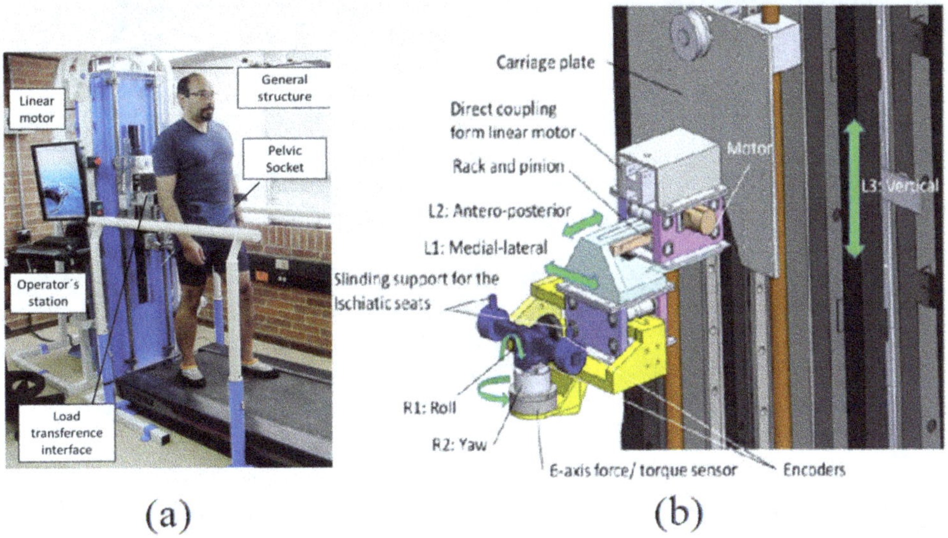

Figure 8. An image of a rehabilitation training session applying the LOKOMAT™ structure (Font: https://exoskeletonreport.com/product/lokomat/).

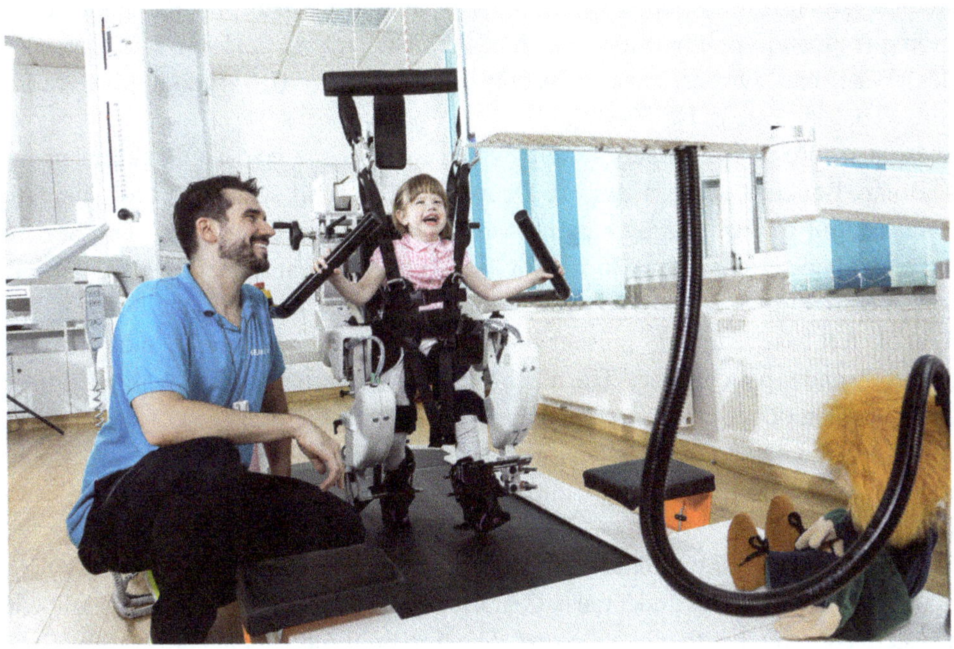

Figure 9. Image of the Gangtrainer GTI structure performing rehabilitation exercises (Hesse & Uhlenbrock, 2000).

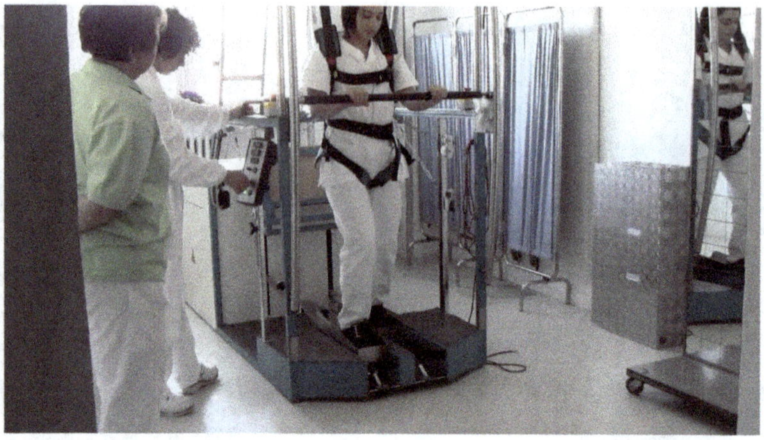

An similar structure, known as G-EO System (Galli et al., 2016) and distributed by RehaTecnology, can also be classified as a programmable feet base rehabilitation structure. Instead of cables, the G-EO System apply robotic arms with the foot plates as terminal elements to reproduce gait movements. Clinical tests of this structure has shown efficient treatment results on Parkinson Disease patients (Galli et al., 2016). An image of the G-EO System is presented in Fig. 10.

There are also structures developed specifically for ankle treatment, for the ankle sprain is one of the most common human joint impairment, with an incidence rate of 2.15 per 1000 person-years in the United States (Waterman, Owens, Davey, Zacchilli, & Belmont, 2010), and applies a complex bone structure (Sui, Yao, Lin, Yan, & Dai, 2009).

One of these structures is the Rutgers system, which consists on two 6-DOF parallel pneumatic actuated robots (Boian, Bouzit, Burdea, Lewis, & Deutsch, 2005). This structure uses kinematic variables to monitor and control movements on patients' ankles, while also interacting with a virtual graphic interface and electronic serious games to improve the entertainment during the sessions. The Rutgers system is presented In Fig. 11.

Programmable feet base structures present simpler control methods overall, once the movements are reproduced directly under patient's feet. The rehabilitation exercises provided by these structures rely on an indirect transmission of the gait movements from the foot plates to the rest of the limb. This method of gait training has shown good results on similar structures like the MIT-Skywalker, which can be defined as an hybrid structure, combining treadmill and programmable feet bases (Susko, Swaminathan, & Krebs, 2016).

The MIT-Skywalker, one of the main references of the HOPE-G structure, applies two basculation treadmills activated by linear cams under the patient's feet, while supporting its bodyweight with a passive suspension system combined with a bicycle seat. The treadmill movements are controlled by an image tracking system based on a high-speed camera and infrared markers (Rogerio S Gonçalves, Hamilton, Daher, Hirai, & Hermano, 2017). Recently, a markerless system is being evaluated to replace the high-speed camera (Gonçalves, Hamilton, & Krebs, 2017). An image of this structure is presented in Fig. 12.

Figure 10. The G-EO System
(www.rehatecnology.com).

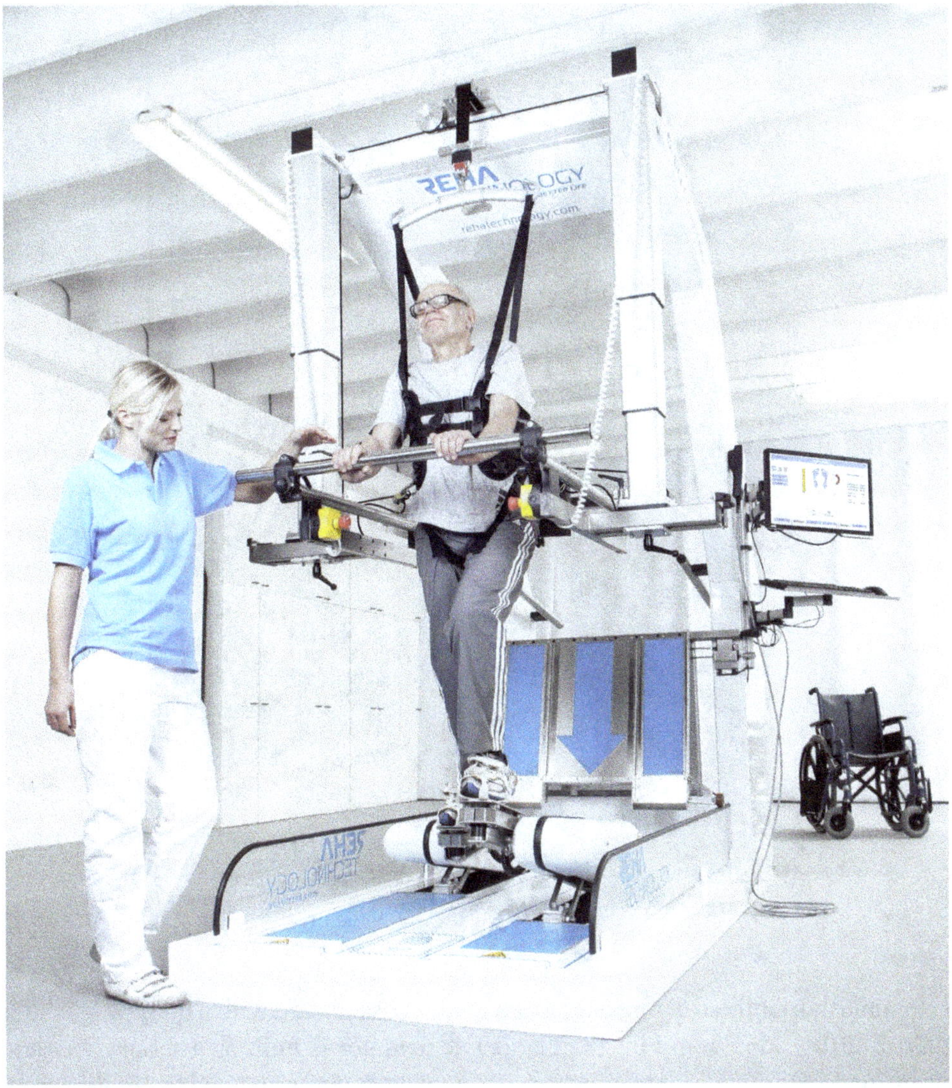

Mobile Rehabilitation Structures

The third type of rehabilitation structures is mainly characterized by a mobile bodyweight support that allow patients to perform overground exercises on large areas instead of a limited workspace of a conventional robotic structure. Some examples found in literature of these structures are presented next.

Some mobile structures can also be applied to assist daily tasks of people with low mobility. An example of this concept is the exoskeleton under patent owned by Goffer (US7153242 B2, 2006) and sold commercially on United States as ReWalk®. This structure is based on two robotic arms coupled directly on user lower-limbs, and pelvic actuators that assist balance during gait. An image of a user with a ReWalk® robot is seen in Fig. 13.

Figure 11. Image of a patient operating the Rutgers system
(Boian et al., 2005).

Another example of mobile structure uses a pelvic robot associated with a bodyweight support (Mun, Yu, Zhu, & Cruz, 2014), shown in Fig. 14. The pelvic actuator is built as a serial robot and assist the user along overground exercises, controlling the bodyweight suppression and facilitating movements. A study with this structure shows significant gait increases for different levels of bodyweight suppression (Mun, Lim, Guo, & Yu, 2017).

There are also mobile structures that applies cable-based mobile bodyweight supports. Examples of these structures are the named "The Float" (Bannwart, Rauter, Curt, & Bolliger, 2015) and the "Vector Gait & Safety", developed by Bioness™. Both examples are mobile robots installed near the ceiling of large exercise rooms, using trails to follow patients along the exercises routines and using cables connected to motors and actuators to suppress part of the bodyweight during the exercise sessions. Images of these structures are shown in Fig. 15.

Based on the features of the presented structures, the authors selected important characteristics to consider on the development of a rehabilitation structure, named HOPE-G, and used the MIT-Skywalker (Susko, 2015), and the bodyweight support proposed by Roberts (Roberts, 2004), as main references.

Figure 12. An image of a gait rehabilitation training session with the MIT-Skywalker (Font: https://tnsre.embs.org/).

Figure 13. The ReWalk robotic structure (Font: www.rewalk.com).

Figure 14. (a) Schematic of Mun's structure; and (b) its built prototype (Mun et al., 2014).

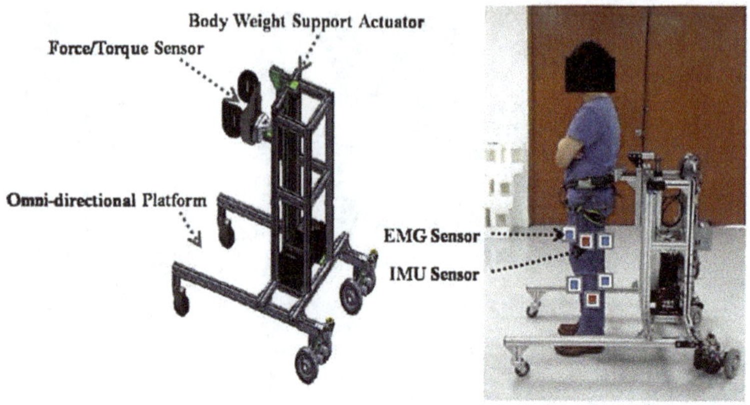

(a) (b)

Figure 15. (a) "The Float" robot (Bannwart et al., 2015); and (b) Vector Gait & Safety (www.bioness.com).

(a) (b)

Importance of the Pelvic Movements

The lower-limbs of the human body rely on its three joints, hip, knee and ankle, to perform gait properly. Although the main kinematics are actuated by these joints, other body movements have an important role in the regular gait. Some of these movements are performed on the pelvis, which is commonly immobilized during robotic rehabilitation exercises.

The pelvic movements are generated by the sacroiliac joint, which is a connection joint between the sacrum and ilium bones (Rafei et al., 2018). This joint has 3-DOF of rotation on the frontal, transversal and sagittal human reference planes, with small angular amplitudes (Roberts, 2004).

Some studies, as listed below, has shown good results on active assistance of pelvic movements with robotic rehabilitation structures and setbacks of limiting them.

There are several papers dealing with the effects of bodyweight supports that constrain the movements of pelvis and thorax on robotic rehabilitation treadmill structures, describing a significant balance variation on individual when compared with non-assisted traditional therapy (Hesse et al., 2003; Swinnen, Baeyens, Knaepen, Michielsen, Hens et al., 2015; Swinnen, Baeyens, Hens et al., 2015; Swinnen, Baeyens, Knaepen, Michielsen, Clijsen et al., 2015).

Another work shows that use of active bodyweight support that allow pelvic motion during the rehabilitation exercises allow some patients to perform a "pelvic step", using its rotation to assist the leg swing on gait. This method helps low hip mobility patients to increase their functionality gain (Liang et al., 2014).

There is also evidence that limiting pelvis movements can interfere significantly on the step length and body rotations on frontal and sagittal planes, recommending keep the joint free for a more efficient training (Veneman, Menger, van Asseldonk, van der Helm, & van der Kooij, 2008).

Although the sacroiliac joint features 3 rotation degrees-of-freedom, only frontal and transverse rotation are directly related to human gait (Roberts, 2004). Thus, the sagittal rotation does not require active assistance during training in order to improve balance recovery.

Therefore, it is evident that allowing and assisting pelvic movements can benefit the rehabilitation results of patients assisted by robotic devices. When applying active bodyweight supports, the system must deal with a 5-DOF motion, being 3 spatial translations related to position adjustments during gait, and 2 rotations on the sacroiliac joint, leaving the sagittal rotation unactuated.

HOPE-G: A NOVEL REHABILITATION STRUCTURE

After building a consistent background research on gait rehabilitation robotics, the authors decided to develop a gait rehabilitation structure suitable to small clinical applications and cost efficient. To take advantage of low space requirement, and to follow the good results obtained with the MIT-Skywalker (Susko, 2015), the proposed structure will be a hybrid type, applying a dual treadmill system to indirectly assist human gait and an active bodyweight support with a pelvic assistance.

This structure, named by the authors as "HOPE-G", is subdivided into three modules to facilitate the design process, listed below:

- **Bodyweight Support Module:** The main module of the HOPE-G structure, focus of this chapter. This module is designed as a 5-DOF parallel robotic structure actuating on patient's pelvic joint, while providing dynamic bodyweight suppression and assisting balance maintenance along training sessions.

- **Dual Treadmill Movement Module:** This module is described as two planar parallel structures with 1-DOF. The system applies pneumatic cylinders to perform basculation movements on the treadmills on order to induce the gait movements on patients' lower-limbs.
- **Markerless Control & Monitoring Module:** In order to synchronize the actions of the other modules with the lower-limb movements, this system applies an image processing system based on Microsoft Kinect® 2.0 to track specific points of patients' legs and send control signals to the respective actuators. The Markerless Control & Monitoring Module is already developed, and its first prototype built. Preliminary experimental tests with the prototype provided consistent evidence of its capability to control the other HOPE-G modules (Salim, 2018).

An illustrative scheme of the HOPE-G structure is presented in Fig. 16.

In order to control the lower-limb induced trajectories of the HOPE-G structure, the design is oriented in accordance with a theoretical framework of the lower-limb behaviors (Hogan & Sternad, 2013), dividing human gait into three primitives: discrete, rhythmic and balance.

The rhythmic primitive describes the regular human gait movements. The discrete primitive, also nominated as deambulation, deals with the gait learning process, where initial and final steps can be defined. This is an important phase in the rehabilitation process of patients with severe stroke impairments, as balance and muscular coordination need to be relearned in order to achieve a rhythmic gait again.

When properly assisted, the transition between discrete and rhythmic happens quickly, evidencing the importance of preparing HOPE-G structure to deal with patients in discrete gait primitive (Susko, 2015).

Figure 16. Conceptual model of the HOPE-G structure

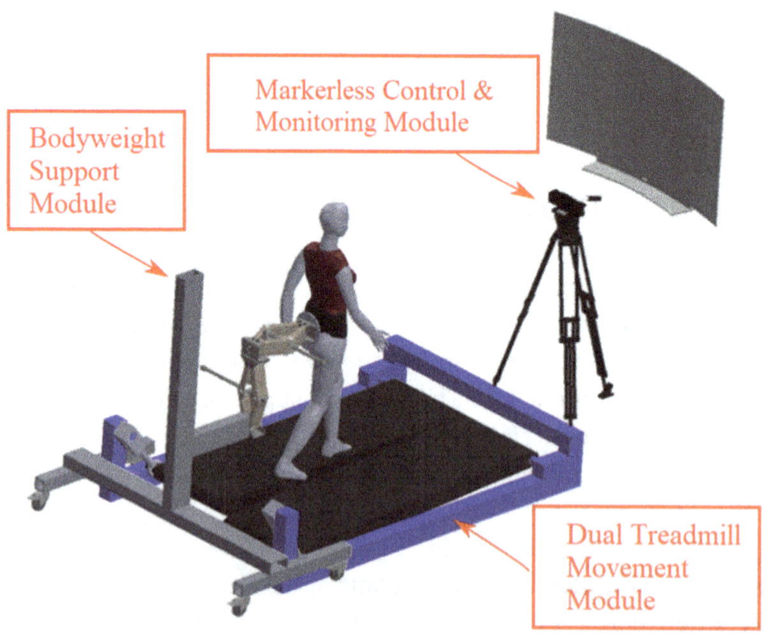

Balance, as the very name indicates, deals with the balance maintenance during the regular human gait. Neurological diseases as stroke and cerebral palsy significantly affects victims ability to keep the regular gait due to balance disorders (Nadeau et al., 2016). Thus, the bodyweight support will also allow balance training actuating on frontal and coronal planes, leaving only the sagittal rotation due its low influence under treatment results (Roberts, 2004).

In this chapter, the authors present the geometrical model, as well as the dynamic model for the bodyweight support module, as it represents the main innovation of the HOPE-G project. The development of the dual treadmill movement module is scheduled to be part of further work research of the authors.

KINEMATICS AND DYNAMICS OF THE BODYWEIGHT SUPPORT MODULE

Once the conceptual model of the HOPE-G structure is obtained, the next step on its development is to obtain the mathematical models of the proposed structure. Since the objective of this chapter is strict to the bodyweight support module, in this section all the proceedings are systematically presented.

In order to simplify the design process of this module, a method for parallel structures that minimizes articular coordinates coupling is required. Thus, the bodyweight support was firstly idealized as a 4-PRRU+PRRS parallel manipulator, using the "multipteron" mechanism family as its main reference (Gosselin et al., 2007). The schematics drawing of the structure is presented on Fig. 17.

Figure 17. (a) Schematic drawing the bodyweight support, representing the reference frames; and (b) detail of the position of the point P on the mobile platform

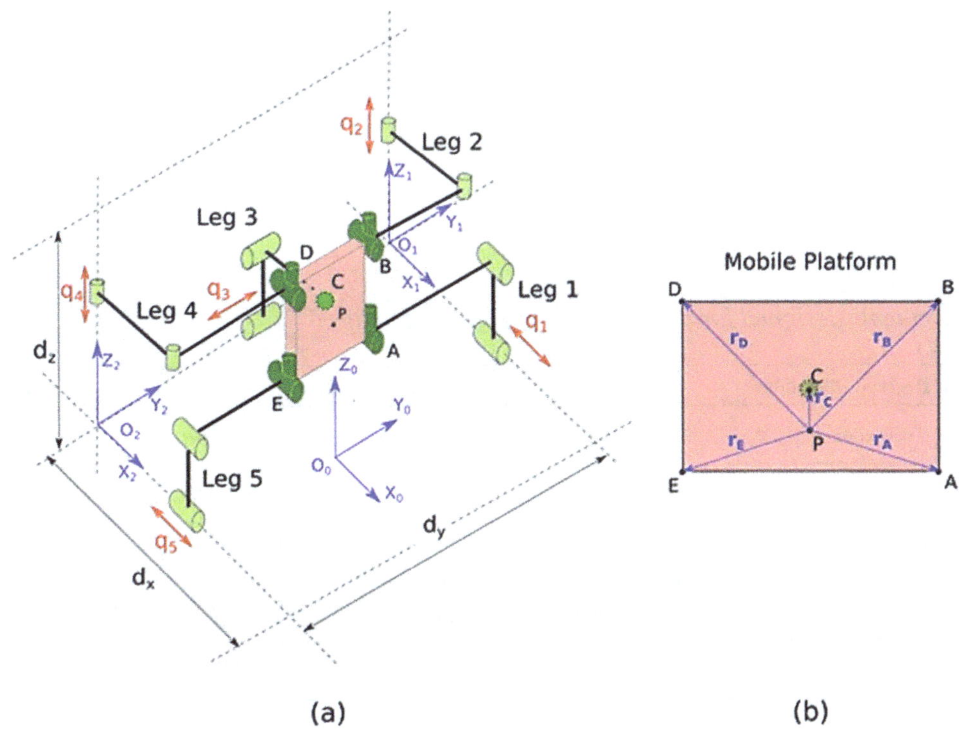

The mobile platform is represented by a pink rectangular solid, and its connected to the fixed platform by 5 legs, each one composed of two bars, two rotational joints, and one universal joint. The legs are identified on Fig. 17 and its bars are represented with black lines.

The rotational joints are represented by light green cylinders, where the ones crossed by dashed lines also includes a translation joint. The universal joints are represented by dark green cross-cylinders, and the spherical joint as a dashed green circumference, placed behind the mobile platform.

Conceptually, this parallel structure applies a fixed platform with 5 articular coordinates, q_1 to q_5, to manipulate a mobile platform containing the terminal point P, Fig. 17 (b), where a seat for the patient will be installed.

The reference frames are also represented in blue in Fig.17 (a). The inertial reference $X_0Y_0Z_0$ plane is placed below the initial position of the mobile platform, and the auxiliary reference frames, $X_1Y_1Z_1$ and $X_2Y_2Z_2$, are placed on the corners of the fixed platform. Five reference vectors, r_A to r_E, are also shown on Fig. 17(b), which represents a frontal view of the mobile platform. These vectors are defined in order to facilitate the geometric model of this structure.

The initially proposed prismatic active joints were replaced, applying a 3 bar-linkages with DC motors to reduce costs with precision linear rails and servomotors. The preliminary model considering the linear actuators is kept, as the mathematical model for the linkages can be directly associated on the final model.

The "multipteron" structures features geometrical characteristics that enhance the independence between the joint coordinates, reducing coupling and singularities on the workspace of the structure (Gosselin et al., 2007).

As one of the main objectives of this module is the pelvic motion assistance, the workspace of the manipulator needs to match with the amplitudes of the pelvis in order to actuate properly on it. Thus, the recommended amplitude values for each joint coordinate proposed by Roberts (2004) were applied on this project, as described on Table 1.

The translation movements of the structure are necessary to correctly adjust the mobile platform with respect to patients' proportions and gait characteristics, while the rotation movements must match the same angular amplitude, skipping the sagittal motion due to its low influence on gait rehabilitation training (Roberts, 2004).

Table 1. Movements designed for the bodyweight support module

Type of Degree-of-Freedom	Direction	Pelvic Amplitude	Structure Minimum Amplitude
Translation	Frontal	-	$\pm 200mm$
Translation	Lateral	-	$\pm 100mm$
Translation	Vertical	-	$\pm 150mm$
Rotation	Transversal	$\pm 25°$	$\pm 25°$
Rotation	Coronal	$\pm 5°$	$\pm 5°$

Source: (Roberts, 2004).

Inverse Geometric Model

In order to obtain the inverse geometric model of the module, equipollent coordinates (Selig, 2000) were applied to the reference frames, where $X_0Y_0Z_0$ represents the inertial reference. The frames were placed in order to align each articular coordinate q_i with one of the unit vectors of each $X_iY_iZ_i$ frame.

In addition, for each leg i, let u_i be a vector linking the joint coordinate q_i to the terminal point of leg i on the mobile platform, and contained on a plane perpendicular to the unit vector aligned with q_i.

Based on the definition given above, the inverse geometric model is obtained analyzing separately each leg i, and grouping all equations on a matrix relating the articular and operational coordinates.

Leg 1

The first analysis is done on the leg that links the articular coordinate q_1 to the mobile platform on point A, applying a comparison between the position of point A directly with respect to the inertial frame $X_0Y_0Z_0$ and its position including reference frame $X_1Y_1Z_1$.

This vector analysis is represented on Fig. 18 and formalized on Eq. (1), where ${}^0r_p$ is the position of point P with respect to $X_0Y_0Z_0$, r_A is the relative position of point A with respect to point P, and $\boldsymbol{i}_1$ is the unit vector of the X direction of reference frame $X_1Y_1Z_1$.

$$ {}^0r_p + r_A = {}^0r_1 + q_1 i_1 + u_1 \tag{1} $$

As previously defined, vector u_1 is parallel to plane Y_1Z_1. Therefore, it is perpendicular to i_1, and as the model is built considering equipollent coordinates, $i_1 = i_0$. Thus, applying the scalar product of i_0 on both sides of Eq. (1), it is obtained the following sentence, where φ is the angle between vectors r_A and i_0, as shown in Fig. 19(b), which also represents orientation around Z_0 axis of the mobile platform.

The schematic presented in Fig. 19 represents the orientation of the platform, using the angles obtained with the unit vectors of the reference frame, and the reference vectors defined on Fig.17(b). Angles θ and φ represents the rotations around X_0 and Z_0, respectively. Angles ϕ and ψ represent the angular difference with respect to θ.

$$ q_1 = \left({}^0r_p + r_A - {}^0r_1\right) \cdot i_0 \Rightarrow $$

$$ \Rightarrow q_1 = \left(\begin{Bmatrix} {}^0r_{px} \\ {}^0r_{py} \\ {}^0r_{pz} \end{Bmatrix} + \begin{Bmatrix} r_{Ax} \\ r_{Ay} \\ r_{Az} \end{Bmatrix} - \begin{Bmatrix} -\dfrac{dx}{2} \\ \dfrac{dy}{2} \\ 0 \end{Bmatrix} \right) \cdot \begin{Bmatrix} 1 \\ 0 \\ 0 \end{Bmatrix} \Rightarrow $$

Figure 18. Position analysis of point A. In blue, directly with respect to the inertial frame $X_0Y_0Z_0$. *In red, through the referential frame* $X_1Y_1Z_1$

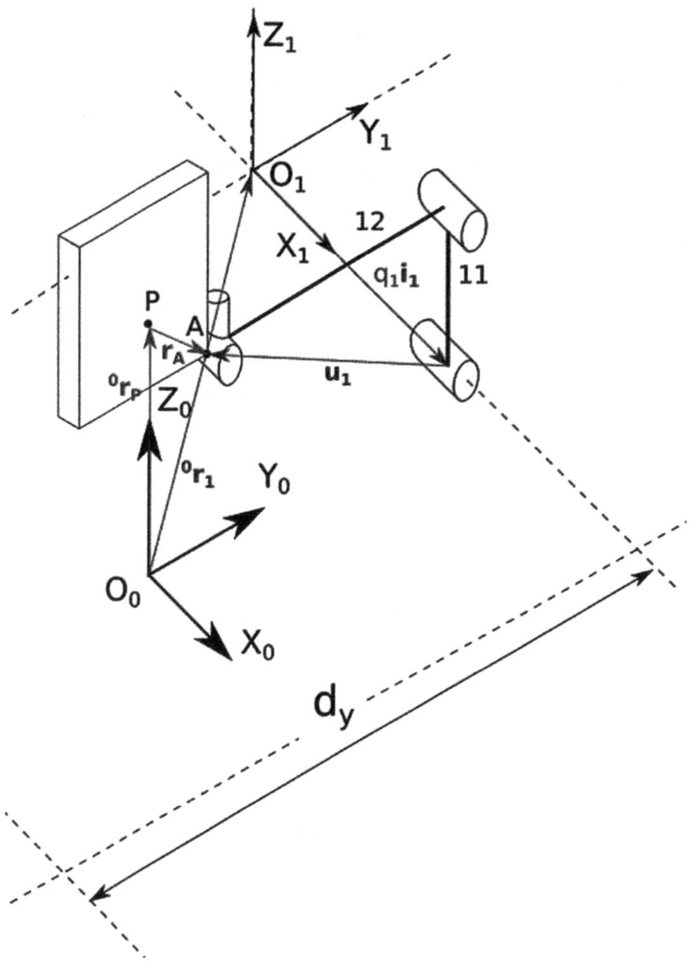

$$\Rightarrow q_1 = {}^0r_{px} + \frac{dx}{2} + r_A \cos\varphi \tag{2}$$

Leg 2

Applying the same process to the leg between joint coordinate q_2 and point B, it is obtained the vector expression shown in Fig. 20 and Eq. (3).

$${}^0r_p + r_B = {}^0r_1 + q_2k_1 + u_2 \tag{3}$$

Figure 19. Angles between unit vectors, in redm and position vectors, in blue, with respect to point P: (a) Angles around X_0 axis; and (b) around Z_0 axis.

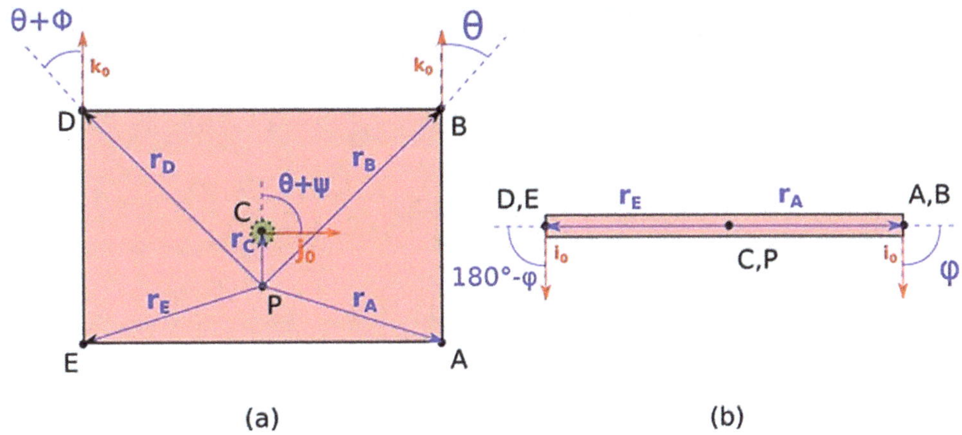

As vector u_2 is parallel to plane X_2Y_2, and consequently perpendicular to $k_1 = k_0$, the Eq. (4) relating q_2 with the operational coordinates can be obtained with the dot product of k_0 on Eq. (3).

$$q_2 = \left({}^0r_p + r_B - {}^0r_1\right)\cdot k_0 \Rightarrow\Rightarrow q_2 = \left(\begin{Bmatrix} {}^0r_{px} \\ {}^0r_{py} \\ {}^0r_{pz} \end{Bmatrix} + \begin{Bmatrix} r_{Bx} \\ r_{By} \\ r_{Bz} \end{Bmatrix} - \begin{Bmatrix} -\dfrac{dx}{2} \\ \dfrac{dy}{2} \\ 0 \end{Bmatrix}\right)\cdot\begin{Bmatrix} 0 \\ 0 \\ 1 \end{Bmatrix} \Rightarrow q_2 = {}^0r_{pz} + r_B\cos(\theta) \tag{4}$$

With θ as the angle between unit vector k_0 and the vector r_b, as shown in Fig. 19(a).

Leg 3

Repeating the analysis to the leg between joint coordinate q_3 and point C, the next vector expression is obtained as shown in Fig. 21 and Eq. (5). Again, taking the advantage of the perpendicularity between vector u_3 and the unit vector j_2, and applying the scalar product with $j_0 = j_2$, joint coordinate q_3 can be written as described in Eq. (6), where angle $\theta + \psi$ that comes from dot product r_c and j_0 also represents the mobile platform around X_0 axis, as shown in Fig. 19(a).

$${}^0r_p + r_C = {}^0r_2 + q_3 j_2 + u_3 \tag{5}$$

$$q_3 = \left({}^0r_p + r_C - {}^0r_2\right)\cdot j_0 \Rightarrow$$

Figure 20. Position analysis of point B. In blue, directly with respect to the inertial frame $X_0Y_0Z_0$. *In red, through the referential frame* $X_1Y_1Z_1$

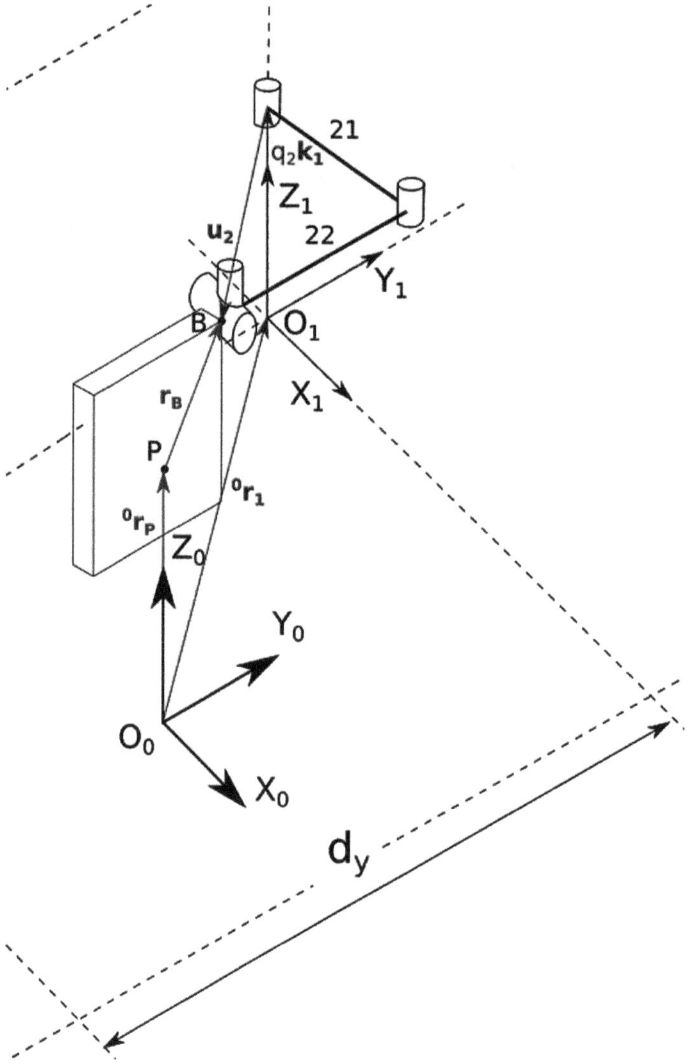

$$\Rightarrow q_3 = \left(\begin{Bmatrix} {}^0r_{px} \\ {}^0r_{py} \\ {}^0r_{pz} \end{Bmatrix} + \begin{Bmatrix} r_{Cx} \\ r_{Cy} \\ r_{Cz} \end{Bmatrix} - \begin{Bmatrix} -\dfrac{dx}{2} \\ -\dfrac{dy}{2} \\ 0 \end{Bmatrix} \right) \cdot \begin{Bmatrix} 0 \\ 1 \\ 0 \end{Bmatrix} \Rightarrow$$

$$\Rightarrow q_3 = {}^0r_{py} + r_C \cos\left(\theta + \psi\right) + \frac{dy}{2} \quad (6)$$

Figure 21. Position analysis of point C. In blue, directly with respect to the inertial frame $X_0Y_0Z_0$. In red, through the referential frame $X_2Y_2Z_2$

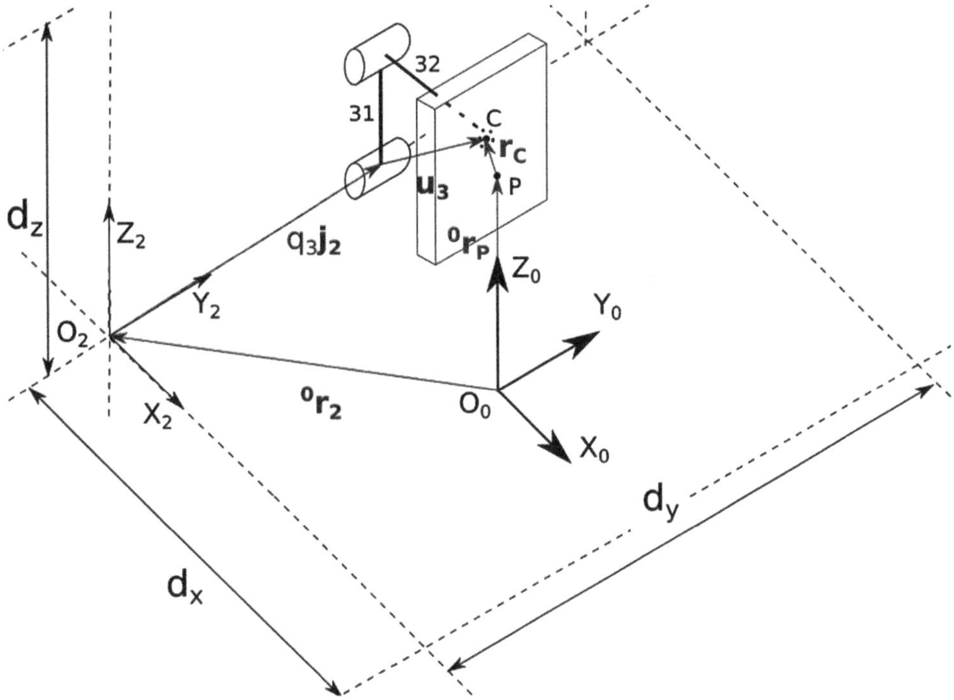

Leg 4

Next, leg 4 is analyzed using point D to obtain its vector relation between the inertial reference frame and $X_2Y_2Z_2$, represented on Fig. 22 and described on Eq. (7). This expression is then dot product by $K_2 = K_0$ to obtain q_4 expression, shown in Eq. (8), where angle $\theta + \phi$ represents the angular difference between K_0 and r_D.

$$^0r_p + r_D = {}^0r_2 + q_4k_2 + u_4 \tag{7}$$

$$q_4 = \left({}^0r_p + r_D - {}^0r_2\right) \cdot k_0 \Rightarrow$$

$$\Rightarrow q_4 = \left(\begin{Bmatrix} {}^0r_{px} \\ {}^0r_{py} \\ {}^0r_{pz} \end{Bmatrix} + \begin{Bmatrix} r_{Dx} \\ r_{Dy} \\ r_{D_z} \end{Bmatrix} + \begin{Bmatrix} -\dfrac{dx}{2} \\ -\dfrac{dy}{2} \\ 0 \end{Bmatrix}\right) \cdot \begin{Bmatrix} 0 \\ 0 \\ 1 \end{Bmatrix} \Rightarrow$$

Figure 22. Position analysis of point D. In blue, directly with respect to the inertial frame $X_0Y_0Z_0$. In red, through the referential frame $X_2Y_2Z_2$

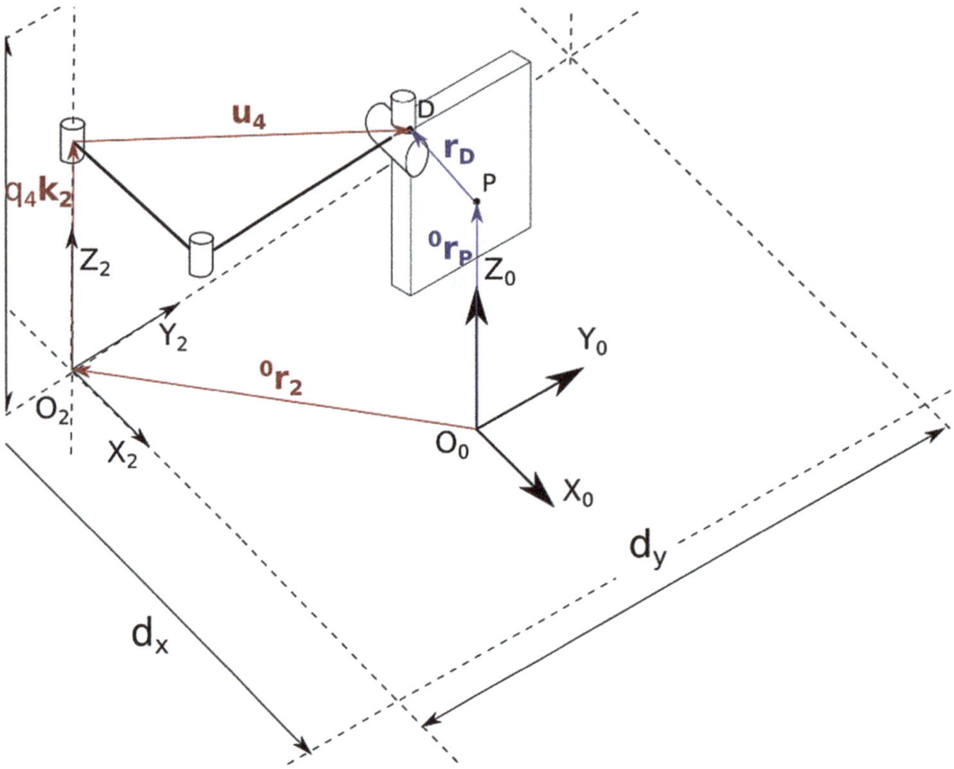

$$\Rightarrow q_4 = {}^0r_{pz} + r_D \cos\left(\theta + \phi\right) \tag{8}$$

Leg 5

Finally, the last leg analysis is performed on point E, Fig. 23 and Eq. (9). Using perpendicularity of u_5 with respect to unit vector i_2, q_5 is then described in terms of the operational coordinates on Eq. (10).

$${}^0r_p + r_E = {}^0r_2 + q_5 i_2 + u_5 \tag{9}$$

$$q_5 = \left({}^0r_p + r_E - {}^0r_2\right) \cdot i_0 \Rightarrow$$

Figure 23. Position analysis of point E. In blue, directly with respect to the inertial frame $X_0Y_0Z_0$. *In red, through the referential frame* $X_2Y_2Z_2$

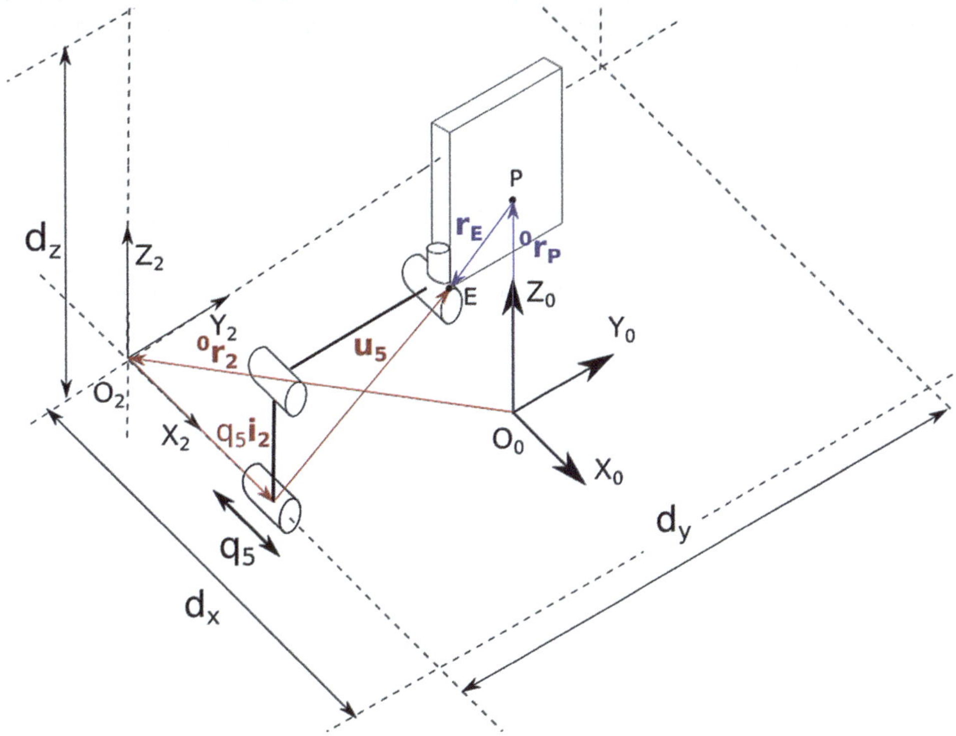

$$\Rightarrow q_5 = \left(\begin{Bmatrix} {}^0r_{px} \\ {}^0r_{py} \\ {}^0r_{pz} \end{Bmatrix} + \begin{Bmatrix} r_{Ex} \\ r_{Ey} \\ r_{E_z} \end{Bmatrix} - \begin{Bmatrix} -\dfrac{dx}{2} \\ -\dfrac{dy}{2} \\ 0 \end{Bmatrix} \right) \cdot \begin{Bmatrix} 1 \\ 0 \\ 0 \end{Bmatrix} \Rightarrow$$

$$\Rightarrow q_5 = {}^0r_{px} + r_E \cos\left(180^\circ - \varphi\right) + \frac{dx}{2} = {}^0r_{px} - r_E \cos\left(\varphi\right) + \frac{dx}{2} \tag{10}$$

Concluding the inverse geometrical model, Eq. (2), (4), (6), (8) and (10) presents all relations between the operational coordinates ${}^0r_p$, θ and φ, and the joint coordinates q_1, q_2, q_3, q_4 and q_5. The relations presented on the mentioned equations are used in next subtopic to calculate the jacobian matrix of the module, which is used to verify the existence of possible singularity positions.

Singularity Analysis

Singularities can be defined as an impossibility of solving the kinematics of a mechanism through mathematical methods. Physically, singularities can be comprehended as positions that instantaneously change the mobility of a robotic structure which can lead to control loss in one or more joints during its moving cycle, and must be avoided (Altuzarra, Pinto, Avilés, & Hernández, 2004). In this work, the determination of the singular positions of the bodyweight support is performed analyzing the jacobian matrix of the structure (Selig, 2000).

To determine possible singularity positions on the proposed parallel manipulator, it is necessary to analyze the geometrical model, applying a time differentiation on the expressions for the joints coordinates q_1, q_2, q_3, q_4 and q_5. The derivatives of this process are presented on Eq. (11) to (15).

$$\dot{q}_1 = {}^0\dot{r}_{pz} - r_A \sin(\varphi)\dot{\varphi} \tag{11}$$

$$\dot{q}_2 = {}^0\dot{r}_{pz} - r_B \sin(\theta)\dot{\theta} \tag{12}$$

$$\dot{q}_3 = {}^0\dot{r}_{py} - r_C \sin(\theta + \psi)\dot{\psi} \tag{13}$$

$$\dot{q}_4 = {}^0\dot{r}_{pz} - r_D \sin(\theta + \phi)\dot{\theta} \tag{14}$$

$$\dot{q}_5 = {}^0\dot{r}_{pz} + r_E \sin(\varphi)\dot{\varphi} \tag{15}$$

With the equations differentiated in time, the jacobian matrix can be obtained building the matrixial form of Eq. (11) to (15), as shown in Eq. (16). Separating the joint and operational velocity coordinates vectors on different sides of the matrixial equation, to match the form $[J_q]\{\dot{q}\} = [J_x]\{\dot{x}\}$, it is evident that $[J_q] = [I]$, where $[I]$ is the identity matrix, which means that singular positions can only exist if $\det([J_x]) = 0$.

$$\Rightarrow \begin{Bmatrix} \dot{q}_1 \\ \dot{q}_2 \\ \dot{q}_3 \\ \dot{q}_4 \\ \dot{q}_5 \end{Bmatrix} = \begin{bmatrix} 1 & 0 & 0 & 0 & -r_A \sin(\varphi) \\ 0 & 0 & 1 & -r_B \sin(\theta) & 0 \\ 0 & 1 & 0 & -r_C \sin(\theta+\psi) & 0 \\ 0 & 0 & 1 & -r_D \sin(\theta+\phi) & 0 \\ 1 & 0 & 0 & 0 & r_E \sin(\varphi) \end{bmatrix} \begin{Bmatrix} {}^0\dot{r}_{px} \\ {}^0\dot{r}_{py} \\ {}^0\dot{r}_{pz} \\ \dot{\theta} \\ \dot{\varphi} \end{Bmatrix} \tag{16}$$

Thus, $\det([J_x]) = 0$ is solved as described in Eq. (17).

$$\begin{aligned} &\det([J_x]) = r_A r_D \sin(\phi+\theta)\sin(\varphi) - r_B r_E \sin(\theta)\sin(\varphi) - r_A r_B \sin(\theta)\sin(\varphi) \\ &+ r_D r_E \sin(\phi+\theta)\sin(\varphi) = 0 \Rightarrow (r_A + r_E) r_D \sin(\phi+\theta)\sin(\varphi) \\ &-(r_A + r_E) r_B \sin(\theta)\sin(\varphi) = 0 \Rightarrow r_D \sin(\phi+\theta) = r_B \sin(\theta) \end{aligned} \tag{17}$$

Therefore, Equation (17) express the only condition that leads to singular positions, where the module is subject to control and mobility loss. Hence, it is important to select dimensions for the structure in order to prevent Eq. (17) being true in any points of its workspace.

Finally, observing the inverse geometrical model, and the equation of singular positions, one can note that the selected architecture for the module successfully presents minimal coupling between joint coordinates, and a low influence of the translation degrees-of-freedom, once they are not present on the singularity equation.

Actuators Design

This project aims to assist medical rehabilitation professionals. However, rehabilitation robotic structures are generally expensive equipment, especially for small clinics. This fact motivates the authors optimize all structural characteristics of the HOPE-G in order to make it affordable for rehabilitation clinics and hospitals.

Therefore, the original linear actuators described on the inverse geometric model q_1 to q_5 were replaced by 3-bar linkages, based on the crank-slider mechanism. With this replacement, the required gear for the prototype construction had its cost reduced, complying with the affordability objective of the HOPE-G project.

The design of the new actuators is presented on Fig. 24. This mechanism receives the angular input θ_i and transform it into a linear output q_i, designed to match with the articular coordinates previously presented on the geometrical model. The relation between q_i and θ_i is described on Eq. (18).

$$q_i = r_{2i} \cos\theta_i - r_{3i} \cos\theta_i \tag{18}$$

Figure 24. (a) CAD model of the actuator based on the crank-slider mechanism; and (b) the schematic drawing evidencing its mechanical parameters

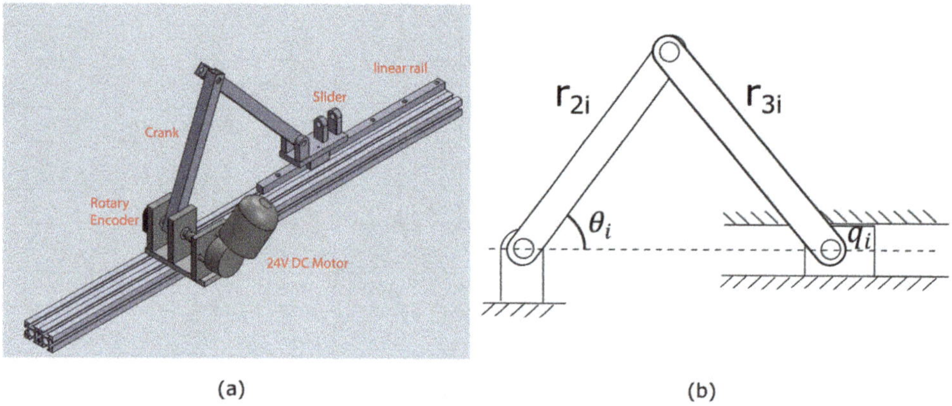

Concluding this section, the expressions obtained during the construction of the inverse geometric model, especially the singularity analysis, will be applied on a computational model associated with an evolutional algorithm in order to achieve optimum dimensions for the structure.

DYNAMICAL SPECIFICATIONS AND OPTIMIZATION

To define correctly the bar dimensions and the DC Motors for the structure, a computational model, considering both geometrical and dynamical model, is constructed. This model deals with the geometrical characteristics presented on the previous section, along with dynamical specifications of the actuators and the geometry of the linking bars.

In this section, the parameters for these specifications are firstly detailed. Then, the algorithm applied on the optimization process is described, ending with the obtained results for the bar dimensions and motor specifications.

Actuators Dynamical Specifications

In order to simplify this process and avoid computational overload on the optimization process, the dynamical model of the bodyweight support module considered only the dynamical forces directly applied on the actuators. This premise is valid due the low velocity and accelerations performed on the mobile platform during gait rehabilitation training (Rodrigues, 2017). Therefore, four directives for the dynamical model are specified, prioritizing the patient safety during the rehabilitation exercises. The directives are:

- Bars that link the mobile platform with the sliders must support the maximum static load, specified as the maximum weight capacity of the module.
- The sliders must be able to go through its workspace under the stride time of the rhythmic gait.
- Bars that link a slider to a DC motor must support the maximum dynamical load.
- Actuators 2 and 4 will feature a spring passive system to support the static weight of the patient, as shown in Fig. 25, leaving the actuators free to deal only with the dynamical load.

Figure 25. Passive spring system on sliders q_2 and q_4 to support the static weight

The maximum weight supported by the module is specified referencing to the Brazilian populational data, selecting a critical mass $M_p = 150Kg$, which covers up to 95% of the population (IBGE, 2010; Moreira et al., 2018).

Another important parameter is the time limit for the system to respond. As stated on the second directive, the stride time, which is the required time to complete the gait cycle, have a direct influence on the dynamical requirements of the structure. To proceed with the specification, an average time $t_c = 0,4s$ is selected (Kharb, Saini, Jain, & Dhiman, 2011).

Thus, the sliders must be able to start the movement without an initial velocity, go through its workspace, and stop before reaching its rail limit. Based on these assumptions and the mechanical theory, the maximum dynamical force can be calculated as stated on Eq. (19), where F_i is the maximum dynamical load on the slider q_i, and d_i is its workspace.

$$F_i = \alpha_i M_p \frac{d_i}{t_c^2}$$
$$\begin{cases} i = 1,2,4,5 \Rightarrow \alpha_i = 0.5 \\ i = 3 \Rightarrow \alpha_i = 1 \end{cases} \tag{19}$$

The parameter α_i corresponds to a correction factor to consider the distribution of the load on actuators 1, 2, 4 and 5. Since the slider q_3 is the only one that actuates on the j_0 direction, it does not need to correct the critical mass value.

Applying the specifications in Eq. (19) on the entire structure, the expected dynamical loads are obtained, and the results are presented in Table 2.

Next, in order to calculate the required power and angular velocity for the DC motors on the crank-slider mechanisms, a free body diagram is constructed, as seen in Fig. 26, where T_i represents the torque input of motor M_i, and F_i the force applied on the slider q_i. Note that inertia and mass of the bars were not considered on the model, due its low influence on the system compared with patient's bodyweight load.

Applying the dynamical equilibrium conditions to the free body diagram, toque T_i can be written as described on Eq. (20).

$$T_i = F_i r_{2i} \sin\theta_i \tag{20}$$

To calculate the angular velocity of the motors, it is necessary to obtain an expression for $\dot{\theta}_i$ To build this expression, first the Cosine Law is applied on the Crank-slider mechanism, and θ_i is place in evidence, as shown in Eq (21).

$$\theta_i = \cos^{-1}\left(\frac{r_{3i}^2 - r_{2i}^2 - q_i^2}{2r_2 q_i}\right) \tag{21}$$

Then, it is possible to obtain the angular velocity applying a time derivation on Eq. (21):

Table 2. Simulation Optimization Parameters

Setup Parameter	Value Applied
Number of replications	10
Population Size	50
Maximum iterations	200
Crossover rate	0.8
Mutation probability	0.005

Figure 26. Free body diagram of the crank-slider mechanisms

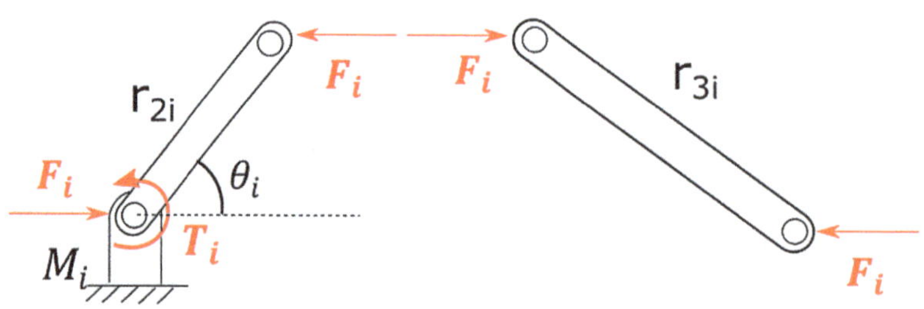

$$\dot{\theta}_i = \frac{r_{2i} \sin \theta_i}{r_{3i}\sqrt{1 - \frac{r_{2i}^2 \cos^2 \theta_i}{r_{3i}^2}}} \dot{q}_i \tag{22}$$

Finally, the required power, in watts, can be obtained applying Eq. (23).

$$P_i = \frac{\left(\max\left(\dot{\theta}_i\right) \max\left(T_i\right)\right)}{9.55} \tag{23}$$

Bar Geometry Specifications

In the previous subsection, it is stated that mass and inertia of the linking bars of the structure are ignored. To ensure that these parameters will be low enough to be disregarded from the analysis, bars should be built with a light material and with a transversal profile as thin as possible.

In order to obtain the optimum dimensions of the bars, the tension on the transversal section is analyzed applying the Material Resistance Theory (Uicker, Pennock, Shigley, & McCarthy, 2003). To calculate the tension σ_{ij} of each bar, the same mechanical characteristics of the free body diagram shown in Fig. 26 is used. The expression obtained for the tensions σ_{ij} of each bar r_{ij} is presented on Eq. (24), where I_{ij} represents the moment of inertia of bar r_{ij}.

$$\sigma_{ij} = \frac{F_j r_{ij}}{4I_{ij}} \tag{24}$$

To simplify the selection of the dimensions, all bars were considered as square tubular profiles, and the computational algorithm were able to change length, thickness and transversal section side. The same material was also applied to all bars. In order to keep a low influence of the bars on the dynamical model, an aluminum alloy was selected as the raw material to the bars, with a maximum transversal tension of $\sigma_{\max} = 180MPa$.

Evolutional Algorithm

A computational model was constructed on MATLAB® and associated with an evolutional algorithm to obtain an optimized solution for the dimensions of the structural elements.

The algorithm applied is denominated "differential evolution algorithm" (Lobato, Steffen Jr., & Oliveira-Lopes, 2007). This method combines characteristics of genetic algorithms with the classic analytical optimization methods, presenting faster solution convergences and less vulnerability to local minimum or maximum solutions.

This method works with lateral limits of the input variables, which represents minimum and maximum values allowed on the optimization process, and configuration parameters similar to genetic algorithms (Lobato et al., 2007).The algorithm also features setup parameters to configure the initial population size, maximum number of iterations, crossover rate, mutation probability and convergence strategy.

The input variables applied on the evolutional algorithm represented bars length, transversal section side, and thickness. Bars r_{2i}, r_{3i} represents the crank-slider bars of the leg i, while r_{4i}, r_{5i} are the bars connecting slider q_i to the mobile platform. As the optimization process of the algorithm proceeds, the evolving geometry of the bars will have a direct effect on the presence of the singular positions on workspace, tension parameters and the required motor power. Therefore, the maximum motor power requirement is selected as a minimization objective function, while the other parameters are applied on the optimization as constraints.

The Simulation Optimization process was applied with the parameters listed in Table 2.

After running the Simulation Optimization process for the specified number of replications, the first raw results are obtained. These values were approximated to match products available on the local market, in order make further prototype construction easier and reduce fabrication costs. The optimum dimensions of the bars and required motor power for the actuators are detailed in Tables 3 and 4, respectively.

Table 3. Optimized bar dimensions

Bar	Length (mm)	Transversal Section (mm)	Thickness (mm)
r_{2i}	250	25.4	5
r_{3i}	300	25.4	5
r_{4i}	300	25.4	5
r_{5i}	300	25.4	5

Table 4. Power and velocity specifications for the DC Motors

Motor	Required Power (W)	Required Angular Velocity (RPM)
M_1	35	20
M_2	20	15
M_3	20	10
M_4	20	15
M_5	35	20

Note that the same transversal section was able to be applied on all bars after rounding the values calculated by the optimization algorithm, and only the segment r_{2i} of structure legs needed a smaller length. This fact makes the building process easier and reduce the material cost of the prototype.

CAD SIMULATIONS

The preceding sections provided all the information regarding the construction of the bodyweight module, from its initial conception and motivation to the geometrical and mechanical requirements. After gathering this data, was possible to construct a CAD model of the module and simulate the specified movements in order to validate the manipulator and prepare for the prototype construction and experimental tests.

Beside that, the CAD model of the structure also consider a mobile base, designed to fit to width dimensions of the dual treadmill movement module, to allow the application of this module as a stand-alone rehabilitation device, performing overground training, and to facilitate the access to the mobile platform for the patients.

The assembly of the bodyweight support module, along with all motion simulation was developed with SolidWorks® CAD/CAE software. The complete assembly of the module is shown in Fig. 1. Overall, four main simulations were executed with the CAD/CAE model.

First, the elementary movements of the workspace were tested, simulating the expected 5-DOF of the structure. The main objective of this simulation is to visualize the aspect of the movements and ensure if it reach the required minimum amplitude. A sequential illustration of this simulation is shown in Fig. 27.

Next, the authors made a simple simulation to give an example of the expected setup, considering the proceedings to connect the patient from a wheelchair to the mobile platform. The simulation does not represent the connection with details but demonstrates that the setup process does not require a previous full lifting of the patient. A sequence of this proposed proceeding is shown in Fig.28.

The last two simulations present the behavior of the module during an overground rehabilitation training session, and its application with a dual treadmill movement module, representing the full assembly expected to the HOPE-G structure. Both simulations are sequenced on images, as seen on Fig. 29.

Additionally to the image sequences here presented, a demonstrative video featuring all the animations generated by the simulations is available on https://tinyurl.com/HOPE-G-bodyweight-support.

FUTURE RESEARCH DIRECTIONS

The CAD model of the bodyweight support module demonstrated the expected behavior and operation of the designed device. Based on these results, the project is now ready for prototype building and experimental tests.

The first crank-slider mechanism prototype has been already built on the Robotics and Automation Laboratory of the Federal University of Uberlândia, and its control system is at the final stage of development. The authors are performing load-less trajectory tests on this prototype in order to validate the geometrical model and ensure position and velocity control.

Figure 27. Sequence of images showing the elementary movements simulation

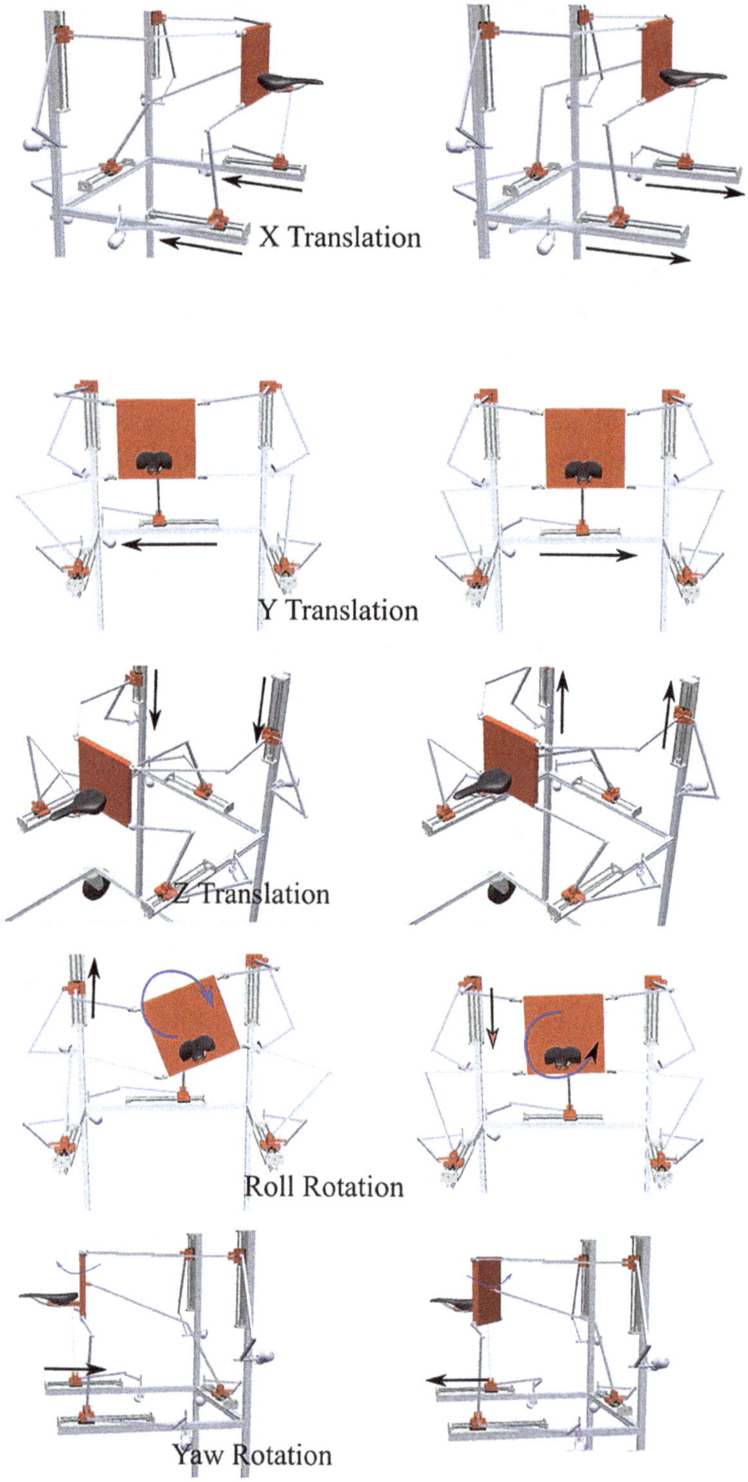

Figure 28. Sequence of images showing a setup proceeding simulated

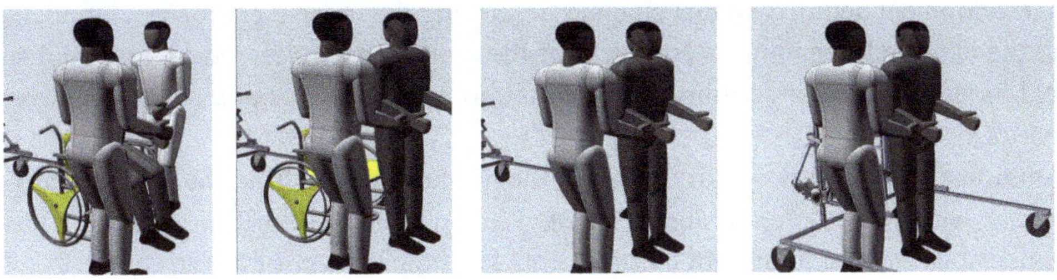

Figure 29. Sequence of images showing the simulations of overground rehabilitation exercises, and dual treadmill rehabilitation exercises

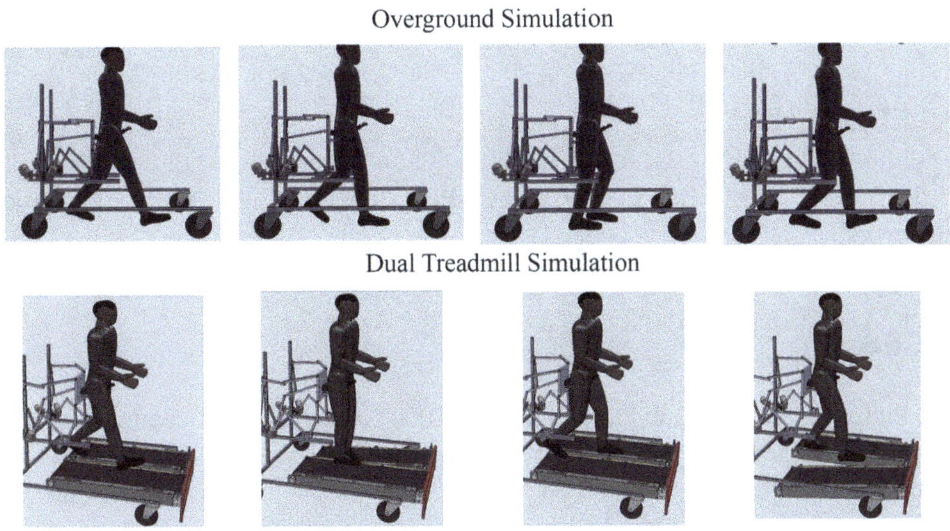

The next steps of this research start with the complete construction of the bodyweight support module, along with full control validation and preliminary experimental test on an anthropometric wooden puppet.

After concluding the bodyweight support module, the research proceeds with the development of the dual treadmill movement module and the construction of its prototype, followed by the integration with the other modules. The last phase of the project, the complete HOPE-G structure will be submitted to clinical tests in order to and measure systematically the robot contributions to the rehabilitation training sessions.

CONCLUSION

This chapter has presented the proceeding on the development of a novel robotic rehabilitation structure. From its very first concept, reviewing the recent literature and proposing a solution combining important characteristics observed on previous work.

The proposed structure was named "HOPE-G", and one of three modules of this rehabilitation device had its development detailed on this chapter, describing the inverse geometrical model, singularity analysis and the expected dynamics. Regardless of the module been defined as a parallel structure, its mathematical model shows low coupling between the joint coordinates and only a single equation for singular positions.

A computational model constructed combining the results obtained from the other models was implemented together with an evolutional algorithm, and a simulation optimization process has generated optimum dimensions and the minimum requirements for motor power of the structure.

A CAD/CAE model based on these optimum parameters was built and used to develop graphic simulations, which presents strong evidence of the correct functionality of the proposed structure.

As the next steps of this research go ahead, the authors look forward in performing tests with a complete prototype of the module and compare results with the CAD simulations in order to validate the computational model experimentally.

ACKNOWLEDGMENT

The authors thank UFU, FAPEMIG, CNPQ, and CAPES - Finance Code 001 for the partial financial support to this work.

REFERENCES

Altuzarra, O., Pinto, C., Avilés, R., & Hernández, A. (2004). A practical procedure to analyze singular configurations in closed kinematic chains. *IEEE Transactions on Robotics*, *20*(6), 929–940. doi:10.1109/TRO.2004.832798

Aoyagi, D., Ichinose, W. E., Reinkensmeyer, D. J., & Bobrow, J. (2004). Human Step Rehabilitation Using a Robot Attached To the Pelvis. In *2004 ASME International Mechanical Engineering Congress and Exposition* (p. 7).

Babaiasl, M., Mahdioun, S. H., Jaryani, P., & Yazdani, M. (2015). A review of technological and clinical aspects of robot-aided rehabilitation of upper-extremity after stroke. *Disability and Rehabilitation. Assistive Technology*, 1–18. doi:10.3109/17483107.2014.1002539 PMID:25600057

Bannwart, M., Rauter, G., Curt, A., & Bolliger, M. (2015). The FLOAT: Interactive BWS training of everyday walking challenges in 4 dimensions. Retrieved from https://www.research-collection.ethz.ch/handle/20.500.11850/105182

Barbosa, A. M. (2013). *Desenvolvimento de uma estrutura robótica paralela atuada por cabos para reabilitação dos movimentos do ombro*. Universidade Federal de Uberlândia.

Boian, R. F., Bouzit, M., Burdea, G. C., Lewis, J., & Deutsch, J. E. (2005). Dual stewart platform mobility simulator. In *Proceedings of the 2005 IEEE 9th International Conference on Rehabilitation Robotics* (Vol. 2005, pp. 550–555). https://doi.org/10.1109/ICORR.2005.1502023

Brasil. (2013). *Diretrizes de Atenção à Reabilitação da Pessoa com Acidente Vascular Cerebral.*

Brasil. (2017). Acidente Vascular Cerebral - AVC. Ministério da Saúde. Retrieved from http://portalms.saude.gov.br/saude-de-a-z/acidente-vascular-cerebral-avc

Camargos, A. C. R., Lacerda, T. T. B., Taise, V. B., Silva, G. C., Parreiras, J. T., & Vidal, T. H. J. (2012). Relação entre independência funcional e qualidade de vida na paralisia cerebral Relationship between functional independence and quality of life in cerebral palsy, *25*(1), 83–92.

El Rafei, M., Badr, S., Lefebvre, G., Machuron, F., Capon, B., Flipo, R.-M., & Cotten, A. (2018). Sacroiliac joints: Anatomical variations on MR images. *European Radiology*, *28*(12), 5328–5337. doi:10.100700330-018-5540-x PMID:29876707

Galli, M., Cimolin, V., De Pandis, M. F., Le Pera, D., Sova, I., Albertini, G., ... Franceschini, M. (2016). Robot-assisted gait training versus treadmill training in patients with Parkinson's disease: A kinematic evaluation with gait profile score. *Functional Neurology*, *31*(3), 163–170. doi:10.11138/FNeur/2016.31.3.163 PMID:27678210

Goffer, A. (2006). *US7153242 B2.*

Gonçalves, R. S., Hamilton, T., Daher, A. R., Hirai, H., & Hermano, I. (2017). MIT-Skywalker : Evaluating Comfort of Bicycle / Saddle Seat *. In *2017 International Conference on Rehabilitation Robotics (ICORR)* (pp. 516–520). London, UK.

Gonçalves, R. S., Hamilton, T., & Krebs, H. I. (2017). MIT-Skywalker: On the use of a markerless system. In *IEEE International Conference on Rehabilitation Robotics* (pp. 205–210). https://doi.org/10.1109/ICORR.2017.8009247

Gonçalves, R. S. (2010). Desenvolvimento de uma estrutura robótica paralela atuada por cabos para reabilitação dos movimentos do ombro. In *VI Congresso Nacional de Engenharia Mecânica*. Campina Grande.

Gosselin, C. M., Masouleh, M. T., Duchaine, V., Richard, P.-L., Foucault, S., & Kong, X. (2007). Parallel Mechanisms of the Multipteron Family: Kinematic Architectures and Benchmarking. In *Proceedings 2007 IEEE International Conference on Robotics and Automation* (pp. 555–560). IEEE. https://doi.org/10.1109/ROBOT.2007.363045

Habib, M. (2007). Mechatronics - A unifying interdisciplinary and intelligent engineering science paradigm. *IEEE Industrial Electronics Magazine*, *1*(2), 12–24. doi:10.1109/MIE.2007.901480

Habib, M. K. (2008). Human adaptive and friendly mechatronics (HAFM). In *2008 IEEE International Conference on Mechatronics and Automation* (pp. 61–65). IEEE. https://doi.org/10.1109/ICMA.2008.4798726

Hesse, S., Schmidt, H., Werner, C., & Bardeleben, A. (2003). *Upper and lower extremity robotic devices for rehabilitation and for studying motor control. Current Opinion in Neurology*. Lippincott Williams & Wilkins. doi:10.1097/00019052-200312000-00010

Hesse, S., & Uhlenbrock, D. (2000). A mechanized gait trainer for restoration of gait. *Journal of Rehabilitation Research and Development*, *37*(6), 701–708. Retrieved from https://www.rehab.research.va.gov/jour/00/37/6/pdf/Hesse.pdf PMID:11321006

Hogan, N., & Sternad, D. (2013). Dynamic primitives in the control of locomotion. *Frontiers in Computational Neuroscience*, *7*. doi:10.3389/fncom.2013.00071 PMID:23801959

IBGE - Instituto Brasileiro de Geografia e Estatística. (2010). *Antropometria e Estado Nutricional no Brasil 2008-2009*. Retrieved from https://doi.org/978-85-240-4131-0

Jezernik, S., Colombo, G., & Morari, M. (2004). Automatic gait-pattern adaptation algorithms for rehabilitation with a 4-DOF robotic orthosis. *IEEE Transactions on Robotics and Automation*, *20*(3), 574–582. doi:10.1109/TRA.2004.825515

Kharb, A., Saini, V., Jain, Y. K., & Dhiman, S. (2011). *A REVIEW OF GAIT CYCLE AND ITS PARAMETERS. IJCEM International Journal of Computational Engineering & Management* (Vol. 13). Retrieved from www.IJCEM.orgIJCEMwww.ijcem.org

Lackland, D. (2017). *Heart Disease and Stroke Statistics — 2017 Update A Report From the American Heart Association*. https://doi.org/ doi:10.1161/CIR.0000000000000485

Lloyd-Jones, D., Adams, R. J., Brown, T. M., Carnethon, M., Dai, S., & De Simone, G. … Wylie-Rosett, J. (2010). Executive summary: Heart disease and stroke statistics-2010 update: A report from the American Heart Association. *Circulation*. doi:10.1161/CIRCULATIONAHA.109.192667

Lobato, F. S., Steffen, V. Jr, & Oliveira-Lopes, L. C. (2007). *AN EVOLUTIONARY APPROACH BASED ON SEARCH CHAOTIC PATTERN ASSOCIATED TO DIFFERENTIAL EVOLUTION ALGORITHM. In 17º* (p. 9). Uberlandia: Simpósio do Programa de Pós-graduação em Engenharia Mecânica.

Louie, D. R., & Eng, J. J. (2016). Powered robotic exoskeletons in post-stroke rehabilitation of gait: A scoping review. *Journal of Neuroengineering and Rehabilitation*, *13*(1), 53. doi:10.118612984-016-0162-5 PMID:27278136

Lum, P., Reinkensmeyer, D., Mahoney, R., Rymer, W. Z., & Burgar, C. (2002). Robotic Devices for Movement Therapy After Stroke: Current Status and Challenges to Clinical Acceptance. *Topics in Stroke Rehabilitation*, *8*(4), 40–53. doi:10.1310/9KFM-KF81-P9A4-5WW0 PMID:14523729

Mekki, M., Delgado, A. D., Fry, A., Putrino, D., & Huang, V. (2018). Robotic Rehabilitation and Spinal Cord Injury: A Narrative Review. *Neurotherapeutics; the Journal of the American Society for Experimental NeuroTherapeutics*, *15*(3), 604–617. doi:10.100713311-018-0642-3 PMID:29987763

Moreira, N. F., Luz, V. G., Moreira, C. C., Pereira, R. A., Sichieri, R., Ferreira, M. G., … Rodrigues, P. R. M. (2018). Self-reported weight and height are valid measures to determine weight status: Results from the Brazilian National Health Survey (PNS 2013). *Cadernos de Saude Publica*, *34*(5). doi:10.1590/0102-311x00063917 PMID:29768583

Mun, K. R., & Lim, S. Bin, Guo, Z., & Yu, H. (2017). Biomechanical effects of body weight support with a novel robotic walker for over-ground gait rehabilitation. *Medical & Biological Engineering & Computing*, *55*(2), 315–326. doi:10.100711517-016-1515-8 PMID:27193227

Mun, K. R., Yu, H., Zhu, C., & Cruz, M. S. (2014). Design of a novel robotic over-ground walking device for gait rehabilitation. In *IEEE 13th international workshop - Advanced Motion Control(AMC)*. https://doi.org/10.1109/AMC.2014.6823325

Nadeau, S. E., Dobkin, B., Wu, S. S., Pei, Q., & Duncan, P. W., & Team, for the L. I. (2016). The Effects of Stroke Type, Locus, and Extent on Long-Term Outcome of Gait Rehabilitation. *Neurorehabilitation and Neural Repair*, *30*(7), 615–625. doi:10.1177/1545968315613851 PMID:26498434

Nunes, W. M. (2012). *Desenvolvimento de uma estrutura robótica atuada por cabos para reabilitação/ recuperação dos movimentos do ombro humano*. Universidade Federal De Uberlândia. Retrieved from https://repositorio.ufu.br/bitstream/123456789/14900/1/d.pdf

Peurala, S. H., Airaksinen, O., Huuskonen, P., Jäkälä, P., Juhakoski, M., Sandell, K., ... Sivenius, J. (2009). Effects of intensive therapy using gait trainer or floor walking exercises early after stroke. *Journal of Rehabilitation Medicine*, *41*(3), 166–173. doi:10.2340/16501977-0304 PMID:19229450

Pietrusinski, M., Cajigas, I., Mizikacioglu, Y., Goldsmith, M., Bonato, P., & Mavroidis, C. (2010). Gait rehabilitation therapy using robot generated force fields applied at the pelvis. *2010 IEEE Haptics Symposium, HAPTICS 2010*, 401–407. https://doi.org/10.1109/HAPTIC.2010.5444624

Roberts, M. (2004). A robot for gait rehabilitation. Retrieved from http://dspace.mit.edu/handle/1721.1/34562

Rodrigues, L. A. O. (2017). *DESENVOLVIMENTO DE DISPOSITIVOS NÃO MOTORIZADOS PARA REABILITAÇÃO DO MEMBRO INFERIOR HUMANO*. Universidade Federal de Uberlândia.

Salguero-Beltran, A., Yamhure, G., Manrique, M., Jimenez, L. C., Hernandez, A. M., & Cotrino, C. (2012). On the design of an ischiatic body weight support system (IBWS) for gait rehabilitation. *Proceedings of the IEEE RAS and EMBS International Conference on Biomedical Robotics and Biomechatronics*, (5), 1434–1439. https://doi.org/10.1109/BioRob.2012.6290940

Salim, V. V. (2018). *DESENVOLVIMENTO DE UM SISTEMA SEM MARCADORES PARA CONTROLE DE UM DISPOSITIVO ROBÓTICO PARA REABILITAÇÃO DA MARCHA HUMANA*. Universidade Federal de Uberlândia.

Selig, J. M. (2000). *Geometrical Foundations of Robotics*. WORLD SCIENTIFIC. doi:10.1142/4257

Sui, P., Yao, L., Lin, Z., Yan, H., & Dai, J. S. (2009). Analysis and synthesis of ankle motion and rehabilitation robots. In *2009 IEEE International Conference on Robotics and Biomimetics, ROBIO 2009* (pp. 2533–2538). IEEE. https://doi.org/10.1109/ROBIO.2009.5420487

Susko, T., Swaminathan, K., & Krebs, H. I. (2016). MIT-Skywalker: A Novel Gait Neurorehabilitation Robot for Stroke and Cerebral Palsy. *IEEE Transactions on Neural Systems and Rehabilitation Engineering*, *24*(10), 1089–1099. doi:10.1109/TNSRE.2016.2533492 PMID:26929056

Susko, T. G. (2015). *MIT Skywalker: A novel robot for gait rehabilitation of stroke and cerebral palsy patients.*

Swinnen, E., Baeyens, J. P., Hens, G., & Knaepen, K., Beckwee, D., Michielsen, M., … Kerckhofs, E. (2015). Body weight support during robot-assisted walking: Influence on the trunk and pelvis kinematics. *NeuroRehabilitation*. doi:10.3233/NRE-141195

Swinnen, E., Baeyens, J. P., Knaepen, K., Michielsen, M., & Clijsen, R., Beckwee, D., & Kerckhofs, E. (2015). Robot-assisted walking with the Lokomat: The influence of different levels of guidance force on thorax and pelvis kinematics. *Clinical Biomechanics (Bristol, Avon)*, *30*(3), 254–259. doi:10.1016/j.clinbiomech.2015.01.006 PMID:25662678

Swinnen, E., Baeyens, J.-P., Knaepen, K., Michielsen, M., Hens, G., Clijsen, R., ... Kerckhofs, E. (2015). Walking with robot assistance: The influence of body weight support on the trunk and pelvis kinematics. *Disability and Rehabilitation. Assistive Technology*, *10*(3), 252–257. doi:10.3109/17483107.2014.888487 PMID:24512196

Uicker, J. J., Pennock, G. R., Shigley, J. E., & McCarthy, J. M. (2003). Theory of Machines and Mechanisms. *Journal of Mechanical Design*. Oxford Press. https://doi.org/ doi:10.1115/1.1605769

Veneman, J. F., Kruidhof, R., Hekman, E. E. G., Ekkelenkamp, R., Van Asseldonk, E. H. F., & Van Der Kooij, H. (2007). Design and evaluation of the LOPES exoskeleton robot for interactive gait rehabilitation. *IEEE Transactions on Neural Systems and Rehabilitation Engineering*, *15*(1), 379–386. doi:10.1109/TNSRE.2007.903919 PMID:17894270

Veneman, J. F., Menger, J., van Asseldonk, E. H. F., van der Helm, F. C. T., & van der Kooij, H. (2008). Fixating the pelvis in the horizontal plane affects gait characteristics. *Gait & Posture*, *28*(1), 157–163. doi:10.1016/j.gaitpost.2007.11.008 PMID:18207406

Waterman, B. R., Owens, B. D., Davey, S., Zacchilli, M. A., & Belmont, P. J. (2010). The epidemiology of ankle sprains in the United States. *Journal of Bone and Joint Surgery - Series A*, *92*(13), 2279–2284. doi:10.2106/JBJS.I.01537 PMID:20926721

Wu, Y., Hwang, M., Ren, Y., Gaebler-spira, D., & Zhang, L. (2011). Combined Passive Stretching and Active Movement Rehabilitation of Lower-Limb Impairments in Children With Cerebral Palsy Using a Portable Robot. https://doi.org/ doi:10.1177/1545968310388666

ADDITIONAL READING

American Heart Association, American Stroke Association. (2017). Heart Disease and Stroke Statistics: 2017 Update. Retrieved from http://www.acc.org/latest-in-cardiology/ten-points-to-remember/2017/02/09/14/58/heart-disease-and-stroke-statistics-2017

Carvaho, J. C. M., Ibrahim, R. C., & Coelho, T. A. H. (2017). *Mecanismos, Máquinas e Robôs: uma abordagem unificada para análise e síntese*. São Paulo, Brazil: Elsevier.

Carvalho, J. C. M., & Martins, D. (2017). Multibody Mechatronic Systems: *Proceedings of the MUSME Conference*, Florianópolis, Brazil, Springer.

Diaz, I., Gil, J., & Sanchez, E. (2011). Lower-Limb Robotic Rehabilitation: Literature Review and Challenges. *Journal of Robotics*, (11): 1–11. doi:10.1155/2011/759764

Hussain, S., Xie, S., & Jamwal, P. (2013). Adaptive impedance control of a robotic orthosis for gait *rehabilitation. IEEE Transactions on Cybernetics*, *43*(3), 1025–1034. doi:10.1109/TSMCB.2012.2222374 PMID:23193241

Hussain, S., Xie, S., Quan, S., & Liu, G. (2011). Robot assisted treadmill training: Mechanisms and training strategies. *Medical Engineering & Physics*, *33*(5), 527–533. doi:10.1016/j.medengphy.2010.12.010 PMID:21216650

Skidmore, J., Barkan, A., & Artemiadis, P. (2015). Variable Stiffness Treadmill (VST): System Development, Characterization, and Preliminary Experiments. *IEEE/ASME Transactions on Mechatronics*, *20*(4), 1717–1724. doi:10.1109/TMECH.2014.2350456

Wu, S., Duncan, P., & Sullivan, K. (2007). Protocal for the Locomotor Experience Applied Post-stroke (LEAPS) trial: A randomized controlled trial. *BMC Neurology*, *7*(1), 1–23. doi:10.1186/1471-2377-7-39 PMID:17204141

KEY TERMS AND DEFINITIONS

Degree-of-Freedom (DOF): Any of a limited number of ways in which a body may move or in which a dynamic system may change. Defines the possible elementary movements of a robotic structure.

Frontal/Coronal Plane: A hypothetical plane defined in human anatomy that divides the body into ventral and dorsal sections.

Genetic Algorithm: A metaheuristic algorithms inspired on the biological process of evolution and natural selection. Genetic algorithms are known to generate high-quality solutions with low influence of local minimums or maximums, relying on computational equivalents of natural processes such as crossover, mutation and environment fitness.

Parallel Manipulator: A robotic structure that connects a static platform to a mobile platform using two or more legs, where each leg may contain an active actuator.

Rehabilitation: The act of restoring something to its original state. In this work, it applies to the lower-limb motion rehabilitation.

Sagittal Plane: A hypothetical plane defined in human anatomy that divides the body into left and right sections.

Simulation Optimization: A computational technique that applies the simulation of a real system, based on a mathematical model, integrated with an optimization algorithm. This process works applies recursive iterations between the computational simulation and the optimization algorithm directly connecting inputs and outputs.

Transversal Plane: A hypothetical plane defined in human anatomy that divides the body into upper and lower sections.

Chapter 4
Mechatronic Design of Low-Cost Control Systems for Rehabilitation and Assisting Devices

Pierluigi Rea
University of Cassino and Southern Lazio, Italy

Erika Ottaviano
https://orcid.org/0000-0002-7903-155X
University of Cassino and Southern Lazio, Italy

ABSTRACT

In this chapter, authors focus on mechatronic solutions for low-cost control devices for the homecare of elderly and people with reduced mobility, having the goal of assisting people in daily-life activities. The development of such systems can be exploited in the form of a toolkit to be flexible and applied either to rehabilitation or assistive systems, in order to aid movements with controlled position, force/torque, and acceleration, possibly introducing reliability, repetitiveness, and low cost in the related sector. In particular, the development of a low-cost control system design allows the applicability of the solution to a broad range of devices for the home-care. The authors aim to develop low-cost technologies for the homecare of elderly and people with motor impairments trying to reduce significantly the overall costs to facilitate wider use of assisting and rehabilitation systems. In particular, a Sit-To-Stand (STS) assisting device is considered as a paradigmatic example to illustrate design considerations.

DOI: 10.4018/978-1-7998-0137-5.ch004

INTRODUCTION

Recent studies report that the number of elderly and people with motion disabilities that cannot perform Activities of Daily Life (ADL) without the help of other people is increasing. In order to face this problem, nurses, caregivers or relatives are involved for continuous aiding at home, alternatively, a number of commercial solutions have been developed to help the execution of some tasks. Among several ADL, such as walking, feeding, bathing, toileting, we are interested in the Sit-To-Stand (STS), which is one of the most frequent and, despite its apparent simplicity, it is a mechanically demanding functional task undertaken daily, requiring a strong coordination between posture and movement. Some commercial solutions were developed (Hari Krishnan & Pugazhenthi, 2014). Most of these devices act like end-effector robot, i.e. they provide assistive forces only at one body segment, while the trunk is either rigidly supported by a back support of a seat or through use of the arms. All these approaches may or surely result in unnatural kinematics. The purpose of this Chapter is to discuss solutions of mechatronic design for the control of the actuation of assisting devices designed for the STS, as evolution of the ones developed by the authors (Rea et al., 2013; Rea & Ottaviano 2018). They can be either used as rehabilitation devices, as the one proposed in (Rea et al., 2013). The mechatronic design is particularly important when the motion or force have to be controlled (Figliolini et al., 2010, Borras et al., 2009; Ottaviano et al, 2004). Simulation results are reported to show the feasibility of the proposed solution together an experimental test bed of the mechatronic design. This Chapter reports perspectives on mechatronics integrated in home environment and enhancements and of research activities presented in (Ottaviano et al., 2019) as a paradigmatic case of study.

One of the main challenges in the design of assisting and/or rehabilitation systems is their portability and cost, which are in most of the cases prohibitive, even for the solutions developed and commercially available.

Nevertheless, in the last decades, the cost of electronics and microcontrollers has been decreasing, but software in mechanical products have increased. In addition, trends in new products ranging from mechanical solutions with added functionality realized by electronics and software to electronic products with embedded software with the need of mechanical encapsulation. The multi-technological products pose a fundamental challenge for companies joining development processes, different in nature, into a synergistic process to produce successful products. While some companies see new market potentials in the joining of solutions from several engineering disciplines, other companies sees it as a necessity to stay competitive on already captured market segments. The added complexity caused by going from single- to multi-disciplinary development can have various consequences, bringing potential danger to the project. When designing a new mechatronic product or solution, these issues should be faced.

The engineering disciplines mainly considered in the mechatronic design are mechanical engineering, electronics engineering and software engineering. Thus, control engineering might be considered as an engineering discipline on the same level as mechanical, electronics and software engineering. The concept of mechatronics was born in Japan in 1969, (Fukuda et al., 2010). Research efforts continuously demand for fruitful and seamless integration between the involved engineering disciplines (Andreasen & McAloone, 2001, Gausemeier et al., 2009; Tomiyama et al., 2007) to create success in the development of mechatronic products. Mechatronics can be defined in several ways. A common consensus would be to describe mechatronics as an integrative discipline based on mechanics, electronics and information technology to provide enhanced products, processes and systems. It commonly integrates the classical fields of mechanical and electronic engineering, and computer science/information technology since the design stage of a product or a system. Mechatronics cannot be considered a new branch of engineering, but a newly concept that underlines the necessity for integration and intensive interaction among different branches of the engineering science. Mechatronics is based undoubtedly on the core disciplines set by the name, i.e., *mechanics* and *electronics*. This should not be considered as a boundary or limit of mechatronics; *mech* should be understood as the widest aspects of the physical embodiments of mechanical engineering, including optical elements, whilst *tronics* should be understood to embrace all aspects of microelectronics and information technology including control. In addition, it should be understood that most the engineered products or processes have moving parts and require manipulation and control of their dynamic constructions with a given accuracy. This may involve the use of enabling technologies related to sensors, actuators, software, communications, optics, electronics, structural mechanics, and control engineering. If fact, a basic concept of mechatronics philosophy is the integration of microelectronics and information technology into mechanical systems. Design of such products and processes, therefore, has to be the outcome of a multidisciplinary activity rather than an interdisciplinary one. Therefore, mechatronics challenges traditional engineering thinking because the selection of means to the contemplated functional ends must involve crossing the boundaries between traditional engineering disciplines.

Mechatronics represents a unifying interdisciplinary and intelligent engineering science paradigm which fuses, permeates, and comprehends modern engineering science and technologies, as reported in the wok (Habib, 2007) in which it was introduced the concept of Human Adaptive and Friendly Mechatronics (HAFM). This concept was better explained in (Habib, 2008) being a pioneer work in which it was introduced that a key topic in the new mechatronics era was the design of human-oriented machines that implies intelligent and cooperative coexistence (beyond time and physical constraints) between technical systems and biological ones (human) within their natural environment.

The Mechatronics integration can be pursued by considering the so-called parallel configuration or through continuous modelling, as displayed in the schemes of Figure 1. A prerequisite is that the project plan should favor the possibility of collaboration among the domains even in the case where the development is closer to a serial than a parallel configuration. If the configuration is of the parallel type, the creation of the Mechatronic Integration Concept (MIC) can be used as baseline descriptions needed at

Figure 1. Possible ways for mechatronic integration: a) base lines; b) 'continuous modelling

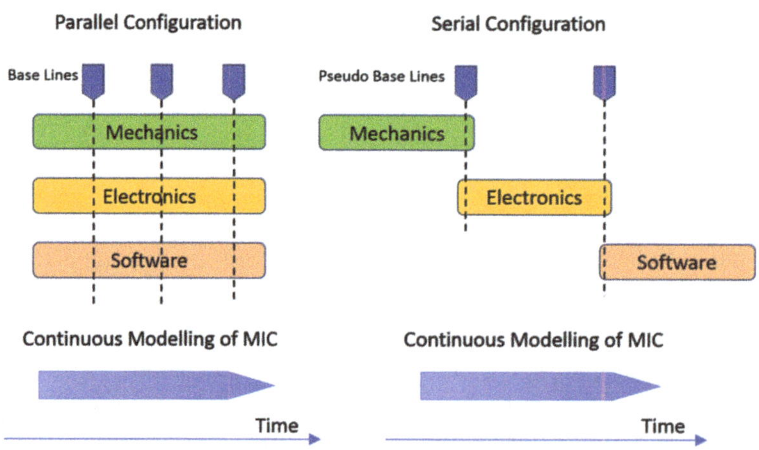

e.g. decision points or gates or milestones among the project phases. It can also be used as a continuous modelling of the dependencies in the product concept as the concept evolves and the detail level increases-

In this case, the problem of integration can be solved thanks decision points that have been taken. In the serial configuration, base lines are more difficult to establish because development is ongoing at the same time for the three domains. Instead, a continuous modelling of the product concept in the form of the Mechatronic Integration Concept (MIC) can be achieved to reveal dependencies as early as possible.

In addition, it might be advantageous to create the Mechatronic Integration Concept in the transition between two domains; e.g., where the product concept is handed from the mechanical development to the electronics development. In the case of a serial configuration the main objective would be to avoid integration problems and to a lesser degree to obtain synergistic solution finding.

A report on some challenges related to the integration in mechatronic development shows that the it is still a source for problems in development (Cabrera et al., 2010). The stated challenges are specifically directed at modelling the product concept. Encountered difficulties in developing mechatronic products are related to:

- Multi-disciplinary modelling.
- Design models and data exchange.
- Cooperation and communication among the design engineers.
- Simultaneous consideration of designs from different disciplines.
- Early testing and verification.
- Persistence of a sequential design process.
- Lack of tools and methods supporting multi-disciplinary design.
- Support of the design of control software.

A typical mechatronic system picks up signals, processes them, and, as an output, generates forces and motion. Mechanical systems are integrated with sensors, microprocessors and controllers. The fact that such a system detects environmental or parametrical changes by sensors and, after suitably processing this information, reacts to them, makes it quite different from conventional mechanical systems. For example, robots, automatic machined, digitally controlled engines, automated guided vehicles, electronic cameras, telefax machines, and photocopiers can be mentioned as typical mechatronics products.

Main characteristics of mechatronic systems according to Buur (1989) are: functional interaction among mechanical, electronic, and information technologies; physical integration of several subsystems in one unit; intelligence related to the control functions of the mechatronic system; flexibility, ability of mechatronic products being modified to fit changing requirements and situations; multifunctionality attributed to the software defined functions of the microprocessor;; and technology dependence, closely related to available manufacturing technologies. Increased flexibility, versatility, intelligence level of products, safety, and reliability as well as lower energy consumption and relatively low-cost are the gains achieved through applying mechatronics concepts to the product design. These advantages translate into a product with more customer appeal, produced quickly at reduced cost and serving larger markets.

Mechatronics Integrated in Home Environment

Robotics and Mechatronics play an important role in building Cyber-Physical Systems (CPS), which can be defined as physical and engineered systems whose operations are monitored, coordinated, controlled, and integrated by a computing and communication core. The intimate coupling between the cyber and physical will be evident from the Nano world to large-scale wide area of systems. CPS will transform how we interact with the physical world just as the Web transformed people's interaction (Ragulskis et al., 2014). CPS are built and used to closely interact also with people in human environment, therefore, their main characteristics play a crucial role to understand and interact with physical world.

Since last two decades, several researchers have started to develop solutions for the robotic service systems to support Activities of Daily Living (ADLs), as reported in (Linner et al., 2012). Proposed approached combines and applies strategies of performance-oriented architecture, ubiquitous computing and service robotics. Researches have been developed solution for embedding robotics and mechatron-

Table 1. Examples of embodiment of robotics in architectural elements

sub-system	**wall**	**ceiling**	**bedroom**	**kitchen**
project	Age UK stairlift (https://www.ageuk.org.uk/products/)	Transfer Lifts & Ceiling Mounted Lifts (https://www.ehls.com)	Stryker-InTouch (https://patientcare.stryker.com/en/products/beds/critical-care-beds/intouch)	OBI feeding system (https://meetobi.com/)
product	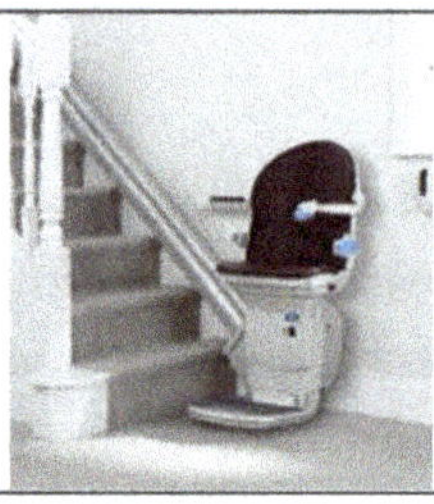	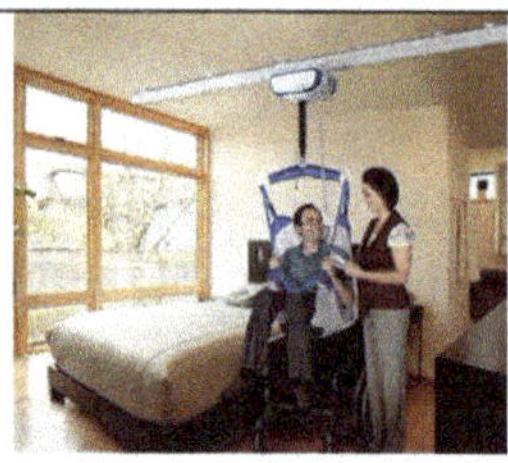	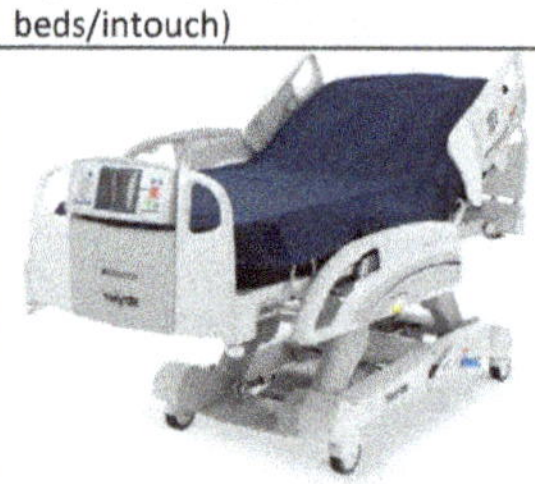	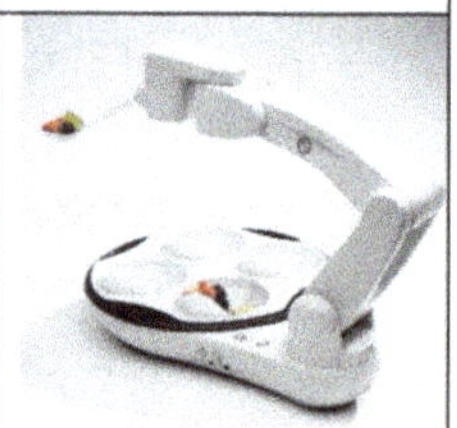

sub-system	**seat**	**in-house STS and mobility**	**bathroom/shower**	**home assisting robots**
project	Up-Lift Seat Assist Chair Lift (https://www.caregiverproducts.com/up-lift-seat-assist-chair-lift.html)	Struzzo (http://www.lyncare.ie/product/struzzo-3420/)	I-SUPPORT (Dometius et al., 2016)	CES 2019: Robotic assistants (https://www.jwtintelligence.com/2019/01/ces-2019-robotic-assistants/)
product		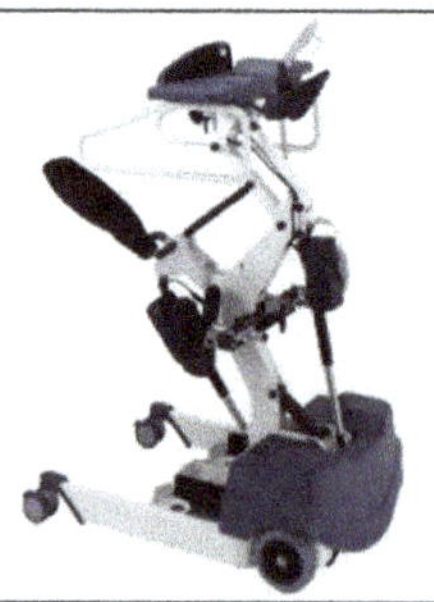	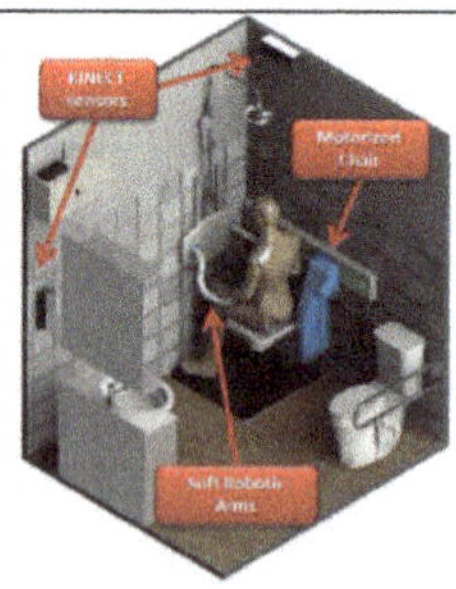	

ics in architectonic elements. These interesting concepts can be further developed to get an integrated vision of the rehabilitation and assisting systems, which can be thought as embedded systems highly developed for aiding ADL. Table I reports examples of embodiment of robotics in architectural elements at different levels, environments and functions.

There has been a growing interest in the recent years in the research on human-robot interactions, human-machine interfaces and intelligent devices that are centered around human application, however, these works mostly lacked in focus on how to harmonize the interactions between mechatronic systems and users in the loop (Zhang et al., 2018). This is one of the key areas for evaluating the success of any mechatronic system implementation on human. Examples in Table I show how mechatronics is the core of the proposed solutions, most of them are commercially available to demonstrate the awareness and spread of the technology in the sector of assisting and healthcare systems.

Advanced human-machine interfaces (HMIs) are the key to harmonizing mechatronic systems with human beings; many developed systems fail because they lack of good interfaces for interpreting user motion intention/states and providing tactile feedback in the systems. The level of harmonization between mechatronic systems and human beings could not be assessed without validating these devices under experiments with human subjects. There are several studies that focused on experimental trials to prove the efficacy of their devices.

A Mechatronic Design of the Actuation Control for the Sit-To-Stand Application

In modern societies, several factors may contribute to the need for rehabilitative and assistive services, as complement and support to surgical and medical treatments. The main of them are the increasing longevity of the population, the trend towards reducing the duration of hospitalization, the use of therapies that can treat highly progressive debilitating diseases, the increased incidence of severe and moderate disabilities resulting from the activities at risk of injury and trauma, the use of advanced techniques of resuscitation. The need for appropriate rehabilitative therapies has an increasing importance in many motor disorders of neurological origin, referring to neurorehabilitation. A large number of people worldwide suffer from motor disorders associated with neurological problems such as stroke, brain injuries, spinal cord injuries, multiple sclerosis, Parkinson's disease.

The World Health Organization estimates that each year 15 million people worldwide are affected by a stroke and 5 million of those are living with a permanent disability (WHO). The majority of post-stroke patients is able to recover an independent walking, but many of them fail to obtain a functional use of upper limbs, even after a prolonged rehabilitation treatment: these functional limitations are responsible for a significant reduction quality of life. Several robotic devices for rehabilitation have been developed to overcome the disadvantages, they require a manual interaction by the therapists that must be provided on a daily basis for several weeks, meaning time consuming with high costs. Robotics may provide a safe and intensive therapy to patients with different degrees of motor impairment. Furthermore, the training with the support of the robot can be extremely precise, intense and prolonged. The robotic systems can also measure the progress of the patient in an objective way, increase the effectiveness of treatment and reduce the costs associated with the healthcare system. A mechatronic approach can show several benefits such as great flexibility, performances, high quality, wide areas of applications and are less expensive in two main research fields: functional assessment and movement analysis.

Several reviews have shown the robot-assisted sensorimotor treatments and task-orients repetitive movements can improve muscle strength and motor coordination in patients with disabilities after stroke or trauma. In the paper (Vazquez-Santacruz et al., 2015) a mechatronic approach has been used to design an assistive system. The mechanical structure was synthesized according to the identified needs. The full design needs to integrate mechanical and electronic components together with industrial design, resulting in an affordable device. The intelligent system applied demonstrated a strategy to endow robotic assistant's ability to detect risk scenarios to patients. The mechatronics approach that was used involved the mechanical and electronic design, the control design and the use of a Graphical Users Interface (GUI), which allows the user getting real information to observe the scenario and handle the variables in the system.

A mechatronic tool for arthroscopy, which is at the same time a smart tool for traditional arthroscopy and the main component of a system for computer-assisted arthroscopy, was proposed in (Dario et al., 2000).

In the case of surgical instrumentation for minimally invasive surgery, research efforts are devoted to enhance the dexterity and sensory capabilities, to improve visual and tactile rendering to the surgeon, and to develop complete systems able to integrate multimodal data (images, models, planned procedures, sensors signals). The objective of Computer-Assisted Surgery CAS is to provide the surgeon with benefits deriving from the integration of advanced image processing techniques and computer graphics with new robotic and mechatronic surgical instruments, in order to achieve more effective planning and more accurate execution of surgical interventions. Surgical tools have been classified in (Dario et al., 2000) into one of three main types, based on their technical characteristics and on the aspect of surgeon performance they aim to enhance: a) robotic tools; b) tele-operated dexterous tools; and c) manuall-operated mechatronic tools.

According to general design consideration above, the design of mechatronic device should take into account three main phases: the first one deals with the acquisition/monitoring task, in which the motion is captured and recorded by a MoCap system. Alternatively, it can be simulated by CAD based design with a model interacting with a human body model, (Castelli & Ottaviano, 2010). The second part of the control deals with the kinetostatic/dynamic model of the system. Data obtained by the acquisition are used into the math model to get the motion law of the actuation system, which is sent to the controller (which may be Arduino based) to drive the actuation. Figure 2 shows a general scheme for the architecture of the control system of the assisting device referring on a scaled prototype here, or alternatively it can be sent to a real prototype.

Figure 3 shows an experimental test-bed realized as illustrative example to demonstrate the effectiveness of the proposed procedure. Two brushless motors and drivers controlled by Raspberry Pi III and Pican2 compose the experimental test-bed. A monitor is used for raspberry-PI programming using Python software. Label 3) in Figure 3 indicates the laptop with caliper software for the set-up of brushless drives. The controller reads the motion laws for the actuation from a memory card, then it commands the two brushless motors. Figure 4 reports a flow-chart for a suitable program to drive the two motors named as M1 and M2.

The use of this kind of low-cost control system solution offers several advantages, which can be summarized as follows:

1. Low-cost - Raspberry boards are relatively low-cost, if compared to other microcontroller platforms. They may enable diffusion and affordable cost of the overall system.

Figure 2. A scheme for the mechatronic architecture for the control of the system

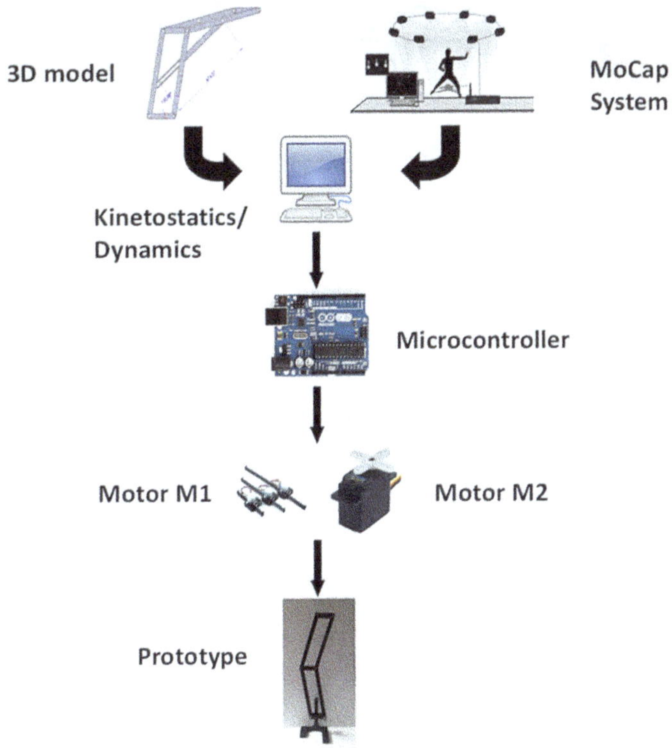

Figure 3. Experimental Test-Bed: 1);Brushless motors and drivers controlled by Raspberry Pi III model B and Pican2, 2); Monitor for raspberry-PI programming, software Python, 3); Laptop with caliper software for the set-up of brushless drive

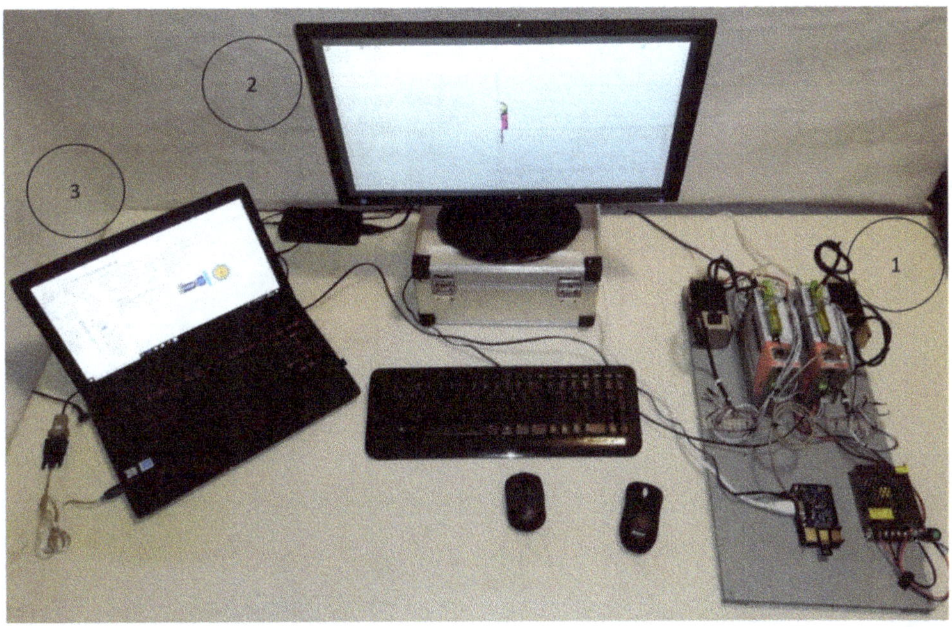

Figure 4. Flow-chart for the programming

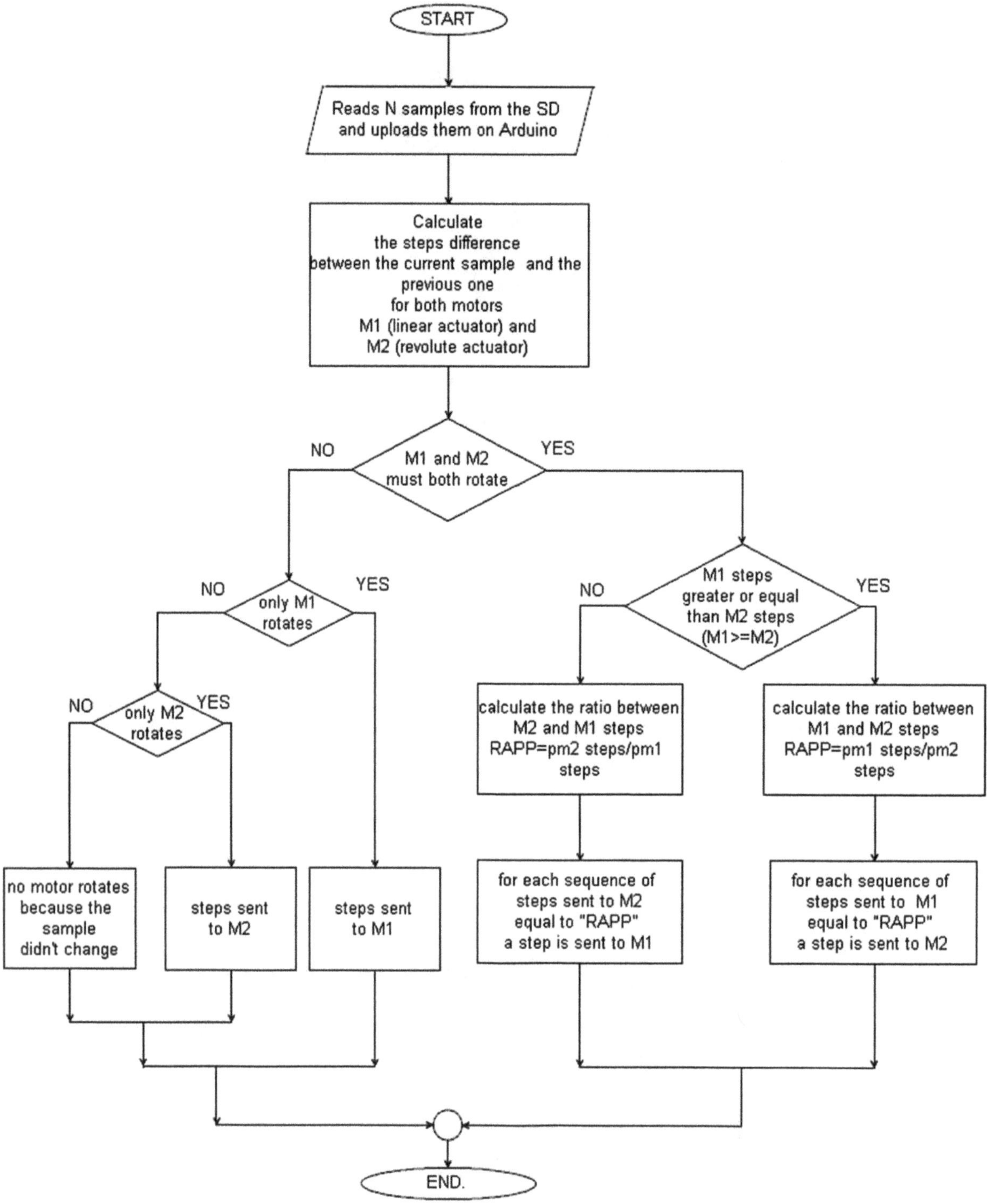

2. The Python software runs on Windows, Macintosh OSX, and Linux operating systems. Most microcontroller systems are limited to Windows.
3. Simple, clear programming environment - The programming environment is easy-to-use for beginners, yet flexible enough for advanced users to take advantage of as well.
4. Open source and extensible software - The software is made available as open source tools, suitable for extension by experienced programmers. The language can be expanded through C++ libraries.
5. Open source and extensible hardware - The plans for the modules are published under a Creative Commons license, so experienced circuit designers can make their own version of the module, extending it and improving it. Even relatively inexperienced users can build the breadboard version of the module in order to understand how it works and save money.

Main limitations are due to its hardware. Because of its processor, some common ones like Windows and Linux distros are not compatible. In addition, some applications, which require high demands on CPU processing, are off-limits.

CANopen is a CAN-based communication system. It comprises higher-layer protocols and profile specifications. CANopen has been developed as a standardized embedded network with highly flexible configuration capabilities. It was designed originally for motion-oriented machine control systems, such as handling systems. Today it is used in various application fields, such as medical equipment, off-road vehicles, maritime electronics, railway applications, or building automation.

In order to obtain this low-cost control system, the advantages of using Raspberry with CANopen for data transmission have been combined. This was possible using a Pican2 breadboard interface card that allows Raspberry to send commands via CANopen.

CASE OF STUDY: THE DESIGN OF A 2-DOF MECHANISM FOR THE SIT-TO-STAND APPLICATION

The so-called "natural" STS movement is composed by a translation in two directions with a rotation of the trunk, as described in (Marlene & Cooper, 1995). It can be defined as a balanced movement of the body's center of mass from a seated position to the standing one (Roebroeck et al., 1994) being a transition between two stable postures (Tully et al., 2005). Several factors can influence the STS, e.g. the anthropometric data, age, environment, strategies (velocity, upper limb configuration) and objective (assistance, rehabilitation). Therefore, the design of a support mechanism, which is devoted to generate the requested motion and support the body of the individual during the STS is a crucial element in the design of the device (Ottaviano et al., 2007). Synthesis procedures can be taken into account for the design of support mechanisms, as described in (Figliolini & Rea, 2012; Figliolini et al., 2016).

Among several solutions, we have chosen to illustrate here a 2-DOF mechanism shown in Fig. 5, which is able of reproducing any trajectory in the sagittal plane with a fixed orientation of the trunk. It is well known that the combination of the motion laws for the two actuators will produce then any trajectory in a region of the sagittal plane. Experiments and numerical simulations were conducted and aimed at clarifying the effects of the motion law of the actuation and speed for the execution of the STS movement supported by the assisting device. CAD based modelling allows the realization of realistic simulations that are effectively used at the design stage (Castelli et al., 2014). Figure 6 shows a motion

Figure 5. Kinematic scheme and planar sketch of the 2-DOF STS mechanism

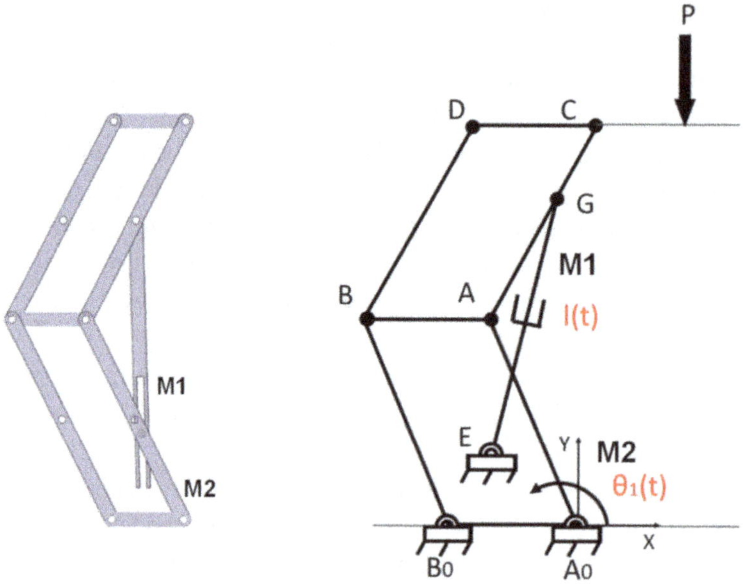

Figure 6. Motion sequence of first version of a scaled prototype obtained with a 3D printing technique

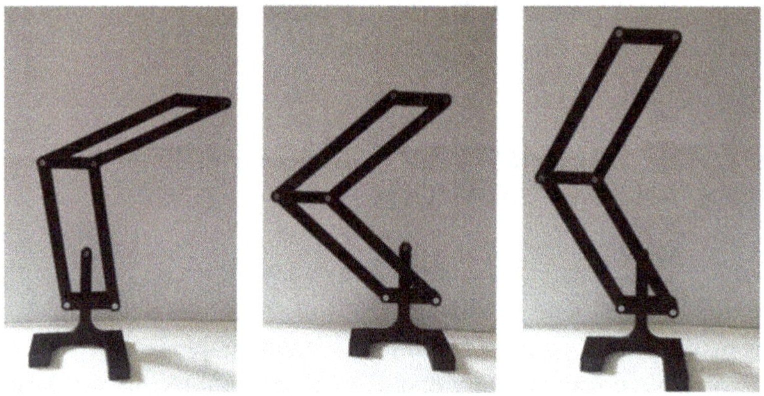

sequence of a first version of a scaled prototype realized with 3D printing technique. Figure 7 shows the motion sequence of the device during the STS motion interacting with a realistic model of the human body, (Castelli & Ottaviano, 2010). . Figure 8 reports the motion laws of the two actuators M1 and M2 and their velocity profiles. During the simulation, the velocity and acceleration components in the sagittal plane have been also displayed to monitor the effect of the STS device on a reference point on the human body model. In particular, a reference point S has been chosen on the center of the interface of the mechanism that supports the armpit. This point has been set to monitor the effects of the movement on the body, kinematics is used to drive M1 and M2.

Figure 7. Motion sequence of the STS with the proposed 2-DOF mechanism

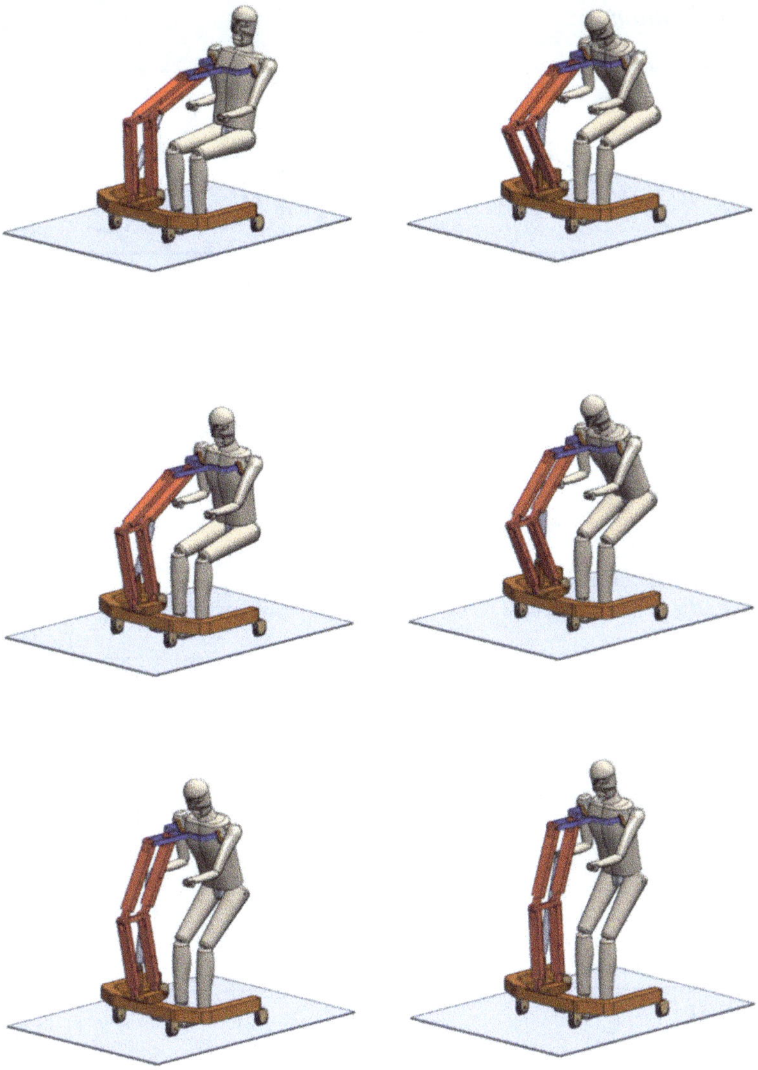

CONCLUSION

In this Chapter, we presented overall discussion and design issues related to a mechatronic design for the actuation control of STS devices with the aim to provide solutions for the homecare of elderly and people with reduced mobility. The mechatronic solution has the aim to test low-cost technologies trying to reduce significantly the overall costs to facilitate interconnection and wider use of assisting systems.

Figure 8. Results for the simulation in Fig. 2: a) displacement the linear actuator M1; b) velocity profile of M1; c) angular displacement of the rotational actuator M2; d) velocity profile of the M2; e) point S velocity; f) point S acceleration

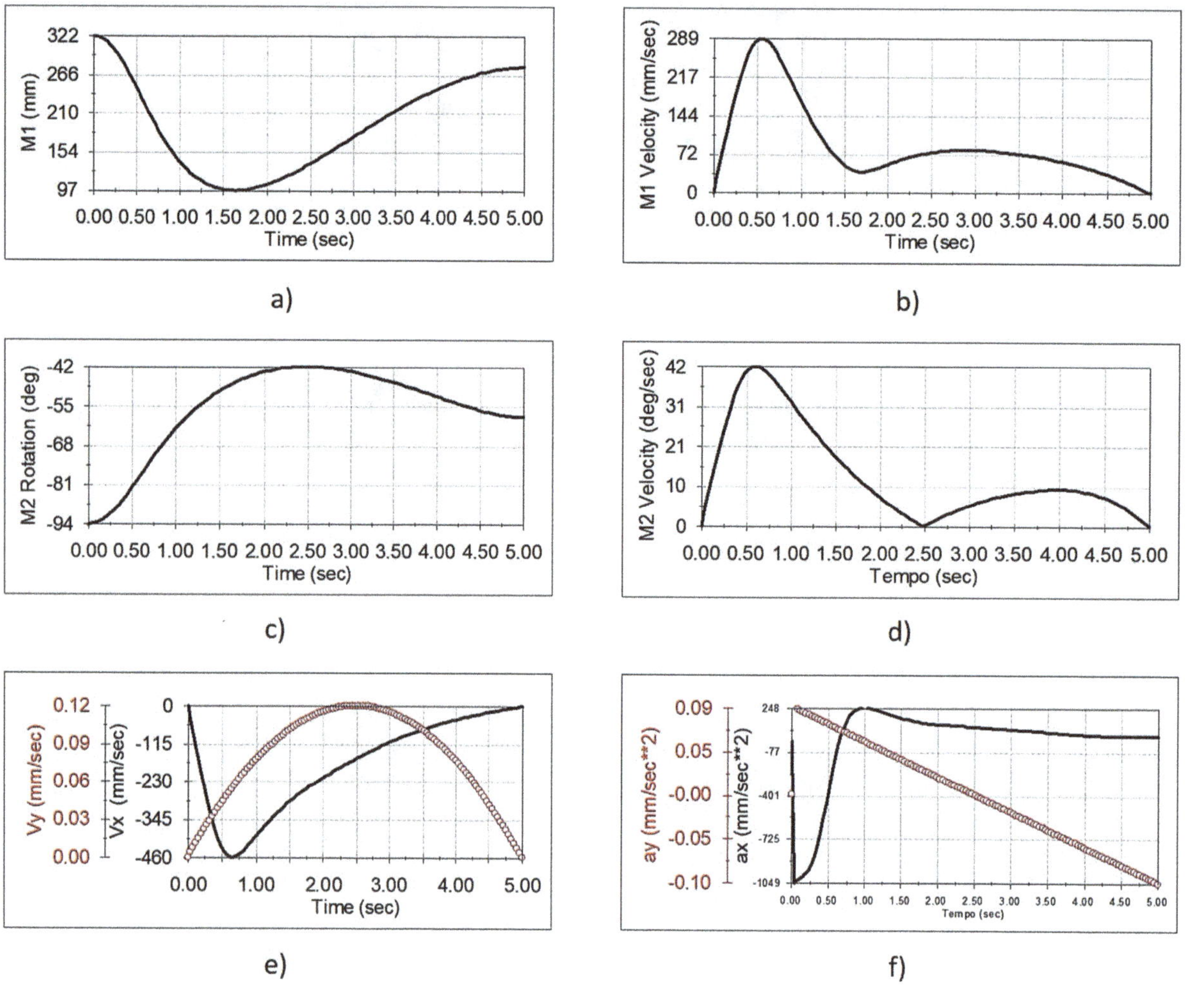

REFERENCES

Andreasen, M. M., & McAloone, T. C. (2001). Joining Three Heads - Experiences From Mechatronic Projects. In *12th Design For X Symposium,* Neukirchen, Germany. pp 151-156.

Borràs, J., Thomas, F., Ottaviano, E., & Ceccarelli, M. (2009). A reconfigurable 5-DoF 5-SPU parallel platform" Proceedings of the 2009 ASME/IFToMM International Conference on Reconfigurable Mechanisms and Robots, London, UK, 5173892, pp. 617-623.

Buur, J. (1989). *Mechatronics design in Japan.* Institute for Engineering Design, Technical University of Denmark.

Cabrera, A. A., Foeken, M. J., Tekin, O. A., Woestenenk, K., Erden, M. S., De Schutter, B., ... Tomiyama, T. (2010). Towards Automation of Control Software: A Review of Challenges in Mechatronic Design. *Mechatronics*, *20*(8), 876–886. doi:10.1016/j.mechatronics.2010.05.003

Castelli, G., & Ottaviano, E. (2010). Modelling, Simulation and Testing of a Reconfigurable Cable-Based Parallel Manipulator as Motion Aiding System. *Applied Bionics and Biomechanics*, *7*(4), 253–268. doi:10.1155/2010/548197

Castelli, G., Ottaviano, E., & Rea, P. (2014). A Cartesian Cable-Suspended Robot for improving end-users' mobility in an urban environment. *Robotics and Computer-integrated Manufacturing*, *30*(3), 335–343. doi:10.1016/j.rcim.2013.11.001

Dario, P., Carrozza, M. C., Marcacci, M., D'Attanasio, S., Magnani, B., Tonet, O., & Megali, G. (2000). A Novel Mechatronic Tool for Computer-Assisted Arthroscopy. *IEEE Transactions on Information Technology in Biomedicine*, *4*(1), 15–29. doi:10.1109/4233.826855 PMID:10761770

Dometios, A. C., Papageorgiou, X. S., Tzafestas, C. S., & Vartholomeos, P. (2016, June). Towards ICT-Supported Bath Robots: Control Architecture Description and Localized Perception of User for Robot Motion Planning, *24th Mediterranean Conference on Control and Automation (MED)*, Athens, Greece. 10.1109/MED.2016.7535954

Figliolini, G., & Rea, P. (2012). Effects of the design parameters on the synthesis of Geneva mechanisms". Proceedings of the Institution of Mechanical Engineers. *PART C, Journal of Mechanical Engineering Science, vol. 2012*, pp. 1-10. doi:10.1177/0954406212466194

Figliolini, G., Rea, P., & Angeles, J. (2016). The synthesis of the axodes of RCCC linkages, *Journal of Mechanisms and Robotics, 8*(2). doi:. doi:10.1115/1.4031950

Figliolini, G., Rea, P., & Conte, M. (2010). Mechanical Design of a Novel Biped Climbing and Walking Robot, ROMANSY 2010, 18th CISM-IFToMM Symposium on Robot Design, Dynamics, and Control, pp.199-206.

Fukuda, T., Nakajima, M., Ahmad, M., Shen, Y., & Kojima, M. (2010). Micro- and Nanomechatronics. *IEEE Industrial Electronics Magazine*, *4*(4), 13–22. doi:10.1109/MIE.2010.938721

Gausemeier, J., Frank, U., Donoth, J., & Kahl, S. (2009). Specification Technique for the Description of Self-optimizing Mechatronic Systems. *Research in Engineering Design*, *20*(4), 201–223. doi:10.100700163-008-0058-x

Habib, M. K. (2007, Summer). Mechatronics: A Unifying Interdisciplinary and Intelligent Engineering Paradigm. *IEEE Industrial Electronics Magazine*, *1*(2), 12-24. doi:10.1109/MIE.2007.901480

Habib, M. K. (2008, August). Human Adaptive and Friendly Mechatronics (HAFM), IEEE International Conference on Mechatronics and Automation 'ICMA2008', Takamatsu-Kagawa, Japan, pp. 61-65. 10.1109/ICMA.2008.4798726

Hari Krishnan, R., & Pugazhenthi, S. (2014). Mobility assistive devices and self-transfer robotic systems for elderly, a review. *Intelligent Service Robotics*, *7*(1), 37-49. doi:10.100711370-013-0142-6

Linner, T., Georgoulas, C., & Bock, T. (n.d.). Advanced building engineering: Deploying mechatronics and robotics in architecture, *Gerontechnology, 11*(2), doi:. doi:10.4017/gt.2012.11.02.158.002012

Marlene, J. A., & Cooper, J. (1995). *Biomechanics of Human Movement*. Brown & Benchmark.

Ottaviano, E., Ceccarelli, M., & Castelli, G. (2004). *Experimental Results of a 3-DOF Parallel Manipulator a an Earthquake Motion Simulator, ASME Design Engineering Technical Conference and Computers in Engineering Conference* (pp. C2004–C57075). Salt Lake City, Utah: Paper DET.

Ottaviano, E., Ceccarelli, M., & De Ciantis, M. (2007). A 4-4 Cable-Based Parallel Manipulator for an Application in Hospital Environment, 15th Mediterranean Conference on Control and Automation – MED07, Athens, Greece. 10.1109/MED.2007.4433839

Ottaviano, E., Rea, P., & Grandinetti, A. (2019). Mechatronic Design of a Low-Cost Control System for Assisting Devices. In J. Machado, F. Soares, & G. Veiga (Eds.), *Innovation, Engineering and Entrepreneurship. HELIX 2018. Lecture Notes in Electrical Engineering* (Vol. 505). Cham, Switzerland: Springer. doi:10.1007/978-3-319-91334-6_21

Papageorgiou, X. S., Tzafestas, C. S., Vartholomeos, P., Laschi, C., & Lopez, R. (2015). ICT-Supported Bath Robots: Design Concepts, 17th International Conference on Social Robotics (ICSR 2015). Retrieved from http://www.i-support-project.eu/web/wp-content/uploads/2018/08/Papageorgiou_et-al_ICSR2015.v.2.pdf

Ragulskis, M., Jiang, H., Quan, Q., Fedaravicius, A., & Xie, G. (2014). Mechatronics and Mechanical Engineering in Cyber-Physical Systems. *Hindawi Publishing Corporation Advances in Mechanical Engineering*. doi:10.1155/2014/591629

Rea, P., & Ottaviano, E. (2018). Functional Design for Customizing Sit-To-Stand Assisting Devices. *Journal of Bionics Engineering, 15*(1), 83–93. doi:10.100742235-017-0006-4

Rea, P., Ottaviano, E., & Castelli, G. A. (2013). Procedure for the Design of Novel Assisting Devices for the Sit-to-Stand, *Journal of Bionic Engineering, 10*(4), 488-496.

Rea, P., Ottaviano, E., Conte, M., D'Aguanno, A., & De Carolis, D. (2013). The design of a novel tilt seat for inversion therapy. *Int. Journal of Imaging and Robotics, 11*(3), 1–10.

Roebroeck, M. E., Doorenbosch, C. A. M., Harlaar, J., Jacobs, R., & Lankhorst, G. J. (1994). Biomechanics and muscular activity during sit-to-stand transfer. *Clin. Biom., 9*(4), 235–244. doi:10.1016/0268-0033(94)90004-3 PMID:23916233

Stryker-InTouch. (2019). Retrieved from https://patientcare.stryker.com/en/products/beds/critical-care-beds/intouch

Tomiyama, T., Amelio, V. D., Urbanic, J., & ElMaraghy, W. (2007). Complexity of Multi-Disciplinary Design. *Annals of the CIRP, 56*(1), 185–188. doi:10.1016/j.cirp.2007.05.044

Tully, E. A., Fotoohabadi, M. R., & Galea, M. P. (2005). Sagittal spine and lower limb movement during sit-to-stand in healthy young subjects. *Gait & Posture, 22*(4), 338–345. doi:10.1016/j.gaitpost.2004.11.007 PMID:16274916

Vazquez-Santacruz, E., Morales-Cruz, C., Portillo-Velez, R., & Gamboa-Zuniga, M. (2015). Mechatronic Approach to an Intelligent Machine: The Case of an Assistive Device. *Research in Computing Science, 100*, 115–127.

Zhang, D., Dubey, V. N., Yu, W., & Low, K. H. (2018). Editorial: Biomechatronics: Harmonizing Mechatronic Systems With Human Beings. *Frontiers in Neuroscience*. doi:10.3389/fnins.2018.00768 PMID:30459543

Chapter 5
Infrared Thermography for Intelligent Robotic Systems in Research Industry Inspections:
Thermography in Industry Processes

Alessandro Massaro
https://orcid.org/0000-0003-1744-783X
Dyrecta Lab srl, Italy

Angelo Galiano
Dyrecta Lab srl, Italy

ABSTRACT

Infrared thermography has been applied in different research projects involving leakages, micro-climate monitoring, energy control, security, and predictive maintenance. The chapter focuses on image and data process techniques improving intelligent inspection systems able to extract hidden information. Image segmentation techniques, data mining, and machine learning approaches have been applied by proposing innovative models suitable for accurate process monitoring. The proposed models can be adopted for the engineering of industry processes involving quality, production performance indicators, security systems, resources management, and intelligent scheduling of activities. Different networks able to transmit data and thermograms are described by discussing system integration in robotic systems. The proposed chapter provides elements and requirements for the design of infrared thermography facilities in industries. Different cases of study of industry projects have been analyzed in order to provide models usable in different application fields.

DOI: 10.4018/978-1-7998-0137-5.ch005

INTRODUCTION: SOME RESEARCH TOPICS REGARDING IR THERMOGRAPHY IN INDUSTRIES

Mechatronics and Robotics are of strong impact in engineering science oriented on innovative ways to synergize multiple disciplines and enhance machine intelligence (Habib, 2008). In particular mechatronics is interdisciplinary and opens new horizons in all field (Habib, 2007). An interesting field for mechatronics is thermography. Thermography is adopted in different industrial research fields. Regarding food and machines to preserve food quality, thermography has been adopted for food quality monitoring (Samanian et al., 2016), for the check of refrigerator thermal behaviour (Björk et al., 2010), and for the identification of thermal regions of food affected by induced mechanical injuries or cold injuries (Samanian et al., 2016). Concerning leakages checking, infrared –IR- thermography has been adopted for gas sensing (Rogalski et al., 2002), for building thermal leakage using data mining approaches (Galiano et al., 2016). Artificial neural network has been applied in literature for the defect classification in electric applications improving predictive maintenance for power substation equipment (Ullah et al., 2017), and K-means clustering algorithms has been implemented for thermal clustering in photovoltaic networks (Massaro et al., 2018). In agriculture thermal images provided important information about moisture stress conditions of plants (Banerjee et al., 2018), soil microwave treatment (Khan et al., 2018), and other agriculture parameters (Ishimwe et al., 2014). Also in manufacturing control processes thermography is adopted for motor monitoring (Manama et al., 2011), degradation assessment control (Zhao et al., 2015), and other real time monitoring manufacturing processes (Usamentiaga et al., 2014). All the mentioned works prove that infrared thermography (IRT) is an important tool for process monitoring and for quality process improvement according with ISO 9001: 2015 for different industry fields. Other aspects correlated with IRT inspections are:

- Worker security (controlling temperature of work environments machine and tolls);
- Crack detection in robotic systems (Ghidoni et al., 2015);
- Defect classification (Grys, 2018);
- Detection of the excessive friction due to improper lubrication or material fatigue (La Rosa et al., 2000);
- Robotic system for electrical distribution inspection in industry (Cheng et al. 2019);
- photovoltaic efficiency (Crişan et al., 2019);
- gas leakage (Yang et al., 2019);
- Night vision enhancement (Raghatate et al., 2013);
- Water building infiltration (Ljungberg, 1994);
- Visibility through smoke, light fog, snow, and rain in defence systems (Akula et al., 2011).

Many of the listed applications can be applied for specific cases of study involving industry monitoring processes, thus providing solution in energy efficiency, security, and production line predictive maintenance. A good implementation of the potential use of IRT is in robotic and automated systems oriented on the inspection time reduction and on the measurement optimization. In this scenario some authors used Unmanned Aerial Vehicle (UAVs) for photovoltaic control (Aghaei et al., 2014), building inspection protocols (Entrop et al., 2017), and vegetation monitoring (Sagan et al., 2019). UAVs has been applied also for surveillance applications involving image fusion between visual and thermal image enabling tracking monitoring (Kumar et al., 2011). In this direction in this chapter will be proposed some industry solution concerning real needs, embedding IRT systems and discussing how the radiometric inspection can be automated and supported by other tools including data mining and image processing. According with the state of the art, the goal of the proposed chapter is to discuss some models and basic protocols suitable for industry research. In this direction the used approaches are useful in order to perform service innovation and production process innovation by means of a knowledge gain provided (Frascati, 2015) by the IRT systems and data and image processing based on innovative algorithms designed for the specific cases of study. It is important to note that in research industry, a research and development project must aim to create new concepts

or ideas that improve existing knowledge (Frascati, 2015). For this purpose the models proposed in this chapter are oriented on the formulation of new concepts to solve and optimize industry problems involving critical issues concerning robotic and automated systems.

The chapter is structured as follows:

- Description of the IRT basic principles useful to understand the electromagnetic phenomena involving IRT measurements;
- Examples of measurement approaches and measurements protocols;
- Presentation of different cases of study where IRT systems providing a knowledge gain about industrial and production processes, including data flow automatism;
- Some aspects regarding IRT implementation in robotic systems and inspection protocols involving image processing and data mining post-processing approach;
- Future research directions of IRT;
- Conclusions.

In the table reported below are indicated the acronyms and variables discussed into the chapter:

IR THERMOGRAPHY: BASIC PRINCIPLE

In this section are provided different notions useful in order to understand the basic physical principles of thermography useful for measurements approaches. Below is an explanatory concept of thermal radiation.

The molecules of a target are always in motion. The heater molecules move faster than the cold ones, and the molecular velocity corresponds to a defined temperature level. The radiated temperature is therefore the heat of the object. Heat and thermal energy are measured in:

- Joule (J);
- Watt-second (Ws) or Kilowatt-Hour KWh;
- Newton-meter (Nm).

A thermogram shows the temperature values of the target in each its parts.

According with the scheme of Fig. 1, thermograms can be acquired as passive detection or in active modality by exciting the target object by a heater or a light source such as a pulsed laser (Ciampa et al., 2018). The total radiation received by an infrared thermal camera comes mainly from three sources: the emission of the target object (W_o), the emission of other emitting objects (W_r), and the emission of the environment (W_e):

Table 1. Acronyms and variables

Variables	Description
W_T	Total thermal radiation
W_o	Object thermal radiation
W_e	Environment thermal emission
α	Absorption coefficient
τ	Trasmittance coefficient
NIR	Near-infrared band
IR	Infrared band
IRT	Infrared Thermography
SWIR	Short-wavelength infrared
MWIR	Mid-wavelength infrared
LWIR	Long-wavelength infrared
FDTD	Finite Difference Time Domain
FEM	Finite Element Method
FOV	Field of View
IFOV	Instant Field of View
3D	Three-Dimensional
RGB	Red, Green, Blue
E_{int}	Internal energy of pixel clusters
E_{Snake}	Functional energy associated to the snake contour
E_{image}	Image energy representing image forces
E_{con}	Constraint forces
$v(s)$	Snake pattern
k	Cluster numbers of the k-means algorithm
CE	Contrast enhancement
R&D	Research and Development
DB	Database
CAN	Controlled Area Network
ODB	On Board Diagnostics
POE	Power over Ethernet
UNI	Italian National Unification
ISO	International Standards Organization
GPS	Global Positioning System
UAV	Unmanned Aerial Vehicle
HD	High Definition
IEC	International Electrotechnical Commission
DIN	Deutsches Institut für Normung
ASTM	American Society for Testing and Materials International
MIL-STD	Military Standard
VdS	Verband der Sachversicherer
NFPA	National Fire Protection Association
NMS	National Master Specification
UKTA	United Kingdom Thermography Association
RESNET	Residential Energy Services Network

Figure 1. Basic principle of thermography

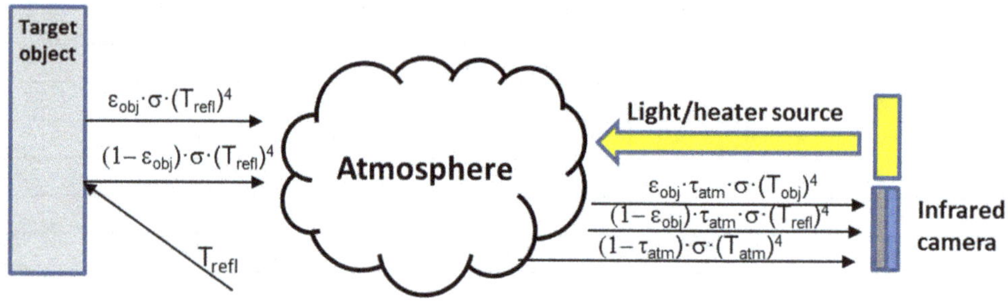

$$W_T = W_o + W_r + W_e \tag{1}$$

, where

$$\begin{aligned} W_o &= \varepsilon_{obj} \cdot \tau_{atm} \cdot \sigma \cdot (T_{obj})^4 \\ W_r &= (1-\varepsilon_{obj}) \cdot \tau_{atm} \cdot \sigma \cdot (T_{refl})^4 \\ W_e &= (1-\tau_{atm}) \cdot \sigma \cdot (T_{atm})^4 \end{aligned} \tag{2}$$

being σ is the Stefan–Boltzmann constant, ε is the emissivity (ratio of how well a material radiates infrared energy, compared to a perfect radiator; emissivity values fall between 0 and 1; an object that measures 1.0 is considered a perfect radiator and is called black body), τ is the transmittance, and T is the temperature. The same equation (1) can be written as:

$$\alpha + \tau + \rho = 1 \tag{3}$$

being α the absorption, τ the transmission and ρ the reflection coefficient, respectively.

In order to estimate correctly the temperature of a target it is important to know the emissivity of materials: a material's emissivity depends on its nature, its surface finish and its temperature (ratio of the energy radiated by an emission source to the energy radiated by a black body at the same temperature). In the table 2 are indicated the emissivity values of different materials:

The emissivity can changes for the same material versus the working wavelength (Rakrueangdet et al., 2016). There are the following three methods to determine emissivity of a target characterized by an unknown material to set on thermal camera:

- **Method 1 (Using a Thermocouple Characterized by a Good Temperature Contact and by a Low Heat Dissipation Loss):** By a thermocouple is measured, simultaneously to the radiation, the real temperature of target surface is measured (the emissivity is subsequently adjusted on the thermal camera so that the temperature measurement of the infrared thermometer corresponds to the value shown by the contact measurement).
- **Method 2 (Creating a Black Body With a Test Object)**: By a drilled hole having a ratio diameter to drilling depth ≤ ⅓ in thermal conducting material reacts similarly to a black body with an

Table 2. Emissivity values

Material	Emissivity
Aluminium	0.03
Aluminum, rough surface	0.07
Brick	0.95
Concrete	0.9
Chalk	0.34
Tin	0.06
Gold	0.02
Lead	0.28
Human skin (perfect radiator)	0.98
Polished Copper	0.01
Water	0.98
Ice	0.97
Electrical tape (black plastic)	0.95
Black paper	0.94
White paper	0.9
Rubber	0.93
Iron, oxidized	0.74
Steel, rusty red	0.69
Iron, sheet galvanized, burnished	0.23
Brass, dull, tarnished	0.22
Cast iron, polished	0.21
Steel, sheet, nickel plated	0.11

emissivity near 1; emissivity can be subsequently determined moving camera from the hole to the target object.

- **Method 3 (Applying Reference Emissivity):** An electrical black tape having a known emissivity is applied to the measurement object; the emissivity must simultaneously be adjusted until the same temperature measurement of the electrical black tape is displayed.

The transmittance of the environment can be calculated through the distance from the object to the camera and the relative humidity (generally this value is very close to one). The temperature of the environment can be obtained using a common thermometer both for indoor and outdoor applications.

The filters of the thermal camera can be select different wavelengths ranges such as:

- Near-infrared –NIR-: 0.8 μm <NIR< 1.7 μm;
- Short-wavelength infrared –SWIR-: 1 μm <SWIR< 2.5 μm;
- Mid-wavelength infrared –MWIR-: 2 μm <MWIR< 5 μm;
- Long-wavelength infrared -LWIR-: 8 μm <LWIR< 14 μm.

Typically commercial cameras works in the LWIR range (spectral range between 7.5μm – 14 μm). The thermography can be classified as:

- Passive (without any external heat source);
- Active (with an external pulsed heat source impinging on the target).

Other thermography application concern the visibility in cases of fog and rain (transport security). Specifically, the penetration of visibility in the fog is higher for the LWIR (range between 7 and 14 microns) than in the band MWIR (range between 3 and 5 microns). LWIR and MWIR behave similarly in the presence of rain. In the table 3 are indicated different IR wavelengths useful for thermography filtering application (Vollmer et al., 2010).

In agriculture different wavebands and IR sensors, including satellite sensors, are able to acquire different information such as sea surface temperature, lower-level water vapor, winds, SO_2, total water for stability, cloud phase, dust, rainfall, total ozone, turbulence, clouds, ash, air temperature, cloud heights and amounts, surface temperature, night cloud mapping, oil spill detection, mapping and detection of fire and geologic areas, water algae, forest inventory, water suspended sediment, soil moisture, canopy reflectance, water vapor and aerosol, etc. (Ishimwe et al., 2014). In table 4 are summarized these applications:

SOME MEASUREMENT APPROACHES TO USE IN MONITORING PROTOCOLS

Measurement approach in thermography is relevant in order to perform a good thermal detection especially when a robotic handling is involved. The measurement methodology is function of the adopted technology. In the proposed first example it is highlighted how the punctual measurement is important. The experimentation has been executed into a room by measuring the punctual point of three different walls characterized by three different external environments. The IR thermal instruments adopted for the measurements is a Parkside detector detecting temperature in a range between -50 °C and + 380 °C. four time steps of the same day have been considered during the measurement (at 8.30 a.m., at 11.00

Table 3. Standard IR filters for thermography.

Wavelength (μm)	Application
2.3	Measurements through glass
3.42	Measurements of plastics temperature (polyethylene)
3.6-4	Reduction of atmospheric influences at longer measurement distances
3.6	Reduction of influence of sun reflections
3.7-4	Measurements through flames
3.9	Measurements of high object temperatures (reduces incident radiation on the detector)
4.25	Measurements of flame temperatures
5.0	Measurement of glass temperature
7.5	Measurements only in the LWIR spectral region
8.5	Measurements of Teflon temperatures

Table 4. IR wavelengths and applications

Wavelengths (μm)	Application
10.8, 12, 11.5-12.5, 3.66-3.84, 10.3-11.3, 3.55-3.93	sea surface temperature
7.34	lower-level water vapor
7.34, 10.8	winds
7.34, 8.5	SO_2
8.5, 10.8, 10.35	Cloud/cloud phase
8.5	dust
8.5,11.2	rainfall
10.78-11.28, 9.61	total ozone
9.61	turbulence
12.3	ash
13.3	air temperature
13.3	cloud heights and amounts
11.5-12.5, 8.12-9.27, 10.25-10.95, 10.95-11.65, 10.3-11.3	surface temperature
10.3-11.31, 3.50-3.93	night cloud mapping
3.0-5.4, 8.5-12.5	oil spill detection
11.7-12.27, 3.0-5.4, 8.5-12.5, 9.58-9.88, 11.77-12.27	mapping and detection of fire and geologic areas
8.5-12.5	water algae
8.5-12.5	forest inventory
8.5-12.5	water suspended sediment
10.4-12.5, 10.3-11.3, 10.4-12.5	soil moisture
10.4-12.5	canopy reflectance
7.34, 10.3-11.3, 11.512.5, 3.55-3.93	water vapor and aerosol

a.m, 1.30 p.m, and at 17.00 p.m). According with the performed measurements it is observed that for each wall the temperature trend changes with the incremental time step: by increasing the dairy time the outdoor temperature increases together with the time step. The adopted protocol for the acquired measurement is the following one:

- Four main time steps;
- Five measurements for each wall and for each time steps;
- Interval between three time steps of about three hours;
- For each time step measurement the thermal camera or the infrared laser sensor is switched on before (five minutes before).

In Fig. 2 are illustrated all the phases of the infrared temperature measurements.

The adopted protocol can be used in other applications where the analysis during the time is important in order to check the thermal behaviour of different objects allocated in the same environment. This approach it also useful to characterize the effects of near objects reflecting thermal radiated energy in order to understand how the whole ambient it is influenced by each object. This approach can be executed by analysing the thermal contour conditions of each measured object. In this context can be analysed for example the heating effect of a part of a machine on the other machine components by comparing for each part the correct operative temperature condition and by overcoming interrupts of the production line where the machine works. The same approach can be followed by acquiring different thermograms in different time steps, and by analysing the thermal inducted effect on each target object. Concerning an automated system, the protocol can be implemented by programming the acquisition time of the robot enabling infrared measurements. A global view of the thermal measurements can be obtained by a full radiometric image. In Fig. 3 is illustrated an example of a thermograms, where each pixel represents a radiometric measurement. The information contained into the thermogram is representative of the whole environment behaviour. For this reason it is important to distinguish during the measurements:

- Possible thermal background reflections (noise covering the thermal information);
- The energetic sources and their distances from the target object;
- The object to distinguish or the defect to analyse;

According with the point described above, the next step of the image processing is to select the area where are identified the defects: in Fig. 3 is shown the box where are included point defects and where it is possible to analyse the values of each pixel (see inset of Fig. 3 representing punctual values of the box matrix extracted by FLIR tools software). An important aspect concerning defect checking is the

Figure 2. (a) Indoor measurements performed in different time steps on three different room walls; (b) IR laser pointer (the measurement is the average value of eight measured closed points); (c) experimental indoor layout

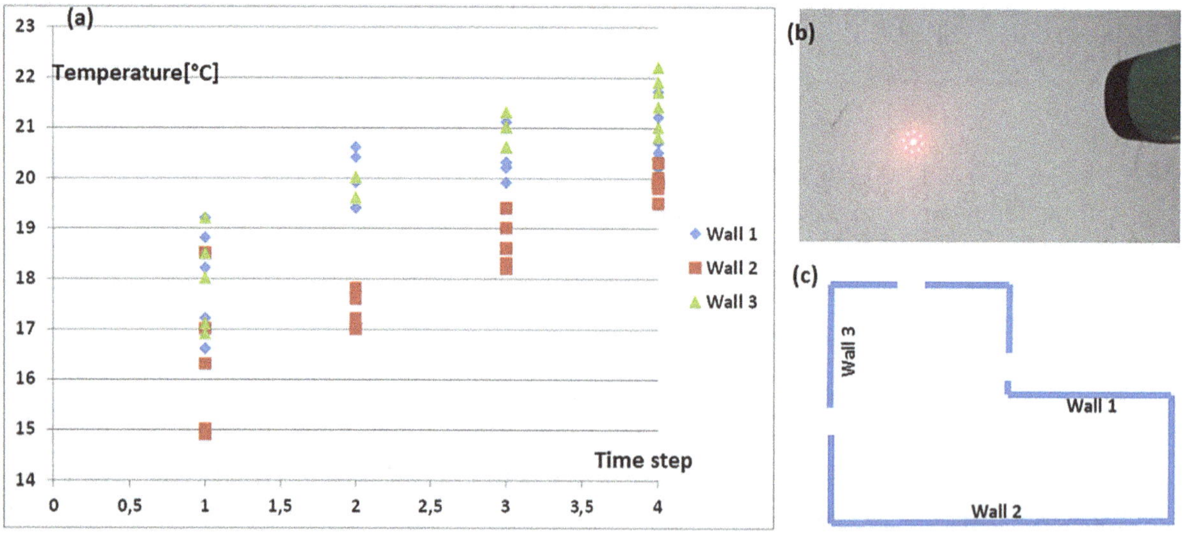

Figure 3. Thermogram of a photovoltaic panel and radiometric measurements of a region delimited by a box

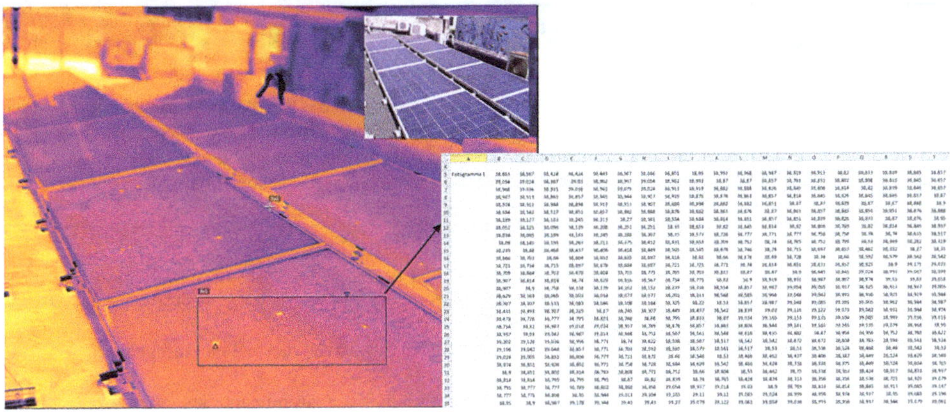

thermal and the image resolution of the camera. For example in order to detect little temperature gradients (as in hydric leakages for liquids having a temperature similar to the surrounding environment) has been adopted a FLIR T1020 having the following main specifications: thermal sensitivity <0,02 °C at 30 °C, IR sensor resolution of 1024 x 768 pixel, temperature range of -40°C to 2000°C, frame rate of 30 Hz, spectral range between 7.5 μm – 14 μm. Each measured pixel of the box is characterized by a measure value: major is the image resolution, major will be the number of the measured pixels, major is the thermal resolution, minor will be the temperature gradient that will be possible to measure.

Other protocols can be defined into an inspection process. In Fig. 4 is illustrated an example of infrared (IR) measurement protocol integrated into a complete inspection process involving the following phases:

- Process design and simulation (the inspection process is planned and engineered by analyzing accurately the case of study; simulation of the physical behavior can be performed by different methods such as Finite Difference Time Domain –FDTD-, and finite element method –FEM-);
- Analysis (the analysis is related to the simultaneous data processing of data measured by sensors, thermograms, and design specifications);
- Output formulation (the output is formulated by the analysis output enabling predictive maintenance procedures and optimization of quality processes).

All the listed phases will be repeated in order to update the analysis output.

Measurement protocol are different for outdoor and indoor applications. in the table 5 are indicated the main procedures to be considered for

For the best measurement operative condition it is preferable to wait the exact moment when the sun terminates directly to excite the target object (beginning of shadow condition): in this way a big quantity of energy it is released into the external environment and measurement is easier. For the camera setting it is important to check also the field of view –FOV- and the instant field of view –IFOV.

Table 5. Thermal measurement procedures for indoor and outdoor applications

Application Typology	Step to Follow
Indoor	1- Check camera hardware (battery, true optics, etc.); 2- Check of internal thermal source (heater, conditioner,); 3- Disactivation of internal thermal sources; 4- Analysis of possible noise sources which can disturb the thermal measurement of the target (reflection of a window or of a metallic body, presence of people in proximity of the object, etc.); 5- Setting of the measurement parameters (emissivity, environment temperature, etc.); 6- Setting of the true distance to detect the target object (according with the true optics); 7- Measurement acquisition.
Outdoor	1- Check camera hardware (battery, true optics, etc.); 2- Check of meteorological conditions in order to choose the best external condition (no rain, no fog, etc.); 3- Choice of the best time of the day to perform the measurements (it is preferable an hour without direct solar exposition of the object); 4- Analysis of possible noise sources which can disturb the thermal measurement of the target (reflection of a other body, cars in motion, people, etc.); 5- Setting of the measurement parameters (emissivity, environment temperature, etc.); 6- Setting of the true distance to detect the target object (according with the true optics); 7- Measurement acquisition.

Figure 4. Infrared –IR- measurement protocol workflow as a part of an inspection process

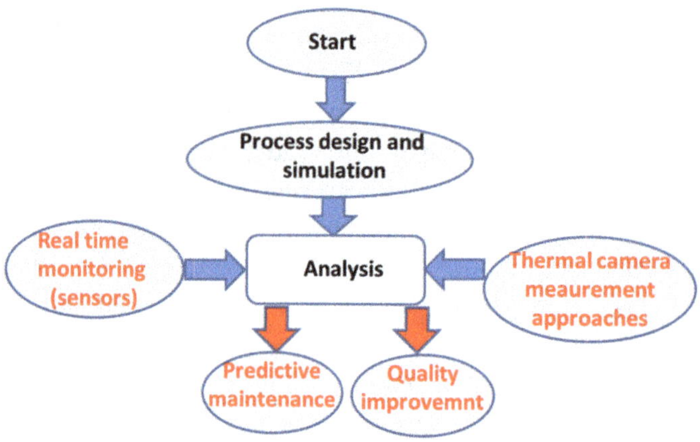

IMAGE PROCESSING APPROACHES ASSOCIATED WITH THERMOGRAPHY

In this section are discussed two image processing approaches concerning image processing segmentation and data mining clustering. Different image segmentation tools can be applied to the thermograms in order to highlight and to find hidden information. ImageJ tool (Abràmoff et al., 2004) can be applied for image segmentation and 3D processing. In Fig. 5 are illustrated different image processing involving 3D surface plot function and filtering, 3D RGB colour representation of the analysed image, and active segmentation by means of active contour snake approach (Massaro et al., 2012). Concerning this last approach we observe that the image processing is dynamic: snakes are active contour models converging in different time steps on edges having a minimum energy. Parametrically the snake contour is $v(s) = (x(s), y(s))$. The functional energy associated with the snake is defined as (Kass et al., 1987):

Figure 5. (a) Example of underground hydric loss acquired by thermal camera; (b)-(d) 3D surface processing and filtering; (e) color distribution; (f) RGB color percentage;(g) snake active contour processing; (h) 3D image processing performed by setting image processing parameters

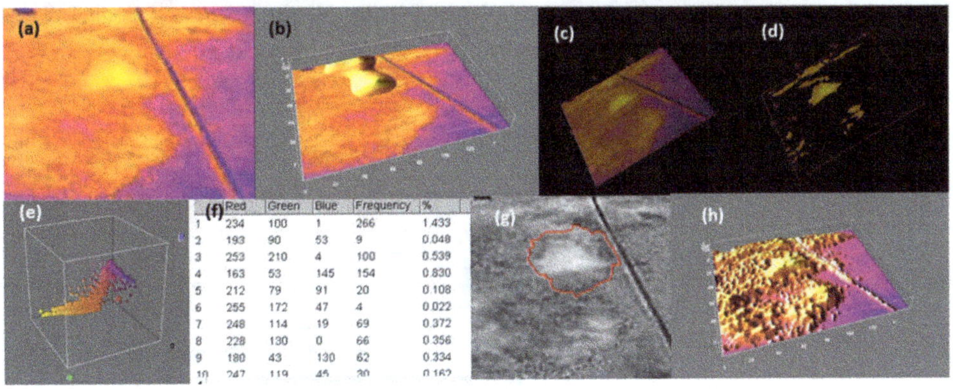

$$E_{snake} = \int_0^t E_{snake}\left(v\left(s\right)\right)ds = \int_0^t \left[E_{\text{int}}\left(v\left(s\right)\right) + E_{image}\left(v\left(s\right)\right) + E_{con}\left(v\left(s\right)\right)\right]ds \tag{3}$$

Being E_{int} the internal energy of the spline due to the bending, E_{image} the image forces, and E_{con} the constraint forces. Matching all the information of the image processing it is possible to distinguish possible geolocalization of the underground leakage (represented for example by the high level of intensity of Fig. 5 (b)),

Another approach to post-process radiometric data is the threshold filtering: in order to highlight the exact leakage region it is possible to define the temperature threshold indicating a filtered alerting condition which could activate automatic alarm conditions. In Fig. 6 (a) is shown an example of a radiometric image indicating an underground water leakage of a hydric pipeline, and in Fig. 6 (b) is illustrated the color features extraction (filtering mask obtained with Flir Tools software) of values major than 24 °C (black shape) indicating the exact leakage detected geolocalization. The filtering can be set on real time thermal images thus accelerating the repair operations. For the experimental setting of Fig. 6 has been realized an hydric system having a buried pipeline at about 1m from the external surface, and characterized by an hole simulating the real water leakage. The measurements of Fig. 6 has been performed during an hot day. For this reason the leakage appears as a darker shape (in the winter period the shape could appear colder than the surrounding surface). Observing measurements it is clear how the thermograms should be classified in function of the environment temperature.

The real time radiometric processing it is important to find quickly find an alerting condition but could not provide more information about the leakage evolution or about induced thermal effects of defected parts. In this direction it is useful to post-process data in order to find hidden information explaining phenomena evolution. An algorithm which can be applied is based on k –Means approach able to calculate cluster of radiometric pixels having similar features (Massaro et al., 2018). In Fig. 7 is illustrated the k-means post processing for different k values (k=7, k= 10, k=15, k=25). The choice of the k values can be optimized in order to find hidden information. In the case of hydric leakage of Fig. 6 the analysis provides information about the plume extension and direction: by analyzing Fig. 7 (c) and

Figure 6. (a) Radiometric image of water leakage; (b) Threshold filtering applied on the radiometric image

(a)

(b)

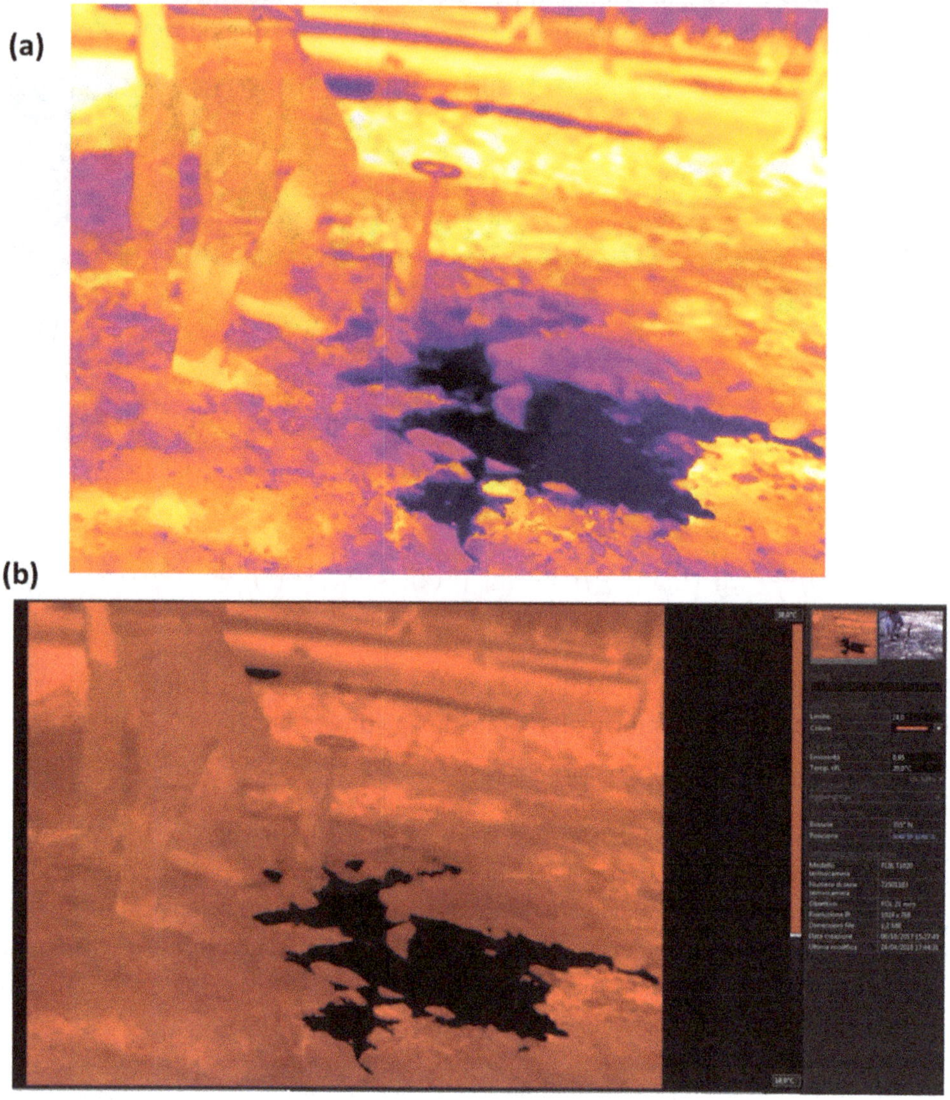

Fig. 7 (d) it is possible to observe for the main shape indicating the underground leakage two different cluster of colors representing two temperatures (the external cluster indicates the extension direction of the leakage). We note that for k minor then 15 it is no possible to distinguish the plume direction from the whole shape indicating the main effect of the temperature gradient variation. By detecting different thermograms in different periods will be possible also to understand better how evolves the leakage thus providing also information about volumetric losses.

In order to analyze correctly the thermal images it is important to pre-process properly the detected radiometric image: the image contrast enhancement could allow to decrease the image processing error by neglecting image noise. In this direction some authors (Versaci et al., 2017) adopted a new fuzzy

Figure 7. K-means clustering processing of the radiometric image: (a) k=7, (b) k=10, (c) k=15, k=25 (b). The different cluster coloration indicates different underground water concentrations

procedure for adaptive gray-level image contrast enhancement (CE). Other studies highlighted the image pre-processing procedures (Petkov et al., 2015) thus confirming that the image pre-process is an important step to acquire hidden information. According with k-means algorithm, a good image contrast setting could:

- Decrease the computational error;
- Decrease the iteration numbers for a convergent solution;
- Decrease the computational cost;
- Add more clear information such as the leakage plume evolution.
- Add more information about image clusters for high *k* values.

EXAMPLES OF APPLICATIONS OF THERMOGRAPHY IN INDUSTRY PROCESSES

It is known that data processing is not a Research and Development -R&D- activity unless it does part of a project to develop new data processing methods. In this scenario are described how IRT and data processing can be a part of an industry research project, thus providing a knowledge gain. Specifically, in this section we discuss about some applications of thermography in industrial sectors. The proposed examples are useful in order to understand how thermography could be improved in robotic systems oriented in industrial applications. In particular in this section is discussed:

- An industrial application in fresh food monitoring and control (project 1);
- An industrial application in predictive maintenance (project 2);
- An industrial application in transportation security (project 3).

The first industry project embeds the following modules:

Infrared Thermography Module

This module takes into account of the use of an infrared camera for monitoring fresh product environments (micro-climate defined by areas and volumes where fresh products are stored and exposed in the store). By means of this energy control device, it will also be possible to monitor both the energy dispersions of the equipment and the temperature of the individual fresh products within the store environment. The camera has an ethernet connection for remote real time control. The thermograms are into a physical server and are processed in order to analyze the energy efficiency of the refrigerator and the classifications of defects (defects of non-homogeneity cooling, deterioration of products classified according to thermal inhomogeneities found on the individual products, etc.).

QR Code Tracking Form

Warehouse products, including fresh ones, are tracked using QR code technology improving traceability of the entire production chain (mapping of quality processes).

Data Mining Engine

Data mining algorithms will be applied with the aim of optimizing stock levels and optimally managing the sale of fresh and non-fresh products (forecast sales calculation). The temperature data recorded with the thermography are processed by this engine in order to draw predictive control maps relating to the quality of the exposure refrigeration environments and to the levels risks of deterioration.

Electronic Labels

electronic labels represent a positive element for product facing and inventory management (analysis of quantitative verification of products on the shelf compared to those available in the warehouse).

In Fig. 8 is illustrated the main architecture of the project 1.

As second example is discussed how IRT system could be integrated into a quality process according with ISO 9001:2015.

Below are listed the topics and the requirements of the second industry project.

Figure 8. Project 1: 'Traceability/IR Thermography Fresh Food'

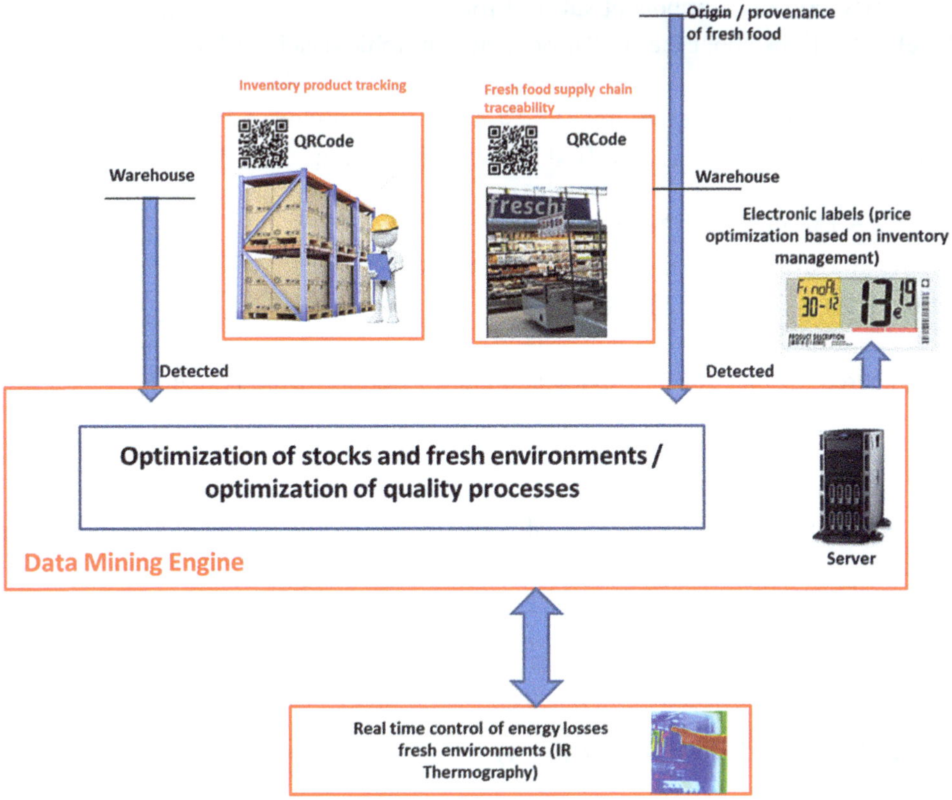

Predictive Maintenance Module

The thermogram acquisition procedures is designed and engineered to predict the maintenance of plants and particular engines/machines. Furthermore, at least two types of systems and machines will be experimentally analyzed, and the acquisition procedures will be established according to:

- Specific boundary conditions (temperature, humidity, etc.);
- Inspections at different times;

The Predictive Maintenance concerns both the verification of the electrical and mechanical parts, and allows to verify and to compare the situations of absence of load with different loading conditions.

Furthermore, data mining algorithms are applied by:

- Clustering algorithms (for example k-Means);
- Classification algorithms;
- Predictive algorithms (for example neural networks);
- Correlation algorithms (correlation of radiometric data with other data and measurements).

The module provides the following outputs:

- Risk maps with preventive indications for the replacement of plant or of machine components;
- Predictive prediction of component substitution;
- Graphical indications about the risk of breakage (graphic dashboards);

In addition, a combination of other image post-processing engines for the features extraction of the radiometric images are useful in order to find hidden information.

Quality Testing Module of Components

This module includes the following facilities:

- New engineered verification processes applied on quality of components (according with ISO 9001: 2015);
- Methods of thermographic verification and analysis of the individual components defined through appropriate image processing tools;
- Methods of thermographic verification of networks (photovoltaic, electrical panels, heating systems, etc.) ;
- Test procedures defined by checking the basic functionalities of the prototype platform (thermographic testing workflow, thermographic testing report, statistical analysis of thermographic tests, etc.).

Database System Module (DB System)

Different data collection tables are implemented for the various modules; these tables collects data into an unique DB system which is interfaced with the data mining engine.

Data Mining Engine Module

A data mining engine reads the data of the database -DB- system as inputs and processes the information of the different modules; as data mining tools can be adopted Orange Canvas, or RapidMiner, or Weka, or KNIME Studio, or other applications that can be integrated into a backend/frontend platforms such as Keras, TensorFlow, Theano, or others; the outputs of the engine will be the platform system outputs.

The proposed platform potentially provides:

- A systemic study of the external conditions effects on radiometric measurements (effect of the environment where measurements are performed, reflection effects, etc.);
- Best parameter settings (material emissivity, background, etc.);
- Procedures for calibrating parameters for quantitative measurements;
- Qualitative and quantitative acquisition methods (acquisition contexts based on the application field);
- Defect classification (defects detected by the IR thermal camera);
- Relevant procedures for thermographic operators (UNI EN ISO 9712:2012);
- Definition of alert thresholds of the image processor;
- Excitation methodologies (flash light or by heating sources) of the mechanical system to control.

In Fig. 9 is illustrated the main architecture of the project 2.

The third proposed project is related to transport sector, concerning thermography in security bus systems. Specifically, the proposed research project discusses the design and the development of an integrated platform aimed at the safety of the bus fleet. The project provides for the integration of the following main functionalities into a prototype platform such as:

- Enhancement of the driver's nightly visibility by means of thermography,
- Real-time analysis of the positioning of the vehicle (tracking),
- Reliability analysis of the vehicle,
- Adoption of smart medical devices (defibrillators) improving passengers security on board.

The prototype platform consists of the following modules:

- **Control unit based on CAN** (Controller Area Network)/OBD-II (On Board Diagnostics) communication standards: the data, extracted from the control units using cadenced procedures and after the deposit of the vehicles, are stored into appropriate databases, and are processed by predictive data mining algorithms (artificial intelligence) in order to predict malfunctions that may compromise the safety of the travelers;
- **A thermographic module** for enhancement of visibility: this module involves the use of a particular compact and smart camera, able to significantly reduce the risks of night driving, allowing drivers to see much further, and with greater clarity, than the visibility provided from standard vehicle headlights; drivers can see better (by means of thermographic detection) pedestrians (risk of serious injury or death of the pedestrian), animals (potential damage of the vehicle or risk of

Figure 9. Project2: Infrared predictivemaintenance/quality'

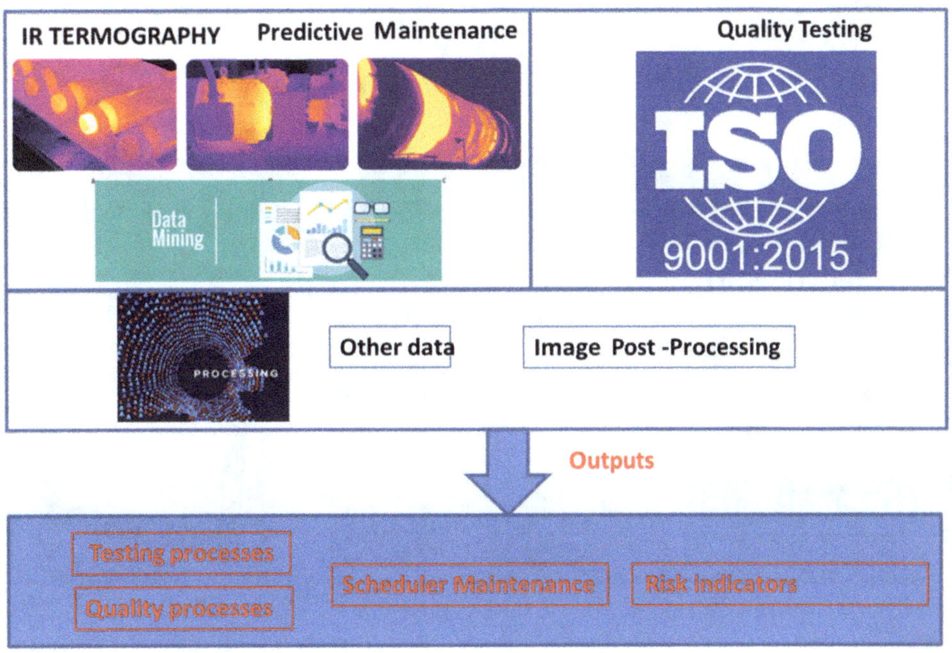

the accident) or objects on the road or on the edge of the same, having more time to react to any potential danger; thermography therefore helps to detect and recognize potential risks in total darkness, and in smoke (limited visibility by smoking); the automatic recognition feature with automatic alerting function will facilitate the real time video reading.

- **A GPS tracking module:** this module allows you to view the positioning of the buses in real time and from the operations center (remotely connected) to areas of danger (risk of weather warning areas, risk of blockage in traffic due to accidents, etc.), and therefore to provide information on the safety of passengers, guaranteeing the correct travel time for journeys.
- **Semi-automatic defibrillators on board** which automatically identify the energy dose thus facilitating the aid intervention.

In Fig. 10 is illustrated the main architecture of the third project involving thermography. We observe that in this last case thermograms are detected in quasi real time, because it is important for the driver to see imminent dangers.

Figure 10. Project 3: Security based integrated platform'

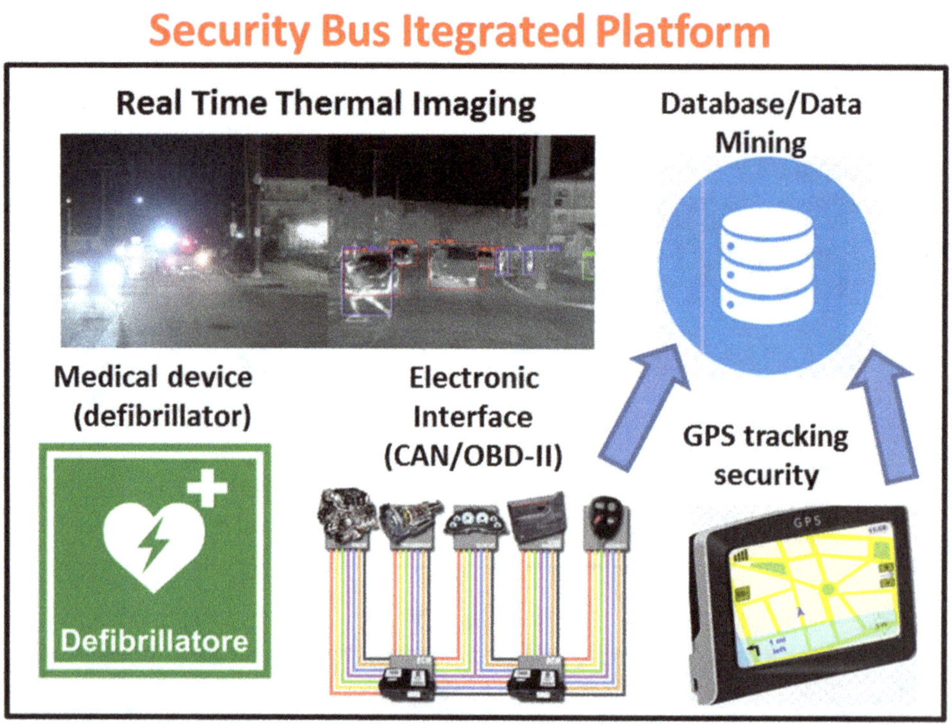

SOME SYSTEM INTEGRATION ASPECTS IN AUTOMATISM ORIENTED ON ROBOTICS

In this section is discussed how IRT systems discussed in the previous sections could be embedded into a robotic system improving automatism. In particular in the table 6 are summarized the main features of the automatism:

From the automatism discussed it is clear how the concept of "robotic automatism" can be extended to the:

- The data acquisition;
- The data flow;
- The data processing (real time, pre-processing and post-processing) ;
- The alerting system;
- Mechanical commands and motion;

Concerning hydric leakage detection executed by UAV system, the automatism can be implemented by programming the UAV trajectory according to pipeline underground positions and to total length to

Table 6. Automatism in IRT systems

Project	IRT System	Automatism and Data Mining Facilities
1	Different thermal cameras are connected by POE cables to an information infrastructure made by a switch and by a router system (hub) transmitting data to an external database. Each thermal camera control a single refrigerator environment and each product contained in it.	Every 30 minutes the thermograms are detected and sent in cloud for data processing. The images will be automatically processed by a data mining engine able to classify product defects and to send alerting signal in case of uncontrolled temperatures. The region of product defects are classified and stored into a database useful to predict the life cycle of the product. Other information coming from traceability could help to find possible causes of deterioration.
2	Different thermal cameras are connected by POE cables to an information infrastructure made by a switch and by a router system (hub) transmitting data to an external database. Each thermal camera control a single machine of a production line. The production line is monitored by different thermal cameras.	Every minute the thermograms are sent to an external database linked to a data mining engine (Industry 4.0 facility). An automated data pre-processing based on threshold analysis provides information about imminent breakage risks. The data mining post- processing performed by artificial neural networks could provide a detailed plan of predictive maintenance which will be automatically updated. The post- processing is computed and updated at the end of the production activity. Data mining automatically process data of different thermograms by finding possible correlations and causes of machine failures (the malfunction of a single machine can be repercussed on the other machines) or product defects. Data mining classificators can classify the machine malfunctions thus predicting failures or production stop. Also machine log errors can provide data to process for the function forecasting.
3	Thermal camera detects quasi real time thermal images of possible dangers on the road.	The alerting system could be connected to brake circuit in order to automatically slow down the vehicle speed. The guide position can be potentially slowly adjusted by a controlled arm connected to the steering (important aspect for automotive industry sector).

monitor. In Fig. 11 is illustrated the octocopter used for water leakage inspection and the related real time thermal image detecting leakage.

Below are indicated the main characteristics of the UAV system adopted for radiometric measurements:

- Video transmitter 25 / 200 / 600mW 40ch 5,8 Ghz+ 2 antennas (circular polarization);
- Optris PI 450 thermal camera (IR resolution = 382 x 288; frequency acquisition = 80 Hz; precision = ± 2 °C; operation temperature = 0° C - 50 °C; FOV = 30 ° x 23 ° f =17mm/ 13 ° x 10 ° f =40mm; temperature range = -20°C - 100°C);
- Balanced carbon propellers;
- Telemetry until 500 m;
- Lithium power supply 5100mAh 4S;
- Monitor station HD 7'';
- Independent terminator on 869MHz frequency for power cut-off emergency;
- Gimbal two-axis gyro stabilized;
- GoPro Hero 5 (for primary checking or inspection).

The acquired thermograms can be furthermore automatically processed by snake active contour algorithm able to extract information useful to estimate plume extension and volumetric losses. The aerial pattern could be programmed automatically after the data post-processing results. By defining properly alerting thresholds, the UAV can come closer to the plume region in order to acquire more details according with its optical characteristics. As alternative, the thermal camera can automatically zoom the image according with its image and thermal resolution. The same behavior can be attributed from drones moving on the ground or for thermal camera fixed on robotic arms following by a controlled motion the object to detect. Other important aspects about the automated IR detections are:

- Auto-calibration;
- Auto-focus;
- Automatic overcoming of obstacles;
- Automatic detection corresponding with particular gps coordinates;
- Object tracking (motion tracking);
- Automatic adjustment of motion trajectory;
- Automatic data post-processing;
- Automatic thermal thresholding according with the environmental conditions.

Figure 11. (a) Flying Eye octocopter equipped by an Optris PI450 thermal camera; (b) Optris radiometric detection of a water leakage.

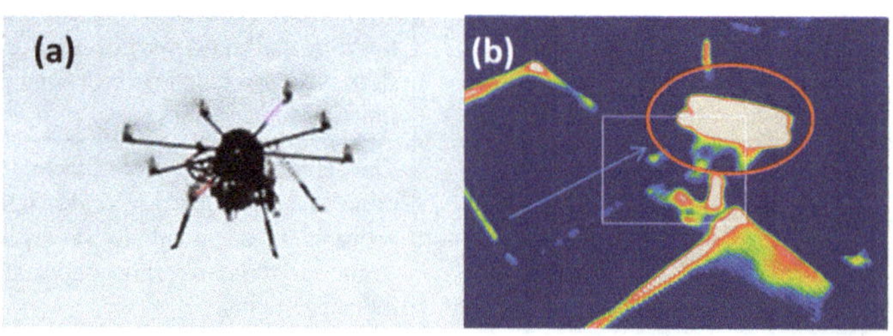

In intelligent robotic systems the output of the data processing can used for the optimization of the robot handling and of thermograms acquisition.

REGULATORY FRAMEWORK

Below are listed the regulatory framework concerning IRT systems suitable for industry applications:

- UNI EN 16714-1: 2016 "Non-destructive tests-Thermographic tests - Part 1: General principles";
- UNI EN 16714-2: 2016 "Non-destructive tests-Thermographic tests - Part 2: Instrumentation";
- UNI EN 16714-3: 2016 "Non-destructive tests-Thermographic tests - Part 3: Terms and definitions";
- UNI 10824-1: 2000 "Non-destructive tests - Infrared thermography - Terms and definitions";
- UNI EN 473: 2008 "Non-destructive testing - Qualification and certification of personnel involved in non-destructive testing - General principles";
- ISO 9712 - Third edition - 2005 "Non-destructive testing - Qualification and certification of personnel";
- ISO 18436-1 "Monitoring and diagnostics of machines - Requirements for training and certification of personnel - Part 1: Requirements for certification and the certification process";
- ISO 18436-3 "Monitoring and diagnostics of machines - Requirements for personnel qualification - Part 3: Requirements for training and education institutions";
- ISO 18436-7 "Monitoring and diagnostics of machines - Requirements for qualification and assessment of personnel - Part 7: Thermography";
- ISO 18434-1: 2008 Condition monitoring and diagnostics of machines — Thermography — Part 1: General procedures.
- ISO 6781: 1983 Thermal insulation — Qualitative detection of thermal irregularities in building envelopes — Infrared method.
- EN 13187 Thermal performance of buildings. Qualitative detection of thermal irregularities in building envelopes. Infrared method.
- ISO / TR 13154: 2009 Medical electrical equipment — Deployment, implementation and operational guidelines for identifying febrile humans using a screening thermograph.
- IEC 80601-2-59: 2008 Medical electrical equipment — Part 2-59: Particular requirements for basic safety and essential performance of screening thermographs for human temperature screening.
- DIN 54190-1 ZerstörungsfreiePrüfung — ThermografischePrüfung — Teil 1: Allgemeine Grundlagen.
- DIN 54190-2 Non-destructive testing — Thermographic testing — Part 2: Equipment.
- DIN 54190-3 ZerstörungsfreiePrüfung — ThermografischePrüfung — Teil 3: Begriffe.
- DIN 54191 ZerstörungsfreiePrüfung — ThermografischePrüfungelektrischer Anlagen.
- DIN 54192 ZerstörungsfreiePrüfung — AktiveThermografie.
- VdS 2858en Thermography in electrical installations, a contribute to loss prevention and operational reliability.
- MIL-STD-2194 Military standard: Infrared thermal imaging survey procedure for electrical equipment.

- ASTM E1934-99a (2010) Standard guide for examining electrical and mechanical equipment with infrared thermography.
- UNI EN 13187: 2000 "Thermal performance of buildings - Qualitative detection of thermal irregularities in building envelopes - Infrared method";
- UNI 9252: 1988 "Thermal insulation - Survey and qualitative analysis of thermal irregularities in building envelopes - Infrared thermography method";
- CAN / CGSB 149-GP-2MP: Manual for Thermographic Analysis of Building Enclosures;
- ASTM C1060: Standard Practice for Thermographic Inspection of Insulation Installations in Envelope Cavities of Frame Buildings;
- ASTM E1186: Standard Practice for Air Leakage Site Detection in Building Envelopes and Air Barrier Systems;
- ASTM C1153: Standard Practice for Locating of Wet Insulation in Roofing System Using Infrared Imaging;
- ASTM E1316: Terminology for Non Destructive Examinations;
- ASTM E1213: Standard Test Methods for Minimum Resolvable Difference for Thermal Imaging System;
- ASTM E1311: Standard Test Methods for Minimum Detectable Temperature Difference for Thermal Imaging System;
- ASTM E1862: Standard Test Methods for Measuring and Compensating for Reflected Temperature Using Infrared Imaging Radiometers;
- ASTM E1897: Measuring and Compensating for Transmittance and Using Infrared Imaging Radiometers;
- ASTM E1933: Standard Test Method for Measuring and Compensating for Emissivity Using Infrared Imaging Radiometers;
- ASTM D4788: Standard Test Method for Detecting Delaminations in Bridge Decks Using Infrared Thermography;
- MIL-STD-2194: Military Standard - Infrared thermal Imaging Survey Procedures for electrical Equipment;
- Canada NMS Section 022713-2008: Thermographic Assessment: Building Envelope;
- Canada NMS Section 022719-2008: Thermographic Assessment: Mechanical Equipment;
- Canada NMS Section 022723-2008: Thermographic Assessment: Electrical Systems;
- Building Research Establishment (BRE) - BR 176: A Practical Guide to Infra-Red Termography For Building Surveys;
- UKTA Code of Practice Number 2. Assessing thermal bridging and insulation continuity;
- RESNET standard 802: General Provisions for Thermographic Inspection of Buildings;
- NFPA 70B: Recommended Practice for Electrical Equipment Maintenance;
- DIN 54 190-1 Non-destructive tests - Thermographic examination - Part 1: General principles;
- DIN 54 190-2 Non-destructive tests - Thermographic examination - Part 2: Equipment;
- DIN 54 190-3 Non-destructive tests - Thermographic examination - Part 3: Conditions;
- DIN 54 191 Non-destructive tests - Thermographic examination of electrical systems;
- DIN 54 192 Non-destructive tests - Active thermography (draft standard).

CONCLUSION

The goal of the chapter is to provide an overview about scientific research topics involving thermography. Specifically, some procedures have been proposed to acquire IR measurements and some approaches about data pre-processing and data post-processing. The discussions about industry project specifications allow to understand how IRT systems can be integrated into an information system concerning different applications such as transport, quality process, predictive maintenance, product quality monitoring and leakage detection. By taking into account the project examples has been analysed the automatisms involving IRT implementation.

The attention has been focused on radiometric data post processing by means of image segmentation and data mining algorithms. In particular has been applied active snake contour and clustering k-means algorithms. The analysis of measurement procedures, and of the data algorithms could help to formulate a research industry project to apply on a specific application.

Finally, in order to complete the discussion, the chapter proposes a full list of standards defining possible regulatory frameworks. The whole discussion is oriented on the choose of facilities suitable for industry monitoring, industry production processes and security. Different IR thermography applications are commented including predictive maintenance, prediction of machinery failures, production quality control, food temperature control, energetic indoor monitoring, energy efficiency, non–destructive liquid leakage detection, real time security systems. Concerning this last point, a useful application is in transport security where thermography could detect in real time obstacles on the roads and could support the night visibility and through smoke. The chapter highlights the potential applications of thermography in research projects involving data processing, data mining and artificial intelligence, by addressing the reader to define research topics adopting thermograhy facilities compatible with fundamental pillars in research industry and in general in research and development projects. All the proposed applications are suitable for automated processes and for integration in robotic systems. The example of thermography in UAV systems proves that a good development can be the management of the control of the thermal inspections which can be performed in real time. The data post-processing by data mining algorithms could increase the inspection efficiency.

ACKNOWLEDGMENT

The work has been carried out during the development of the following different industry projects: *"Procedure Innovative e Processing Intelligente delle Misurazioni Radiometriche per la Predictive Maintenance degli Impianti Tecnologici e il Quality Control di Testing dei Componenti: 'Infrared PredictiveMaintenance/Quality'* " [Innovative Procedures and Intelligent Processing of Radiometric Measurement for Predictive Maintenance of Technological Systems and for Component Testing Quality Control 'Infrared PredictiveMaintenance / Quality']; *"Piattaforma di tracciabilità integrata ed engine di data mining, applicata ai freschi con monitoraggio real time degli ambienti controllati in temperatura mediante termografia ad infrarossi ed orientata all'ottimizzazione delle scorte / prezzi ed al controllo energetico ottimale degli ambienti: 'Traceability/IR Thermography Fresh Food'* " [Integrated traceability platform and data mining engine applied to the fresh food, with real time temperature monitoring of the controlled environments by infrared thermography, and oriented to the optimization of the stocks / prices and to the optimal energy control of the environments]; *"Sistema tecnologico non invasivo per il rilevamento e la localizzazione intelligente delle perdite idriche mediante immagini termiche automatizzate, sistema di monitoraggio/alerting, image processing e sistema decisionale di gestione interventi con tecnologie alternative 'ThermRadLeack'*" [Non-invasive technological system for the intelligent detection and localization of water leackage by means of automated thermal images, monitoring / alerting system, image processing, and decision-making systems for managing interventions with alternative technologies 'ThermRadLeack']; "*Piattaforma di sicurezza integrata per flotte bus orientate al enhancement di visibilità, all'analisi dell'affidabilità del mezzo, al tracking real time e alla dotazione di defibrillatori 'Security Based Integrated Platform'*" [Integrated security platform for bus fleets oriented on enhanced visibility, vehicle reliability, real time tracking and healthcare by defibrillators: "Security Based Integrated Platform"].

REFERENCES

Abràmoff, M. D., Magallhas, P. J., & Ram, S. J. (2004). Image Processing with ImageJ. *Biophotonics International*, *11*(4), 36–42.

Aghaei, M., Quater, P. B. F., Grimaccia, F., Leva, S., & Mussetta, M. (2014). Unmanned aerial vehicles in photovoltaic systems monitoring applications. *Proceedings of 29th European Photovoltaic Solar Energy Conference and Exhibition*, *1*(1), 2734-2739.

Akula, A., Ghosh, R., & Sardana, H. K. (2011). Thermal imaging and its application in defence systems. *Optics: Phenomena, Materials, Devices, and Characterization*. In *AIP Conference Proceedings*, *1391*(1), 333–335.

Banerjee, K., Krishnan, P., & Mridha, N. (2018). Application of thermal imaging of wheat crop canopy to estimate leaf area index under different moisture stress conditions. *Biosystems Engineering*, *166*(1), 13–27.

Björk, E., Palm, B., & Nordenberg, J. (2010). A thermographic study of the on-off behavior of an all-refrigerator. *Applied Thermal Engineering*, *30*(14-15), 1974–1984.

Cheng, M., & Xiang, D. (2019). The Design and Application of a Track-type Autonomous Inspection Robot for Electrical Distribution Room. *Robotica*, *1*(22).

Ciampa, F., Mahomoodi, P., Pinto, F., & Meo, M. (2018). Recent analysis in active infrared thermography for non-destructive testing of aerospace components. *Sensors (Basel)*, *18*(609), 1–37.

Crişan, L.-M., & Borza, I. (2019). Management of Maintenance for Photovoltaic Systems. *Nonconventional Technologies Review*, *23*(1), 40–45.

Entrop, A. G., & Vasenev, A. (2017). Infrared drones in the construction industry: Designing a protocol for building thermography procedures. *Energy Procedia*, *137*(1), 63–68.

Frascati Manual. (2015). *The Measurement of Scientific, Technological and Innovation Activities- Guidelines for Collecting and Reporting Data on Research and Experimental Development*. OECD.

Galiano, A., Massaro, A., Barbuzzi, D., Pellicani, L., & Birardi, V. (2016). Heat Monitoring Plan in Building for Energy Efficiency using Thermal Imaging Analysis. [JNIT]. *Journal of Next Generation Information Technology*, *7*(4), 8–18.

Ghidoni, S., Antonello, M., Nanni, L., & Menegatti, E. (2015). A thermographic visual inspection system for crack detection in metal parts exploiting a robotic workcell. *Robotics and Autonomous Systems 74*(B), 351-359.

Gonçalves, B. J., de Oliveira Giarola, T. M., Pereira, D. F., de Barros Vilas Boas, E. V., & de Resende, J. V. (2016). Using infrared thermography to evaluate the injuries of cold-stored guava. *Journal of Food Science and Technology*, *53*(2), 1063–1070.

Grys, S. (2018). Determining the dimension of subsurface defects by active infrared thermography- experimental research. *Journal of Sensors and Sensor System*, *7*(1), 153–160.

Habib, M. K. (2007). Mechatronics: A Unifying Interdisciplinary and Intelligent Engineering Paradigm. *IEEE Industrial Electronics Magazine*, *1*(2), 12–24.

Habib, M. K. (2008, August) Human Adaptive and Friendly Mechatronics (HAFM). *IEEE International Conference on Mechatronics and Automation 'ICMA2008'*, Takamatsu-Kagawa, Japan, 61-65.

Ishimwe, R., Abutaleb, K., & Ahmed, F. (2014). Applications of thermal imaging in agriculture – a review. *Advances in Remote Sensing*, *3*(3), 128–140.

Kass, M., Witkin, A., & Terzopoulos, D. (1987). Snakes: Active contour models. *International Journal of Computer Vision*, *1*(4), 321–331.

Khan, M. J., Brodie, G., Gupta, D., & Foletta, S. (2018). Microwave soil treatment improves weed management in Australian dryland wheat. *Transactions of the ASABE*, *61*(2), 671–680.

Kumar, K. S., Kavitha, G., Subramanian, R., & Ramesh, G. (2011). Visual and Thermal Image Fusion for Based Target Tracking, U. A. V. MATLAB - A Ubiquitous Tool for the Practical Engineer, Prof. Clara Ionescu (Ed.): ch. 16.

La Rosa, G., & Risitano, A. (2000). Thermographic methodology for rapid determination of the fatigue limit of materials and mechanical components. *International Journal of Fatigue*, *22*(1), 65–73.

Ljungberg, S. A. (1994). *Infrared techniques in buildings and structures: operation and maintenance. Infrared Methodology and Technology, 7, 211.* Gordon and Breach Science Publishers.

Manama, M., Arrayo, A., Ortiz, A., Renedo, C. J., Perez, S., & Delgado, F. (2011). Field winding fault diagnosis in DC motors during manufacturing using thermal monitoring. *Applied Thermal Engineering*, *31*(5), 978–983.

Massaro, A., Ekuakille, A. L., Caratelli, D., Palamara, I., & Morabito, F. C. (2012). Optical performance evaluation of oil spill detection methods: Thickness and extent. *IEEE Transactions on Instrumentation and Measurement*, *61*(12), 3332–3339.

Massaro, A., & Galiano, A., Meuli, G., & Massari, S. F. (2018). Overview and application of enabling technologies oriented on energy routing monitoring, on network installation and on predictive maintenance. *International Journal of Artificial Intelligence and Applications*, *9*(2), 1–20.

Petkov, N., & Azzopardi, G. (2015). *Lecture Notes in Computer Science*, *9256*, Springer International Publishing.

Raghatate, R. P., Rajurkar, S. S., Waghmare, M. P. & Ambakar, P. V. (2013). *International Journal of Modern Engineering research*, *3*(2), 816-820.

Rakrueangdet, K., Nunak, N., Suesut, T., & Sritham, E. (2016). Emissivity measurements of reflective materials using infrared thermography. *Proceedings of the International MultiConference of Engineers and Computer Scientists 2016* Vol I.

Rogalski, A., & Chrzanowski, K. (2002). Infrared devices and techniques. *Opto-Electronics Review*, *10*(2), 111–136.

Sagan, V., Maimaitijiang, M., Sidike, P., Eblimit, K., Peterson, K. T., Hartling, S., ... Mockler, T. (2019). UAV-Based High Resolution Thermal Imaging for Vegetation Monitoring, and Plant Phenotyping Using ICI 8640 P, FLIR Vue Pro R 640, and thermoMap Cameras. *Remote Sensing*, *11*(1), 330.

Samanian, N., & Mohebbi, M. (2016). Thermography, a new approach in food science studies: A review. *MOJ Food Processing & Technology*, *2*(3), 110–119.

Ullah, I., Yang, F., Khan, R., Liu, L., Yang, H., Gao, B., & Sun, K. (2017). Predictive mainyenance of power substation equipment by infrared thermography using a machine-learning approach. *Energies*, *10*(12), 1–13.

Usamentiaga, R., Venegas, P., Guerediaga, J., Vega, L., Molleda, J., & Bulnes, F. G. (2014). Infrared thermography for temperature measurement and non-destructive testing. *Sensors (Basel)*, *14*(7), 12305–12348.

Versaci, M., Morabito, F. C., & Angiulli, G. (2017). Adaptive Image Contrast Enhancement by Computing Distances into a 4-Dimensional Fuzzy Unit Hypercube. *IEEE Access: Practical Innovations, Open Solutions*, *5*(1), 26922–26931.

Vollmer, M. (2010). *Mollmann. K.-P.* Infrared Thermal Imaging. WILEY-VCH Verlag GmbH & Co.

Yang, H., Yao, X.-F., Wang, S., Ke, Y.-C., & Liu, Y.-H. (2019). Simultaneous determination of gas leakage location and leakage rate based on local temperature gradient. *Measurement*, *133*(1), 233–240.

Zhao, Y., Mehenen, J., Xu, W., Alrashed, M., Abineri, S., & Roy, R. (2015). Degradation assessment of industrial composite using thermography. *Procedia CIRP*, *38*(1), 147-152.

KEY TERMS AND DEFINITIONS

Data Clustering: A set of multivariate data analysis techniques aimed at the selection and grouping of homogeneous elements in a data set.

Field of View (FOV): Is the largest area that your imager can see at a set distance.

Infrared Radiation (IR): Infrared radiation (infrared light) is electromagnetic radiation with longer wavelengths than those of visible light.

Infrared Termography: The use of radiometric thermograms to study heat distribution in structures or regions.

Infrared Thermography (IRT): Infrared thermography technique performed by IR thermal camera.

Instant Field of View (IFOV): Represents the spatial resolution otherwise known as Spatial Resolution), is the smallest detail within the FOV that can be detected or seen at a set distance.

Predictive Maintenance: Data mining techniques designed to help determine the condition of in-service equipment in order to estimate when maintenance should be performed.

Thermogram: Radimetric record made by thermography.

Chapter 6
Safe Development Environments for Radiation Tracing Robots

Kai Borgeest
https://orcid.org/0000-0003-4289-0083
TH Aschaffenburg, Germany

Daniel Kern
TH Aschaffenburg, Germany

ABSTRACT

Robots can substitute for men in radioactively-contaminated areas. This is a suitable field to deploy robots for measurements, repair, or clearance, but development and test of such robots could be dangerous, because radiation sources need to be handled. To avoid these hazards in development or public demonstrations, safe alternatives to radiation samples have been sought using an already existing robot (EtaBot). One proposed solution is an optical substitution ("light follower"), the other one a fully-digital simulation of the contaminated area and the robot movement inside it using a hardware-in-the-loop simulator (HiL).

INTRODUCTION

After a nuclear disaster (e.g. Chernobyl, Fukushima) or in some military scenarios, it is too dangerous for people to enter a contaminated area. This is a suitable field to deploy robots for measurements, repair or clearance. Several fully autonomous or remote operated robots have been already built for this purpose. Before operation the robot must be developed, built and tested safely for critical missions. So it is desirable to perform these activities without handling dangerous radiation sources. A further advantage of simulated environments is a higher flexibility. This paper describes two ways of safe development. For the work a self-built, wheeled, multi-purpose robot "EtaBot" has been used. Radiation tracing and optional radiation-based navigation have been implemented in ROS.

DOI: 10.4018/978-1-7998-0137-5.ch006

One simple way has been dubbed "light follower". The radiation sensor has been substituted by a photo diode as a light sensor. For the light sensor an interface circuit emulates the signals which in true operation would come from the radiation sensor. The robot can run a given path and just measure radiation on this path, but additionally an algorithm has been implemented to follow actively the radiation source. In this mode a directive radiation source could be represented e.g. by a simple torch light, a distributed scenario by diffuse lamps.

A second, more sophisticated approach is a complete digital simulation of the contaminated environment using a hardware-in-the-loop simulator (HiL). This implies two technical core questions, on the one hand the representation of the contaminated environment and the robot motion in the HiL, on the other side the interfacing between the robot and the HiL. Depending on the interface further functionality needs to be implemented in the simulator. In the particular case of this article, the robot has been reduced to its computation platform. The ROS computer receives radiation signals from the HiL, passes motion commands to an additional microcontroller board which would control steering and the motors. The robot is jacked with its wheels free from ground, the PWM signals are intercepted and fed into the HiL to simulate the robot motion, its location and posture in the contaminated environment. With a graphical map creator, scenarios can be defined and uploaded to the HiL. Radiation exploring robots and the presented approaches are good examples where deep interaction of mechanics and electronics (mechatronics) solves problems (Habib, 2007). A very similar application is exploration for chemicals where sensors may be more complicated, but exploration is easier, because in contrast to the stochastic nature, chemical sensing delivers continuous values (Ishida, 2012).

BACKGROUND

Properties of Radioactive Radiation

Decomposition of atomic nuclei releases α-rays (particles consisting of two protons and two neutrons), β-rays (electron rays, rarely positron rays) and γ-rays (energy rich photons). The wave-particle duality allows considering these kinds of radiation also as waves. In particular γ-radiation is often considered as an electromagnetic wave with higher frequency, higher energy and shorter wavelength as X rays. All kinds are capable to damage living cells and to cause cancer and birth defects. Whereas α-rays from outside the body are absorbed by the outer skin layers, γ-rays have the deepest penetration through the body. X rays have similar properties as γ-rays, they have a longer wavelength, so the related energy of an X ray photon is lower (there is no clearly defined limit between X-rays and γ-rays, often the limit is assumed at 1 MeV or below, sometimes X-rays and γ-rays are even used synonymously).

Radiation interacts with living or dead matter by ionization. Ionization is also used to detect radioactive radiation. Typical detectors are Geiger-Müller Tubes; a typical implementation is a tube with an easily penetrated end window in which a high voltage electrode separates ions and causes a measurable current. So Geiger-Müller Tubes have a maximum sensitivity in the window direction, whereas sensitivity to radiation from the side is much lower. Further radiation sensors (not considered here) are semiconductor sensors including radiation sensitive image sensors and scintillators which turn radiation into light.

Radioactive decay is a stochastic process, so in particular with weak radiation a sufficiently long averaging time must be scheduled which limits the velocity of moving sensors. There are several criteria and derived from these criteria, there are different units to characterize radiation. Biologically relevant is the equivalent dose in Sievert with 1 Sv = 1 J/kg. Often the time derivative in μSv/h (equivalent dose power) is measured.

Artificial sources which may be used in laboratory are radioactive minerals (α, β, γ), radionuclides in a windowed shield (α, β, γ) and X-ray tubes (as γ-radiator substitutes). An advantage of X-ray tube under laboratory safety consideration is safe storage; radiation is only emitted with a high voltage applied. Although α-rays do not penetrate the body, ingestion or incorporation of α-radiating matter (e.g. dust) is dangerous.

Radiation Tracing Robots

After the nuclear accident in Harrisburg (USA) there have been first efforts in the 1980es to send robots into contaminated areas. The first INES7 nuclear accident (major accident on International Nuclear and Radiological Event Scale, the highest rating) has been 1986 in Chernobyl in the former Soviet Union, the second one 2011 in Fukushima (Japan). In Chernobyl, immediate disaster recovery has been still manual with many immediate or late casualties. Later first efforts to send robots into the plant have come up (Potemkin et. al., 1992; Abouaf, 1998). In Fukushima robots have been deployed very quickly. The debris of the Fukushima reactors can be considered now as a large test field. A good survey on realized recovery robots is given by Tsitsimpelis, Taylor, Lennox, and Joyce (2019).

Radiation tracing robots need radiation sensors and depending on their task they must either map the radiation or they navigate depending on navigation. Beyond these mission capabilities they need a sufficient energetic autonomy, the electronics need to be radiation hardened and in some cases they also need the capability to climb over debris. Soft-shaped robots may have better capabilities to move through difficult terrain (Oshiro, 2018). Often parts of the sites are under water, so there are also diving robots such as sunfish (Peko et al., 2018), under some circumstances also flying robots may be useful. Very often multiple robots accomplish a task together with either central coordination or distributed intelligence.

If the robot must not be tethered by supply cables energy autonomy requires sufficiently large batteries/accumulator and low power consumption.

HiL

A hardware-in-the-loop simulator (HiL) is a box which simulates the environment of an electronic control. The interfaces of electronic controls to their environment are sensors and actors. For a HiL simulation the signals intended for the actors are fed into a simulation model inside the HiL system, an adequate system response is calculated. The response from the simulated system is translated into electrical signals which would be received by sensors, these artificial sensor signals are fed back into the physically existing electronic control. This way closed control loops including hardware can be realized in laboratory without the need of the true operating environment of a control.

Figure 1 shows such a box consisting out of a chassis with power supply, cooling fans and a backplane (in this case a PXI backplane, from other manufacturers there are other types), input/output boards on the right side and the main board running the real time model on the left side.

Figure 1. Example of a HiL-simulator

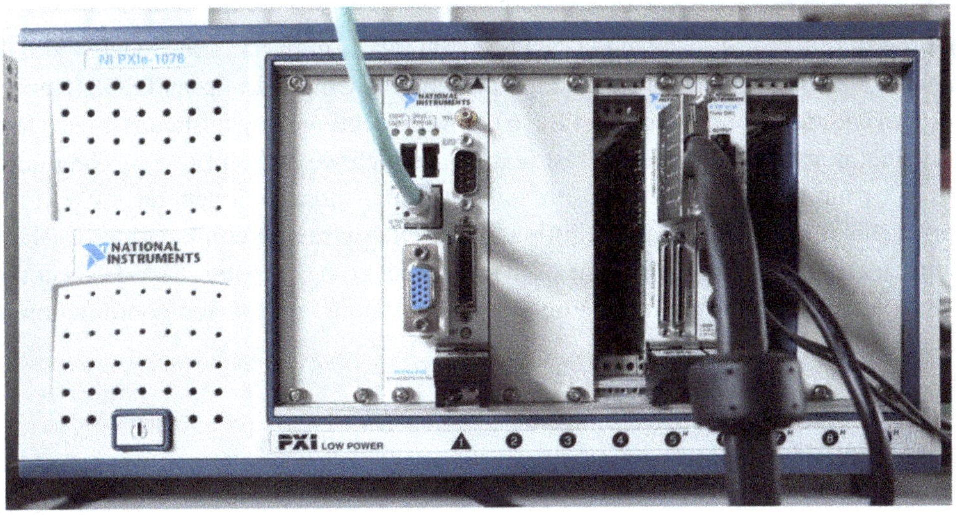

For the connection between the HiL simulator and the embedded control a break-out box is useful, which gives access to all signals passing between the control and the environment which is simulated on the HiL system. This approach allows more flexibility than a dedicated harness with closed connectors, is cheaper and offers the additional possibility to integrate measurement gear or signal generators.

ROS

Robotic applications consist of many standard tasks. Software coding should not be the principal task in robotic development. This requirement is addressed by robotic frameworks. A robotic framework is a modular system out of code modules for common robotic tasks and a software infrastructure to establish communication between the different tasks and several development tools, in particular to integrate own code if necessary. One framework which is frequently used in academia, but also in industry is ROS. Although ROS stands for Robot Operating System, it is not an operating system. It is a robotics framework as described and it needs an operating system on which ROS can run. ROS is developed to run on Ubuntu Linux, a common PC operating system; many ROS releases have also been ported to other operating systems. Due to the underlying platform, ROS does not fulfill hard real time requirements, but in practice the combination of Linux and ROS is suitable for soft real time requirements, i.e. for application in which a reaction time cannot be granted, but promised with high probability. It is in particular suitable for the task of radiation searching and mapping if tasks such as obstacle avoidance are delegated to a microcontroller. For hard real time requirements, the use of an additional microcontroller is recommended.

A typical ROS application consists of several small programs called 'nodes'. They communicate among each other via updated messages called 'topics' which have publisher nodes and subscriber nodes. Detailed information can be retrieved from ROS (2019).

EtaBot

The EtaBot used in this project is intended as a research platform. It has a low energy small PC (in its second release an Intel NUC) for complex calculations running Ubuntu Linux and ROS (Robot Operating System) atop. An additional microcontroller (C167, now discontinued by Infineon) is used for tasks with hard real-time requirements, in particular fast reactions which we call "reflexes". The microcontroller is connected to six ultrasound distance sensors for collision avoidance; it generates PWM (pulse width modulation) signals for the drive motors (optionally an energy saving combination of BLDC and DC motor) and the steering servo. A CAN bus is used to connect both computers. A WLAN connection can connect to an external PC for even more complex computations, centralized coordination of multiple robots or to ease function development.

LIGHT FOLLOWER

The light follower is an optical radiation sensor substitute. It senses from the front of the robot (as the radiation sensor does, Figure 2). It consists of a light sensor, a microcontroller and a USB interface which is compatible to the interface of a radiation sensor (in our particular case a "Gamma Scout"). The USB interface has been chosen to obtain complete exchangeability with the radiation sensor including hardware and software; otherwise a CAN bus may be an even better choice, for which a simpler CAN transceiver instead of the USB interface would suffice.

A photo diode BPX61 with a pull-down resistor is attached to an analogue input of the microcontroller (PIC18F4550-I/P). A serial controller port connects to an USB converter card which fits into a second IC socket on the light follower board. Furthermore, there are a 5-V-regulator, a few passive components and a programming interface for the controller.

The software of the light follower controller consists of a part to initialize the microcontroller including the AD converter followed by an infinite loop waiting for an interrupt from the interface. There are few control commands which can be passed to the light follower, i.e. to go online/offline, to query data and to set sampling intervals. The controller then executes the requested command.

The ROS computer in the robot takes over the signals and in follow mode it reacts accordingly to approach to the radiation source. For this purpose, there is a ROS node which publishes the signal received from the interface as the ROS topic 'rad_val' (radiation value) as in Figure 3.

The publisher node has some further tasks such as establishing communication to the sensor. There is a subscribing node 'sp_gammascoutreader' with an algorithm to follow the source (the name is derived from the radiation sensor "Gamma Scout"); alternatively a mapping algorithm can subscribe rad_val. The follower algorithm needs signals from different directions, so with only one sensor it needs to try out by zigzagging where the radiation comes from. It generates setpoints for the real time motion control on the 167 controller which is put by the ROS mode 'cc_etabot_can_tx' on the CAN bus and then sent to the controller.

'sp_gammascoutreader' with the follower algorithm is typically invoked each period T of 2 s with a new reading from 'gammascout_node'. The core of the follower code is a state machine. Initial state 1 compares if the sensor value exceeds a base value representing the environment level. If so, it tries to approach to the source running during the period T straight and returning again into state 1 the next time. Otherwise the follower sends a command to steer right to the microcontroller. The next time state

Figure 2. Front view of EtaBot equipped with light follower including shield around the photo diode. Below, the ultrasonic sensors for collision avoidance are visible.

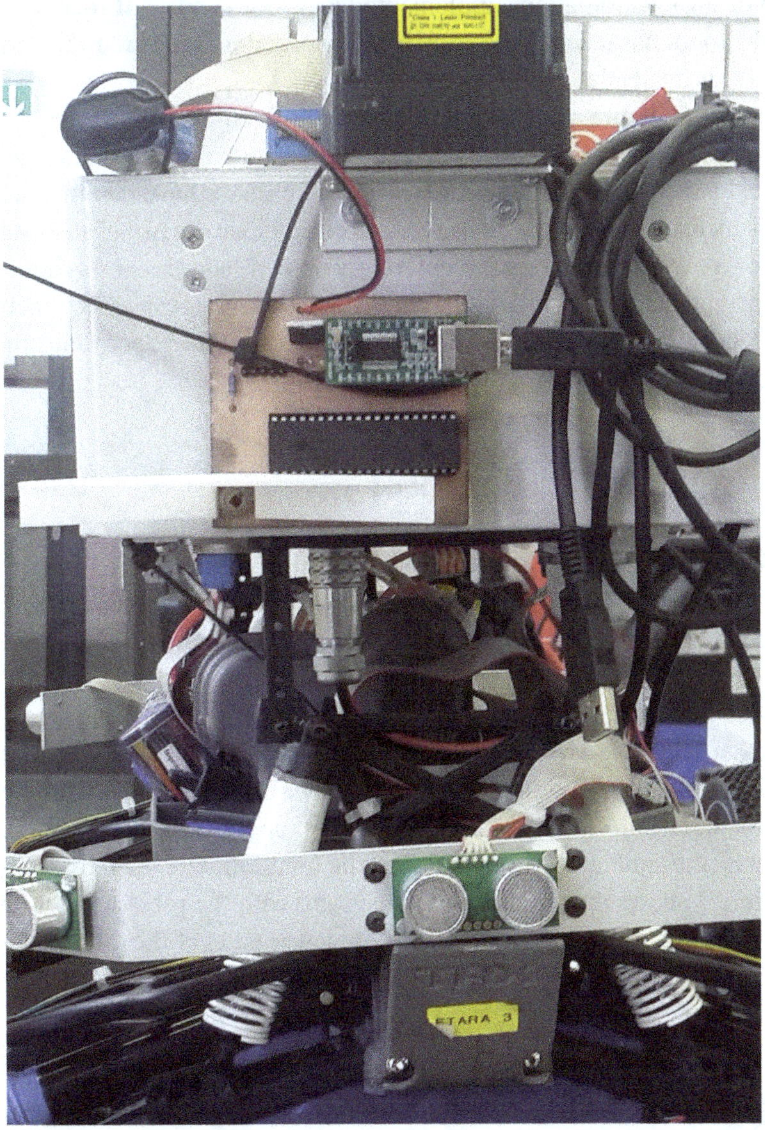

Figure 3. Overview of ROS nodes and their exchanged topics involved in radiation values processing

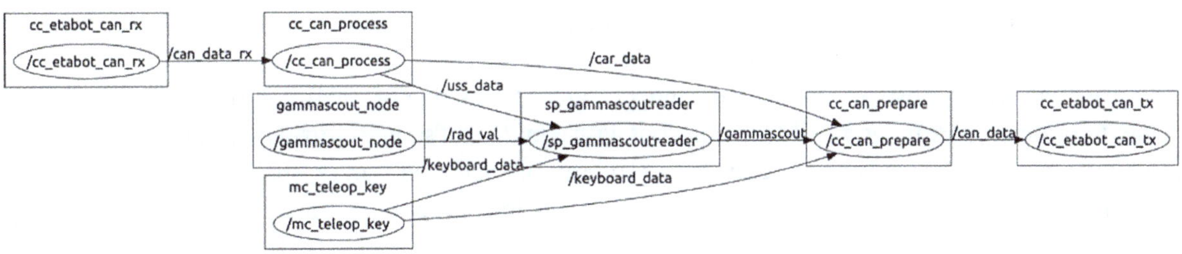

2 (steer one period *T* left) will be executed, so the robot runs a zig-zag and acquires two sensor readings from the left the right side. If the robot notices that it has missed the source it turns back and steers towards the direction of stronger radiation (implemented in states 3 and 4). If there is no further approach possible (because the robot has reached the source), it stops autonomous motion and reports a message if a remote control PC is available.

A practical issue is the optimum motion speed, with light emulation 0.45 m/s; with true radiation which may cover a wide range it needs to be adapted, in most of cases a suitable speed would be lower.

A further practical issue is the steering accuracy. In our case we have obtained a sufficient accuracy by mechanical adjustments, in more general use cases a closed loop control of steering may be necessary.

The follower works well in a laboratory environment with additional daylight or ceiling light, it reaches its limit in bright sunlight. Sensitivity to the simulation source has been increased with a three-dimensionally printed stray light protector on the photo diode. With paper tubes the directivity of the radiation sensor can be simulated.

HARDWARE-IN-THE-LOOP-SIMULATION

An approach which allows more complex scenarios than the relatively simple light follower is a digital simulation. For this purpose, the PWM drive signals for the drive motor and the steering servo are intercepted and fed via a break-out box into the HiL simulator. The motor and steering servo may stay connected, but the robot is lifted, so wheels have no contact to ground.

The motion model calculates how the robot moves through the virtual test area including obstacles and the radiation model generates a radiation value for each position. The signals from virtual ultrasound sensors and from the virtual radiation/light sensor are fed back to the ROS PC in the robot. In our case we decided to use the light follower board as interface for the radiation/light signal, so we replace the analog sensor signal by the simulator output and then the light follower board passes the signal via USB to the ROS PC. Figure 4 shows the structure of the system with the robot core on the left side and the simulator with the simulated environment (and also simulated parts of the robot, i.e. sensors and actors) on the right side.

To represent maps in the model, there are several possibilities, in particular (Berns & von Puttkamer, 2009):

- Metrical maps.
- Grid maps.
- Topologic maps.

Metrical maps as known from geographical maps show all relevant features with map locations which are linearly scaled from true locations, so exact distances and angles can be read out of the map. Feature extraction from a metrical map is computationally intensive.

A grid map is more abstract, it divides the mapped area in a number of small, equidistant squares. Other shapes than squares are theoretically possible, but with equidistant squares a grid map can be stored easily as a matrix where each coefficient describes a property of each map square, such as radiation intensity or occupancy by an obstacle or by the robot itself. If sufficient memory is available for each property, separate matrices with identical mapping from the mapped plain to local coefficients can be used, e.g.

Figure 4. Connecting the robot to the HiL simulator

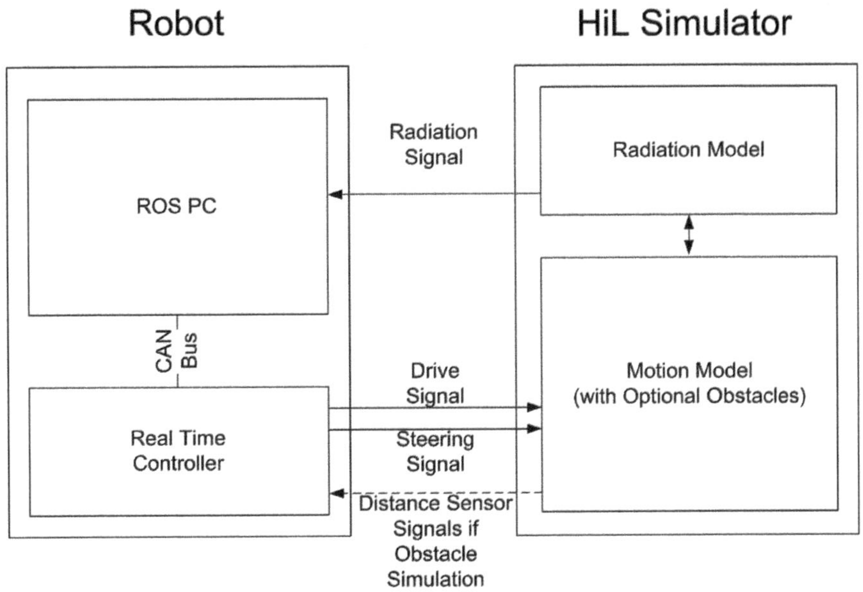

one matrix for radiation intensity and another occupancy matrix to mark obstacles. The occupancy map uses 3 different numerical values to mark obstacles, free area and the recognition area around obstacles in which virtual ultrasound sensors in the model recognize the obstacle. Matrices with different mapping scale are possible, in some cases they may save memory, but they are more difficult to handle.

The highest level of abstraction is achieved with topographical maps which are useful e. g. for long distance navigation, but less in the present application.

The motion model in the simulator initially places the robot in a grid map and depending on the drive and steering signals it tracks the virtual motion in the map and the radiation model takes the local radiation value from the radiation matrix. For this purpose the motion model evaluates the PWM signals from the real time controller which normally drive the motors (with a power stage in between) and the steering servo.

The obstacles are fixed, the robot moves and although in typical scenarios the radiation sources do not move, they are movable here, so the light follower can also be tested with the HiL system.

There are many possibilities, how simulation models can be implemented on HiL simulators, depending on the supplier of the HiL system, available software and also experience. In this particular application a HiL simulator from National Instruments together with a realtime environment from the same supplier (VeriStand) has been used. The models have been implemented in Simulink. They easily drive the simulator into full load, so a resource-saving implementation is necessary.

Figure 5 gives an overview of the Simulink model on the HiL system. On initialization (represented by the two respective if conditions) the block 'Fahrzeug_erstellen' ('Build_vehicle') is executed to enter the robot geometry into the map and the block 'Start_Auswertung' ('Start_evaluation') is executed for further initializations. 'Karten_erstellen' ('Generate_Maps') is invoked continuously and delivers two square maps ('Karte' and 'Lichtkarte') for the motion model and the radiation/light model with a side length of 10 m in this version.

Figure 5. Overview of Simulink model on the HiL simulator (original German annotation, relevant translations in the text)

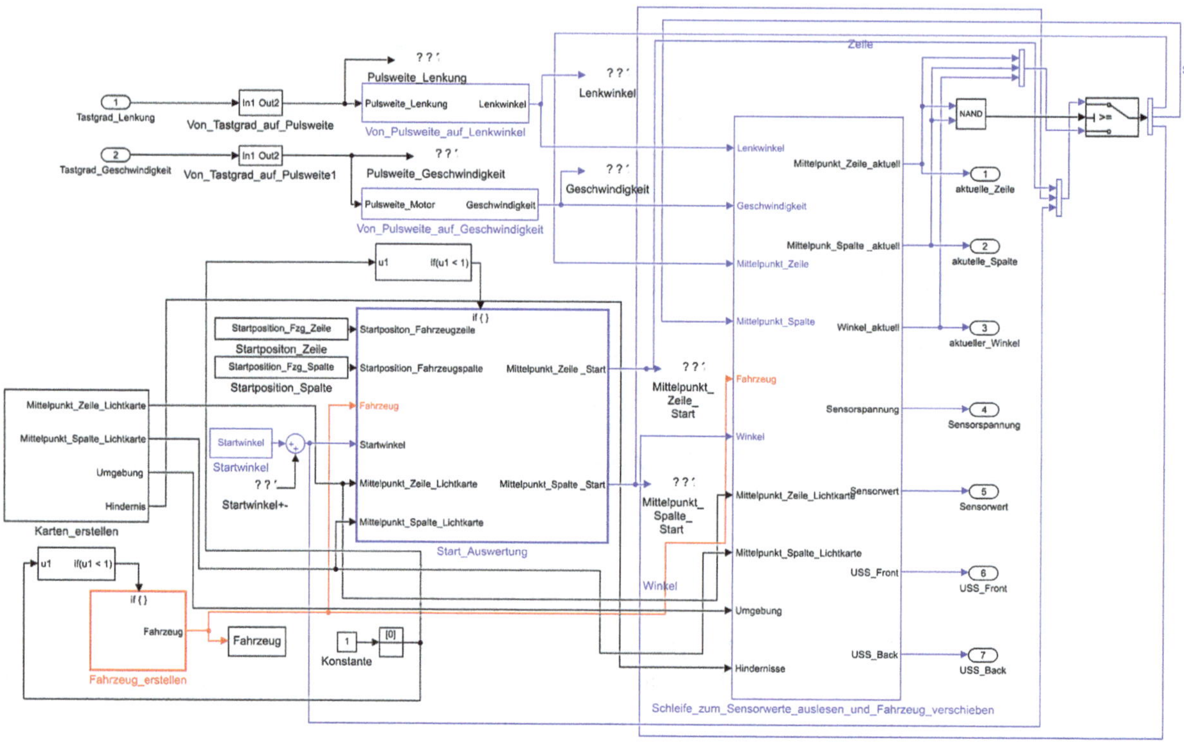

The small blocks on the top convert raw values of the steering angle ('Tastgrad_Lenkung') and velocity ('Tastgrad_Geschwindigkeit') into useful values of steering angle ('Lenkwinkel') and velocity ('Geschwindigkeit'). The core simulation is implemented in the block 'Schleife_zum_Sensorwerte_auslesen_und_Fahrzeug_verschieben' ('Loop_to_read_sensor_ and_move_vehicle'). This block contains further subblocks and Matlab functions (the only difference between subblocks and Matlab functions is the graphical representation versus representation as textual code) to move the robot, to extract data from the radiation map in order to update simulated sensor readings, to recognize obstacles and to translate physical motion parameters into map coordinates.

FUTURE RESEARCH DIRECTIONS

As mentioned in the introduction, EtaBot is a research platform. Whereas our principle goal is reached, it still takes a long path to commercialize a robot to be used in nuclear disaster recovery. Besides validation under different conditions for commercialization there are some very basic technical issues, in particular:

- Principle of motion.
- Ray hardening.
- Power reduction.

- Cooperation between robots.

Furthermore some improvement of HiL simulation may be possible.

Principle of Motion

Most mobile robots are wheel robots. A wheel drive is technically well established and rugged, but it requires an even ground which cannot be expected in disaster areas. For this reason, tracked vehicles are frequently used, often in combination with wheels. Some robots wind along like snakes. The best principle for rough territory has been invented by nature: walking. However, walking requires complex mechanics and also control and keeping balance are still problems associated with walking. At present time robotic walking is by a factor around 10 less efficient than natural walking and still less nimble.

Ray Hardening

Electronic equipment is sensitive to radioactive radiation. In particular memory and cameras are concerned. Ray hardening is on the one hand a practical necessity apart from research with well-known methods, on the other hand it is an area of fundamental research.

The practical part includes in particular lead shielding and preference to low integration semiconductors. The scientific part is the optimization of the semiconductors themselves.

Power Reduction

If not tethered to a cable, power management is crucial to achieve a long autonomous operation time. Power consumers are drives, computers, wireless communication and a laser scanner. Debor, Borgeest, and Imhof (2015) show an approach to optimize driving with respect to consumption whereas Pissulla and Borgeest (2016) show a combined DC/brushless drive with the goal to use each motor in its optimum operating range. The progress of smartphone technology may deliver interesting approaches to reduce power demand for computing and communication. A remaining problem is the energy consuming electromechanical laser scanner which is important for robot navigation, in future pure solid-state laser scanners may bring relief.

Cooperation Between Robots

Searching radioactive matter across a large surface is a task in which multiple equal or also different robots with individual special features can be deployed (Bischoff & Bruhm, 2016). All robots may cooperate on the same level or there may be one master and several slaves (theoretically a hierarchy with multiple levels is possible, but hardly reasonable). Humans could be involved, too.

Improvement of HiL Simulation

In typical applications obstacles and radiation sources are fixed, whereas the robot moves. One may imagine special situations in which also obstacles move or radiation sources vary in activity (actually only their location can move during simulation in the model). Basically, it is possible to extend the model in this respect; after using all possibilities to reduce computational load, a more powerful hardware may extend the bottleneck of computational power additionally.

A further useful improvement would be the inclusion of radiation sensor directivity in the model.

CONCLUSION

Two platforms have been established in which algorithms for radiation search or mapping can be deployed without hazards for the experimenters; this is in particular important for universities where students work on problems. The light follower is also suitable for public demonstrations and lectures; it does not need additional knowledge for a layman to imagine a radiation source instead of the light source, whereas it is difficult to grasp the digital simulation quickly in this case. The digital simulation is more versatile, of course its application needs a high initial work effort, in particular for people who are not familiar with HiL simulators, although tools such as map editors have been developed and short manuals have been written, to get started relatively easily. Application is not limited to radioactive radiation, even for mapping or searching applications without comparable hazards (e.g. electromagnetic fields), it is easy to build any test scenario in the simulation, whereas some scenarios would require a much larger effort with true physical quantities.

REFERENCES

Abouaf, J. (1998). Trial by Fire: Teleoperated Robot Targets Chernobyl. *IEEE Computer Graphics and Applications*, *18*(4), 10–14. doi:10.1109/38.689654

Berns, K., & von Puttkamer, E. (2009). Autonomous land vehicles: Steps towards service robots, Springer-Vieweg, Wiesbaden doi:10.1007/978-3-8348-9334-5

Bischoff, A. & Bruhm, H. (2016). *GIS-basierte Cloud-Services für die Geländeexploration mit kooperierenden autonomen Systemen*, GIS Talk, Munich, Oct. 25, 2016 (German Presentation)

Debor, S., Borgeest, K. & Imhof, Chr. (2015). Evolutionäre Algorithmen bei der Optimierung und beim Test mobiler Roboter (in German language). *12. AALE-Konferenz*, Jena

Habib, M. K. (2007). Mechatronics: A Unifying Interdisciplinary and Intelligent Engineering Paradigm. *IEEE Industrial Electronics Magazine*, *1*(2), 12–24. doi:10.1109/MIE.2007.901480

Ishida, H., Wada, Y., & Matsukura, H. (2012). Chemical Sensing in Robotic Applications: A Review. *IEEE Sensors Journal*, *12*(11), 3163–3173. doi:10.1109/JSEN.2012.2208740

Open Source Robotics Foundation. (2019). *ROS documentation.* Retrieved from http://wiki.ros.org/Documentation

Oshiro, T. (2018). *Soft Robotics in Radiation Environments: A Prospective Study of an Emerging Automated Technology for Existing Nuclear Applications*, (Master thesis, Oregon State University, Oregon).

Peko, D., Basu, S., Kraft, S., Mizokami, S., & Rempe, J. (2018). Working together to enhance nuclear reactor safety; Information from examinations at the affected reactors at Fukushima Daiichi is being used to enhance reactor safety in U.S. operating plants, plants under construction, and future nuclear power plants, Nuclear News, American Nuclear Society, 36-42.

Pissulla, D., & Borgeest, K. (2016). *Energieeffizienter Antrieb eines Mobilroboters mit einem kombinierten DC/BLDC-Motor (in German). 13.* Lübeck, Germany: AALE-Konferenz.

Potemkin, E., Astafurov, P., Osipov, A., Malenkov, M., Mishkinyuk, V., & Sologub, P. (1992). Remote-controlled robots for repair and recovery in the zones of high radiation levels, *Proceedings of the 1992 IEEE International Conference on Robotics and Automation*, Nice, France 10.1109/ROBOT.1992.220331

Tsitsimpelis, I., Taylor, C. J., Lennox, B., & Joyce, M. J. (2019). A review of ground-based robotic systems for the characterization of nuclear environments. *Progress in Nuclear Energy, 111*, 109–124. doi:10.1016/j.pnucene.2018.10.023

ADDITIONAL READING

Bashyal, S., & Venayagamoorthy, G. K. (2008). Human Swarm Interaction for Radiation Source Search and Localization, *Proceedings of the 2008 IEEE Swarm Intelligence Symposium*, St. Louis, MO. 10.1109/SIS.2008.4668287

Iborra, A., Pastor, J. A., Álvarez, B., Fernández, C., & Fernández Meroño, J. M. (2003). Robots in Radioactive Environments; Operational Experiences Using Robotics During Maintenance Services in PWR Nuclear Power Plants. *IEEE Robotics & Automation Magazine, 10*(4), 12–22. doi:10.1109/MRA.2003.1256294

Krishnanand, K. N., Amruth, P., Guruprasad, M. H., Sharschchandra, V. B., & Ghose, D. (2006). Glowworm-inspired Robot Swarm for Simultaneous Taxis towards Multiple Radiation Sources, *Proceedings of the 2006 IEEE International Conference on Robotics and Automation*, Orlando, FL. 10.1109/ROBOT.2006.1641833

La, H. M., Sheng, W., & Chen, J. (2015). Cooperative and Active Sensing in Mobile Sensor Networks for Scalar Field Mapping. *IEEE Transactions on Systems, Man, and Cybernetics. Systems, 45*(1), 1–12. doi:10.1109/TSMC.2014.2318282

Siegwart, R. (2011). *Introduction to Autonomous Mobile Robots.* Cambrige, MA: MIT Press.

KEY TERMS AND DEFINITIONS

AD Converter: Converter from analogue signals to digital signals for further processing in the controller.

BLDC: BrushLess Direct Current, BLDC motors are supplied with DC, but run internally with multiple phases.

CAN: Controller Area Network, a digital bus which is common in many industries such as automotive industry.

Geiger-Müller Tube: Tube in which ions are generated by radioactive radiation. An applied voltage causes a measurable discharge.

HiL: Hardware in the Loop.

INES: International Nuclear and Radiological Event Scale (7 is the maximum category of nuclear accidents).

LIDAR: Light detection and ranging.

PWM: Pulse Width modulation.

PXI: PCI eXtensions for Instrumentation, a PCI based backplane bus.

Radioactivity: Nuclear decomposition releasing radiation.

ROS: "Robot Operating System", a platform for robotic applications.

Transceiver: Transmitter and receiver, an integrated circuit which physically attaches a controller to the CAN bus.

USB: Universal serial Bus.

WLAN: Wireless LAN (Local Area Network).

Chapter 7
Membrane Micro Electro-Mechanical Systems for Industrial Applications

Mario Versaci
University "Mediterranea" of Reggio Calabria, Italy

Francesco Carlo Morabito
University "Mediterranea" of Reggio Calabria, Italy

ABSTRACT

The objective of this chapter is to provide the analytical-numerical tools for the simplified rewriting of the most important mathematical models of MEMS membrane devices for Mechatronics, exploiting advanced concepts and results in the theory of curves and surfaces. Moreover, when the solution in closed form could not be obtained (that is, it is impossible to obtain the membrane deflection analytically), some consolidated techniques will be described both to obtain conditions ensuring existence/uniqueness of the solution, and the most suitable approaches for obtaining numerical solutions in the absence of ghost solutions. Finally, some practical examples will illustrate the approaches presented.

INTRODUCTION

Mechatronics engineering studies mechanical, electronic and computer interactions for the analysis and design of fully automated control systems (Bolton, 2003), (Jouaneh, 2012). This term is composed of the words "mechanics" and "electronics", historically, the company that used this word for the first time was Yaskawa, in Japan, in the 1960s. The first works concerned the development of systems requiring knowledge on both domains. Since, the development of this discipline has been exponential, requiring analysis/synthesis techniques assisted and supported by constant technological innovation with the main purpose of creating devices on an increasingly smaller scale. At the same time, the most recent industrial guidelines direct both researchers and designers to the development of low-cost devices able to combine the physical properties of the problem with low-level machine languages. Hence, the need to design

DOI: 10.4018/978-1-7998-0137-5.ch007

sensors and actuators able to meet the multiple needs required by the most widespread industrial, civil and biomedical applications. In such a context, static and dynamic Micro-Electro-Mechanical-Systems (MEMS) technology has matured, especially in all of those domains where modern miniaturized and integrated electromechanical systems are required. Specifically, in the general knowledge space, MEMS microsystems are considered as very important devices for mechatronics foundational and core knowledge (Habib, Mechatronics: A Unifying Interdisciplinary and Intelligent Engineering Paradigm, 2007), (Habib, Human Adaptive and Friendly Mechatronics (HAFM), 2008). Moreover, thanks also to modern numerical techniques capable of simulating extremely realistic industrial scenarios, MEMS represent one of the most important achievements of engineering on an industrial scale (Gad-el Hak, 2006), (Pelesko & Bernstein, 2003) (Phan & et al., 2018). Currently, the industrial applications of MEMS devices in mechatronics are extremely varied, from applications in the biomedical domain, such as surgical-diagnostic-therapeutic microsystems (Khoshnoud & de Silva, 2012), (Wang & Soper, 2006) to recent developments in the opto-electronic field (compact sensors systems, photodiode) (Warren, 2008); thermally driven MEMS actuator for mechatronics (Kimura, 2009) to modeling vibration elastic structures (Pelesko & Bernstein, Modeling MEMS and NEMS, 2003). Additionally, modeling MEMS has gained wide acclaim both coupled thermal-elastic systems (Pelesko & Bernstein, 2003) and electrostatic-elastic systems (Pelesko, 2004) making MEMS devices extremely attractive for mechatronics applications, due to the fact that their small size as well as the easy production with relatively low costs (Gad-el Hak, 2006) (Mistry & Nahapatra, 2012), (Warren, 2008). If on the one hand the demand for MEMS devices in Mechatronics is very strong, on the other hand, it is not always possible to formulate analytical mathematical models easily implemented. The MEMS technology makes it possible to integrate both electronic circuits and opto-mechanical devices on the same silicon substrate (Plander & Stepanovsky, 2017), employing manufacturing technologies similar to those used for the realization of integrated circuits. The dimensions of a MEMS device are generally variable between a few *μms* and 1 *mm*, while the individual components of which it is composed vary between 1 and 100 *μms*. The most widely used MEMS devices in mechatronics belong to the class of the general electrostatic-elastic systems. This is due to the biggest problem of the mass-spring model of the electrostatic actuators is the inability to capture real geometry and elastic effects as well. Usually, an electrostatic-elastic MEMS device consists of two parallel plates, one of which is fixed and the other deformable. A difference potential V is applied and the deformable plate moves. Often, however, many applications require that a membrane replace the deformable plate in order to reduce the inertial effects. Usually, the physical-mathematical models use the deflection of the membrane, u, as an independent variable. Furthermore, these analytical formulations hardly allow resolution in closed form (Angiulli, Jannelli, Morabito, & Versaci, 2018) (Di Barba, Fattorusso, & Versaci, 2017). As a result, one must be satisfied with obtaining conditions that guarantee both existence and uniqueness of the solution, often using techniques based on fixed-point theorems formulated in suitable functional spaces with evident aggravation in terms of computational complexity (Esposito, Ghoussoub, & Guo, 2010). The alternative, therefore, is the numerical approach, which, although of great practical interest, do not prevent us from being present for ghost solutions interest, do not prevent us from being present for ghost solutions (Angiulli, Jannelli, Morabito, & Versaci, 2018), (Esposito, Ghoussoub, & Guo, 2010). Thus, a viable path is the joint use of analytical approaches assisted by numerical techniques in order to obtain numerical solutions that respect the analytical conditions that guarantee existence and uniqueness of the solution by eliminating the possibility of obtaining ghost solutions. In this context, the experience of the authors in the field of modeling electrostatic MEMS membrane devices with strong non-linearity of interest to Mechatronics has matured. In particular, by combining the physics of the problem with important results of differential geometry (do Carmo, 2016), in some cases, analytical results have been obtained in terms of the existence and uniqueness of the solutions (Di Barba, Fattorusso, & Versaci, 2017) and, in other cases, numerical solutions in the absence of ghost solutions (Angiulli, Jannelli, Morabito, & Versaci, 2018). However, the conditions that guarantee existence and uniqueness are often independent of the electromechanical properties of the membrane material, making the models studied unattractive from an industrial point of view (Di Barba, Fattorusso, & Versaci, 2017). The purpose of this chapter is to provide the analytical-numerical tools for the simplified rewriting of the most important mathematical models of electrostatic-elastic membrane MEMS devices for Mechatronics, exploiting advanced concepts and results in the theory of curves and surfaces. Moreover, when it is impossible to obtain the membrane deflection analytically[1], some well-established techniques will be described obtaining both existence/uniqueness conditions of the solution depending on the electromechanical properties of the membrane material.

The remainder of the chapter is structured as follows: Sections one and two, briefly recall the one-dimensional physical-mathematical model here studied. In particular, starting from the modeling general electrostatic-elastic systems constituted by two parallel plates one of which is deformable (see, Figure 2), and introducing appropriate scale factors, we obtain model (4) as a simplification of the three-dimensional electrostatic-elastic model. Then, framing the problem in a suitable functional space and exploiting relation (5), the authors rewrite the problem in terms of curvature K of the membrane and exploit the result of existence of the solution (Di Barba, Fattorusso, & Versaci, 2017) (Fattorusso & Versaci, 2018). Following, a new condition ensuring the uniqueness of the solution, and depending on the electromechanical property of the material of the membrane, will be presented in order to achieve a new condition of both existence and uniqueness dependent on the electromechanical characteristics of the material constituting the membrane. Following, a brief description of the usual numerical resolution techniques for BVPs is presented. After, shooting techniques for solving the problem at hand is presented motivating its choice and showing the main results of convergence in absence of ghost solutions. Furthermore, new ranges of values of σ (mechanical tension of the material constituting the membrane), V and $sup|\mathbf{E}|$ are shown to highlight the applicability of the device under study in the mechatronics domain. Finally, after to have shown an overview of the most important applications of membrane MEMS in mechatronics, some due comments and useful conclusions will conclude the chapter.

MEMBRANE ELECTROSTATIC MEMS DEVICES: A STARTING POINT

Membrane electrostatic-elastic MEMS devices of interest to Mechatronics have often a circular geometry (Bolton, 2003). It is, therefore, obvious to structure the analytical models in order that the radius, r, of the membrane itself (starting from the center of the membrane towards the edges) represents the independent variable. The models obtained, in these cases, is a second-order nonlinear ordinary differential equation in which boundary conditions must be imposed (Boundary Value Problems, BVPs). For example, a typical dimensionless model is as follows (Pelesko & Bernstein, 2003):

$$\begin{cases} \dfrac{d^2u}{dr^2}+\dfrac{1}{r}\dfrac{du}{dr}=\dfrac{\lambda^2}{u^2} \\ u(1)=1 \end{cases} \tag{1}$$

Figure 1. Membrane electrostatic mems device with axial symmetry: From 3D to 1D modeling

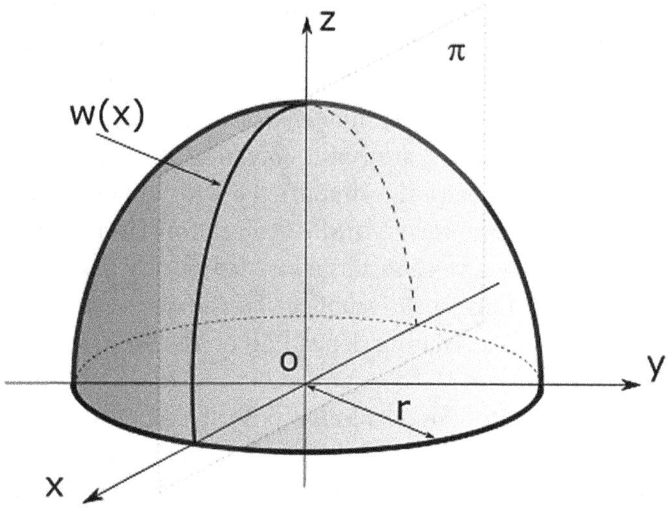

Figure 2. A simplified representation of the general electrostatic-elastic system in which both elastic plate (at potential V) and fixed ground plate (at potential 0) are shown

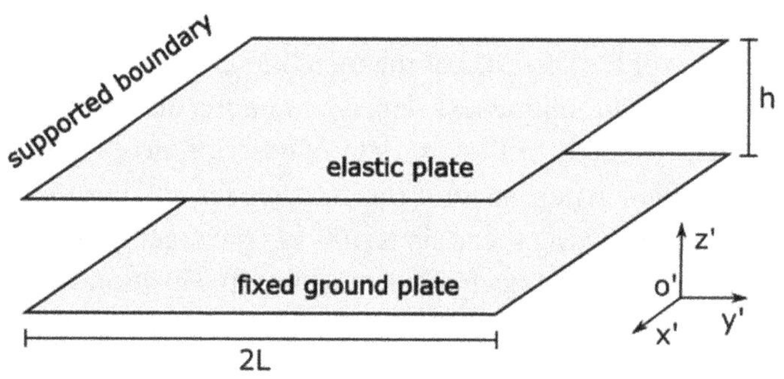

In addition, a second condition must be considered, such as:

$$\frac{du}{dr}(0) = 0 \qquad (2)$$

in order to ensure that solutions be symmetric about the point *r=0*. Most analytical models for circular electrostatic-elastic MEMS are invariant under change of variables containing a free parameter. Hence, these models can be considered as invariant under a one-parameter family of transformations (the so-called Lie groups) (Esposito, Ghoussoub, & Guo, 2010), (Pelesko & Bernstein, 2003). Then, at least theoretically, such analytical models can be solved by using Lie symmetries that maps any solutions to another solution of the model. Such transformations are algebraic groups depending on continuous parameters and consist of either point transformations acting on the so-called contact transformations (Bluman & Anco, 2002). This procedure, even if very interesting from a theoretical viewpoint, needs specific software based on symbolic calculus and, however, the obtained solutions are often too complicated to be exploited for real time applications. However, by subjecting the device to a given *V*, the membrane deforms symmetrically with respect to the central axis passing through *r=0* which becomes an axes of symmetry (in a three-dimensional Cartesian space where *r=0* is the axis *z*). The problem becomes an axial symmetric one (with respect to *r=0*) so that it is sufficient to consider the profile, *u*, generated by the intersection between the deformed membrane and a not specified plane, π, whose support is *r=0* admitting, in this way, a 1*D* modeling of the problem.

The most simple way to obtain a *1D* model of a membrane electrostatic-elastic MEMS consists in starting from a model of MEMS with two parallel plates, one of which is fixed and the other deformable, imposing the relaxation of mechanical constraints. One of the most interesting dimensionless physical-mathematical models of electrostatic-elastic MEMS devices with parallel plates, whose just one of them is deformable but anchored to their edges, is (Esposito, Ghoussoub, & Guo, 2010):

$$\begin{cases} \alpha\Delta^2 u = \left(\beta_1 \int_\Omega \left|\nabla u\right|^2 dx + \gamma\right)\Delta u + \dfrac{\lambda_1 f_1(x)}{\left(1-u\right)^{\sigma_1}\left(\chi \int_\Omega \dfrac{dx}{\left(1-u\right)^{\sigma_1 - 1}}\right)} \\ u(x) = 0, x \in \partial\Omega; 0 < u(x) < 1, x \in \Omega \end{cases} \tag{3}$$

in which *u(x)* represents the deflection of the deformable plate and the $f_1(x)$, bounded function, represents the dielectric characteristic of the material constituting the deformable plate. In addition, λ_1 depends on the applied voltage, while α_1, β, γ and χ are physical parameters identifying the electrical and mechanical properties of the material constituting the deformable plate (for details, see Table 1). Model (3), parameter σ_1 can take into account general electrostatic potentials; however, for all purposes of this chapter, attention is limited to Coulombian potentials $\sigma_1 \geq 2$. To adapt the model (3) to membrane electrostatic-elastic MEMS devices, one must neglect both the thickness of the deformable plate and the mechanical effects due to inertial phenomena. This means that, choosing $\sigma_1 = 2$, $\alpha_1 = 1$, $\beta = 0$, $\gamma = 0$ and $\chi = 0$ and indicating by ' the derivative with respect to *x*, model (3) can be written as follows (Di Barba, Fattorusso, & Versaci, 2017):

$$\begin{cases} u''(x) = \dfrac{\lambda_1 f_1(x)}{\left(1+u(x)\right)^2}; 0 < u(x) < 1 \, in \, \Omega \\ u = 0, on \, \partial\Omega \end{cases} \tag{4}$$

and it is a model related to a membrane electrostatic-elastic MEMS device. Since in model (3) the deformable plate is the lower one, for reasons of opportunity it is preferable to consider the lower plate as a non-deformable one, so that it is necessary to invert the orientation of the ordinate axis.

With these premises, model (4) assumes the following form:

$$\begin{cases} u''(x) = -\dfrac{\lambda_1 f_1(x)}{\left(1-u(x)\right)^2} \, in\,\Omega \\ u(-L_1 = -0.5) = u(L_1 = 0.5) = 0. \end{cases} \quad (5)$$

The applied voltage *V*, which usually is closely related to the parameter λ_1, can be formulated in terms of the amplitude of the electrostatic field, |**E**(*x*)| (Esposito, Ghoussoub, & Guo, 2010), while **E**(*x*) can be thought as a vector locally orthogonal to the tangent direction of the membrane (Di Barba, Fattorusso, & Versaci, 2017; Voltmer, 2007). Increasing *V*, |**E**(*x*)| increases so that the membrane deforms and, locally, increases its curvature *K*. Moreover, for physical reasons, *K*, besides being dependent on *x*, also depends on the deflection of the membrane at the same point, $K\left(x, u(x)\right)$. Thus, it seems natural to consider the direct proportionality between |**E**(*x*)| and *K* as follows:

$$\left|\boldsymbol{E}(x)\right| = \mu\left(x, u(x), V\right) K\left(x, u(x)\right) \quad (6)$$

where $\mu\left(x, u(x), V\right)$ represents the function of proportionality between $\left|\boldsymbol{E}(x)\right|$ and $K\left(x, u(x)\right)$. In such a context, by means of both (5) and (6), we obtain a model in which the singularity, *1-u(x)*, was eliminated. In the work of (Di Barba, Fattorusso, & Versaci, 2017), important results have been obtained regarding the conditions ensuring both existence and uniqueness of the solution for the model (5), by means the exploitation of (6). However, their work demonstrates that the condition ensuring the existence of the solution depended strongly on the electromechanical characteristics of the material constituting the membrane, but the condition governing the uniqueness of the solution was independent of such electromechanical characteristics. Since in the field of mechatronics the use of the device is strongly conditioned by the electromechanical properties of the membrane, it is evident to have conditions of existence and uniqueness of the solution that are both dependent on such properties. In the first part of this chapter a new condition ensuring the uniqueness of the solution depending on the characteristics of the membrane is presented. Moreover, the presence of *V* in such new condition of uniqueness here obtained, provides the possibility to establish more precisely the domain of applicability of the device. The second part of the chapter deals with solving the problem from a numerical point of view. This is due to the fact that the model under study does not allow an explicit solution for which the most attractive approach, from an industrial point of view, is a numerical one. In particular, being the model under study a boundary value problem (BVP), the technique most interesting (and computationally not expensive) is the shooting approach able to transform the BVP into an equivalent initial value problem (IVP). The profile of the membrane obtained by this approach put in evidence the values of operational parameters of the membrane, *V* and *sup*|**E**|, ensuring the absence of ghost solutions (numerical solutions that do not satisfy the analytical conditions of existence and uniqueness). In this way, numerical solutions can be considered more realistic because they are obtained taking into account the new condition of uniqueness, here presented, depending on the characteristics of the membrane constituting the device.

Table 1 provides a list of symbols used throughout the chapter in order, first the electrostatic parameters; then, the mechanical ones. In the following, the geometrical parameter describing the MEMS device. Finally, some electromechanical parameters and some quantities defined mathematically

Table 1. A list of symbols used throughout the chapter

Symbol	Description
Electrostatic Parameters	
	Electrostatic field inside the MEMS device
E	The usual electrostatic permittivity of the vacuum (or free space)
	Coulomb exponent
	External applied electric voltage generating **E**
	Electrostatic potentials (the second one is in dimensionless conditions)
V	Voltage to be applied externally to the MEMS device to overcome the mechanical inertia of its membrane
	Pull-in voltage
	Bounded function indicative of electrical properties of the membrane
	Dielectric strength of the membrane
Mechanical Parameters	
	Mechanical tension of the material constituting membrane
D	Flexure rigidity of the material constituting the membrane
	Relative importance of tension and rigidity
Electromechanical Parameters	
$\alpha_1, \beta, \gamma, \chi$	Electromechanical quantities of the material constituting the membrane
Geometrical Parameters	
2L, $2L_1$	Lengths of the MEMS device (the second one is in the dimensionless conditions
h	Height of the MEMS device
u	Deflection of the membrane in the MEMS device (dimensionless conditions)
*d**	Critical distance
K	Curvature of the membrane
Parameters Defined Mathematically	
M	
H	
G(x,s)	Green's function exploited to transform the differential problem into its corresponding integral formulation

SOME PHYSICAL-MATHEMATICAL BACKGROUNDS FOR MEMBRANE ELECTROSTATIC-ELASTIC MEMS

In order to understand how a membrane electrostatic MEMS device works, let us consider an usual orthonormal system of Cartesian axes in $\mathbb{R}^3$, whose origin and axes are labeled by *O'x'y'z'*. An electrostatic-elastic MEMS constituted by a pair of parallel plates, one plate fixed and the other one elastic and deformable but fixed along its edges, is located normally to the axis *z'*. Moreover, the fixed plate is located at *z'=0* and the deformable one is at *z'=h* (mutual distance between the plates). The length of the device is *2L* while its width is negligible (see Figure 1). If an electrostatic voltage, *V*, is applied to the MEMS device, the plates are subject to two different electrostatic potentials: the fixed plate is at zero potential, $\phi = 0$, while the deformable one has potential $\phi = V$, so that inside the MEMS device the Laplace's equations yields, $\Delta\phi = 0$ (Di Barba, Fattorusso, & Versaci, 2017). If we indicate as dependent variable the elastic deflection of the deformable plate, *w*, and considering the mechanical tension of the material constituting both the plates, σ, the well-known nonlinear coupled partial differential equations which modeling the problem is as follows (Di Barba, Fattorusso, & Versaci, 2017):

$$\begin{cases} \in^2 \Delta_\perp \Phi + \dfrac{\partial^2 \Phi}{\partial z^2} = 0 \\ -\Delta_\perp w + \delta \Delta_\perp^2 w = -\lambda^2 \left[\in^2 \left| \nabla_\perp \Phi \right|^2 + \left(\dfrac{\partial \Phi}{\partial z} \right)^2 \right] \\ \Phi = 1 \, on \, the \, elastic \, plate; \Phi = 0 \, on \, the \, fixed \, plate \end{cases} \tag{7}$$

where *D* indicates the flexural rigidity of the plates. It worth noting that model (7) is a dimensionless one and it is obtained by exploiting a suitable set of scaling factors. Moreover, $\Delta_\perp = \dfrac{\partial^2}{\partial x^2} + \dfrac{\partial^2}{\partial y^2};$; δ and $\in$ are computed as $\dfrac{D}{(2L)^2 \sigma}$ and $h/(2L)$, respectively. Finally, the ratio of a reference electrostatic force to a reference elastic force, λ^2, is computed as follows (Pelesko & Bernstein, 2003):

$$\lambda_1 = \lambda^2 = 0.5 h^{-3} \sigma^{-1} \in_0 V^2 (2L)^2 = \beta V^2 \tag{8}$$

which represent the link between model (7) and model (4) where β, defined as:

$$\beta = 0.5 h^{-3} \sigma^{-1} \in_0 (2L)^2 \tag{9}$$

is the physical parameter that takes into account the electro-mechanical properties of the material constituting the membrane. It is worth observing that β, in dimensionless conditions (indicated by β_1), satisfies the following condition:

$$\beta_1 = 0.5 \sigma^{-1} \in_0 \geq 10^{12} \tag{10}$$

However, experimentally, the values of β_1 are greater than 10^{12}. Both modern industry and technologies exploit the materials up to a limit state of mechanical resistance; so that *D* can be neglected in the physical-mathematical modeling. This means that the thickness of the deformable plate can be neglected. Moreover, in order to get a 1*D* model, also the width of the device, compared to its length, can be neglected. Finally, to get a membrane electrostatic-elastic MEMS device, we replace the deformable plate with an elastic membrane whose edges are anchored to a fixed plate (which serves only as a support for the membrane itself). Under these conditions, and in steady state conditions, model (7) becomes model (5) where $f_1(x)\lambda_1$ is replaced by λ^2 and axes z is reversed, $L_1 = 0.5$ represents dimensionless L and u is the new deflection of the membrane. It is well-known that the solution is expressible in terms of the energy, En, defined as

$$En = 0.5\left(u'(x)\right)^2 + \lambda^2\left(1-u(x)\right)^{-1}$$

obtaining the following implicit solution (Pelesko, 2004):

$$\sqrt{\frac{\left(u(x)-1\right)\left(u(x)-1-\frac{\lambda^2}{En}\right)}{2En}} + \frac{\lambda^2}{En\sqrt{2En}}\tanh^{-1}\sqrt{\frac{u(x)-1-\frac{\beta}{En}}{u(x)-1}} = x. \tag{11}$$

where the symmetry of the solutions allows us to eliminate the second constant of integration. (11) can be solved numerically obtaining that, for $\lambda \in \left[0,\lambda^*\right]$ with λ^* critical value for λ (the so-called pull-in voltage), there exist two values of En for each value of λ^2, while beyond λ^* no solutions exist. So that, the two solutions found for $\lambda < \lambda^*$ correspond to two different solutions for $u(x)$. Following this approach, the existence of the solution is surely guaranteed, while we certainly lose the certainty of guaranteeing the uniqueness of the solution. Keeping in mind the intended aim, that is to express |E| in terms of the curvature K of the membrane (see, Eq. (6)), reformulate the problem under study by observing that λ^2 can be considered as proportional to V^2, so that $\lambda^2\left(1-u(x)\right)^2$ it is "proportional" to $\left|\boldsymbol{E}\right|^2$. Then, model (5) can be write as follows:

$$\begin{cases} -u'' = \theta\left|E\right|^2 \; in\, \Omega \subset \mathbb{R}^3 \\ \theta \in \mathbb{R}^+; u\left(-L_1\right) = u\left(L_1\right) = 0. \end{cases} \tag{12}$$

where θ represents the function of proportionality between $\lambda^2\left(1-u(x)\right)^2$ and $\left|\boldsymbol{E}\right|^2$ (for details, see (Di Barba, Fattorusso, & Versaci, 2017)). In addition, since λ is proportional to V, (6) assumes the following functional dependence:

$$\left|E(x)\right| = \mu\left(x, u(x), \lambda\right) K\left(x, u(x)\right) \tag{13}$$

with

$$\mu\left(x,u\left(x\right),\lambda\right)=\frac{\lambda}{\left(1-u\left(x\right)-d^{*}\right)}.$$

It is worth observing that

$$\mu\left(x,u\left(x\right),\lambda\right)\in C^{2}\left(\left[-L_{1},L_{1}\right]\times\left[0,1\right)\times\left[\bar{\lambda},M\right]\right)$$

in which $\bar{\lambda}^{2}$ is the minimum voltage for the device in order to overcome the inertia of the membrane, while M is the $\sup\left\{\lambda\right\}$. Finally, $d^{*}=\lambda\left(\in_{t}\right)^{-1}$, with $\in_{t}$ the dielectric strength of the material of the membrane, is the critical distance to be imposed on the model in order to avoid contact between the membrane of the device and the non-deformable upper plate (in other words, the singularity present in the model is avoided). On the other hand, taking into account that *u=u(x)* can be considered as the graph of a function, and using the symbol "'" instead of $\frac{d}{dx}$, the 1*D* Cartesian curvature can be computed as

$$K\left(x,u\left(x\right)\right)=\left|u''\left(x\right)\right|/\sqrt{\left(1+\left|u'\left(x\right)\right|^{2}\right)^{3}}$$

(for more details, see APPENDIX). Then, taking into account the curvature as below defined, problem (12) can be easily rewritten as follows:

$$\begin{cases} u''\left(x\right)=-\dfrac{\left(1+\left(u'\left(x\right)\right)^{2}\right)^{3}}{\beta_{1}\mu^{2}\left(x,u\left(x\right),\lambda\right)}\, in\,\Omega=\left[-L_{1},+L_{1}\right] \\ u\left(-L_{1}\right)=u\left(L_{1}\right)=0 \\ 0<u\left(x\right)<1-d^{*}=\alpha \end{cases} \tag{14}$$

where it is evident that the singularity $\left(1-u\left(x\right)\right)^{2}$ is no longer present. Moreover, taking into account that

$$\mu\left(x,u\left(x\right),\lambda\right)=\frac{\lambda}{\left(1-u\left(x\right)-d^{*}\right)},$$

problem (13) assumes the following final form:

$$\begin{cases} u''(x) = -\dfrac{\left(1+\left(u'(x)\right)^2\right)^3}{\theta\mu^2\left(x,u(x),\lambda\right)} = -\dfrac{1}{\theta\lambda^2}\left(1+\left(u'(x)\right)^2\right)^3\left(\alpha-u(x)\right)^2 \; in\,\Omega \\ u(-L_1) = u(L_1) = 0; 0 < u < \alpha = 1 - d^* \end{cases} \tag{15}$$

in which $u \in C^2(\Omega)$, because the membrane in the device does not allows any tears, so that we impose that the slope, point-by-point, must vary continuously. Mathematically, we need to impose that

$$\mu = \mu\left(x,u(x),\lambda\right) \in C^0\left(\Omega \times [0,1],[\bar{\lambda},M]\right)$$

and $\mu = \lambda\left(\alpha - u(x)\right)^{-1}$ so that (Di Barba, Fattorusso, & Versaci, 2017):

$$\begin{cases} u''(x) = -\dfrac{\left(1+\left(u'(x)\right)^2\right)^3}{\theta\mu^2\left(x,u(x),\lambda\right)} = -\dfrac{1}{\theta\lambda^2}\left(1+\left(u'(x)\right)^2\right)\left(\alpha-u(x)\right)^2 \; in\,\Omega \\ u(-L_1) = u(L_1) = 0 \\ 0 < u < \alpha = 1 - d^* \\ \mu = \mu\left(x,u(x),\lambda\right) \in C^0\left(\Omega \times [0,1],[\bar{\lambda},M]\right) \\ \mu = \dfrac{\lambda}{\alpha - u(x)}. \end{cases} \tag{16}$$

At this point, some technical clarifications are necessary. Firstly, we note that increasing $\theta\lambda^2$, $u''(x)$ decreases. In other words, this means that geometrically the concavity of the membrane is reduced by increasing $\theta\lambda^2$. Secondly, as will been seen, $\theta\lambda^2$ is the main responsible for the convergence of the numerical method used to reconstruct the membrane profile. In other words, higher values of $\theta\lambda^2$ ensure the convergence of the exploited numerical procedure. So that, the greater $\theta\lambda^2$ is, the flattened the membrane profile will be, but the convergence of the numerical method will be more certain. Finally, since high values of $\theta\lambda^2$, avoid ghost solutions (see below), flattened membrane profiles ensure convergence of the numerical solution which is not a ghost solution.

ON THE CONDITION OF EXISTENCE OF THE SOLUTION FOR THE PROBLEM UNDER STUDY: A RESULT KNOWN IN LITERATURE

In the work (Di Barba, Fattorusso, & Versaci, 2017), it is not possible to derive the solution of (16) explicitly, the authors sought conditions that guaranteed both the existence and uniqueness for it. The approach used exploited known fixed point results performed by Schauder-Tychonoff, according to which a completely continuous operator was defined by a closed, limited and convex subset of a Banach space in itself has at least one fixed point. The need to use Banach spaces stems from the fact that each Banach space is both normed (defining a norm in it, inducing a distance) and complete (each Cauchy sequence converges within it, and therefore the risk of going out of the space is avoided). Two functional spaces, P and P_1 were defined in $\Omega = \left[-L_1, L_1\right]$ (Ω, as it is defined, it turs out to be a closed and limited set) as follows:

$$P = \left\{C_0^2\left(\Omega\right) : 0 < u\left(x\right) < \alpha = 1 - d^*, \left|u'\left(x\right)\right| < H < +\infty\right\} \tag{17}$$

$$P_1 = \left\{C_0^2\left(\Omega\right) : 0 < u\left(x\right) < \alpha_1, \left|u'\left(x\right)\right| < H < +\infty\right\} \tag{18}$$

where $H=sup\{|u'(x)|\}$. As known, it is possible to get an equivalent integral formulation of the problem (16) by considering a suitable Green function $G(x,s)$ (Bailey, Shampine, & Waltman, 1968) so you can write (Di Barba, Fattorusso, & Versaci, 2017):

$$u\left(x\right) = \int_{-L_1}^{L_1} G\left(x,s\right) \frac{\left(1 + \left(u'\left(s\right)\right)^2\right)^3}{\theta\mu^s\left(s, u\left(s\right), \lambda\right)} ds. \tag{19}$$

Then, by this new formulation, the existence of the solution for (19) is equivalent to the existence of the solution of the operator equation $T(u)=w$ where $u \in P$. The uniqueness of the solution for $T(u)=w$ has been proven in (Di Barba, Fattorusso, & Versaci, 2017) by the Schauder-Tychonoff fixed point approach, with $T : P \rightarrow P$. This condition guaranteeing the uniqueness of the solution for $T(u)=w$, depended on H (that is, $\sup\left\{\left|u'\right|\right\}$. and also on the parameters characterizing the physical and electrical properties of the material constituting the membrane of the MEMS device. As reported in (Di Barba, Fattorusso, & Versaci, 2017), this condition is:

$$1 + H^6 < \left(4\alpha L_1\right)^{-1} H\theta\overline{\lambda}^2 \tag{20}$$

which turns out to be valid for any value of H, and therefore also for its upper extremity, $sup\{H\}$. It is worth observing that in (Di Barba, Fattorusso, & Versaci, 2017) inequality **(20)** was numerically studied by standard Newtonian iterative techniques proving that, in conditions of existence and uniqueness of the solution,

$$\sup\left\{H\right\} = \sup\left\{\left|u'\left(x\right)\right|\right\} = 99\,.$$

This value corresponds, in fact, 88.92° (dimensionless conditions). This value, however, is considered high from an experimental point of view. This is caused by the fact that, to obtain inequality **(20),** in (Di Barba, Fattorusso, & Versaci, 2017), a large number of increases were used which produced the above-mentioned value (99° - dimensionless). Value certainly high but correct from an analytical point of view.

A NEW CONDITION OF UNIQUENESS BASED ON ELECTROMECHANICAL PROPERTIES OF THE MATERIAL CONSTITUTING THE MEMBRANE

The proof of the uniqueness of the solution of the problem under study **(16)** proposed in (Di Barba, Fattorusso, & Versaci, 2017) exploited the joint use of the Gronwall Lemma and Poincaré inequality. The condition obtained, although mathematically very simple, was not dependent on the electro-mechanical properties of the material constituting the membrane of the device. Therefore, it appears evident to undertake a new demonstrative technique that highlights this functional dependence, which is particularly useful for industrial applications in Mechatronics. In other words, this new condition of uniqueness presented here, as a function of the electro-mechanical properties of the membrane will be particularly useful for establishing the intended use of the MEMS device. Then, the theorem presented in this section of the chapter formalizes the uniqueness of the solution to the problem (16). The technique used here for the demonstration is by contradiction: we assume two different solutions for (16), both belonging to *P*, labelled by $u_1\left(x\right)$ and $u_2\left(x\right)$ to then reach an absurd condition. With this goal in mind, we need to introduce two particular lemmas useful for obtaining the condition of uniqueness of the solution for (16). For both Lemmas, we present just a sketch of the proof.

Lemma 1

$\forall u_1$ and u_2 both belonging to *P*, the following inequality yields:

$$\left|\left(1+\left(u_2'\left(x\right)\right)^2\right)^3-\left(1+\left(u_1'\left(x\right)\right)^2\right)^3\right| \leq 24H^5\left|u_2'\left(x\right)-u_1'\left(x\right)\right|. \tag{21}$$

Proof. As we defined H, we observe that $H > 1$ and, $\forall x \in \Omega = \left[-L_1, +L_1\right]$, we can write:

$$\left|\left(1+\left(u_2'(x)\right)^2\right)^3-\left(1+\left(u_1'(x)\right)^2\right)^3\right|==\left|\left[\left(u_2'(x)\right)^2-\left(u_1'(x)\right)^2\right]\left[\begin{array}{l}\left[1+\left(u_1'(x)\right)^2\right)^2\\+\left(1+\left(u_2'(x)\right)^2\right)\left(1+\left(u_1'(x)\right)^2\right)\\+\left(1+\left(u_2'(x)\right)^2\right)^2\end{array}\right]\right|\leq$$

$$\leq\left|u_2'(x)-u_1'(x)\right|2H\mid\left(1+H^2\right)^2+\left(1+H^2\right)\left(1+H^2\right)+\left(1+H^2\right)^2\ )=$$
$$=\left|u_2'(x)-u_1'(x)\right|\left(6H+6H^5+12H^3\right)\leq 24H^5\left|u_2'(x)-u_1'(x)\right|. \tag{22}$$

In addition, the following Lemma yields:

Lemma 2

Being $\alpha=1-d^*<1$, $\forall u_1,u_2\in P$ and $\forall x\in\Omega=\left[-L_1,+L_1\right]$, it makes sense to write:

$$\left|\left(1+\left(u_2'(x)\right)^2\right)^3\left(\alpha-u_2(x)\right)^2-\left(1+\left(u_1'(x)\right)^2\right)^3\left(\alpha-u_1(x)\right)^2\right|\leq 216H^5\left|u_2'(x)-u_1'(x)\right|$$
$$+24\left(1+H^6\right)\left|u_2(x)-u_1(x)\right|. \tag{23}$$

Proof. In fact, we can write:

$$\begin{aligned}
&\left|\left(1+\left(u_2'(x)\right)^2\right)^3\left(1-d^*-u_2(x)\right)^2-\left(1+\left(u_1'(x)\right)^2\right)^3\left(1-d^*-u_1(x)\right)^2\right|= \\
&\left|\begin{array}{l}\left(1+\left(u_2'(x)\right)^2\right)^3+d^*\left(1+\left(u_2'(x)\right)^2\right)^3-2d^*\left(1+\left(u_1'(x)\right)^2\right)^3-2u_2(x)\left(1+\left(u_2'(x)\right)^2\right)^3 \\
+2u_2(x)d^*\left(1+\left(u_2'(x)\right)^2\right)^3-\left(1+\left(u_1'(x)\right)^2\right)^3-d^*\left(1+\left(u_1'(x)\right)^2\right)-u_1^2(x)\left(1+\left(u_1'(x)\right)^2\right)^3 \\
+2d^*\left(1+\left(u_1'(x)\right)^2\right)^3+2u_1(x)\left(1+\left(u_1'(x)\right)^2\right)^3-2u_1(x)d^*\left(1+\left(u_1'(x)\right)^2\right)^3\end{array}\right| \\
&\leq\left|\left(1+\left(u_2'(x)\right)^2\right)^3-\left(1+\left(u_1'(x)\right)^2\right)^3\right|+\left|u_2^2(x)\right|\left|(1+\left(u_2'(x)\right)^2)^3-\left(1+\left(u_1'(x)\right)^2\right)^3\right| \\
&+\left|u_2^2(x)-u_1^2(x)\right|\left|\left(1+\left(u_1'(x)\right)^2\right)^3+\left(d^*+2d^*\right)\right|\left(1++\left(u_2'(x)\right)^2\right)^3-\left(1+\left(u_1'(x)\right)^2\right)^3\right| \\
&+\left(d^*+2d^*\right)\left|\left(1+\left(u_2'(x)\right)^2\right)^3-\left(1+\left(u_1'(x)\right)^2\right)^3\right| \\
&+\left(2+2d^*\right)\mid u_2(x)\left(1+\left(u_2'(x)\right)^2\right)^3-u_1(x)\left(1+\left(u_1'(x)\right)^2\right)^3 \\
&+u_2(x)\left(1+\left(u_1'(x)\right)^2\right)^3-u_1(x)\left(1+\left(u_1'(x)\right)^2\right)^3\mid \\
&\leq 216H^5\left|u_2'(x)-u_1'(x)\right|+24\left(1+H^6\right)\left|u_2(x)-u_1(x)\right|
\end{aligned} \tag{24}$$

Exploiting both Lemma 1 and Lemma 2, we are now able to achieve a condition ensuring the uniqueness of the solution for problem **(16)** depending on the electromechanical properties of the material constituting the membrane. Formally, the following theorem for uniqueness holds.

Theorem 1

If the inequality

$$1+H^6<\theta\lambda^2\left(24L_1\left(1+L_1\right)\right)^{-1} \tag{25}$$

holds, then the solution for problem (16) is unique.

Proof. Inequality (25) is demonstrated by contradiction. In other words, we suppose two different solutions for the problem (16), u_1 and u_2, both belonging to *P*. Then, introducing the operator *T*, as above defined, we can consider equations $u_1=T\left(u_1\right)$ and $u_2=T\left(u_2\right)$. Then, taking into account (19) and $\mu=\dfrac{\lambda}{\alpha-u(x)}$, it makes sense to write $u_1(x)$ and $u_2(x)$ as in (26) and (27):

$$u_1(x) = \int_{-L_1}^{+L_1} (\theta\lambda^2)^{-1} G(x,s)\left(1 + \left(u_1'(s)^2\right)^3\right)(\alpha - u_1(s))^2\, ds \tag{26}$$

$$u_2(x) = \int_{-L_1}^{+L_1} (\theta\lambda^2)^{-1} G(x,s)\left(1 + \left(u_2'(s)^2\right)^3\right)(\alpha - u_2(s))^2\, ds \tag{27}$$

from which, it is easy to obtain $u_1'(x)$ and $u_2'(x)$ as follows (Bailey, Shampine, & Waltman, 1968):

$$u_1'(x) = \int_{-L_1}^{+L_1} (\theta\lambda^2)^{-1} G_x(x,s)\left(1 + \left(u_1'(s)^2\right)^3\right)(\alpha - u_1(s))^2\, ds \tag{28}$$

$$u_2'(x) = \int_{-L_1}^{+L_1} (\theta\lambda^2)^{-1} G_x(x,s)\left(1 + \left(u_2'(s)^2\right)^3\right)(\alpha - u_2(s))^2\, ds. \tag{29}$$

In (28) and (29) we indicated by $G_x(x,s)$ the derivative of $G(x,s)$ with respect to *x*. Therefore, in the functional space $C^1([-L_1.+L_1])$, we must consider $u_1(x) - u_2(x)_{C^1([-L_1.+L_1])}$ as follows (Bailey, Shampine, & Waltman, 1968):

$$u_1(x) - u_2(x)_{C^1([-L_1.+L_1])} = \sup_{x\in[-L_1,+L_1]} |u_1(x) - u_2(x)| + \sup_{x\in[-L_1,+L_1]} |u_1'(x) - u_2'(x)| \tag{30}$$

Then, we can write the following chain of inequalities:

$$T\left(u_1(x)\right)-T\left(u_2(x)\right)=\left(\theta\lambda^2\right)^{-1}\sup_{x\in[-L_1,+L_1]}\times\left|\begin{array}{l}\int_{-L_1}^{+L_1}G(x,s)\left[\left(1+\left(u_1'(s)\right)^2\right)^3\right]\left(\alpha-u_1(s)\right)^2ds\\-\int_{-L_1}^{+L_1}G(x,s)\left[\left(1+\left(u_2'(s)\right)^2\right)^3\right]\left(\alpha-u_2(s)\right)^2ds\end{array}\right|$$

$$+\left(\theta\lambda^2\right)^{-1}\sup_{x\in[-L_1,L_1]}\left|\begin{array}{l}\int_{-L_1}^{+L_1}G_x(x,s)\left(\left(1+\left(u_1(s)\right)^2\right)^3\left(\alpha-u_1(s)\right)^2ds\\-\int_{-L_1}^{+L_1}G\_x(x,s)\left(\left(1+\left(u_2'(s)\right)^2\right)^3\left(\alpha-u_2(s)\right)^2ds\end{array}\right| \tag{31}$$

$$\leq 0.5\left(\theta\lambda^2\right)^{-1}\left(1+L_1\right)\sup_{x\in[-L_1,+L_1]}\left|\int_{-L_1}^{+L_1}\left[\begin{array}{l}-\left(1+\left(u_1'(s)\right)^2\right)^3\left(\alpha-u_1(s)\right)^2\\+\left(\left(1+\left(u_2'(s)\right)^2\right)^3\right)\left(\alpha-u_2(s)\right)\end{array}\right]ds\right|.$$

Considering (24), we can write:

$$\begin{aligned}&T\left(u_1\right)-T\left(u_2\right)_{C^1([-L_1,+L_1])}\leq 0.5\left(\theta\lambda^2\right)^{-1}\left(1+L_1\right)\sup_{x\in[-L_1,+L_1]}\\&\times\left|\int_{-L_1}^{+L_1}\left(216H^5\mid u_2'-u_1'+24\left(1+H^6\right)\left|u_2-u_1\right|\right)ds\right|\\&=L_1\left(\theta\lambda^2\right)^{-1}\left(1+L_1\right)\left(216H^5L_1\right)\sup_{s\in[-L_1,+L_1]}\left|u_2'(s)-u_1'(s)\right|\\&+24L_1\left(\theta\lambda^2\right)^{-1}\left(1+L_1\right)\left(\left(1+H^6\right)L_1\right)\sup_{s\in[-L_1,+L_1]}\left|u_2(s)-u_1(s)\right|.\end{aligned} \tag{32}$$

Since $u_1=T\left(u_1\right)$ and $u_2=T\left(u_2\right)$, taking into account both (30) and (32), we achieve the following system:

$$\begin{cases}216H^5<\theta\lambda^2L_1\left(1+L_1\right)\\24\left(1+H^6\right)<\theta\lambda^2L_1\left(1+L_1\right).\end{cases} \tag{33}$$

From system (33), we firstly focus our attention to its first inequality and, taking into account that H^6 can be write as $H^5H,$ it makes sense to write

$$H^6=H^5H<(216\left(L_1\left(1+L_1\right)\right)^{-1}H.$$

So that the following inequality holds:

$$1+H^{6}<1+\left(216L_{1}\left(1+L_{1}\right)\right)^{-1}\theta\lambda^{2}H. \tag{34}$$

Then, considering again system (33), it can be easily to write as follows:

$$\begin{cases} 1+H^{6}<1+\left(216L_{1}\left(1+L_{1}\right)\right)^{-1}\theta\lambda^{2}H \\ 1+H^{6}<\left(24L_{1}\left(1+L_{1}\right)\right)^{-1}\theta\lambda^{2} \end{cases} \tag{35}$$

from which, observing that $1+H^{6}$ is the left side of both inequality. In addition, the following Remark allows us to solve the system (35).

Remark 1

$$\left(24L_{1}\left(1+L_{1}\right)\right)^{-1}\theta\lambda^{2}1+\left(216L_{1}\left(1+L_{1}\right)\right)^{-1}\theta\lambda^{2}H \tag{36}$$

Proof. Let us suppose, by contradiction, the following inequality:

$$\left(24L_{1}\left(1+L_{1}\right)\right)^{-1}\theta\lambda^{2}+\left(216L_{1}\left(1+L_{1}\right)\right)^{-1}\theta\lambda^{2}H \tag{37}$$

from which, we achieve:

$$H<9-216L_{1}\left(\theta\lambda^{2}\right)^{-1}\left(1+L_{1}\right)<0. \tag{38}$$

Inequality (39) shows that $H=\sup\left|u'\left(x\right)\right|<0$. This condition, however, represents a condition physically impossible; so that, inequality (36) holds.

Then, taking into account the *Remark 1,* we can say that system (33) is rightly considered as equivalent to the inequality (36). It is worth observing that in inequality (20) (condition ensuring the existence of the solution for the problem under study) the inertia $\bar{\lambda}^{2}$ if the membrane compares, while in condition ensuring the uniqueness of the solution, it does not appear. This last observation, from an experimental point of view, is very interesting as it confirms, firstly, the fact that, if the external voltage V is applied, the membrane starts to move if only if the inertia $\bar{\lambda}^{2}$ of the membrane is overcomes. Then, $\bar{\lambda}^{2}$ must influences the existence of the solution for (16) ($\bar{\lambda}^{2}$ appears in inequality (20)). Secondly, once V wins the inertia of the membrane, it is not necessary that $\bar{\lambda}^{2}$ appears in the condition ensuring the uniqueness of the solution. The following step, then, concerns the search of a unique condition ensuring both existence and uniqueness of the solution for the problem (16).

A CONDITION ENSURING BOTH EXISTENCE AND UNIQUENESS OF THE SOLUTION FOR THE PROBLEM UNDER STUDY

As described above, condition ensuring the existence of the solution was the inequality (20). On the other hand, to satisfy inequality (25), it is equivalent to ensure the uniqueness of the solution. To achieve both existence and uniqueness of the solution for the problem (16), it is sufficient to solve the following system:

$$\begin{cases} 1+H^6 < \left(4\alpha L_1\right)^{-1} H\theta\overline{\lambda}^2 \\ 1+H^6 < \theta\lambda^2\left(24L_1\left(1+L_1\right)\right)^{-1}. \end{cases} \tag{39}$$

As observed in system (35), $1+H^6$ is the left side of each inequality in system (39). Thus, to gain the solution of (39), we exploit the statement of the following Remark.

Remark 2

The following inequality yields:

$$\theta\lambda^2\left(24L_1\left(1+L_1\right)\right)^{-1} < \left(4\alpha L_1\right)^{-1} H\theta\overline{\lambda}^2 \tag{40}$$

Proof. As for the Remark 1, let us suppose, absurdly, that the following condition is true

$$\theta\lambda^2\left(24L_1\left(1+L_1\right)\right)^{-1} > \left(4\alpha L_1\right)^{-1} H\theta\overline{\lambda}^2 \tag{41}$$

achieving, for $\boldsymbol{L}_1 = 0.5$, the following inequality

$$H < 9^{-1}\alpha\left(\overline{\lambda}^2\right)^{-1} \tag{42}$$

that, exploiting the usual scaling factors and the inequality (42), we carry out:

$$H = x^{-1}z = 2z'L_1\left(hx'\right)^{-1} < 9^{-1}\alpha\left(\overline{\lambda}^2\right)^{-1} \tag{43}$$

Finally, indicating by H' the dimensionless value of H e, if we denote the dimensionless value of H by H', the following condition is obtained:

$$H' < \left(18L_1\overline{\lambda}^2\right)^{-1} h\alpha. \tag{44}$$

In other words, according to (44), the values of H' should be lower than $\left(18L_1\bar{\lambda}^2\right)^{-1} h\alpha$ which represents a quantity too small highlighting that condition (44) is physically incompatible. Then, inequality (40) is true.

Therefore, according to Remark 2, we can infer that system (39) is equivalent to the unique following inequality:

$$1 + H^6 < \theta\lambda^2 \left(24L_1\left(1 + L_1\right)\right)^{-1}. \tag{45}$$

In other words, to ensure both existence and uniqueness for the problem (16), *H* must satisfy condition (45). Moreover, according condition (45), to satisfy the condition ensuring the uniqueness of the solution, ensures also its existence; so that, both existence and uniqueness of the solution for (16) depend on the electro-mechanical properties of the material constituting the membrane. In addition, ±. does not appear in (45); so, both existence and uniqueness of the solution for problem (16) does not depend on d^*. It is worth noting that in (Di Barba, Fattorusso, & Versaci, 2017) the uniqueness of the solution for (16) was always ensured, while its existence had to be evaluated. Then, from an industrial point of view, condition (45) is interesting because links each electrostatic-elastic membrane MEMS devices to a specific intended use. As before mentioned, it is not possible to get a solution directly from (16). In other words, it is not possible to recover the membrane profile starting from model (16). Then, we need to exploit suitable numerical techniques able to solve the problem at hand. In particular, since model (16) is an BVP, the numerical shooting approach is one of the best candidates because, besides the fact that transforms a BVP to an equivalent IVP, convergence management does not meet many requirements. In the case of differential problems, the model then requires imposing conditions on the edge, i.e. conditions imposed in several points, rather than initial conditions.

NUMERICAL TECHNIQUES FOR SOLVING BVPS

Many application problems give rise to mathematical models in which the quantities involved depend on the position rather than on time. For our purposes, we observe that model (16) can be considered as a special case of the following general differential model:

$$y''\left(x\right) = f\left(x, u\left(x\right), u'\left(x\right)\right), a < x < b \tag{46}$$

with the following boundary conditions:

$$u\left(a\right) = \alpha, u\left(b\right) = \beta. \tag{47}$$

Unlike the IVPs (Cauchy problems) the regularity of the data may not be sufficient for the existence and uniqueness of the solution of a BVP. The theory is, in reality, more fragmentary, in the sense that it is various but generally unnecessary conditions are known for the existence of the solution. The lack of an adequate general theory, if on the one hand it can prevent a complete justification of the relative numerical approximation, on the other hand it may be precisely a reason for the development and application of numerical methods. The statement is justified considering the possible suggestions, even theoretical ones, that can come from adequate numerical experimentation. Numerical methods for BVPs can be classified, in the one-dimensional case, according to the following criterion:

1. Methods based on more general ideas, valid regardless of dimension;
2. Methods specific to the one-dimensional case or that are generalized with difficulty to multidimensional cases.

The first category of methods is based on a local approximation of the solution, in a set of points (finite differences) or through an approximation with piecewise polynomials on subintervals (finite elements). In the second category there are, for example, methods based on the techniques of initial values in which the solution is approximated by solutions of IVPs. Due to their characteristics, these methods are also known as shooting methods. In both of cases, the problem is reduced to the resolution of a system of equations, in general, non-linear. In the finite element method (FEM) the problem is formulated in variational terms such as the search for the minimum of a functional (Rayleigh-Ritz method) or, more generally, as the resolution of a variational equation (Galerkin method). As an alternative way to the FEM approach, we have, for example, the spectral methods in which the solution is approximated by polynomials, algebraic or trigonometric, over the whole interval. The idea that underlies the methods of the first category is, essentially, independent of the dimension of the problem; however, this does not mean that the difficulties are the same. The passage from the one-dimensional case to the multidimensional case represents, in reality, a qualitative leap both from the technical and theoretical point of view. However, since the problem under study (16) is a classic $1D$ model, we will focus on specific methods for one-dimensional problems and in particular on the shooting method because it has fewer problems regarding convergence. Anyway, for completeness, some detailed information on some general methods by now consolidated in scientific literature are quickly presented.

Finite Difference Methods

It is a strategy used to solve numerically differential equations. The procedure is based on the approximation of derivatives with finite difference equations. It is mainly used for ordinary differential equations, even if the method is used as a time progression scheme for partial derivative problems (PDE). The finite difference method allows to transform a differential problem into an approximate algebraic one. Limiting the problem to the case of an unknown function to a single variable, this function is represented by the set of points of the domain. These values represent the unknowns of the approximate algebraic problem. The method can be defined as simple and even intuitive. It is in fact a question of replacing the derivative, defined as the limit of an incremental relationship, with the incremental relationship itself. Thus, in the differential equation the derivatives are replaced by the incremental ratios calculated through the values of the unknown function in some points of the domain. To define in a rational way the approximation of the derivative of the function f in the point of the domain x, we determine in an approximate way the value of $f(x_{i+1})$ developing the function $f(x)$ by Taylor series, starting from point x_i and stopping at the linear term:

$$f\left(x_{i+1}\right)=f\left(x_i\right)+\left.\frac{\partial f}{\partial x}\right|_{x=x_i}\Delta x \tag{48}$$

From this relation it is possible to obtain the approximate formula of the derivative $\left.\frac{\partial f}{\partial x}\right|_{x=x_i}$ that provides the so-called forwards finite difference. Similarly, the value of the backward numerical derivative can be determined simply

$$f\left(x_{i-1}\right)=f\left(x_i\right)+\left.\frac{\partial f}{\partial x}\right|_{x=x_i}\left(-\Delta x\right) \tag{49}$$

To improve the approximation of the method, it is possible to carry out the formula to the centered differences by developing the function *f(x)* by Taylor series, starting from x_i, and stopping at the quadratic term, from which it is obtained:

$$\left.\frac{\partial f}{\partial x}\right|_{x=x_i}=\frac{f\left(x_{i+1}\right)-f\left(x_{i-1}\right)}{2\Delta x};\left.\frac{\partial^2 f}{\partial x^2}\right|_{x=x_i}=\frac{f\left(x_{i+1}\right)+f\left(x_{i-1}\right)-2f\left(x_i\right)}{\Delta x^2} \tag{50}$$

The goodness of the discrete solution obtained depends mainly on two factors: a) the number of k information used to construct the approximation formula of the derivative; b) the step of discretization Δx. It is clear that the order of accuracy of the truncation error (and, therefore, also of the convergence order) depends on k. The price to pay is that we have to solve a more expensive computational system, so the matrix to reverse is less sparse. On the other hand, reducing Δx does not improve the order of accuracy, but reduces the error.

Finite Element Methods (FEM)

The finite element method (FEM) is a numerical technique aimed at finding approximate solutions of problems described by differential equations (ODE & PDE) by reducing them to a system of algebraic equations. Although it competes in some limited areas with other numerical strategies (for example, the finite difference method and the spectral method), FEM maintains a dominant position in the panorama of numerical approximation techniques and represents the kernel of most analysis codes. Generally, FEM lends very well to solving PDEs whose domain has a complex contour, when the domain is variable, when the accuracy required for the solution is not homogeneous on the domain and when the solution is lacking sufficient regularity. This method applies to physical bodies that can be subdivided into a certain number of elements of defined shape and contained dimensions. In the continuum, each individual element is considered a field of numerical integration of homogeneous characteristics. The main feature of the method is the discretization through the creation of a grid (mesh) composed of primitives (finite elements) of coded shape (triangles and quadrilaterals for *2D* domain, tetrahedrons and hexahedra for *3D* domains). On each element characterized by this elementary form, the solution of the problem is assumed to be expressed by the linear combination of functions called basic functions or shape functions. It should be noted that sometimes the function is approximated, and not necessarily, the exact values of the function will be those calculated in the points, but the values that will provide the least error overall solution (for example, polynomial functions, so the overall solution of the problem is approximated with a piecewise polynomial function). The number of coefficients that identifies the solution on each element is therefore linked to the degree of the chosen polynomial. This, in turn, governs the accuracy of the numerical solution found. FEM belongs to the class of the Galerkin methods, whose starting point is the so-called weak formulation of a differential problem. This formulation, based on the concept of derivative in the sense of distributions, of Lebesgue integral and weighted average (through appropriate test functions), has the great merit of requiring the solution to have realistic regularity characteristics for (almost) all-engineering problems. Galerkin-type methods are based on the idea of approximating the solution of the problem written in weak form by linear combination of elementary functions (the shape functions). The coefficients of this linear combination (also called degrees of freedom) become the unknowns of the algebraic problem obtained by the discretization. Finite elements are distinguished by the choice of piecewise basic polynomial functions. Other Galerkin-type methods such as spectral methods use different basic functions.

Spectral Methods

In a schematic way, spectral methods look for the solution of a differential equation in terms of a series (truncated) of assigned regular functions, for example trigonometric functions, or orthogonal polynomials. They have had a particular development thanks to the fast calculation of Fourier transforms (FFT). Significant applications of the method have occurred, in particular, in the numerical study of the problems of fluid dynamics, in the study of turbulence, in meteorological, problems and in general in those problems where good accuracy is required at low cost. The peculiar characteristic of these methods is, in fact, the highest accuracy in the presence of regular solutions (with respect, for example, to finite difference method or FEM).

THE SHOOTING APPROACH AND AN INTERESTING RESULT OF CONVERGENCE

We introduce the idea of the method, considering the BVP (57) and (47). Formally, we associate the problem (57) and (47) the set of BVPs defined as follows:

$$\begin{array}{c} u''\left(x\right)=f\left(x,u\left(x\right),u'\left(x\right)\right) \\ u\left(a\right)=\alpha, u'\left(a\right)=s \end{array} \Rightarrow s\rightarrow u\left(x;s\right) \tag{51}$$

where s is a parameter to be determined. We immediately observe that problem (51) it is not the only way to associate to the problem (46) and (47) some IVPs. For example, instead of starting from a, we could start from b but there are other possibilities, theoretically equivalent, but not numerically. Assuming the IVP (51) have global solution (that is, on $[a,b]$) for a set of values *s*, we can pose the problem of searching for a "good" value *s*, by solving the following problem of finding a zero of a function:

$$u(b;s) = \beta \Leftrightarrow F(s) = u(b;s) - \beta = 0 \tag{52}$$

If (52) has a solution, then, we have shown that the problem (46) and (47) also has a solution. Therefore, the procedure now indicated can be interesting also from a theoretical point of view. Returning, however, to the numerical aspect, it is a matter of solving a non-linear equation. We underline the fact that the function, of which the zero is sought, is not given in explicit form, but that is the result of the numerical resolution of a problem to the initial values. In this way, we encounter an interesting numerical solution (and common in real problems). The problem to be solved decomposes into sub-problems, which we hope will be simpler (in our case: solving a non-linear equation and solving a problem with initial values). In this situation, for a good numerical calculation, it is necessary not only to choose the most suitable algorithm for the single problem, but to coordinate the algorithms used for the different sub-problems. To solve equation (60) we can use, for example, the Newton method, bisection or secants. In this regard, we observe that if the function *f* is Lipschitzian, then the function $s \to u(x;s)$ is continuous. To apply the method, it is sufficient to find two values of *s* in which *F(s)* has the opposite sign and proceed with bisection (middle point between the two values of *s* obtained). In applying the Newton method, instead, the problem of calculating the derivative of *F(s)* arises. Therefore, we have:

$$F'(s) = \frac{\partial}{\partial s} u(b;s) \tag{53}$$

Supposing *f* sufficiently regular and placing, for brevity:

$$v(x;s) = \frac{\partial}{\partial s} u(b:s) \tag{54}$$

we have, from (51), the following system (the so-called variational system):

$$\begin{gathered} v'' = f_u(x, u(x), u'(x)v + f_u^{'}(x, u(x), u'(x)v' \\ v(a) = 0, v'(a) = 1 \end{gathered} \tag{55}$$

Proceeding in this way, the derivative *F'(s)* is calculated by solving a problem with the auxiliary initial values. We observe that this problem is linear and that the solution $v(x;s)$ provides, for each value of *s*, the sensitivity of the solution *u* with respect to the value of the parameter *s*. The term "shooting" derives from the fact that, positioning itself on one edge of the domain, a set of trajectories are "shot" out in different directions; the correct trajectory is that which intercepts the other extreme of the domain (Kress, 1998). In addition, from a computational complexity viewpoint, the Newton method requires the computation of the Jacobian that usually can be evaluated by means the finite difference method. In addition, after our experiments, more restrictive convergence conditions require the exclusion of the Newton method from the shortlist. Of course, as an alternative to the Newton method, various approaches can be used. An interesting detail may present the secant method, which does not require the explicit calculation of the derivative. We conclude by emphasizing the fact that for a successful numerical calculation of the solution of a BVP by means of the shooting method, it is necessary to appropriately control the various types of errors present in the calculation. We recall that they are: the error due to the method of solving the non-linear equation; the error in the numerical calculation of the solution of the IVPs and, finally, the rounding errors. The control of these errors is essential for the solution found to have meaning and the cost of the calculation to be optimal. In the application of the shooting method there is a further aspect, which we have not yet analyzed, but which may be important. We have already mentioned that there can be various ways of associating a BVP with a family of IVPs. It is possible that the solution obtained with the shooting procedure is extremely sensitive to the parameter *s*. It follows that, in order to have the solution of the BVP, it is necessary to calculate the parameter *s* with an excessive precision. This can occur whenever there is a rapidly growing solution. A possible remedy for these situations is the application of the method on smaller intervals. The use of this idea leads to the so-called multiple shooting method. However, for physical reasons, the problem related to the MEMS under study does not fall into this category. Therefore, we invited interested parties to consult specific texts on the subject (Kress, 1998). To understand how the shooting procedure works on our model (16), let us consider again the following second-order ordinary differential model:

$$y''\left(x\right)=f\left(x,u\left(x\right),u'\left(x\right)\right),u\left(x_0\right)=u_0,u\left(x_1\right)=u_1 \tag{56}$$

which represents a BVP. Denoting by $u\left(x,a\right)$ the solution of the following IVP, we can write:

$$u''\left(x\right)=f\left(x,u\left(x\right),u'\left(x\right)\right),u\left(x_0\right)=u_0,u'\left(x_0\right)=a. \tag{57}$$

If we define *F(a)* a function such that $F\left(a\right)=u\left(x_1,a\right)-u_1$, and *a* is its root, then $u\left(x,a\right)$ of the corresponding IVP is a solution of the BVP. Regarding our model (16) it was solved numerically within both of the works of (Versaci, Angiulli, Fattorusso, & Jannelli, 2019) by the shooting procedure as above described. In particular, (16) was firstly rewritten as the following first-order differential systems:

$$\begin{cases} u_1'\left(x\right)=u_2\left(x\right) \\ u_2'\left(x\right)=\left(-\theta\lambda^2\right)^{-1}\left(1+u_2^2\left(x\right)\right)^3\left(\alpha_1-u_1\left(x\right)\right)^2 \end{cases} \tag{58}$$

which represent a BVP where $u_1\left(x\right)=u\left(x\right)$ and $u_2\left(x\right)=u_1'\left(x\right)=u'\left(x\right)$. Then, it can be rewritten as an equivalent IVP placing $u_1(L_1)$ at $x=L_1$ and $u_2(-L_1)=\eta, \eta\in\mathbb{R}$. as initial condition. In this way, implicitly, the nonlinear equation $F\left(\eta\right)=u_1\left(L_1,\eta\right)=0$ has been defined. Finally, the numerical solution of the system has been achieved starting from two values for η, η_0, and η_1. After, by means of secant method routine implemented in MatLab® R2017a code and exploiting the ODE23 solver (Intel Core 5 CPU 1.6 GHz), the procedure generated a sequence η_k, with $k = 2,3,\ldots$, until $\eta_k \to \eta$, if $k \to +\infty$. It is worth noting that it was exploited the secant method instead of the Newton method in order to get less restrictive convergence conditions (for details, see (Fazio & Jannelli, 2017). It is worth observing that to solve (16) numerically, we can apply a relaxation procedure based on Keller-Box scheme very useful for problems in which solutions require regularities (Fazio & Jannelli, 2014). However, applying relaxation procedures, often, convergence is not always ensured, so that it is necessary to exploit shooting techniques. As reported in (Di Barba, Fattorusso, & Versaci, 2017), the range of values ensuring convergence of the numerical shooting procedure is:

$$\theta\lambda^2 \in \left[0.63, +\infty\right) \tag{59}$$

Given the problem under study in (Di Barba, Fattorusso, & Versaci, 2017) and the model presented in this chapter are the same, the range of convergence of the numerical procedure here exploited is (59). In other words, any values of $\theta\lambda^2$ such that $0 \leq \theta\lambda^2 < 0.63$ do not ensure convergence of the numerical procedure. Certainly, if $\theta\lambda^2 \geq 0.63$ the procedure converges but, we need to know if eventual ghost solutions (that is solution achieved numerically but not satisfying analytical condition (45)) occurs in order to eliminate them. The shooting method, of course, is not the only procedure for treating BVPs $1D$. For example, a possible candidate is the relaxation method exploiting the Keller-Box Scheme (Kress, 1998) . In particular, introducing a mesh of points x_j, $j = 0,1,\ldots,J$, from $-L_1$ to $+L_1$ the numerical approximation of the solution $u(x)$ the Keller-Box scheme is structured as follows (Fazio & Jannelli, 2014, 2017):

$$u\left(x_j\right) - u\left(x_{j-1}\right) = G\left(x_{j-1/2}\frac{u\left(x_j\right) - u\left(x_{j-1}\right)}{2}\right), \text{ J=1,2,...,J} \tag{60}$$

in which $x_{j-1/2}$ in (61) is equal to $\left(x_j + x_{j-1}\right)/2$. Assuming than both $u(x)$ and $G(x,u)$ are sufficiently smooth, the approximated solution is computed by (60) using the Newton method. Finally, a termination criterion must be exploited. However, the problem here concerns the convergence of the Newton method because it can converge locally. Thus, in this chapter, we focus our attention on the shooting procedure.

ACCEPTABLE SOLUTIONS AND GHOST SOLUTIONS

In order to highlight the possible ghost solutions, let us consider inequality (45), which represent the condition ensuring both existence and uniqueness of the solution for the problem (16). In particular, from (45) we can write:

$$H < \left(\theta\lambda^2\left(24L_1\left(1+L_1\right)\right)^{-1} - 1\right)^{1/6}. \tag{61}$$

In order that (61) has a sense, it is imperative that

$$\theta\lambda^2\left(24L_1\left(1+L_1\right)\right)^{-1} - 1 \geq 0$$

and, imposing $L_1=0.5$ we obtain $\theta\lambda^2 \geq = 18$. Then, for $\theta\lambda^2 \in \left[0{,}0.63\right)$ the convergence of the numerical procedure is not ensured and, for $\theta\lambda^2 \in \left[18,+\infty\right)$ the convergence is ensured recovering the profile of membrane in absence of ghost solution. Finally, for $\theta\lambda^2 \in \left[0.63,18\right)$ the numerical procedure converges and recover the profile the profile of the membrane but it is a ghost solution because it does not satisfy condition (46). In other words, if $\theta\lambda^2 \in \left[0.63,18\right)$, from (61) we obtain $H<0$ which represent an inacceptable physical condition (even if the external electrical potential V is applied, the membrane does not deform towards the upper plate but deforms in the opposite direction (see also Figure 3). Table 2 summarizes the results of convergence above presented and Figure 4 displays some profiles recovered by the exploited numerical procedure highlighting the zones of convergence/non convergence and ghost solutions as well.

Convergence does not ensure Convergence ensured and presence of ghost solutions Convergence ensured and absence of ghost solutions

CRITERION FOR CHOOSING THE INTENDED USE OF AN ELECTROSTATIC-ELASTIC MEMBRANE MEMS

In order to choose the intended use of a MEMS device, we have to achieve the values of the external voltage V ensuring both convergence and absence of ghost solutions. For this purpose, observing that $\lambda^2 = \theta V^2$ (see, Di Barba, Fattorusso, & Versaci 2017 condition (45) can be rewritten as follows:

$$1 + H^6 < \left(24L_1\left(1+L_1\right)\right)^{-1}\theta\lambda^2 = 0.055\theta V^2 \tag{62}$$

from which, it is easy to achieve the following:

$$V > \left(18\theta^{-2}\left(1+H^6\right)\right)^{1/2}. \tag{63}$$

In other words, once the material constituting the membrane of the device has been chosen (that is, the parameter θ), inequality (63) selects the range of values for V admissible to ensure convergence and without ghost solutions. And again, from (63), we can obtain Angiulli, et al. (2018); Di Barba, Fattorusso, and Versaci (2017):

Table 2. Summary of the main results of convergence of the exploited numerical procedure based on shooting approach

$\theta\lambda^2 \in [0, 0.63)$	$\theta\lambda^2 \in [0.63, 18)$	$\theta\lambda^2 \in [18, +\infty)$
Convergence does not ensure	Convergence ensured and presence of ghost solutions	Convergence ensured and absence of ghost solutions

Figure 3. Curve of merit showing parameter H vs. $\theta\lambda^2$: if $\theta\lambda^2 \geq 0.63$ the convergence of the numerical procedure is always assured but the absence of ghost solutions is ensured if only if $\theta\lambda^2 \geq 18$

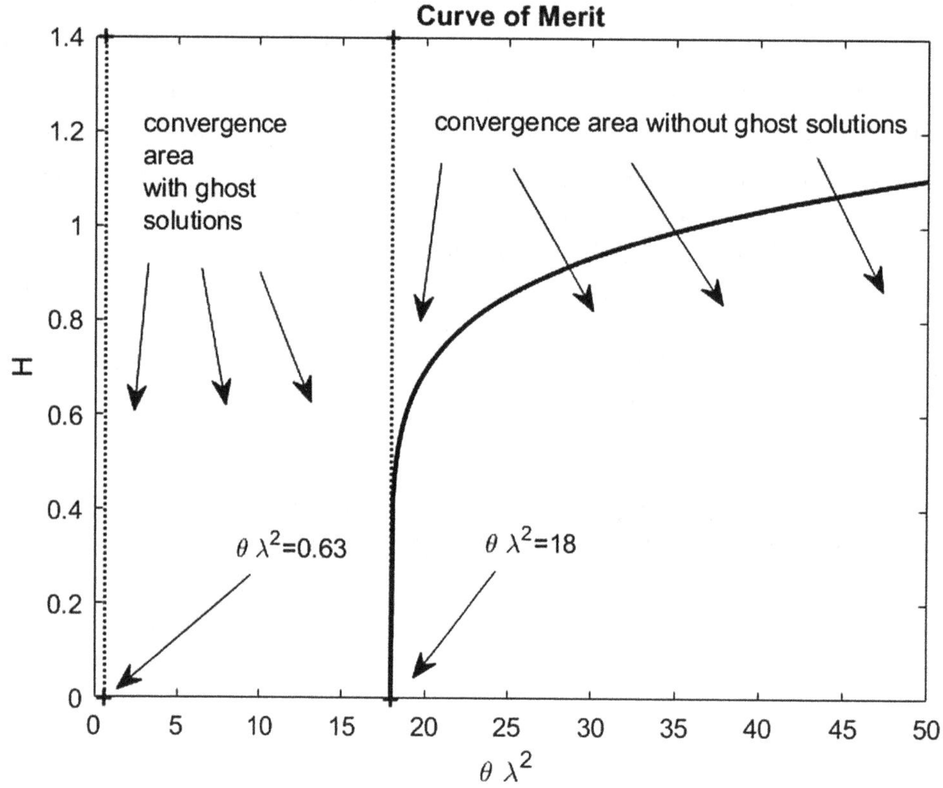

$$H < \left(0.05\theta V^2 - 1\right)^{1/6} \tag{64}$$

which evaluates the range of possible values for H, after to have chosen both the external applied voltage V and the parameter θ (material constituting the membrane). As reported in Angiulli, et al. (2018) as well as Di Barba, Fattorusso, and Versaci (2017) the following condition:

$$\theta E^2 = \left(1 - u(x)\right)^{-2} \lambda^2 = 0.5h^{-3}\left(1 - u(x)\right)^{-2} \in_0 V^2 \left(2L\right)^2 = \left(1 - u(x)\right)^{-2} \beta V^2 \tag{65}$$

Figure 4. Profile recovered by the numerical procedure

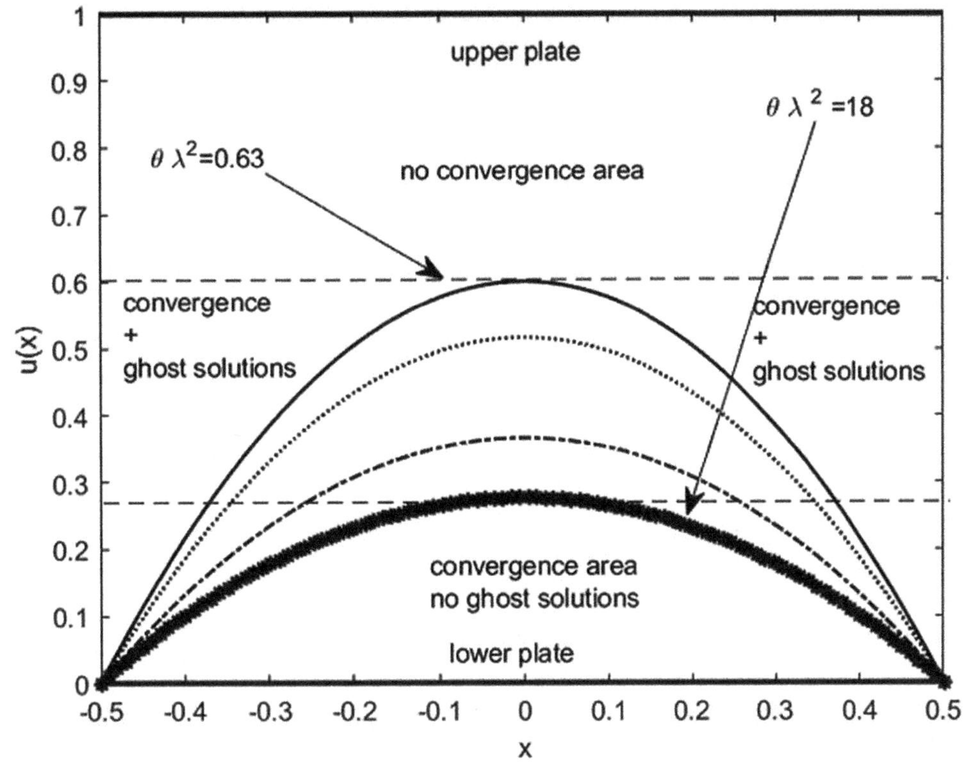

can be considered still valid. Then, we can multiply (66) by the factor λ^2 from which, taking into account that $\left(1-u\left(x\right)\right)^2 < 1$ we achieve the following relations:

$$\theta\lambda^2 E^2 = \left(1-u\left(x\right)\right)^{-2}\beta_1 V^2\lambda^2 = \left(1-u\left(x\right)\right)^{-2}\beta_1 V^4 \tag{66}$$

$$\theta\lambda^2 = 0.25\sigma^{-2}\left(1-u\left(x\right)\right)^{-2} E^{-2} \in_0^2 V^2 > 0.25\sigma^{-2}\left(\sup\left\{E^2\right\}\right)^{-1} \in_0^2 V^4 \tag{67}$$

where $E^2 < \sup\left\{E^2\right\}$, $\beta_1 = 0.5\sigma^{-1}\in_0$ and $\lambda^2 = \beta_1 V^2$. Then, if the numerical procedure converges and, at the same time, there is no trace of ghost solutions, parameter σ (related too the electro-mechanical properties of the membrane) and the pair *V*, $\sup\left\{E^2\right\}$ (the so-called operational parameters) satisfy the following inequality:

$$\sigma^{-2}\left(\sup\left\{E^2\right\}\right)^{-1} \geq 72. \tag{68}$$

Finally, from inequality (63) and (68) we compute σ and V respectively, obtaining the two following conditions:

$$\begin{cases} \sigma \leq \left(72 \sup\left\{E^2\right\}\right)^{-1/2} \in V^2 \\ V > \left(18\theta^{-2}\left(1+H^6\right)\right)^{1/2} \end{cases} \tag{69}$$

that is, both inequalities (69) are the two conditions to which σ, and the pair V and E must meet to guarantee both convergence of the numerical shooting procedure without ghost solutions. Obviously, conditions (69) can be exploited as a criterion for choosing the intended use of the MEMS device, once the electromechanical properties of the material constituting the membrane have been chosen. Conversely, once the intended use of the device is chosen, by (69) we select the material to make the membrane of the device. It is worth highlighting that conditions (69) represent the link, easily verifiably experimentally, between the mathematical model (16), the operational parameters and the electromechanical parameters of the material constituting the membrane.

MEMBRANE MEMS IN INDUSTRIAL APPLICATIONS

Historically, MEMS devices have been exploited in micro-systems dominated by silicon micro-machines (Jouaneh, 2012) (Phan & et al., 2018) (Dorfmeister, Koessl, Schneider, & Schmid, 2019). However, in the recent years and in a natural manner, these technologies have been also exploited for a large number of industrial applications (and more). This is because MEMS, being electro-mechanical micro-devices, are very useful in the applications in which these devices are imperatively required. Mechatronics, as known, represents the fusion among ICT, mechanics and design for producing different object spanning from robots to home automation requiring a set of technologies requiring a multi-disciplinary approach for creating and exploiting materials, devices and systems whose very reduced dimensions can be found in membrane MEMS devices an element of cohesion. Membrane MEMS devices occupy a privileged place in mechatronics because they represent ideal choices in many industrial applications such as healthcare and automotive. Considered by engineers as cutting-edge technology, membrane MEMS sensors have also been embraced by the industry to improve performance, reduce costs and increase reliability. For example, in automotive industry, the crash sensor for airbag control represents the largest automotive use of inertial MEMS sensors. Industrial MEMS sensors offer high performance and the high quality required for applications in robotics, condition monitoring, industrial vehicle driving and stabilization, construction and industry automation, navigation systems, and equipment used in the field of medicine and aerospace and defense. In parallel, Consumers around the world are demanding a more effective means of diagnosing diseases that will guarantee accuracy in less time. Contrary to the high demand for technological advances, the leading membrane MEMS manufacturers in the global medical applications market focus on the production of a single embedded platform with the ability to efficiently perform multiple diagnostic tests. These factors have contributed greatly to the growth of membrane MEMS in the medical applications market. MEMS technology can be used in biomedical applications such as surgical instruments, drug delivery and biosensors, as well as in diagnostic and in vitro applications. With a record of applications, the MEMS micro-interconnection approach offers a powerful and vital tool for design engineers who face the demand for ever-smaller micro-products to meet healthcare needs.

CONCLUSION

In this chapter, a numerical approach to recover the profile of the membrane $u(x)$ for a *1D* electrostatic-elastic membrane MEMS interesting for Mechatronics, is presented. Starting from a well-known membrane MEMS constituted by a fixed plate and a deformable membrane that, after application of an external electrical voltage *V*, moves towards the fixed plate, we have shown a new physical-mathematical *1D* model exploiting differential geometry concepts. Particularly, starting from the idea that, locally, the electrostatic field **E** on the membrane is orthogonal to the tangent straight line, the new formulation here presented considers $|\mathbf{E}|$ proportional to the curvature *K* of the membrane at the same point *x*. The second-order semi-linear elliptic problem achieved, since it is impossible to obtain solutions explicitly, has been treated in order to get both existence and uniqueness conditions depending on the electro-mechanical properties of the material constituting the membrane of the device. Then, the recovering of the profile of the membrane has been obtained numerically by a shooting procedure transforming the problem under study into an equivalent initial value problem. The proposed procedure, implemented by MatLab® solver for ODE, showed that convergence depended on the product $\theta\lambda^2$ whose factors depend on the operational parameters, $\left(V^*, sup\left\{E^2\right\}\right)$, and the electro-mechanical properties, σ, of the material constituting the membrane of the device. The results obtained, firstly, have highlighted the range of values of $\theta\lambda^2$ that ensures the convergence of the proposed numerical procedure highlighting, complementarily, the values of $\theta\lambda^2$, which do not ensure convergence. Then, exploiting the analytical condition that ensure both the existence and uniqueness of the solution for the problem under study, form the range of values of $\theta\lambda^2$, ensuring convergence the values relating to the ghost solutions have been eliminated. Again, under convergence conditions, all the possible combinations among operational parameters and electro-mechanical properties of the material constituting the membrane of the device have been obtained, thus providing a criterion for choosing the intended use of the MEMS device. The results obtained can be considered encouraging as they are interesting from an industrial point of view wince the criterion for choosing the intended use of the device is easy to implement. However, even I, the proposed *1D* model is justified by observations on the symmetry of the problem, its extension to *3D* cases is certainly desirable in order to realistically cover a wider range of cases.

REFERENCES

Angiulli, G., Jannelli, A., Morabito, F. C., & Versaci, M. (2018). Reconstructing the Membrane Detection of a 1D Electrostatic-Driven MEMS Device by the Shooting Method: Convergence Analysis and Ghost Solutions Identification. *Computational & Applied Mathematics*, *37*(September), 4484–4498. doi:10.100740314-017-0564-4

Bailey, P. B., Shampine, L. F., & Waltman, P. E. (1968). *Nonlinear Two Point Boundary Value Problems*. New York, NY: Academic Press.

Bluman, G. W., & Anco, S. C. (2002). *Symmetry and Integration Methods for Differential Equations* (Vol. 154). New York, NY: Springer-Verlag.

Bolton, W. (2003). *Mechatronics - Electronic Control Systems in Mechanical and Electrical Engineering*. Edinburg Gate Harlow Essex CM20 2JE England: Pearson Prentice Hall .

Calcagno, S., La Foresta, F., & Versaci, M. (2014). Independent component analysis and discrete wavelet transform for artifact removal in biomedical signal processing. *American Journal of Applied Sciences*, *11*(1), 57–68. doi:10.3844/ajassp.2014.57.68

Di Barba, P., Fattorusso, L., & Versaci, M. (2017). Electrostatic Field in Terms of Geometric Curvature in Membrane MEMS Devices. *Communications in Applied and Industrial Mathematics*, *8*(1), 165–184. doi:10.1515/caim-2017-0009

do Carmo, M. P. (2016). *Differential Geometry of Curves and Surfaces*. Englewood, NJ: Prentice-Hall.

Dorfmeister, M., Koessl, B., Schneider, M., & Schmid, U. (2019). A Novel Bi-Stable MEMS Membrane Concept Based on a Piezoelectric Thin Film Actuator for Integrated Switching. 1-4. doi:10.3390/proceedings2130912

Esposito, P., Ghoussoub, N., & Guo, Y. (2010). Mathematical Analysis of Partial Differential Equations Modeling Electrostatic MEMS (Vol. 20). American Mathematica Society.

Fattorusso, L., & Versaci, M. (2018). A new one-dimensional electrostatic model for membrane MEMS devices. Lecture Notes in Engineering and Computer Science, 2235. London, UK.

Fazio, R., & Jannelli, A. (2014). Finite Difference Schemes on Quasi-Uniform Grids for BVPs on Infinite Intervals. *Journal of Computational and Applied Mathematics*, *269*, 14–23. doi:10.1016/j.cam.2014.02.036

Fazio, R., & Jannelli, A. (2017). BVPs on Infinite Intervals: A Test Problem, a Nonstandard Finite Difference Scheme and a Posteriori Error Estimator. *Mathematical Methods in the Applied Sciences*, *40*(18), 6285–6294. doi:10.1002/mma.4456

Gad-el Hak, M. (2006). *MEMS: Design and Fabrication*. Boca Raton, FL: Chapman & Hall, CRC Taylor & Francis.

Giusti, G. (2000). *Analisi Matematica*. Turin, Italy: Bollati Boringhieri.

Habib, M. K. (2007). Mechatronics: A Unifying Interdisciplinary and Intelligent Engineering Paradigm. *IEEE Industrial Electronics Magazine*, *1*(2), 12–24. doi:10.1109/MIE.2007.901480

Habib, M. K. (2008). Human Adaptive and Friendly Mechatronics (HAFM). *IEEE International Conference on Mechatronics and Automation 'ICMA2008'*, (pp. 61-65). Takamatsu-Kagawa, Japan. 10.1109/ICMA.2008.4798726

Jouaneh, M. (2012). *Fundamentals of Mechatronics*. CI-Engineering.

Khoshnoud, F., & de Silva, C. W. (2012). *Recent Advances in MEMS Sensor Technology - Biomedical Applications*. IEEE Instrumentation & Measurement Magazine; doi:10.1109/MIM.2012.6145254

Kimura, M. (2009). Micro Temperature Sensors and Their Applications to MEMS Thermal Sensors. In *S. C. Mukhopadhyay, G. S. Gupta, & R. Y. Huang (Eds.), Recent Advances in Sensing Technology. Lecure Notes in Electrical Engineering* (Vol. 49). Berlin, Germany: Springer; doi:10.1007/978-3-642-00578-7_19

Kress, R. (1998). *Numerical Analysis* (Vol. 181). New York, NY: Springer-Verlag; doi:10.1007/978-1-4612-0599-9

Magrab, E. B. (2012). *Vibrations of Elastic Systems with Applications to MEMS and NEMS* (Vol. 184). Springer Netherlands; doi:10.1007/978-94-007-2672-7

Mistry, K. K., & Nahapatra, A. (2012). Design and Simulation of a hermo Transfer Type MEMS Based Micro Flow Sensor for Arterial Blood Flow Measurement. *Microsystem Technologies*, *18*(6), 683–692. doi:10.100700542-012-1515-z

Pelesko, J. A. (2004). Electrostatics in MEMS and NEMS. In *V. M. Harik, & L. S. Luo (Eds.), Miromechanics and Nanoscale Effects* (Vol. 10). Dordrecht, Germany: Springer.

Pelesko, J. A., & Bernstein, D. H. (2003). *Modeling MEMS and NEMS*. Boca Raton, FL: Chapman & Hall, CRC Press Company.

Phan, H. P., Nguyen, T. K., Dinh, T., Iacopi, A., Hold, L., Shiddiky, M. J., & Nguyen, N. T. (2018). Robust Free-Standing Nano-Thin SiC Membranes Enable Direct Photolithography for MEMS Sensing Applications. In *Advanced Engineering Materials, 20(1)*. Wiley Online Library; doi:10.1002/adem.201700858

Plander, I., & Stepanovsky, M. (2017). MEMS Technology in Optical Switching. *IEEE 14th International Scientific Conference on Informatics*. Poprad, Slovakia.

Versaci, M., Angiulli, G., Fattorusso, L., & Jannelli, A. (2019). On the Uniqueness of the Solution for a Semi-Linear Elliptic Boundary Value Problem of the Membrane MEMS Device for Reconstructing the Membrane Profile in Absence of Ghost Solutions. *International Journal of Non-linear Mechanics*, *109*, 24–31. doi:10.1016/j.ijnonlinmec.2018.10.014

Voltmer, D. (2007). *Foundament of Electromagnetics*. Synthesis Lectures on Computational Electromagnetics.

Wang, W., & Soper, S. A. (Eds.). (2006). Bio-MEMS Technologies and Applications. Boca Raton, FL: CRC Press.

Warren, M. E. (2008). Micro-Opto-Mechanical Systems. doi:10.1036/1097-8542.757727

KEY TERMS AND DEFINITIONS

BVP: It is the acronym for boundary value problems. Everyone represents a differential equation whose solution meets certain boundary conditions.

FEM: It is the acronym of Finite Element Method. It is a numerical procedure for solving BVPs for partial differential equations.

FFT: It is the acronym for fast Fourier transform. It is an algorithm for computing the discrete Fourier transform of a sequence converting a signal from its time domain to the frequency domain.

Ghost solutions: Numerical solutions of a problem that do not satisfy the analytical conditions of existence and uniqueness.

IVP: It is the acronym for initial value problems (also called a Cauchy problems). Each IVP is a particular ordinary differential equation whose solution meets a set of initial conditions.

Mechatronics: It studies mechanical, electronic and computer interactions for the analysis and design of fully automated control systems.

MEMS: Also written as Micro Electro-Mechanical System, is a modern technology of devices in microscopic scale. They consist of a central unit processing data and several components interacting with micro-sensors.

MOMS: Also indicated as Micro-Opto-Mechanical Systems. They are special devices made combining MEMS integrated with mechanical and optical systems able to manipulate optical signals.

ODE: It is the acronym of ordinary differential equation.

PDE: It is the acronym of partial differential equation.

Shooting method: In numerical calculus, it is a procedure to solve a BVP by reducing it to an equivalent IVP. Starting from a boundary value, it "shout out" a set of trajectories in different directions until the other boundary value is hit.

Spectral methods: They are particular mathematical procedures to solve (numerically) certain differential equations involving the FFT.

ENDNOTE

1 In other terms, the solution in closed form could not be obtained.

APPENDIX

Arc Length and Curvature of a 2D Line in Explicit Form

Let us consider a *2D* line (in other words, a flat curve) whose equation is $u=u(x)$. We plan to calculate the length of the arc of the curve whose extremes are *P* and *Q*. This length can be calculated knowing the equation of the line and the coordinates of *P* and *Q*. Let us consider two very close points of the arc, *P* and *Q:* let them $A = A\left(x, u\left(x\right)\right)$ and

$B = B\left(x + '' x, u\left(x + '' x\right)\right)$.

Let us consider the righ-angled triangle *ABC,* whose edges measure $AC = '' x$,

$BC = u\left(x + '' x\right) = u\left(x\right) = '' u$.

Obviously, the Pythagorean theorem applies:

$$AB^2 = \left(\Delta x\right)^2 + \left(\Delta u\right)^2 \tag{70}$$

If $'' x$ is small enough, we can write:

$$\left(\Delta s\right)^2 \cong \left(\Delta x\right)^2 + \left(\Delta u\right)^2 \tag{71}$$

from which:

$$\left(\frac{\Delta s}{\Delta x}\right)^2 \cong 1 + \left(\frac{\Delta u}{\Delta x}\right)^2 \Rightarrow \frac{\Delta s}{\Delta x} \cong \sqrt{1 + \left(\frac{\Delta u}{\Delta x}\right)^2} \tag{72}$$

Both sides of (73) are the same if $'' x \to 0$, that is:

$$\lim_{\Delta x \to 0} \frac{\Delta u}{\Delta x} = \mathit{lim}_{\Delta x \to 0} \sqrt{1 + \left(\frac{\Delta u}{\Delta x}\right)^2} \Rightarrow \frac{du}{dx} = \sqrt{1 + \left(\frac{\Delta u}{\Delta x}\right)^2} \tag{73}$$

or

$$ds = \sqrt{1+\left(\frac{\Delta u}{\Delta x}\right)^2} dx = \sqrt{1+\left(u'\left(x\right)\right)^2} dx \tag{74}$$

After integration of (75), we achieve the length of the arc (Giusti, 2000):

$$s = \int_{x_1}^{x_2} \sqrt{1+\left(u'\left(x\right)\right)^2} dx \tag{75}$$

The curvature K of a circle is the ratio between an angle at its center (even infinitesimal), $d\phi$, and the arch (even infinitesimal), ds, that underlies:

$$K = \frac{d\phi}{ds} \tag{76}$$

Being $ds = Rd\phi$, with R the radius of the circle, it easily follows:

$$K = \frac{d\phi}{Rd\phi} = \frac{1}{R}. \tag{77}$$

Relation (77), thanks to the given definition, provides the relationship between curvatures and radius of curvatures of a circle. For any 2D line, its curvature in a point $P_{0,}$ can be defined by (77). On a line, l, we consider three points P_0, P_1, P_2 and the circle C_1 passing through them. The line section $P_1P_0P_2$ of l and the arc $P_1P_0P_2$ of the circle are, generally, different: they coincide with the tendency of P_1 and P_2 to P_0. When P_1 and P_2 are infinitely close to P_0, the circle C_1 becomes a particular circle C whose curvature can be obtained as follows:

$$K = \left|\frac{d\phi}{ds}\right| = \frac{1}{R} \tag{78}$$

where $d\phi$ is the infinitesimal angle to which ϕ tends when P_1 and P_2 tend to $P_{0.}$ ds is the infinitesimal arc to which $P_1P_0P_2$ tends when P_1 and P_2 tend to P_0. We now pass to the effective calculation of the curve at a generic point P of a given line $u = u\left(x\right)$. Obviously, we need to calculate both $d\phi$ and ds. For this purpose, we remember that $\frac{du}{dx}$ in P gives us the slope of the tangent straight line in P

$$\tan\phi = \frac{du}{dx} = u'\left(x\right) \tag{79}$$

from which

$$\phi = a\tan\frac{du}{dx} = a\tan u'(x) \tag{80}$$

The infinitesimal variation of ϕ, that is, the angle between the tangent straight line to the end-points of the arc *ds*, can be achieved by differentiating both sides of (80):

$$d\phi = \frac{1}{1+\left(\frac{du}{dx}\right)^2}\frac{d^2u}{d^2x}dx = \frac{u''(x)}{1+\left(u'(x)\right)^2}dx \tag{81}$$

On the other hand, knowing that *ds* is easily calculated by (75), it follows that (Giusti, 2000):

$$k = \left|\frac{d\phi}{ds}\right| = \left|\frac{u''(x)}{\left(1+u'(x)\right)^{3/2}}\right| = \frac{\left|\frac{d^2u(x)}{dx^2}\right|}{\left[1+\left(\frac{du(x)}{dx}\right)^2\right]} \tag{82}$$

Chapter 8
Analyses on Engineering Mechanics of Robotic Arm for Sorting Multi-Materials

Zol Bahri Razali
Universiti Malaysia Perlis, Malaysia

Mohamed Mydin M. Abdul Kader
Universiti Malaysia Perlis, Malaysia

Mohd Hisam Daud
Universiti Malaysia Perlis, Malaysia

Khor Wen Hwooi Stephen
Universiti Malaysia Perlis, Malaysia

ABSTRACT

The study involves static analysis on the developed robotic arm. Increasing loads are applied to the robotic arm to determine the maximum load that it can hold. Firstly, the robotic arm model is created using CATIA. Then, it is analyzed using the generative structural analysis tool in the same software. Increasing loads are applied to the end of the robotic arm until significant deformation occurs. The same procedure is done for modified designs in the analysis software. The results considered include displacement and stress. Based on the results, the critical stress areas are near to the rotating joints of the robotic arm, the back of the gripper and the sharp edge of the second arm. Proposed modifications include increasing the servo motor shaft radius and edge filleting the affected area, increasing the thickness and reducing the length of the gripper base plate, and implementing a new design for the second arm.

DOI: 10.4018/978-1-7998-0137-5.ch008

INTRODUCTION

A robotic arm is a mechanical arm, usually programmable that resembles with similar functions to a human arm. The arm may be the sum total of the mechanism or may be part of a more complex robot. The links of such a manipulator are connected by joints allowing either rotational motion (such as in an articulated robot) or translational (linear) displacement. The links of the manipulator can be considered to form a kinematic chain. The terminus of the kinematic chain of the manipulator is called the end effector and it is analogous to the human hand.

A serial robot arm can be described as a chain of links that are moved by joints which are actuated by motors. An end-effector, also called a robot hand, can be attached to the end of the chain. As other robotic mechanisms, robot arms are typically classified in terms of the number of degrees of freedom. Usually, the number of degrees of freedom is equal to the number of joints that move the links of the robot arm. At least six degrees of freedom are required to enable the robot hand to reach an arbitrary pose (position and orientation) in three-dimensional space. Additional degrees of freedom allow to change the configuration of some link on the arm (e.g., elbow up/down), while keeping the robot hand in the same pose. Inverse kinematics is the mathematical process to calculate the configuration of an arm, typically in terms of joint angles, given a desired pose of the robot hand in three dimensional spaces.

An industrial robot is a robot system used for manufacturing. Industrial robots are automated, programmable and capable of movement on three or more axis (Mon, Mohd Romlay, & Tamin, 2012). Typical applications of robots include welding, painting, assembly, pick and place for printed circuit boards, packaging and labeling, palletizing, product inspection, and testing; all accomplished with high endurance, speed, and precision. They can assist in material handling.

The end effector, or robotic hand, can be designed to perform any desired task such as gripping, spinning etc., depending on the application. It is the most common manufacturing robot in the industry. In the automotive industry, it is used in assembly lines for welding, parts rotation and placement of parts. It is used in hazardous situations such as bomb disarmament and disposal (Chung & Yi, 2015). Its end effector is used for welding, gripping and spinning of parts. Hydraulic or pneumatic arms are used to lift heavier objects.

It is controlled by a computer program which instructs it to rotate step motors located at each joint. The step motors move in exact increments and is very precise. This is helped by motion sensors which make sure it moves just the right amount. It is useful for operations that require repetition of the same movement. The robotic arm has 7 metal segments which are joined by 6 joints. Typically, it has 6 degrees of freedom. A human hand has 7 degrees of freedom. It usually has a stationary base (Mon, Mohd Romlay, & Tamin, 2012).

Similarly, a mechanical gripper is a robot component that uses movable, finger like levers to grasp objects. It can be used for various types of industrial and household applications. In this case, the mechanical gripper is utilized for industrial pick-and-place purposes. Some mechanical grippers are controlled by servomechanism while others are driven by some linear actuation mechanisms. The most commonly used type of mechanical gripper is the parallel type. This type of mechanical gripper uses two finger-like levers (that move towards each other) to hold objects by applying adequate amount of normal force onto the object. For holding circular objects, a mechanical gripper with three moving levers can be used (Chung & Yi, 2015).

INDUSTRIAL PROBLEM TO BE SOLVED THROUGH THIS PROJECT

Actually, sorting of parts in the production line has to be fast and consistent. There are certain tedious operations that can be done by a pick and place robot to save time and money in hiring operators. However, the design of the robotics system should extremely take care of the fundamental of engineering mechanics, therefore the design of the robotics arm can provide optimum service without faults and failures.

Therefore, among the objectives of this study are to analyze the developed robotic mechanism related to mechanical concepts in static conditions, to conduct static analysis on the overall robotic arm, gripper and the second arm, and to investigate design parameters for improvement.

The study in the scope of robotic arms is very wide, thus the researchers limiting the scope of research, such as: It includes equilibrium force analysis and structural analysis on the robotic system. For the analysis, design of mechanical elements has been done by utilization of CATIA software to analyze the mechanical design.

Literature on Mechanics of the Robotic Arm

This part discusses about the literature relating to the mechanics of the robotic arm. It explains the purpose, assumptions, conclusions, support and qualification, gaps and application to the current research. The literature helps to widen background knowledge on the related topic, includes discussion of literature on statics, dynamics, finite element analysis and mechanical gripper.

Internal Loading

Based on the journal "Analysis of internal Loading at Multiple Robotic Systems" (Chung & Yi, 2015), internal loading is defined as moments or forces that do not affect the motion of the end effector. It analyzes grasping space based internal loading and it does not derive internal loading conditions. However, operational space-based approach is not used. Using reduced row echelon method, the basis of internal loading is as follows:

$$F = [f_{x1}\ f_{y1}\ f_{z1}\ m_{x1}\ m_{y1}\ m_{z1}\ f_{x2}\ f_{y2}\ f_{z2}\ m_{x2}\ m_{y2}\ m_{z2}]^T$$

Statics

According to "Statics of Soft-Fingered Grasping and Manipulation" (Inoue & Hirai, 2009), a local minimum of elastic potential energy with constraints exists during manipulation movement by a soft-fingered robotic hand (Figure 1). It conducts static analysis of overall elastic energy function which includes gravitational force. It assumes that a quasi-continuous motion of both rotational fingers such that each finger moves in keeping with an operation manner. However, it only covers one kind of motion (quasi-continuous motion) for the robot fingers. Numerical method (Nelder-Mead method) is used to search for a force or moment equilibrium point.

Figure 1. Conceptual diagram of force or moment balance on soft fingers

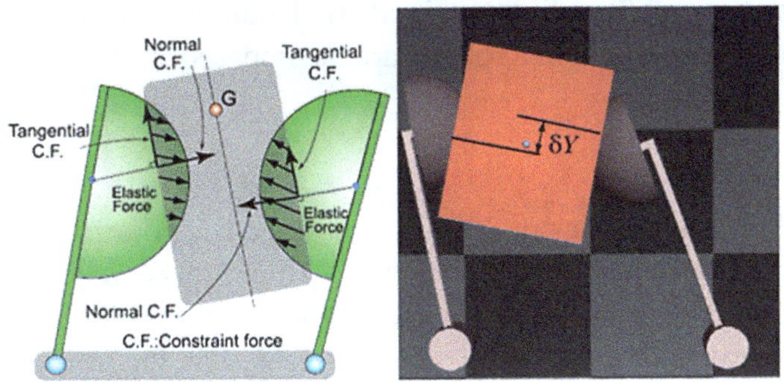

In the book "Statics and Kinematics with applications to Robotics" by Duffy (Duffy, 1996), it was concluded that the Denavit-Hartenberg procedure is suitable to model robot kinematics (Figure 2). All the joints of the robotic arm are assumed to be revolute in the formulation of the Jacobian matrix. This chapter only covers the statics of the robots. Besides the Denavit-Hartenberg convention, it also covers homogeneous transformation, inverse kinematics, Jacobian matrix and differential motion and dual transformation. The Denavit-Hartenberg technique, inverse kinematics, and robotic statics can be applied to this research (Duffy, 1996).

Referring to the book and conference paper "High-End Technology for Effective Glass Manufacturing Industry: Innovative Operating Environment" (Razali & Lee, 2013; Razali & Lee, 2012), it was presented that the operating space is important to the design of the robotic arm. Based on the results of the study, this project is furthering the study on the analyses parts of the robotics arm.

Whilst, in "Automated Mechanical Sorting Device for Mixed Household Waste: Integration Sorting Conveyor with Neural Network" (Razali & Madasamy, 2013; Razali & Madasamy, 2012), the importance of a neural network to the function of the robotic arm is emphasized and an image processing system is studied for integration into the robotic arm design.

Figure 2. Three joint robotic arm hanging a pendulum

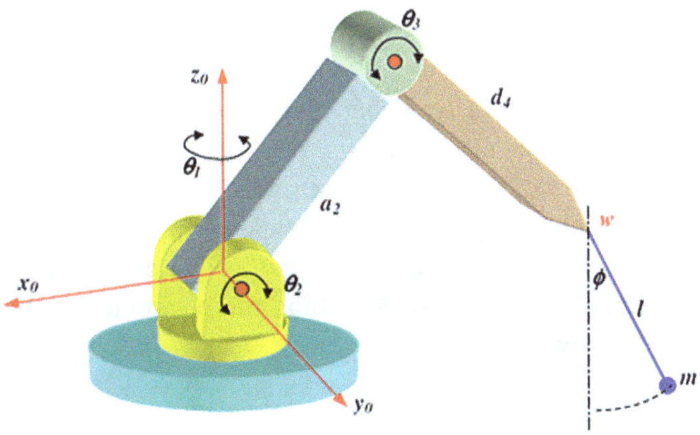

Furthermore in "Automated Segregating Device for Reuse and Recycle Waste: Application on size based, friendly recycling design" (Razali & Nawi, 2013), the importance of automation in the industry is documented and a green way to design a robotic arm is proposed.

Dynamics

According to Purushotham and Anjeneyulu (2013), its method is a powerful way in robot dynamics in that it is comprehensive and efficient. It aims to develop dynamical equations of motion and prepare a solution scheme for space robotic arms. In this journal, virtual displacement is assumed to occur without the passage of time. However, most of the applications presented are for old multi-body systems. It uses the Lagrange form of d'Alembert's principle and the Newton-Euler method (Purushotham & Anjeneyulu, 2013). The Lagrangian method used here can be applied in dynamical systems of robotic arms. Kane's equation is as follows:

$$Fr+Fr^*=0$$

Where

$$F_r=f_r+M_r$$

$$F_r^*=f_r^*+M_r^*$$

f_r and f_r^* - the generalized active and inertia forces respectively
M_r and M_r^* - the generalized active and inertia moments respectively

In the book of "Theory of applied robotics: kinematics, dynamics, and control" (Jazar, 2010), dynamic equations are found by Newton-Euler and Lagrange methods. It assumes no friction in joints, links are made of homogeneous material and bar shaped, no external force on end effector, no gravity, motors are concentrated masses at joints, gravity is the only source of potential energy, mass centre is in the middle of the link and all the product of inertias is zero. It only covers a small part of the statics of the robot. The Lagrange method and robot statics can be used in this research (Jazar, 2010).

Finite Element Analysis

In the book authored by Spentzas and S. A. Kanarachos (Spentzas & Kanarachos 2002), multi-axes positioning is designed using finite element analysis. It is assumed that manipulators are rigid in kinematic analysis and the material model is linear. It was concluded that the finite element method can be used to analyze multi-axes positioning and physical response to load. Denavit-Hartenberg modeling technique and finite element method was used. Inertia and weight effect were ignored in calculations. Denavit-Hartenberg algorithm and finite element method can be applied to this research. The general equation for the system is as follows:

$$[M]\{\ddot{q}\}+[K]\{q\} = \{F\}$$

In "An Incremental Finite Element Analysis of Mechanisms and Robots", calculations are based on the linear theory of rectilinear beams. It was concluded that the incremental analysis of mechanisms and robots by linear finite elements is versatile in that it can be performed on stiff and flexible members. Linear finite element method was used in the analysis. However, non-linear finite element method was not considered. The linear finite element method is suitable for applications here. The system of equations were assembled and written as:

$$M \cdot U + K \cdot U = F$$

Where

M - global mass matrix
U - displacement vector of the mechanism
K - stiffness matrix of the mechanism
F - vector of external loads acting on a beam

Referring to Yuan and Book (Yuan & Book, 1993), aims to provide a way to develop the dynamics of flexible manipulators consisting of two links connected by rotary joints. The link deflection is assumed to be small and the flexible deflection is assumed to be a series of finite separable modes (Figure 3). After the analysis, a transformation between two coordinates consisting of rigid body motion and link deformation has been created in 4 * 4 matrix form. Experimental method, finite element method and Runge-Kutta method was used in the analysis. It does not consider large link deflections. The current research can utilize the experimental and finite element methods used in the journal.

Figure 3. A 2R manipulator carrying a payload

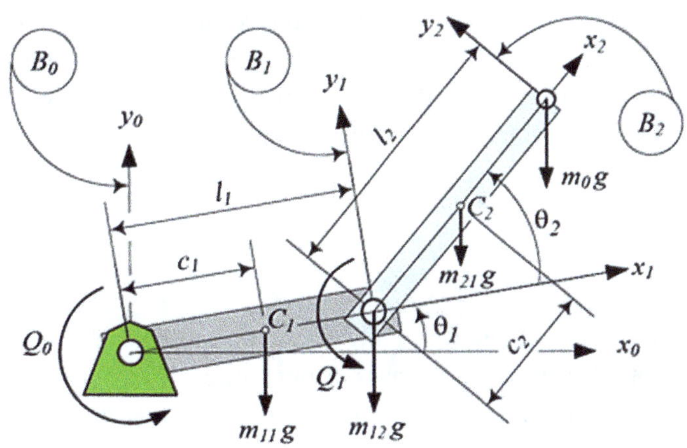

Figure 4. Rigid rotation and link deflection

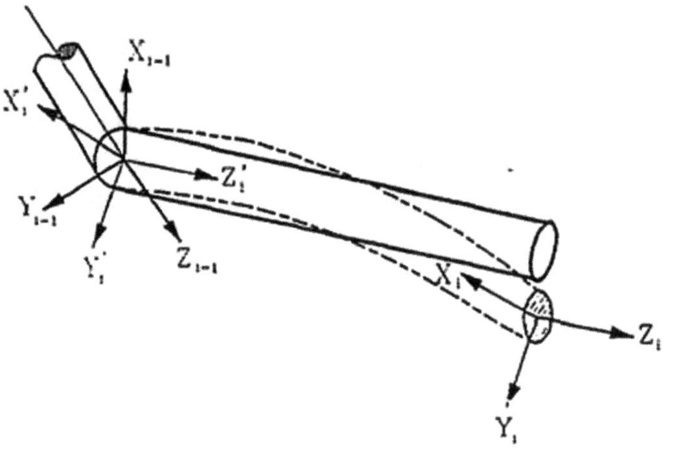

Force and Torque Calculation

To make an extendable arm, a linear actuator can be used provide linear motion along vertical and horizontal axes. To produce such drive at an affordable cost, a combination of belt drive and a motor is more preferable. Nevertheless, selecting a suitable motor for the application requires a very detailed calculation to avoid over-sizing or under-sizing of motor.

According to Panasonic Motor Selection Guide ("Motor Selection", 2010), the general procedure of motor selection must follow the sequence that includes (1) Determination of driving mechanism, (2) Calculation of motor speed, and load, (3) Checking requirement of specifications, (4) Selection of motor model, (5) Temporary selection of motor, and (6) Final determination of the motor and gear head. Load torque, moment of inertia and speed which are converted to those at the motor output shaft have to be calculated. Positioning accuracy, holding of position, speed range and other environmental resistance need to be checked afterwards ("Motor Selection", 2010).

As for the belt-driven vertical axis lifting component, the mechanism resembles conveyor belt carrying loads. Selection of suitable motor must be based on this sort of driven method. According to the literature titled "Motor Torque Calculation" published by Leadshine Technology Co., Ltd. ("Motor Torque Calculation", 2010), load inertia of the system can be calculated using a series of equations that give the separate load inertia of each component. Load inertia exerted by the applied load, the pulleys and the belt can be calculated one at a time. To calculate the load inertia of the applied mass (J_L), equation below should be used. D symbolizes pulley diameter while m_L denotes mass of applied load. All units should be expressed as SI units ("Motor Torque Calculation", 2010).

$$J_L = \frac{1}{4} m_L D^2 \quad (1)$$

Leadshine Technology Co. Ltd stated that pulleys' inertia moment (J_P) should be calculated using the following equation. Note that m_P denotes mass of one pulley. If two identical pulleys are used, the calculated value should be multiplied by two ("Motor Torque Calculation", 2010).

$$J_P = \frac{1}{8} m_P D^2 \tag{2}$$

Belt's load inertia cannot be neglected as belts are mostly reinforced by stronger material such as Kevlar or steel cord. To calculated belt's load inertia (J_B), Leadshine Technology Co. Ltd. suggested the equation stated as follows to relate with mass of belt (m_B) ("Motor Torque Calculation", 2010).

$$J_B = \frac{1}{4} m_B D^2 \tag{3}$$

The total load inertia of the system should be summed and yield the reflected load to the output shaft of selected motor. Thus, the total load inertia of the system (J_T) is expressed as the following equation based on Leadshine Technology Co. Ltd.'s literature ("Motor Torque Calculation", 2010).

$$J_T = J_L + J_P + J_B + J_M \tag{4}$$

Next, acceleration torque (T_a) required to be provided by the motor should be calculated using the value of total load inertia of the system and the desired angular acceleration (a) by utilizing the following equations stated in the literature named "Motor Torque Calculation" published by Leadshine Technology Co. Ltd. ω_0 and ω_1 denote initial and final angular velocity respectively while t is a symbol for time for velocity change ("Motor Torque Calculation", 2010).

$$T_a = J_T a \tag{5}$$

$$T_a = J_T \frac{\omega_1 - \omega_0}{t} \tag{6}$$

As for load torque (T_L), there are two different ways to express the relationship between the applied mass and the load torque required to carry the mass. Leadshine Technology Co. Ltd. ("Motor Torque Calculation", 2010) stated that load torque calculation must consider the inclined angle of the axis of movement of applied mass, as well as the efficiency of the motor. The efficiencies of motors usually vary between 85% and 95%. Equation 7 shows the expression of the relationship. In this case, g, α, and η symbolize gravitational acceleration, angle of inclination of moving plane, and motor efficiency respectively. μ denotes the coefficient of friction of sliding surfaces.

$$T_L = \frac{m_L g D \left(\sin\alpha + \mu \cos\alpha\right)}{2\eta} \tag{7}$$

Nevertheless, Panasonic Motor Selection Guide ("Motor Selection", 2010) expressed the relationship differently. The literature did not consider the efficiency of the motor that drives the pulleys. Also, it expressed the gravitational effects in terms of external forces acting on the applied mass ("Motor Selection", 2010). The equation is given as follows and F denotes the external force acting on the system.

$$T_L = \frac{D\left(F + \mu m_L g\right)}{2} \tag{8}$$

Total calculated torque (T_T) can be obtained by summing up the value of load torque and acceleration torque, according to Leadshine Technology Co. Ltd. ("Motor Torque Calculation", 2010). Mathematically, the relationship between these parameters is expression in the following equation:

$$T_T = T_L + T_a \tag{9}$$

Also, Leadshine Technology Co. Ltd. ("Motor Torque Calculation", 2010) suggested the usage of safety factor when selecting suitable motor for the application. The safety factor (K_S) is merely a constant, user-defined factor to serve as multiple of the total calculated torque. The required motor torque (T_M) is given by the following equation:

$$T_M = K_S T_T \tag{10}$$

Mechanical Gripper

When it comes to gripping mechanism, coefficient of friction between two materials in contact is very important parameters we need to find out. The static friction coefficient between two solid surfaces is defined as the ratio of the tangential force required to produce sliding divided by the normal force between the surfaces. According to Ryan of Pennsylvania State University (Ryan, 2010), rubber on glass has a very high static friction coefficient of 1 to 2. This statement is supported by Haney (2004) in his webpage article titled "Rubber Friction". A table in the article recorded coefficient of friction of rubber on glass under dry condition can achieve a value larger than two. Chen, Magee, and Yakubov (2005) tabled that rubber on glass produces friction coefficient of two or above. The value is way higher than other material combinations. Alexander, et al. (2010) believes that static coefficient of friction of rubber on glass equals the value of two.

Researches were also done on finger type mechanical grippers in the current market. According to El-Kalay et al. (1995), when used as a gripping mechanism, the fingers (generally are two finger-like levers, form two non-jointed tongs) are used to grasp desired parts. The fingers (or jaws) are the appendages that make actual contact with the object. The grip in this case will occur at a point. Alternatively, each of the two pivoted fingers may be made in the form of a parallel linkage, in which the coupler is formed in the shape of a grip head, resulting in a line contact. When grasping, the part is held (constraint) by the gripper by either shaping the fingers so as to enclose the part, or by friction. The latter case yields simpler finger geometry but requires higher grasping forces. Grippers with jointed fingers are practically unknown in industrial robotics.

Actuation of fingers can be by pneumatic or hydraulic actuators, by cable, or by pinion-and-rack arrangements, where the rack may be actuated by a solenoid. Wire-and-pulley mechanisms are used for master-slave type of manipulations. Electric grippers generally use AC/DC motors or solenoids to activate the jaws. However, their use is limited to the weight penalty that they impose. Hydraulic grippers can exert large forces and yet generally carry an excessive weight penalty too.

Linkage-actuated grippers are the most common of all. Fingers may undergo rotation, as in scissors-type grippers, or translation, as in the case of rack-and-pinion drives, or curvilinear translation (CT), as in the case of a parallelogram mechanism. Transformation of linear input from a pneumatic or hydraulic actuator into rotary finger motion is generally carried out by means of lever-type mechanisms, and by rack-and-pinion arrangements (El-Kalay et al., 2010).

Research by Singh and Kumar (2013), the aim was to deduce parametric equations used for the automated kinematic analysis of geared robotic wrists (Figure 5). It was assumed that the arm can rotate about a maximum of two non-parallel intersecting axes. A general displacement equation, which is used in gear trains with gyroscopic complexity, was deduced. However, the parametric equations only work for the lowest gyroscopic complexity. The proposed method of detecting links acting as arms and generating the equation of motion, which is used in this journal, is suitable for this research.

The proposed method of detecting links acting as arms and generating the equation of motion, which is used in this journal, is suitable for this research. The journal assumes that the center of mass of each linkage acts at half of the length. The conclusion was that robotic arms help to make spot welding more flexible, time-oriented and easy. Besides that, the mechanical structure and angle of movement are changeable. However, the design was only based on spot welding applications. The force calculation of joints used in this journal can be applied to the current research.

Figure 5. Components of the gripper

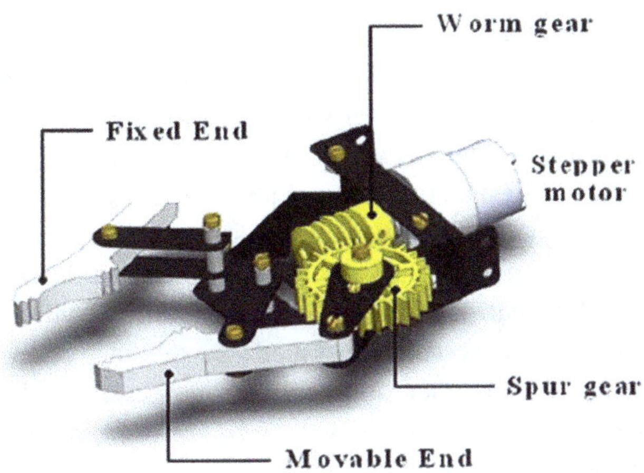

METHODOLOGIES AND MATERIALS

Product Design Process

Generally, the design project follows the workflow of product development process used by EMS companies and other product development departments. The systematic design process starts with gathering design inputs, followed by the stage of design and development. Then, design outputs are generated and design review will be conducted. If the design successfully passed the review, approve documents will be issue if it is required by certain corporations. The design will then be realized by making of prototype. The design should then be tested for verification and if no change is required, design validation (e.g. field trial of medical/healthcare product) will be conducted. If the design passes the trial period, design transfer will be done.

At the beginning of the process, information on requirements for the expected product will be collected. For this case, the material to be picked, the required motion, major dimensions, and other related data regarding the robotic arm will be acquired from supervisor. Design input will be generated, and a series of product design specifications will be proposed to supervisor.

Later on, the design and development stage will be started. Basically, to simplify design process, the expected product will be divided into a few proper modules where concentration and attention of designer will be paid to one module at a time. This is an effective approach to organize design data. By using this method, design data will not be crowded and designer will be able to trace problems easily as they arise. During this stage, researches will be done to gain knowledge or information in order to obtain more understanding on the current design issues. Discussions between designer and advisor will be frequent to avoid misunderstanding and thus misleading to errors and mistakes.

Next, design output will be generated. This is a stage where drawings and 3D models will be produced by designers. All design data must be arranged accordingly and comprehensively to smoothen later processes. Drawings of parts to be custom-made must be provided in order to obtain desired product from vendors/suppliers (if any). All drawings must include comprehensible signs and symbols describing geometric dimensioning and tolerancing (GD&T). This is a way to communicate the desired parts geometry to the fabricator of the parts. Even if the parts are to be fabricated using in-house machining devices, by the designer himself/herself, drawings should be done comprehensively for future development of product (i.e. mass production). Assembly drawings are to be generated for reference of other personnel (if any) involved in the assembly process.

After that, a design review should be conducted thoroughly based on the generated design outputs by independent reviewers. This is to ensure no overlooked factors by the design engineers during the stage of design and development. The reviewer should comment on every doubted area regarding the design and the design engineer is ought to answer to the comments. A meeting should be conducted between the reviewer(s) and the designers to discuss about the issues arose. Reviewer(s) may point out their ideas to improve the design of the product but the designers should always make the final decision by considering all the factors. If there are any changes required, the design process should be brought back to the design and development stage and the consequent stages will be repeated.

If it is required, approve documents should be issued after the stage of reviewing. Documentation and papers describing the design data and their approval information should be kept during this particular stage. This is a record-keeping process and it is done basically by corporate organizations to ensure proficiency of the company and traceability by customers.

Methods

Engineering mechanics analysis will be done on the developed pick and place sorting multi-materials robotics arm. It will be studied in its static conditions. The study covers mechanical aspects such as force analysis, stress, displacement and deformation of the robotic arm. The analysis is done using the computer-aided drawing and analysis software CATIA. The analysis will be done based on the Denavit-Hartenberg convention (Pennestri, 1991). Besides that, the study will involve mechanics topics that have been learned throughout the course such as statics, solid mechanics and mechanics of materials. Next, the results of the analyses are integrated into the improved design. The proposed design must have mechanical design improvements based on the analysis. The proposed concepts will be more efficient and stable than the current design.

Next, the output of the fundamental study of the engineering mechanics is integrated and applied to the conceptual design. Based on the fundamental study and analysis, two improved designs of the robotic arm are proposed. The conceptual designs are created using CATIA. Different variations of developed robotic arm design are analysed based on their suitability to the pick and place application. Static analysis is done on the proposed designs. The analysis outcomes will help to determine the best design. The analysis is done using the generative structural analysis tool in the software. The two proposed concepts are compared and the optimum design is chosen.

From there, two of the best designs are brought forward for comparison based on displacement and stress. Static analysis will be conducted on the robotic arm designs. The analysis will be done on the original design and the improved designs to highlight areas of improvement. The results are reviewed and weaknesses are spotted for revision of the design. Structural imbalances under static conditions are rectified by designing a more stable design.

Based on the analysis of the robotic arm designs, the optimum conceptual design is developed. This design will optimize the stability, power required and strength of the materials during pick and place operations. Lastly, documentation of the analysis results are prepared. The results are obtained from CAD and simulation software. The analysis results include stress, displacements and deformation.

The results are reviewed and checked for their validity and verification. Inconsistencies in the results are spotted. Then, the problem is studied again to troubleshoot the root cause. Besides that, new design concepts are considered for application to the robotic arm mechanism. Applicable concepts are integrated into the final concept design to produce the optimum mechanical design.

Materials

Table 1 shows the material property applied to the overall robotic arm as a general static analysis. The plexiglass material used is in accordance with the material used with the current robotic arm. The material was used because it was the comprised the majority of the robotic arm. This would allow the analysis results to be more accurate compared to using other materials. In order to decrease analysis computation time, the analysis does not consider that washers and screws are made of aluminium, and the hexagonal nuts are made of brass.

Table 1. Properties of the material used for the overall static analysis

Materials	Plexiglass
Young Modulus	2.9e+009N_m2
Poisson Ratio	0.4
Density	1200kg_m3
Thermal Expansion	7e-005 K_deg
Yield Strength	5.5e+007 N_m2

Analyses on the Findings

Engineering mechanics analysis will be done on the developed pick and place sorting multi-materials robotics arm. The analysis will be done based on the Denavit-Hartenberg convention (Habib, 2008). Besides that, the study will involve mechanics topics that have been learned throughout the course such as statics, solid mechanics and mechanics of materials.

Next, the results of the analyses are integrated into the improved design. The proposed design must have mechanical design improvements based on the analysis. The proposed concepts will be more efficient and stable than the current design.

Two designs will be proposed after preliminary analyses on the robotic arm. Different variations of developed robotic arm design are analysed based on their suitability to the pick and place application. From there, two of the best designs are brought forward for comparison based on displacement and stress.

Static analysis will be conducted on the robotic arm designs. The analysis will be done on the original design and the improved designs to highlight areas of improvement. The results are reviewed and weaknesses are spotted for revision of the design. Structural imbalances under static conditions are rectified by designing a more stable design.

Based on the analysis of the robotic arm designs, the optimum conceptual design is developed. This design will optimize the stability, power required and strength of the materials during pick and place operations. Lastly, documentation of the analysis results is prepared. The results are obtained from CAD and simulation software. The analysis results include stress, displacements and deformation.

Results and Discussions

In this part, the discussions on the findings of static analyses are divided into 3 parts:

1. Static Case of Overall Robotic Arm,
2. Static Case of Gripper, and
3. Static Case of Second Arm

The analyses are followed by the proposal of betterment design with some modification on the original Robotics Arm, such as:

1. Increasing the motor shaft radius and its effect on the translational displacement (modification 1)
2. Edge fillet on the motor shaft and its effect on the maximum displacement (modification 2)

3. Increase in thickness of the gripper base plate and its effect on the maximum stress of the gripper (modification 3)
4. Decrease in length of the gripper base plate and its effect on the maximum displacement of the gripper (modification 4)

Static Case of Overall Robotic Arm

Table 2 shows the material property applied to the overall robotic arm as a general static analysis. The plexiglass material used is in accordance with the material used with the current robotic arm. The material was used because it was the comprised the majority of the robotic arm. This would allow the analysis results to be more accurate compared to using other materials. In order to decrease analysis computation time, the analysis does not consider that washers and screws are made of aluminium, and the hexagonal nuts are made of brass.

Figure 6 is the robotic arm drawing created using the CATIA software. All the components of the robotic arm were drawn including the washers, nuts and screws. Furthermore, the dimensions were modeled after the real robotic arm prototype. This was done to ensure that the analysis results reflected accurate results based on the real dimension and shape of the robotic arm. The screw threads were omitted to decrease analysis computation time.

Figure 7 shows the translational displacement of the robotic arm when a vertical 10N downward force applied to the gripper. This was done to determine the effect of an applied force on the maximum displacement. The maximum displacement was found to be 55.585mm located at the gripper. This was due to the large moment created by the force and the distance between the force and the joint which caused deformation at the joint. As shown in the figure, the developed robot arm base is relatively stable with little displacement.

Table 3 shows the translational displacement of the gripper when varying loads are applied on the robotic arm. It was done to determine the relationship between the two variables. It can be seen that the displacement is directly proportional to the applied force. It is important to note that the analysis was done

Table 2. Properties of the material used for the overall static analysis

Applied Force (N)	Displacement (mm)
1	5.5585
2	11.1170
3	16.6755
4	22.234
5	27.7925
6	33.3510
7	38.9095
8	44.4680
9	50.0265
10	55.5850

Figure 6. Detailed CATIA drawing of the robotic arm

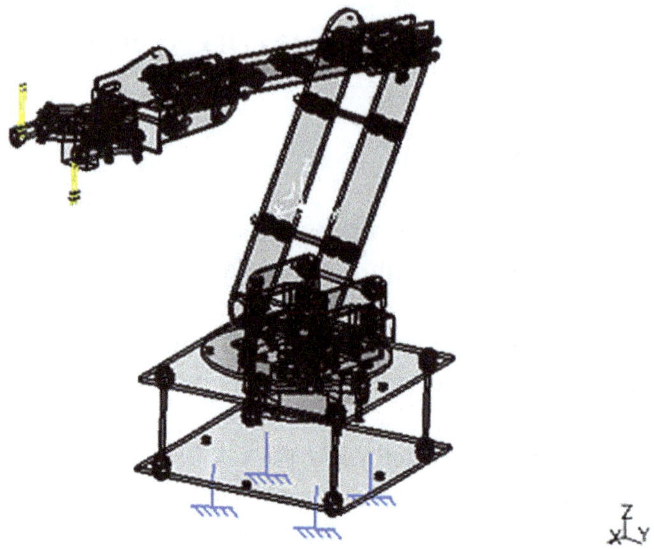

Figure 7. Displacement of the robotic arm

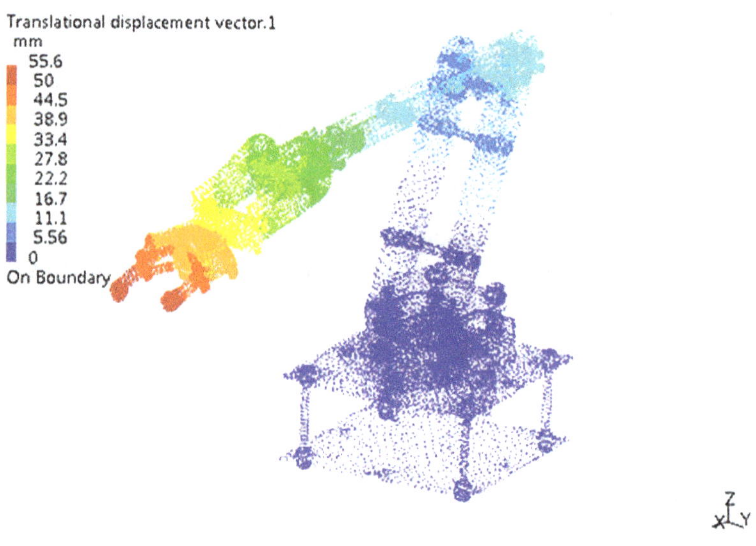

based on the assumption that the whole robotic arm is treated as one part. The maximum displacement is located at the gripper area. This trend is visualized in the graph below which depicts a linear pattern.

Figure 8 shows the displacement of the gripper versus the applied force. This was done to show the effect of varying loads on the robotic arm. The graph shows a linear relationship between displacement and applied force. The maximum displacement was at the gripper with a displacement of 55.585mm. From this graph, larger displacements can be extrapolated based on the gradient of the graph.

Figure 9 shows the location of the maximum stress when a vertical 10N downward force is applied on the gripper. The maximum stress is 1.56424e+008 N/m^2 and located at the servo motor. This is due to

Table 3. Maximum displacement of the robotic arm corresponding to the applied force

Applied Force (N)	Displacement (mm)
1	5.5585
2	11.1170
3	16.6755
4	22.234
5	27.7925
6	33.3510
7	38.9095
8	44.4680
9	50.0265
10	55.5850

Figure 8. Maximum displacement of robotic arm versus applied force

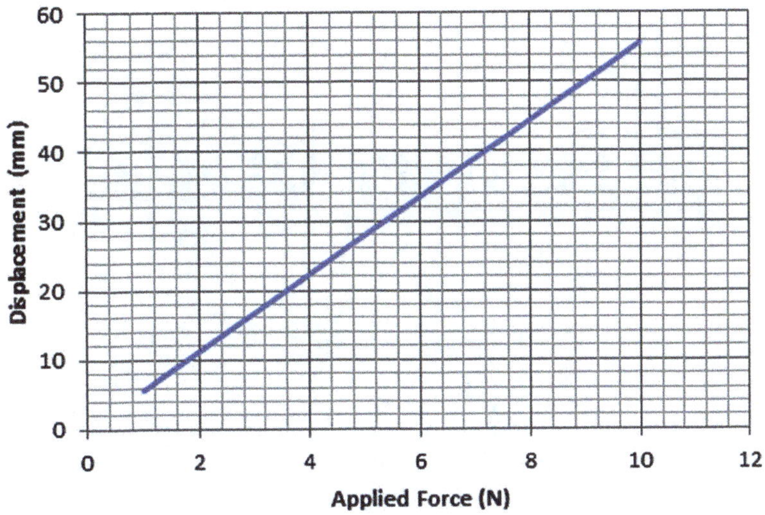

the large moment created on the point when the force is applied. Besides that, the large stress generated is also due the small radius at the maximum stress point. The small radius of the motor shaft makes it a critical point in the structure. The maximum stress area is at the "arm" of the robotic arm.

Figure 10 is a close-up of the maximum stress point which shows that it is located at the motor shaft between the servo motor and its spinner. The motor shaft has a small radius which makes it a weak point in the structure. The large moment created when force is applied at the end of the robotic arm caused the maximum stress to occur at this point. The maximum stress point occurs when a 10N force is applied at the gripper. A closer look at the point shows that it is located at the gap between the servo motor and its spinner which is connected by a shaft.

Table 4 shows the maximum stress when varying loads are applied at the gripper. This was done to determine the relationship between the variables. In other words, it was done to find out the effect of an applied force on the maximum stress of the robotic arm. In the analysis workbench, the load distribution

Figure 9. Stress distribution at the robotic arm

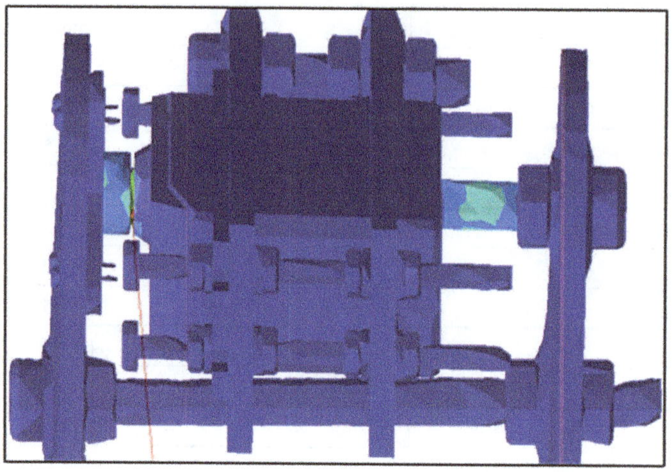

Figure 10. Close-up view of the maximum stress at the robotic arm

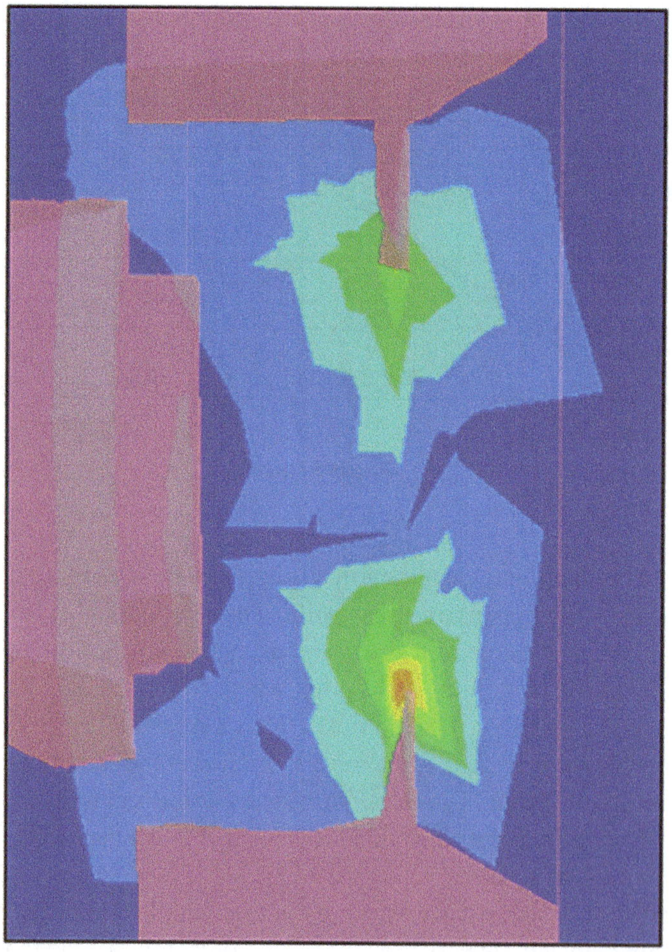

Table 4. Maximum stress of the robotic arm corresponding to the applied force

Applied Force (N)	Maximum Stress (N/m^2)
1	1.56424e+007
2	3.12847e+007
3	4.69271e+007
4	6.25694e+007
5	7.82118e+007
6	9.38542e+007
7	1.09497e+008
8	1.25139e+008
9	1.40781e+008
10	1.56424e+008

parameter was changed to the relevant value. Then, the analysis was computed to determine the corresponding maximum stress. The computation time for each analysis was about 10 seconds. The maximum stress was located at the "arm" of the robotic arm. More specifically, it is located at the servo motor shaft.

Figure 11 shows the relationship between maximum stress and applied force. This was done to determine the effect of applied force on the maximum stress of the robotic arm. The graph shows that the maximum stress is directly proportional to the applied force. The force is applied at the gripper of the

Figure 11. Maximum stress of robotic arm versus applied force

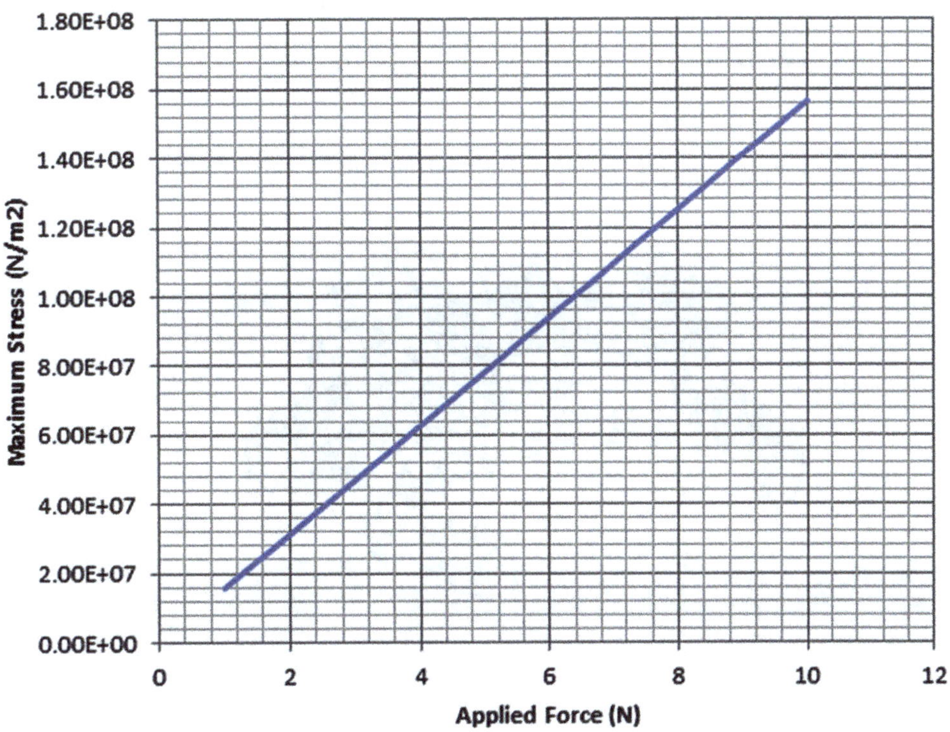

robotic arm. Different values of applied force are applied to the analysis. Then, the analysis is computed to determine the maximum stress. The maximum stress is assumed to be at the same point for each case.

Static Case of Gripper

Table 5 shows the material used in the generative structural analysis on the gripper. It shows the exact numerical values of the properties used in the analysis computation. Aluminium is used as it reflects the material used to manufacture the gripper in the current model. The material property is applied to the whole gripper part in the analysis.

Figure 12 shows the detailed drawing of the gripper of the robotic arm. It shows the dimensions which reflect the true size of the current gripper. The screw threads are not included to reduce analysis computation time. Each part of the gripper is created individually and the overall gripper product is combined through contact constraints. Then, the product is converted into a part for static analysis.

Table 5. Properties of the material used for the gripper analysis

Material	Aluminium
Young Modulus	$7 \times 10^{10} Nm^{-2}$
Poisson Ratio	0.346
Density	$2710 kgm^{-3}$
Thermal Expansion	$2.36 \times 10^{-5} K$
Yield Strength	$9.5 \times 10^{7} Nm^{-2}$

Figure 12. Detailed CATIA drawing of the gripper

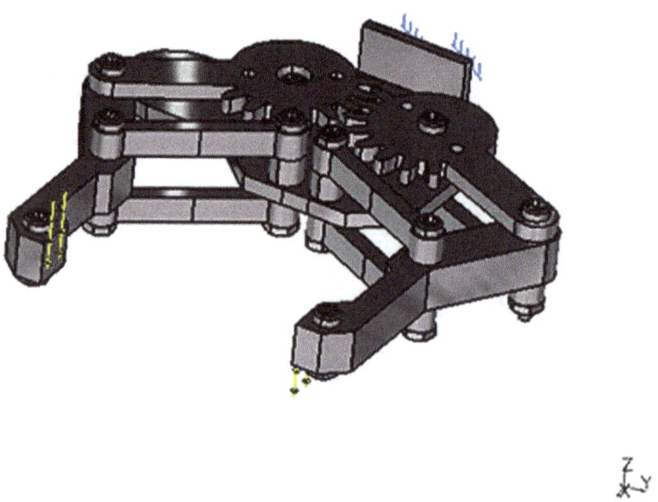

The translational displacement of the gripper when 1 N of force is applied at the two contacting surfaces, the back of the gripper is clamped as it is assumed to be fixed to the end of the robotic arm. It shows the maximum displacement is located at one of two ends of the gripper. The displacement of the gripper decreases as the distance from the clamp decreases. This indicates that the large moment created by the applied force at the other end caused the maximum displacement to be located there.

Table 6 shows the maximum displacement of the gripper when a force is applied at the two contacting surfaces. It presents the effect of increasing applied force on the maximum displacement. The displacement increases as the applied force. The maximum displacement is located near one of the contacting surfaces.

Figure 13 shows the maximum displacement of the gripper versus the force applied at the end of the gripper. It shows the detailed relationship between the two variables. The maximum displacement is directly proportional to the applied force. The minimal displacement of the gripper indicates that it is able to carry much heavier loads. The maximum displacement is near the contacting surface of the gripper.

Figure 14 shows the stress distribution when 1 N of force is applied at the two contacting surfaces. It was done to determine the critical stress areas for further improvement. More stress is experienced by

Table 6. Maximum displacement of the gripper corresponding to the applied force

Applied Force (N)	Maximum Displacement (mm)
1	0.0117372
2	0.0234744
3	0.0352116
4	0.0469488
5	0.0586860
6	0.0704232
7	0.0821604
8	0.0938976
9	0.1056348
10	0.1173720

Figure 13. Maximum displacement of gripper versus applied force

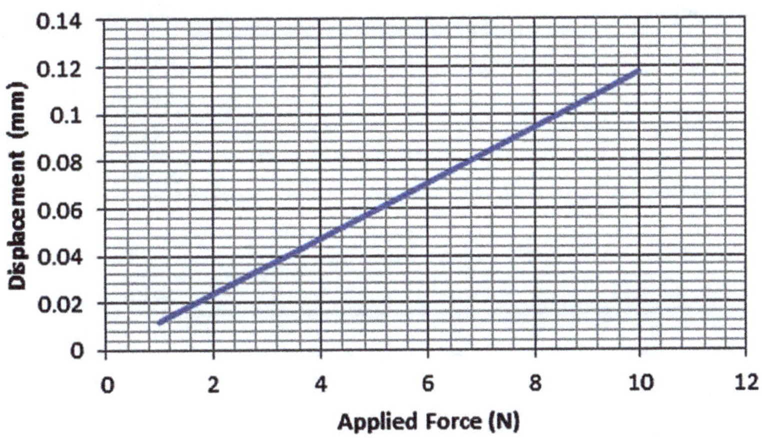

Figure 14. Stress distribution of the gripper

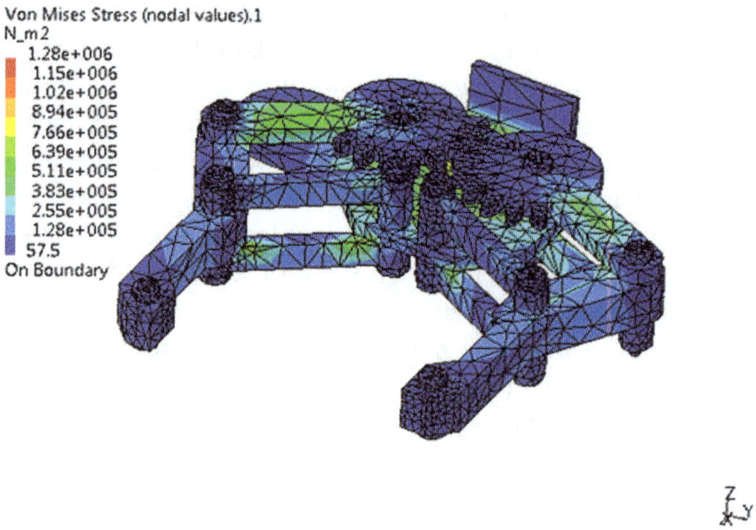

the gripper as the distance from the back of the gripper decreases. The maximum stress area is located near to the clamp at the back of the gripper. More specifically, it is located at the base plate. The stress areas spread out along the surface of the gripper as increasing force is applied.

Table 7 shows the maximum stress when increasing force is applied at the end of the gripper. It shows the effect of the applied force on the magnitude of maximum stress. The maximum stress increases as the applied force increases. The maximum stress is assumed to be at the same point because the structure of the gripper is constant during the analysis. The force is applied at the two contacting surfaces of the gripper. The back of the gripper is clamped to reflect its attachment to the end of the robotic arm.

Figure 15 shows the relationship between maximum stress and applied force at the gripper. It shows the effect of force applied at the end of the gripper. The maximum stress increases linearly with the ap-

Table 7. Maximum stress of the gripper corresponding to the applied force

Applied Force (N)	Maximum Stress (N/m²)
1	1.2770e+006
2	2.5540e+006
3	3.8310e+006
4	5.1080e+006
5	6.3850e+006
6	7.6620e+006
7	8.9390e+006
8	1.0216e+007
9	1.1493e+007
10	1.2770e+007

Figure 15. Maximum stress of gripper versus applied force

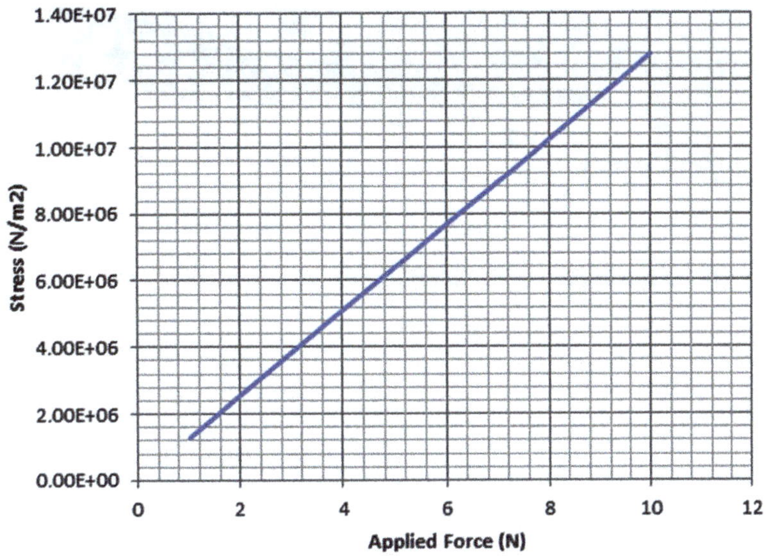

plied force. The force is applied at the two surfaces of the gripper that contact with the specimen to be picked up. In the analysis, all the variables except the applied force is controlled to obtain accurate results.

Static Case of Second Arm

Table 8 shows the material properties used in the static case of the second arm. It shows the values of the properties set before the analysis. The plexiglass material used reflects the real material used for the current model. Since only the static case is considered, only the Young modulus, density and yield strength are used in the analysis.

Figure 16 shows the second arm of the robotic arm drawn through CATIA. It shows the shape of the arm used in the analysis.

Figure 17 shows that the maximum displacement is located at one end. This is because of the moment created across the length of the arm. The displacement increases as the distance from the clamp increases.

Table 8. Properties of the material used for the second arm analysis

Material	Plexiglass
Young Modulus	$2.9 \times 10^{9} Nm^{-2}$
Poisson Ratio	0.4
Density	$1200 kgm^{-3}$
Thermal Expansion	$7 \times 10^{-5} K$
Yield Strength	$5.5 \times 10^{7} Nm^{-2}$

Figure 16. Detailed CATIA drawing of the second arm

Figure 17. Displacement of second arm of current model

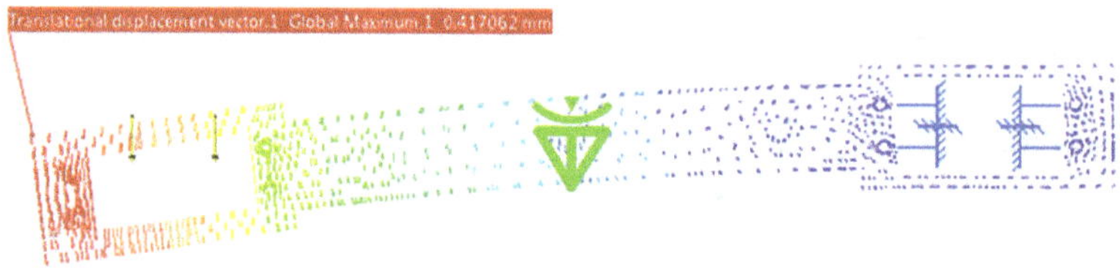

Table 9 shows the maximum displacement when increasing force is applied at one end of the second arm. 1 N of distributed force is applied as shown in the figure above. It shows that the displacement increases as the applied force increases. Besides that, significant displacement occurs when the applied force is 10 N as the maximum displacement exceeds 4 mm.

Figure 18 show the graph of maximum displacement versus applied force for the second arm of the current model.

Table 10 shows the maximum stress corresponding to the applied force at the second arm. It shows the effect of the force on the maximum stress. The maximum stress increases with increasing applied force.

Figure 20 shows the maximum stress when varying force is applied at one end of the second arm. It shows the trend between the maximum stress and applied force. The maximum stress at the second arm increases linearly with the applied force.

Table 9. Maximum displacement of the second arm corresponding to the applied force

Applied Force (N)	Maximum Displacement (mm)
1	0.4170620
2	0.8341240
3	1.2511860
4	1.6682480
5	2.0853100
6	2.5023720
7	2.9194340
8	3.3364496
9	3.7535580
10	4.1706200

Figure 18. Maximum displacement versus applied force for second arm

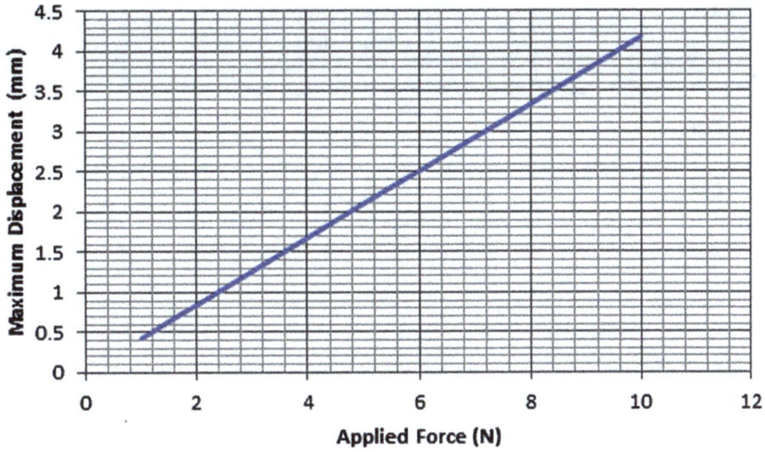

Figure 19. Stress analysis of second arm of current model

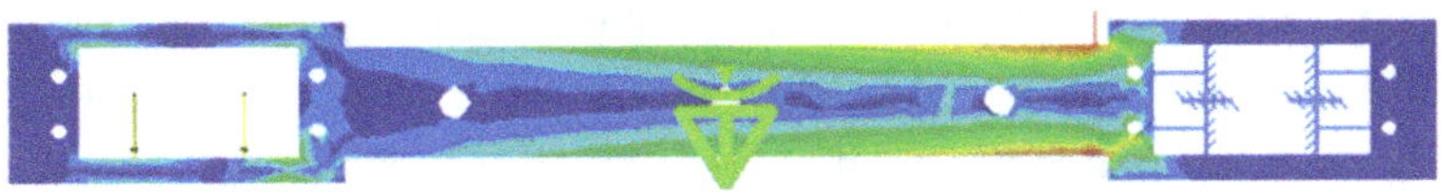

Table 10. Maximum stress at the second arm corresponding to the applied force

Applied Force (N)	Maximum Stress (N/m^2)
1	1.0069e+006
2	2.0138e+006
3	3.0207e+006
4	4.0276e+006
5	5.0345e+006
6	6.0414e+006
7	7.0483e+006
8	8.0552e+006
9	9.0621e+006
10	1.0069e+007

Propose Concepts of Design Improvement

Based on the analyses done as stated earlier, a few modifications have been proposed to get betterment in the design of the robotic arm.

Increasing the motor shaft radius and its effect on the translational displacement (modification 1)

Figure 20. Maximum stress versus applied force for the second arm

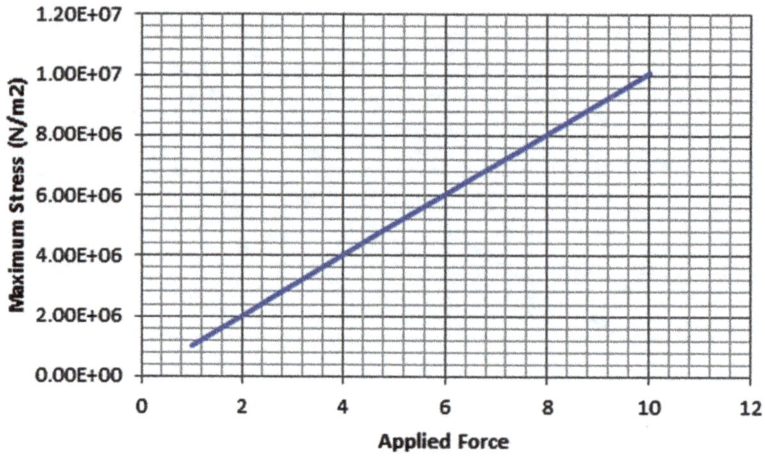

Modification 1 involves an increase in the radius of the motor shaft between the spinner and the base. It was done to increase the ability of the shaft to withstand large forces. The radius of the motor shaft is increased by 1 mm. This modification is applied to all 4 servo motors of the robotic arm.

Table 11 shows the maximum displacement before and after modification as well as the reduction in displacement when a force is applied to the gripper. It was done to determine the effect of the modification on the maximum displacement of the robotic arm. The maximum displacement is assumed to be at the same position for each case since the other parameters are controlled. The maximum displacement is located at the gripper.

Similarly, Table 12 shows the maximum stress before and after modification as well as the reduction in stress when a force is applied to the gripper. It was done to determine the effect of the modification on the maximum stress of the robotic arm. It is shown that the modification of the shaft diameter

Table 11. Maximum displacement comparison for modification 1

Applied Force (N)	Maximum Displacement Before Modification 1 (mm)	Maximum Displacement After Modification 1 (mm)	Reduction in Displacement (mm)
1	5.5585	4.7023	0.8562
2	11.1170	9.4047	1.7123
3	16.6755	14.1070	2.5686
4	22.234	18.8093	3.4248
5	27.7925	23.5116	4.281
6	33.3510	28.2139	5.1372
7	38.9095	32.9163	5.9934
8	44.4680	37.6186	6.8496
9	50.0265	42.3209	7.7058
10	55.5850	47.0232	8.5620

Table 12. Maximum stress comparison and reduction in stress

Applied Force (N)	Maximum Stress Before Modification 1 (N/m²)	Maximum Stress After Modification 1 (N/m²)	Reduction in Stress (N/m²)
1	1.56424e+007	5.39466e+006	1.02477e+007
2	3.12847e+007	1.07893e+007	2.04954e+007
3	4.69271e+007	1.61840e+007	3.07431e+007
4	6.25694e+007	2.15786e+007	4.0991e+007
5	7.82118e+007	2.69733e+007	5.1239e+007
6	9.38542e+007	3.23679e+007	6.1486e+007
7	1.09497e+008	3.77626e+007	7.1734e+007
8	1.25139e+008	4.31573e+007	8.1982e+007
9	1.40781e+008	4.85519e+007	9.2229e+007
10	1.56424e+008	5.39466e+007	1.02477e+008

Figure 21 is a comparison of the maximum displacement before and after modification. Table 13 shows the maximum stress before and after increasing the radius of the motor shaft by 1 mm as well as the reduction in stress.

Figure 22 shows the trend of maximum stress corresponding to the applied force after modification. It shows the effect of the design modification on the reduction in stress. The reduction in stress shows a linear pattern. In other words, the reduction in stress increases linearly with an increase in the applied force.

Edge fillet on the motor shaft and its effect on the maximum displacement (modification 2)

Figure 21. Maximum displacement comparison for modification 1

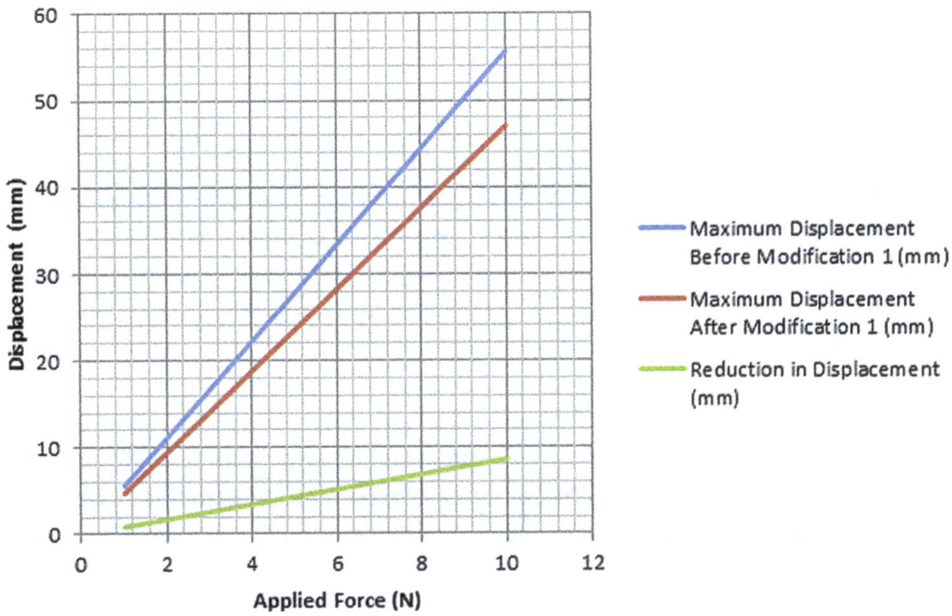

Table 13. Maximum stress comparison for modification 1

Applied Force (N)	Maximum Displacement Before Modification 2 (mm)	Maximum Displacement After Modification 2 (mm)	Reduction in Displacement (mm)
1	5.5585	5.4714	0.0871
2	11.1170	10.9429	0.1741
3	16.6755	16.4142	0.2613
4	22.2340	21.8856	0.3484
5	27.7925	27.3570	0.4355
6	33.3510	32.8283	0.5226
7	38.9095	38.2997	0.6097
8	44.4680	43.7711	0.6968
9	50.0265	49.2425	0.7839
10	55.5850	54.7139	0.8710

Figure 22. Maximum stress comparison for modification 1

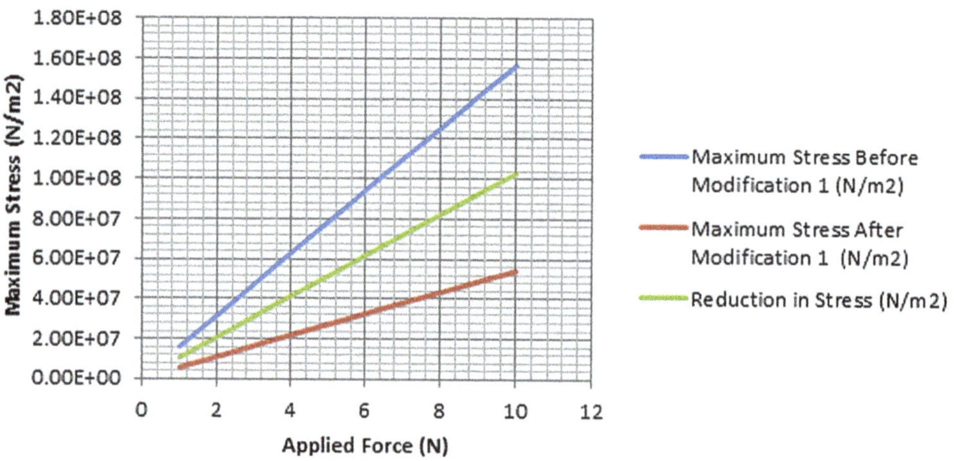

Modification 2 is an edge fillet operation on the motor shaft. It was done to reduce the stress by smoothening out the sharp edges. The radius of the edge fillet at each corner is 0.2 mm. The modification is applied to all the servo motors of the robotic arm to ensure consistency. The difference in maximum displacement after edge filleting the critical stress area shows the difference in displacement is 0.8711 mm. The maximum displacement is located at the gripper. In the analysis, an edge fillet operation is done on the motor shaft to produce smooth and curvy surfaces at the sharp edges. The analysis shows that modification 2 slightly reduces the maximum displacement.

Edge fillet on the motor shaft and its effect on the maximum stress (modification 2)

The maximum stress point when 1 N of force is applied at the gripper after modification 2. It depicts the specific point where the stress is critical. As shown, the maximum stress in this case is 7.79145e+006 N/m^2 which is considerably lower than that of the current design. The maximum stress point is located at the same point, which is the servo motor shaft. However, the stress is more spread out to other parts of the motor and the other side of the joint.

The maximum stress before and after modification 2 as well as the reduction in stress. This was done to determine the effect of modification 2 on the maximum stress. It shows that the reduction in stress increases as the applied force increases. Modification 2 involves an edge filleting operation on the servo motor shaft. In order to identify the maximum stress, the Von-Mises stress visualization is selected in the analysis. Then, the image extreme is selected.

Increase in thickness of the gripper base plate and its effect on the maximum displacement of the gripper (modification 3)

Modification 3 is the increase in thickness of the base plate near the back of the gripper. It was done to increase the capability of the gripper to carry heavier loads. The thickness of the base plate is increased by 1 mm. The displacement when 10N of force is applied at the two contacting surfaces of the gripper. The maximum displacement is shown to be 0.0764882 mm which is lower than the current design with the same force applied. The displacement is reduced by 0.0408838 mm. The maximum displacement is located near one of the two contacting surfaces. The displacement of the gripper reduces going towards the back of the gripper. It shows the effect of increasing the thickness of the gripper base plate, the reduction in displacement increases as the applied force increases.

Figure 23 shows the maximum displacement comparison after modification 3. It shows that modification 3 has significantly reduced the displacement of the gripper. The reduction in displacement is directly proportional to the applied force. All variables except the thickness of the base plate is kept constant to ensure accurate results.

Increase in thickness of the gripper base plate and its effect on the maximum stress of the gripper (modification 3)

The results show that the maximum stress is reduced because the red critical region subsides. The figure is the stress visualization when 10N of force is applied at the two contacting surfaces. Modification 3 has caused the maximum stress to occur at the tip of the gripper. The reduction in stress increases linearly with the applied force. The location of the maximum stress has changed to the tip of the gripper. Modification 3 is the increase in thickness by 1 mm.

Figure 23. Maximum displacement comparison for modification 3

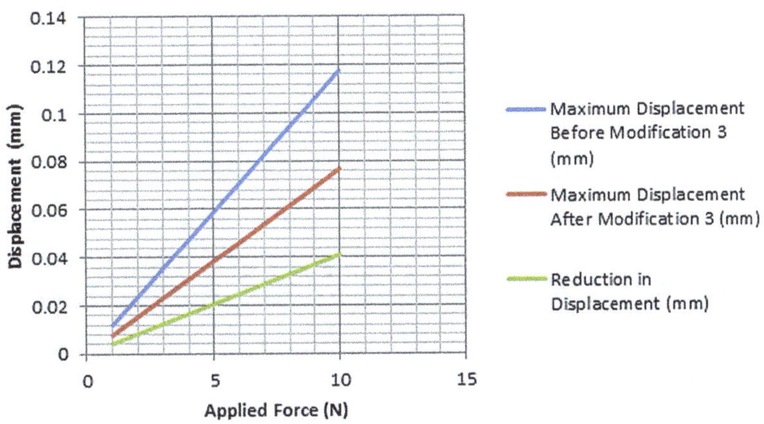

Modification 4 is a decrease in the length of the gripper base plate (Figure 24). It was done to reduce the moment created by the force applied at the end of the gripper. The length of the base plate is decreased by 2 mm from 38 mm to 36 mm.

The analysis shows the effect of reducing the length of the base plate on the maximum displacement numerically. The reduction in displacement is directly proportional to the applied force. The reduction is not as large as that of modification 3. However, modification 4 allows a reduction in material costs.

Figure 25 shows the trend of reduction in displacement after reduction in the length of the base plate. It shows the degree to which the modification reduces the displacement. The reduction in displacement increases linearly with the applied force. All the variables except the applied force is kept constant to ensure the results reflect only the effect of the modification. The displacement is located at the right tip of the gripper.

The analysis of decreasing in length of the gripper base plate and its effect on the maximum stress of the gripper (modification 4) also shows the reduction increases as the applied force increases. Forces varying from 1 to 10 N is applied to the two contacting surfaces. The reduction in stress is directly proportional to the applied force. The length of the base plate is reduced to decrease the moment.

Figure 24. Decrease in length of gripper base plate

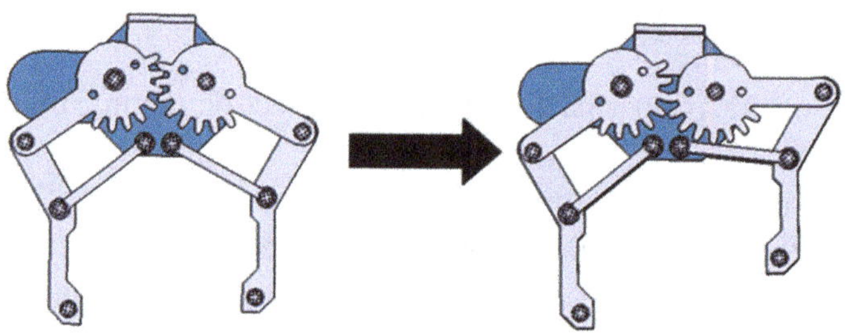

Figure 25. Maximum displacement comparison for modification 4

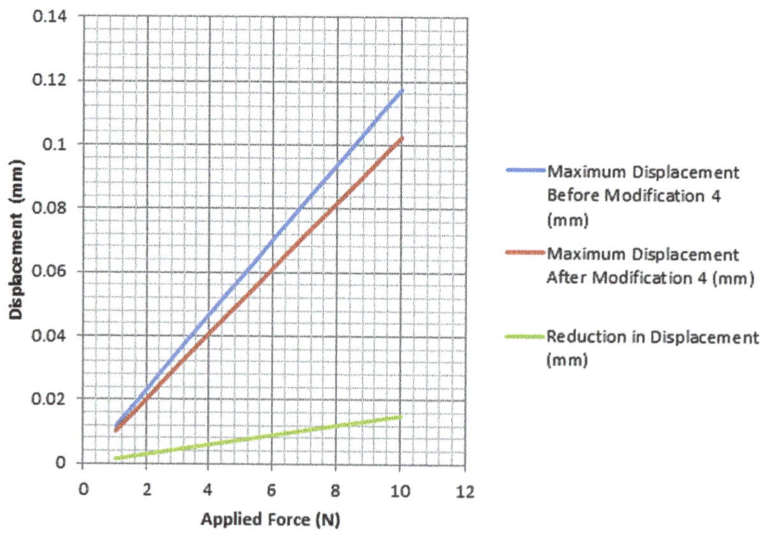

CONCLUSION

The project has successfully identified the critical stress areas and maximum displacement of the robotic arm when it carries out the pick and place operation (Table 14). It has been found that both variables increase linearly with increasing applied force. An increase in the radius and edge filleting of the servo motor shaft, increase in thickness and reduction in the length of the gripper base plate, and the new design of the second arm are proven to increase the capability of the robotic arm to carry heavy loads at extreme angles. The maximum stress is located at the servo motor shaft, the back of the gripper and the top edge of the second arm. The maximum displacement is located near to the applied force. Therefore, the objective to analyse and define critical stress areas as well as determine the location of maximum displacement has been achieved. Other than that, the objective to propose five design improvements on the robotic arm has been achieved since the five modifications have been able to increase the capability of the robotic arm to carry heavier loads. For the overall robotic arm analysis, modification 1 performed better than modification 2. For the case of the gripper, modification 3 is better than modification 4. Modification 5 on the second arm has proven to increase the capability of the robot arm to carry heavier objects.

A possible extension of the project is a proposal of a new mechanism to increase the capability of the robotic arm to carry heavier loads as well as its flexibility of movement. This would increase the functionality of the robotic arm for various industrial applications. For example, it would enable the robotic arm to function in small and confined spaces. The speed of the mechanism can also considered to increase the operation speed of the robotic arm. The location of the mechanism can be analysed to determine the best position for stable operation. The mechanism can include elements of the mechanical system design course studied beforehand. One of the objectives can be to propose new, innovative designs to widen research knowledge on various mechanisms.

The idea of this project is that it offers an increase in the capability of the robotic arm to carry heavier loads and to show the importance of Mechatronics and Robotics and their future impact into heavy industries (Habib, 2008). This is because the increasing needs of Mechatronics and Robotics in the industry to automate its manufacturing process and the increasing of need to design robotic arms

Table 14. Summary of all 5 design changes

Design Change	Result (Reduction)	Improvement in Comparison With Original Design (%)
Increase in the motor shaft radius	Displacement = 8.5620 mm Stress = $1.02 \times 10^8 \, Nm^{-2}$	Displacement = 15.4% Stress = 65.5%
Edge fillet on the motor shaft	Displacement = 0.8710 mm Stress = $7.85 \times 10^7 \, Nm^{-2}$	Displacement = 1.57% Stress = 50.2%
Increase in thickness of the gripper base plate	Displacement = 0.0409 mm Stress = $3.79 \times 10^6 \, Nm^{-2}$	Displacement = 34.8% Stress = 29.7%
Decrease in length of gripper base plate	Displacement = 0.0150 mm Stress = $1.97 \times 10^6 \, Nm^{-2}$	Displacement = 12.8% Stress = 15.4%
Edge fillets on the two long edges at the top of the second arm and increase in thickness of the bottom surface by 4.5mm	Displacement = 1.6062 mm Stress = $1.66 \times 10^6 \, Nm^{-2}$	Displacement = 38.5% Stress = 16.4%

with better capabilities for various applications (Habib, 2007). The design improvement idea can be implemented in the industry to enable the robotic arm to carry heavier loads. The idea can be improved by implementing a mechanism that increases the mechanical advantage of the robotic arm during the pick and place operation.

Future Research Lines

A possible extension of the project is a proposal of a new mechanism to increase the capability of the robotic arm to carry heavier loads as well as its flexibility of movement. This would increase the functionality of the robotic arm for various industrial applications. For example, it would enable the robotic arm to function in small and confined spaces. The speed of the mechanism can also be considered to increase the operation speed of the robotic arm. The location of the mechanism can be analyzed to determine the best position for stable operation. The mechanism can include elements of the mechanical system design course studied beforehand. One of the objectives can be to propose new, innovative designs to widen research knowledge on various mechanisms.

ACKNOWLEDGMENT

The author would like to acknowledge the support from the Fundamental Research Grant Scheme (FRGS) under a grant number of FRGS/1/2014/9003-00442 from the Ministry of Higher Education Malaysia, and Faculty of Engineering Technology, Universiti Malaysia Perlis for the lab facilities. Special thanks to those who contributed to this project directly or indirectly.

REFERENCES

Alexander, E., Chow, Y., Cohen, J., & Jeske, S. (2010). WINROBO Window-Washing Robot. *Docstoc.com*. Retrieved from http://www.docstoc.com/docs/ 23539888/Winrobo -Window-Washing-Robot

Chen, D., Magee, R., & Yakubov, N. (2005). Static and Kinetic Friction. *Revitalizing Achievement by using Instrumentation in Science Education*. Retrieved from http://raise.poly.edu/Information/RAISE_Workshop_PowerPointFiles/Static_&_ _Friction.ppt

Chung, J. H., & Yi, B. J. (2015). Analysis of Internal Loading at Multiple Robotic Systems. *Journal of Mechanical Science and Technology. KSME International Journal*, *19*(8), 14.

Duffy, J. (1996). Statics and Kinematics with application to Robotics. Cambridge University Press, Amazon.com

El-Kalay, A. K., Akyurt, M., Aljawi, A. A. N., & Dehlawi, F. M. (1995). On Gripping Mechanism for Industrial Robots. *King Abdulaziz University Official Website*. Retrieved from http://www.kau.edu.sa/Files/320/Researches/46479_17.pdf

Habib, M. K. (2007, Summer). Mechatronics: A Unifying Interdisciplinary and Intelligent Engineering Paradigm. *IEEE Industrial Electronics Magazine*, *1*(2), 12–24. doi:10.1109/MIE.2007.901480

Habib, M. K. (2008, August). Human Adaptive and Friendly Mechatronics (HAFM), IEEE International Conference on Mechatronics and Automation 'ICMA2008', Takamatsu-Kagawa, Japan, pp. 61-65. 10.1109/ICMA.2008.4798726

Haney, P. (2004). Rubber Friction. *Inside Racing Technology*. Retrieved from http://insideracingtechnology.com/tirebkexerpt1.htm

Inoue, T. & Hirai, S. (2009). *Mechanics and Control of Soft-fingered Manipulation.* Springer.

Jazar, R. N. (2010). Theory of applied robotics: kinematics, dynamics, and control. Berlin, Germany: Springer Science & Business Media. doi:10.1007/978-1-4419-1750-8

Mon, T. T., Mohd Romlay, F. R., & Tamin, M. N. (2012). Role of Finite Element Analysis in Designing Multi-axes Positioning for Robotic Manipulators. In E. Hall (Ed.), Advances in Robot Manipulators, InTech.

Motor selection. (n.d.). *Panasonic Global*. Retrieved from http://industrial.panasonic.com/ww/i_e/25000/fa_pro_sgeard_shing1_e/fa_pro_sgeard_shing1_e/ctlg_geared_e_9.pdf

Motor Torque Calculation. (n.d.). *Selection Guides-Leadshine Technology Co., Ltd.* Retrieved from http://www.leadshine.com/UploadFile/documents_software/ Applications/Motor_Torque_Calculation.pdf

Pennestri, E. (1991). On The Kinematic Analysis Of Geared Robotic Wrists. Springer. *Meccanica*, 1991.

Purushotham, A., & Anjeneyulu, M. J. (2013). Kane's Method for Robotic Arm Dynamics: a Novel Approach. *IOSR Journal of Mechanical and Civil Engineering (IOSR-JMCE) 6*(4): 07-13.

Razali, Z. B., & Lee, G. B. (2012, February) *High-end technology for Effective Operating Environment in a Glass Manufacturing Industry.* Paper presented at the International Conference on Application and Design in Mechanical Engineering (ICADME2012), Penang, Malaysia.

Razali, Z. B., & Lee, G. B. (2013). High-End Technology for Effective Glass Manufacturing Industry: Innovative Operating Environment. Germany: LAP Lambert Academic Publishing.

Razali, Z. B., & Madasamy, G. (2013). Automated Mechanical Sorting Device for Mixed Household Waste: Integration Sorting Conveyor with Neural Network. Germany: LAP Lambert Academic Publishing, Germany.

Razali, Z. B., & Madasamy, S. (2012, February). *An Application of Image Processing for Automated Mixed Household Waste Sorting System.* Paper presented at the International Conference on Man-Machine System (ICoMMS2012), Penang, Malaysia.

Razali, Z. B., & Mat Nawi, N. A. (2013). Automated Segregating Device for Reuse and Recycle Waste: Application on size based, friendly recycling design. Germany: LAP Lambert Academic Publishing, Germany.

Ryan, J. V. (2010). Energy Solutions from Glass Road Surfaces. *Glass Manufacturing Industry Council.* Retrieved from http://www.gmic.org/News/ Joe_Ryan_Energy_Solutions_from_Glass_Road_Surfaces.pdf

Singh, P., Kumar, A., & Vashisth, M. (2013). Design of a Robotic Arm with Gripper & End Effector for Spot Welding. *Universal Journal of Mechanical Engineering*, *1*(3), 6. doi:10.13189/ujme.2013.010303

Spentzas, K. N., & Kanarachos, S. A. (2002). An incremental finite element analysis of mechanisms and robots.

Yuan, B. S., & Book, W. J. (1993). Dynamics of Flexible Manipulator Arms: Alternative Derivation, Verification, and Characteristics for Control. ASME 115.

Chapter 9
Control of a Wing Type Flat-Plate for an Ornithopter Autonomous Robot With Differential Flatness

Elkin Yesid Veslin Díaz
Universidade Federal do Rio de Janeiro, Brazil

Cesar Francisco Bogado-Martínez
Universidad Nacional de Asunción, Paraguay

Max Suell Dutra
Universidade Federal do Rio de Janeiro, Brazil

Luciano Santos Constantin Raptopoulos
Centro Federal de Educação Tecnológica Celso Suckow da Fonseca, Brazil

ABSTRACT

In this chapter, a two-degree-of-freedom controller that exploits the flat properties of a three degree-of-freedom wing type flat-plate for an Ornithopter Autonomous robot is proposed. A set of kinematical patterns inspired by nature is used to simulate the wing's movement around two wingtip trajectories; also, the effects of the aerodynamical forces as a function of the wind velocity and the wing's angle of attack are considered. In order to control the system, the effects of these forces are viewed as disturbs that affect the wing's dynamics. The proposed control scheme drives the device through the desired path by generating a set of desired inputs which are compensated by a feedback loop when the aerodynamical forces actuate upon the system.

DOI: 10.4018/978-1-7998-0137-5.ch009

INTRODUCTION

Mechatronic engineering contributions has been becoming a strong requirement to increase the performance of any device, as the addition of an intelligent processing block increments their capabilities to expand the solution margin. However, it is the synergy concept that turns strong the mechatronics design paradigm, as it is intended as a mechanical structure or any system, that can organize and process their proper information through an efficient architecture, that is able to deal with instruments (sensors and actuators), control system, diagnosis, and communication (Habib, 2008).

In the mechatronics engineer concept, the control system acts like the brain (Rahn, 2002), it becomes from a knowledge base, constructing from it a precise model of the system to be controlled through feedback and feedforward loops. To build a proper control algorithm, it is required to a system's mathematical model, parameter identification and estimation, controller design methods and control performance criteria (Isermann, 2005). Being them, the research of control systems is an important requirement into the design of any mechatronic system, it is intended to deal with statics and dynamics behaviors, meaning that, it is able to compensate uncertainties generated by non-linear external or internal factors, who may cause instability when they are not properly faced.

In this chapter, the study and evaluation of a control scheme for an ornithopter wing is proposed. The control scheme will be able to deal with the proper system's dynamics and the aerodynamic forces that continuously affect the wing's stability. To guarantee the proper model design, biomimetic characteristics will be considered into the dynamic model study.

The design of an aircraft with biomimetic characteristics will lead to developing a device with the runway and reduced takeoffs, energy efficiency during flight, high maneuverability and low noise generation. Diverse academics works focused in this area should be found, some of them especially focused in military applications, others present a research emphasis on specific areas: as the wing design, aerodynamical studies, wing kinematics, drive mechanism design, mathematical modeling, control, and simulation.

For the development of this work, an ornithopter wing is considered as an open chain structure consisting of a flat-plate structure. This structure that constitutes the wing's body is attached to the robot body through a spherical joint, named as the shoulder joint, which performs three movements during flying, they are the upstroke and downstroke movement; the rotation of the wing; and, the abduction and adduction movement.

The device's parametric values are based in morphometric data of a pigeon, with this information, both the kinematical and dynamical model are determined. To model the flight dynamics and wing motion, the hypothesis that the bird is flying at a constant velocity in a horizontal rectilinear trajectory is considered.

This work introduces Differential Flatness as a control solution to obtain an appropriate set of joint's moments (inputs) required to perform the flight movement, for doing this, information based on natural wing movement parameterized as trigonometric functions are used. The results are introduced into a feedback control scheme to ensure path tracking during the flying in the presence or not of disturbs generated by aerodynamical forces. Two wingtip trajectories patterns are implemented in this study: the oval and shape-of-eight (infinite) trajectories, for both cases, the results prove that the Differential Flatness is profitable to determine the wing's inputs to develop the flying movement and control them. A flowchart of the proposed work is presented below.

Figure 1. Work Flowchart.

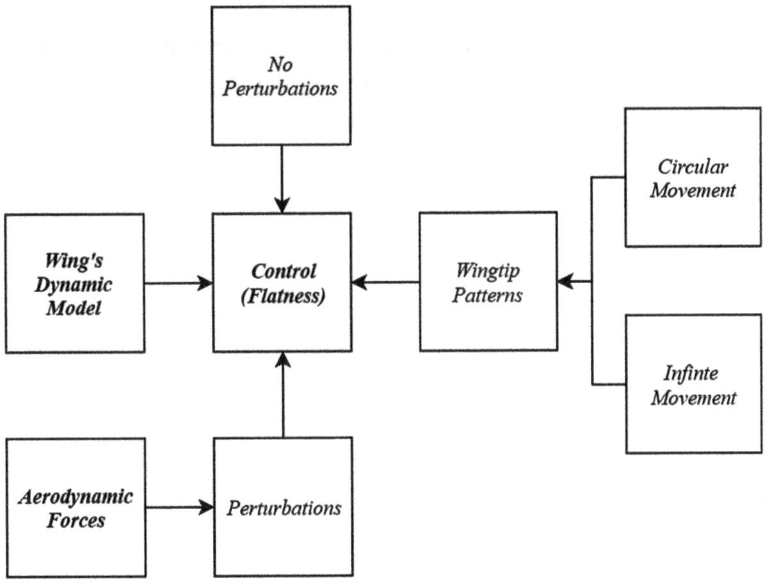

This chapter is organized as follows: first, the Ornithopter Robots state of the technique and the Differential Flatness concepts are presented. Then, the mechanical and dynamical criteria used for ornithopter modeling and simulation made with the Matlab® software are discussed. Afterward, the joint trajectories based in natural movements are introduced; and finally, Differential Flatness and the feedback control rules are introduced to the model. Results of the simulation and the effectiveness of the control concept are discussed. Conclusions and future research end with this chapter.

BACKGROUND

Ornithopters

The designing of aircraft based in ornithopters models provide devices with the following characteristics: runway and reduced takeoffs, energy efficiency in flight, high maneuverability and low noise generation. Current designs are based or inspired by biological models, allowing to gain insight by observation and to explain biological function by means of engineering models (Pons, 2008). Therefore, could be found analysis based in the observation of flying, that allows the understanding of how both birds and insects use their aerodynamical mechanism to gain lift, better than the predicted by the conventional mechanism in stationary aerodynamics (Roccia et al., 2007).

There are a variety of academic works in this area, mainly related to military applications. DeLaurier (1988) designed and built a manned piloted ornithopter that took off and flew for a short time. The author suggested that the importance of research on such devices is that they may be a new type of transport where the flight is clearly bird-like (DeLaurier & Harris, 1993). Park and Yoon (2008) introduces the ornithopter design in small aircraft, inspiring their designing by real insect wings and mimics their flying.

Baek et al. (2011) designed a small flight device with control integrated, introducing a simplified model of ornithopter flight to reduce the order of the control system. The researchers implemented closed-loop feedback with PID controllers to control both, the desired angle and the orientation of the device.

On the other hand, Colorado (2012) considered to mimicking of the bat flight to reproduce their maneuverability, resulting in a flying robot that eliminates the need of flaps and rudders that performs more efficient maneuvers, useful in confined spaces. Other related works based on the study and design of biological inspired flying devices could be found, as tailless flapping-wing micro air vehicles (FM-MAB) (Nguyen & Chan, 2018), hovering flight inspired in fruit flies (Fry et al., 2005), and inspired by birds devices (Bogado Martínez et al., 2018), among others.

These works use a similar approximation to design flying devices. DeLaurier et al. (1993) designed a fixed-wing aircraft based in the projection of the wing and driving mechanism. Park et al. (2008) designed the wing and their driving mechanism. Colorado (2012) designed the kinematic formulation of the wing based in the flight of the bat, projecting posteriorly their dynamics and aerodynamics, and selecting the appropriate materials for the structure design and the control on the actuators. Finally, Fry et al. (2005) studied the insect's movement, replaying their wing kinematics and related aerodynamical forces, allowing him to calculate their flight power.

In general, the flight device design can be summarized as follow: a flight biology study of the species from which it is inspired is made, then, the wing is designed and their aerodynamic is studied. Posteriorly the drive mechanism is designed, a study of kinematics and dynamics of the structure is made, then the wing control is proposed and finally, flight simulations are made.

Fewer researches focus on the development of the entire flight device, instead, their studies are focused into specific parts, such as: wings designs (Liu, 2006), kinematics and aerodynamics (Fry et al., 2005), (Dietl & Garcia, 2008), drive mechanism (Ma et al., 2012), modelling and control of wing movement (Chirarattananon et al., 2014), (Ma et al., 2013), (Armanini et al., 2016), flight control and simulation (Baek et al., 2011), (Pfeiffer et al., 2010), (Verboom et al., 2015) and obstacle avoidance (Lamers et al., 2016)

On the other hand, extended and advanced research could be addressed, as the design and performance of a micro flying device type flapping robot (De Croon et al., 2012) and (De Wagter et al., 2014), whose control system allows the device to control height and obstacle avoidance using stereo vision systems. Forward works of this device will consider adding a multiple degrees-of-freedom wing-structure to gain flying stability, removing the tail. More advances into the research of low-weight flying robots could be found in (Phan et al., 2017), who controlled a four-bar mechanism applying PD feedback law.

In previous works, the authors had been proposed a control system to control a flapping-wing (Bogado Martínez et al., 2018), in these work, a feedback linearization control method with a *PD* controller was proposed, the results allowed to conclude that the control scheme was able to drive the system actuators for both the stabilization of position and the tracking of trajectories. For this work, was considered to test the model with a control system based in differential flatness in a two degree of freedom architecture, with this method, no linearization of the dynamical model is required, and the system response could be stabilized around the desired trajectory using a linear feedback controller.

Differential Flatness

The Differential Flatness concept was firstly introduced by Michel Fliess using differential algebra terminology (Fliess et al., 1995). According to this theory, a system is seen as a differential camp generated by a set of variables (states x and inputs u). Subsequently, Martin et al., (1997) redefined the concept in a geometric context, in which flat systems could be described in absolute equivalence terms.

According to Fliess et al., (1995), a system is defined flat if exits a set of outputs, called flat outputs, that can determine all system states and inputs without a needed of integration. More formally, it means that, a system of states $x \in \mathbb{R}^n$ and inputs $u \in \mathbb{R}^m$ is flat, if can be found a set of outputs $y \in \mathbb{R}^m$ on the form:

$$y = h\left(x, u, \dot{u}, \ldots, u^r\right) \tag{1}$$

Such that,

$$x = f\left(y, \dot{y}, \ldots, y^q\right) \tag{2}$$

$$u = g\left(y, \dot{y}, \ldots, y^q\right) \tag{3}$$

As seen in Equations (2) and (3), in a flat system both states x and inputs u are expressed by the flat outputs y and a finite number q of derivatives. This property has been called by Fliess as diffeomorphism; and it is helpful when explicit trajectory generation is required (P. Martin et al., 1997), it means, when a system requires the desired behavior, it is achieved specifically through a trajectory in the output space.

Considering a system as Flat, through Equation (3) is possible to find a set of inputs u such that the system behavior is related to a trajectory. Several works presents the use of differential flatness, including, mobile robotics (Ryu & Agrawal, 2011),(Nasr et al., 2019), aircraft control (Philippe Martin & Paden, 1994), robotic manipulators (Markus et al., 2017), (Veslin Diaz et al., 2011), among other systems (Murray et al., 1995).

Differential flat systems are useful in situations where explicit trajectory generation is required (Nieuwstadt, 1996). As the flat outputs y determines the behavior of the system, trajectories are planned in the output space, mapping the appropriate inputs, efficiently generating possible trajectories.

Flat Properties of Robotics Systems

According to Levine (2009), a robotic system could be defined as a flat system. Considering that the system is composed of n degrees of freedom with n actuators, the system dynamics is defined as:

$$uQ\left(q, \dot{q}\right) = M\left(q\right)\ddot{q} + C\left(q, \dot{q}\right)\dot{q} + K\left(q\right) \tag{4}$$

Where, **q** stands for the generalized coordinates, **u** the system inputs with $\dim q = \dim u$, where $M(q)\ddot{q}$ is the inertial matrix, $C(q,\dot{q})\dot{q}$ is the centripetal and centrifugal forces matrix, $K(q)$ is the gravitational forces matrix, and $\mathbf{Q}(q,\dot{q})$ is a matrix that characterizes the actuators of the system.

Considering that the system had one single generalized coordinate $q \in \mathbb{R}^{1\times1}$, and considering that $x_1 = q$ and $x_2 = \dot{q}$, equation (4) could be rewritten as a set of EDO's as:

$$\dot{x}_1 = x_2$$

$$\dot{x}_2 = M(x_1)^{-1}\left(uQ(x_1,x_2) - C(x_1,x_2)x_2 - K(x_1)\right) \tag{5}$$

If the dimension of Q is *n*, the input *u* could be defined as

$$u = Q(x_1,x_2)^{-1}\left(M(x_1)\dot{x}_2 + C(x_1,x_2)x_2 + K(x_1)\right) \tag{6}$$

They are x_n the flat output, that are the systems generalized coordinates (angular positions). In both cases, the dynamic system states *x*, and the inputs *u* are defined as a function of the generalized coordinates positions and their derivatives, thus, is it said that the system is flat. This definition could be extended to every Lagrangian system if it is totally actuated.

If an output *y* if differentially flat, it could be labeled as $z \in \mathbb{R}^m$, then, equations (2) and (3), could be rewritten as:

$$x = f\left(z, \dot{z}, \ldots, z^q\right) = f(\overline{z}) \tag{7}$$

$$u = g\left(z, \dot{z}, \ldots, z^q\right) = g(\overline{z}) \tag{8}$$

Where $\overline{z}$, is named *flag flat* and stands for $\overline{z} = \left[z, \dot{z}, \ldots, z^q\right]$.

Two Degrees-Of-Freedom Controller

The two degrees-of-freedom controller (*Figure 2*) is a technique of non-linear system control. A trajectory generator block composes the desired input u_d based in a set of desired state space x_d, the second block of the controller is a feedback compensator which is responsible such in case of perturbations of the trajectory tracking, by stabilizing the system around the desired path by means of the control signal δ_u which compensates the desired input u_d.

Figure 2. Two Degree of Freedom Controller.
Adapted from (Nieuwstadt & Murray, 1997).

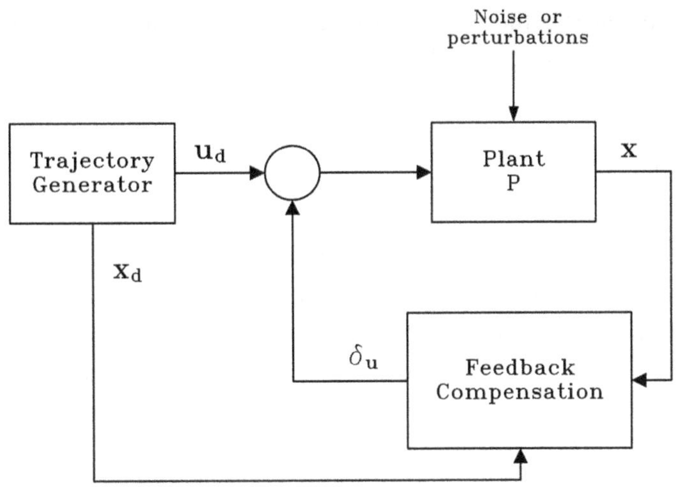

The two degree of freedom controller is considered for this problem, due to it exploits the geometric structure of a differentially flat system without transforming it into a single linear system by separating, in an explicit way, the tracking problem into a trajectory generation block, that generates feasible trajectories in the input space and a feedback compensation block (Nieuwstadt et al., 1997).

According to Nieuwstadt & Murray (1997), differential flatness is equivalent to dynamic feedback linearization using a class of invertible dynamic feedbacks. Flatness generates feasible trajectories rather than transform the system into a single linear system. A linear controller is integrated to maintain the stability of the system around the generated trajectory, linearizing the input/output response of the system and not the full state dynamics, specifically, a portion of the system dynamics, being the remaining (that is well-known) evolving independently, driven by the inputs required to track the output signal.

In order to illustrate this assertion, in Equation (9) is presented a standard control law applied in feedback linearization problem for tracking the desired trajectory *y*. The system output has dimension $y \in \mathbb{R}^n$, so that is full state linearizable. Therefore, their control law is:

$$u = \alpha(x) + \beta(x)\left(y_d^{(n)} + K(\xi - \xi_d)\right) \tag{9}$$

Where $y_d(\cdot)$ is the desired output trajectory, ξ_d is the result of their differentiation, and $\xi = \phi(x)$ is the dynamic system transformed to a linear one with $\xi \in \mathbb{R}^n$.

According to Nieuwstadt et al. (1997), the control law written in Equation (9), can be viewed as a two degree of freedom controller rearranging their terms. Becoming:

$$u = \underbrace{\left(\alpha(x) + \beta(x)y_d^{(n)}\right)}_{feedforward} + \underbrace{\beta(x)K(\xi - \xi_d)}_{feedback} \tag{10}$$

In Equation (10), the feedforward term determines the inputs that are required to track the desired trajectory. Meanwhile, the feedback term is the applied control law, that corrects any error generated by uncertainties or external disturbs in the model. If the feedforward term in Equation (10) is a differentially flat system instead of a feedback linearizable, the controller could be rewritten as:

$$u = \left(\alpha\left(\overline{z_d}\right) + \beta\left(\overline{z_d}\right) z_d^{(n)}\right) + k\left(\overline{z_d}\right)\left(x - x_d\left(\overline{z_d}\right)\right) \tag{11}$$

Where the inputs **u** depend on the flat outputs and their derivatives $\overline{z_d}$, and can be determined in an open-loop form; meanwhile, the control of the feedback loop depends on the error between the non-linear system and the result of the desired state in function of the flat output, the gain **k**, named by Van Nieuwstadt, as a scheduled gain, allows a tradeoff between performance and input magnitude that change depending on the operating conditions.

Therefore, in the absence of perturbations, the controller could drive the system through the desired path with no compensation requirements, due to the system is driven by the feedforward term through a feasible state space trajectory in an open loop fashion.

WING MODELING AND CONTROL

In nature, a bird's wing features a physic structure composed of three joints. The first joint named as *elbow joint* it's a three degrees-of-freedom joint that performs three movements. The ventral to dorsal and dorsal to ventral movements, named also as *upstroke* and *downstroke*, respectively, that move the bird's arm from the front to their belly toward the back, and from the front of the back toward the belly. The rotation in the wing axis, that is named *pronation* that is when the wing changes from downstroke phase to upstroke and, being their inverse action called as *supination*, this movement also is named as the *rotation* of the wing. And, the *abduction* and *adduction* movement, which consists of taking the arm from the legs side toward their head, and carrying the arm from the head side toward the legs, respectively.

The remaining wing joints, the *elbow,* and the *pulse* act like a hinge rotating in one axis. For this study the movements of the second and third joints were neglected, this removal simplifies the model and their results give a suitable performance, as was seen in DeLaurier et al. (1993).

The wing of the bird form is simplified to a rectangle (*Figure 3*) of length $b/2$ and width *c*, where *b* is the total wingspan of the bird and *c* is the length of the mean wing chord. For this work was considered Tenenbaum (2013) that the wing had negligible thickness. For calculations, the wing length will be named L_1, whose value is $b/2$.

Kinematic Model

In Sane & Dickinson (2001) and Roccia et al. (2007), a model of an insect wing kinematic is presented. This model represents the trajectories that both bird and insects perform, hence, there will be adapted to model the system in this work.

Figure 3. Wing model for the study

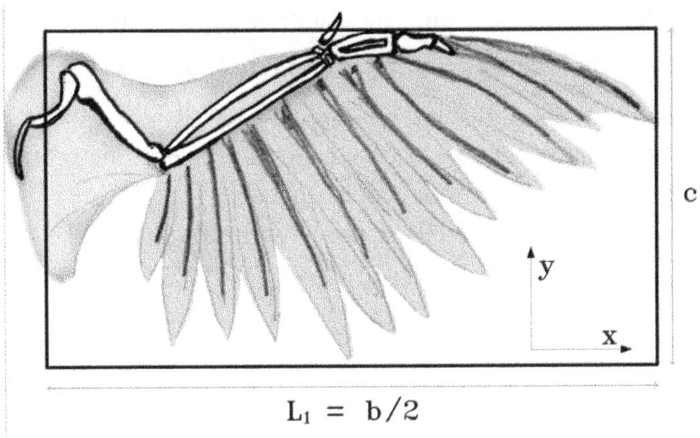

The bird's wing model to determine the position of the wingtip is formulated through Euler angels. In *Figure 4* is presented the configuration of the system, defining a reference inertial frame defined as $\boldsymbol{n} = \begin{bmatrix} n_1 & n_2 & n_3 \end{bmatrix}$ that is placed at the body centers, a second reference frame $\boldsymbol{a} = \begin{bmatrix} a_1 & a_2 & a_3 \end{bmatrix}$ is placed in the bird's elbow where the rotational movements are executed, a third reference frame $\boldsymbol{b} = \begin{bmatrix} b_1 & b_2 & b_3 \end{bmatrix}$ that is fixed to the wing and is located at their beginning which orientation depends on the movements on the reference frame **a**. A third reference frame $\boldsymbol{c} = \begin{bmatrix} c_1 & c_2 & c_3 \end{bmatrix}$, with similar orientation than **b** is used to describe the wingtip position.

The upstroke and downstroke movement or *stroke position* is named as α_1, that is the deviation angle of the wing axis and it is executed in the stroke plane rotating around a_3 axis; the angle that rep-

Figure 4. The configuration of wing kinematics

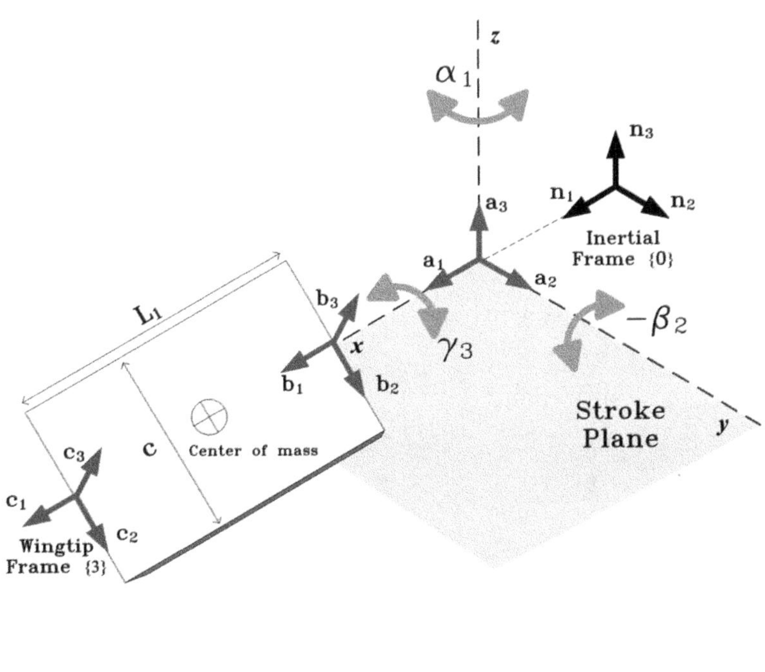

resents the movement of abduction and adduction or *stroke deviation,* named as β_2, that acts normal to the stroke plane. rotating around a_2 axis; and, the rotation angle of the wing or *angle of attack*, that rotates with respect to their wing longitudinal axis a_1, and is named as γ_3.

Being L_1 the length of the wing along the b_1 axis and the distance between frames **b** and **c**, and considering that there is no distance between the inertial reference frame **n** and the elbow frame **a** and neither between frames **a** and **b**, using homogeneous matrices it is defined the set of transformations that leads from the inertial reference frame $\{0\}$ to the wingtip frame in $\{3\}$ as:

$$\boldsymbol{T}_4^0 = \boldsymbol{T}_1^0\left(\alpha_1, a_3\right)\boldsymbol{T}_2^1\left(-\beta_2, a_2\right)\boldsymbol{T}_3^2\left(\gamma_3, a_1\right)\boldsymbol{T}_4^3\left(L_1, b_1\right) \tag{12}$$

Note that the angle β_2 rotates in an inverse way, so that when a positive value in the angle is introduced, the wing rotates in the clockwise direction.

The transformation matrix response in Equation (12) can be represented as:

$$\boldsymbol{T}_4^0 = \begin{bmatrix} n_x & o_x & a_x & p_x \\ n_y & o_y & a_y & p_y \\ n_z & o_z & a_z & p_z \\ 0 & 0 & 0 & 1 \end{bmatrix} \tag{13}$$

Where the position of the wingtip in relation to the inertial reference frame is defined by the vector **p**, the vectors **n, o** and **a** define the direction of the wingtip according to the movements executed by the rotation angles $\alpha_1, -\beta_2,$ and γ_3 and comprise the system's rotation matrix **R**.

The linear velocity of the wingtip related to the inertial frame $\{0\}$ is the derivation of vector **p** respect to time, that is:

$$\boldsymbol{v} = \frac{d}{dt}\begin{bmatrix} p_x \\ p_y \\ p_z \end{bmatrix} = \begin{bmatrix} v_x \\ v_y \\ v_z \end{bmatrix} \tag{14}$$

Thus, ***v*** equals to:

$$\boldsymbol{v} = \begin{bmatrix} -L_1\left(\cos\left(\beta_2\right)\sin\left(\alpha_1\right)\dot{\alpha}_1 + \cos\left(\alpha_1\right)\sin\left(\beta_2\right)\overset{\text{Ù}}{\beta}_2\right) \\ L_1\left(\cos\left(\alpha_1\right)\cos\left(\beta_2\right)\dot{\alpha}_1 - \sin\left(\alpha_1\right)\sin\left(\beta_2\right)\overset{\text{Ù}}{\beta}_2\right) \\ L_1\cos\left(\beta_2\right)\overset{\text{Ù}}{\beta}_2 \end{bmatrix} \tag{15}$$

The angular velocity ω of the wing according to Tenenbaum (2013) is:

$$\omega = J_{\omega}\dot{q} \tag{16}$$

Where J_{ω} is the angular Jacobians matrix of dimension $\mathbb{R}^{3\times3}$, and $\dot{q}$, is the arrangement of time derivatives of the angles. Considering the axis of rotations of the angles, J_{ω} is determined as:

$$\boldsymbol{J}_{\omega} = \begin{bmatrix} \boldsymbol{R}_1^0 \begin{bmatrix} 0 \\ 0 \\ 1 \end{bmatrix} & \boldsymbol{R}_2^0 \begin{bmatrix} 0 \\ 1 \\ 0 \end{bmatrix} & \boldsymbol{R}_3^0 \begin{bmatrix} 1 \\ 0 \\ 0 \end{bmatrix} \end{bmatrix} \tag{17}$$

Where $\boldsymbol{R}_n^0$ is the rotation matrix ($\begin{bmatrix} n & o & a \end{bmatrix}$) of the *n* degree-of-freedom seen from the inertial frame {0}, therefore, from Equation (16) ω is obtained as:

$$\boldsymbol{\omega} = \begin{bmatrix} \sin(\alpha_1)\dot{\beta}_2 + \cos(\alpha_1)\cos(\beta_2)\dot{\gamma}_3 \\ -\cos(\alpha_1)\dot{\beta}_2 + \sin(\alpha_1)\cos(\beta_2)\dot{\gamma}_3 \\ \dot{\alpha}_1 + \sin(\beta_2)\dot{\gamma}_3 \end{bmatrix} \tag{18}$$

Dynamical Model

The dynamical model of the wing system is determined using a Lagrangian conception, in this case:

$$\frac{d}{dt}\left(\frac{\partial L}{\partial \dot{q}_n}\right) - \frac{\partial L}{\partial q_n} = Q_n \tag{19}$$

Where q_n stands for the generalized coordinates, corresponding to the angular displacements of the wing movement, being $q_1 = \alpha_1$, $q_2 = \beta_2$, and $q_3 = \gamma_3$. *L* is the Lagrangian model, and Q_n is the resultant torque, considering that exists external disturbs originating from the aerodynamical forces that generate moments on the wing, Equation (19) takes the form of:

$$Q_n = u_n + \vec{\tau}_{a,n} \tag{20}$$

Where $\vec{\tau}_{a,n}$ correspond to the aerodynamic torque, resulting from the aerodynamical forces that actuate on the wings and are actuating upon the *n* generalized coordinate axis. This aerodynamical forces will be addressed in the next section. Finally, the term u_n stands for the external torque applied to the n generalized coordinate.

Considering that the wing has its center of mass at $L_1 / 2$, has mass m, and its form is equivalent to a rectangular area of $L_1 \times c$ that rotates with respect to $\{0\}$, using the Steiner Theorem the wing inertia $\boldsymbol{I}_0$ equals to:

$$\boldsymbol{I}_0 = \begin{bmatrix} \frac{c^2 m}{12} & 0 & 0 \\ 0 & \frac{L_1^2 m}{3} & 0 \\ 0 & 0 & \frac{m\left(4L_1^2 + c^2\right)}{12} \end{bmatrix} \tag{21}$$

Thus, the Lagrangian model of the wing is written as:

$$L = T + V = \left(\frac{1}{2}\omega I_0 \omega^T + \frac{1}{2} m v v^T\right) + \left(mg\frac{p_z}{2}\right) \tag{22}$$

Being T the kinetical energy and V the potential energy of the wing. Note that gravity actuates on the z-axis, p_z is the wingtip position upon that axis obtained from (12). Finally applying Equation (22) into (19), there are obtained three models with respect to the three generalized coordinates n being:

$$M\left(q\right)\ddot{q} + C\left(q,\dot{q}\right)\dot{q} + K\left(q\right) - \vec{\tau}_{a,n} = u_n \tag{23}$$

Aerodynamical Forces and Torques

The aerodynamic forces calculation was done considering a quasi-static approach. The aerodynamic hysteresis coefficients were considered negligible and only the area of linear behavior of these coefficients was considered; thus, a linear relationship with the variation of the angle of attack was proposed.

The calculation of the angle of attack α is an important item, as the tables of the aerodynamics coefficients depend directly on its value. The angle of attack (*Figure 5*) is defined as the angle existing between the wind chord c and the total flow incident on the wing V_R, that is the velocity seen by the wing at a distance $c / 4$ from the leading edge (Muijres et al., 2012), that is estimated as $V_R = V_\infty + v_a$, where: V_∞ is the flow velocity far ahead of the body (Anderson, 2001), and $\mathbf{v}_a$ is the linear velocity of the wing at $c / 4$ from the leading edge.

Using (*Figure 5*) as a reference, and having that the angles α and θ are complementary, using the scalar product definition is obtained:

Figure 5. Calculation of the angle of attack from the vectors v_∞*, flow velocity incident on the wing, and* $\hat{\eta}_{or}$*, normal unit vector to the wing*

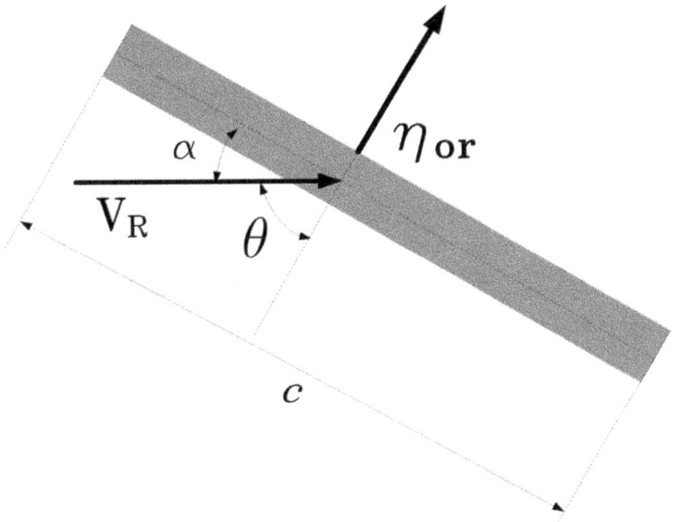

$$\cos\left(\theta\right) = \frac{\hat{\eta}_{or} \cdot V_R}{\left|V_R\right|} \tag{24}$$

Where $\hat{\eta}_{or}$ is normal to the wing, which has the same value that vector **a** in Equation (13). From Equation (24), and taking into consideration the complementarity of the angles, Equation (25) is obtained, this expression relates the angle of attack α with vectors V_R and $\hat{\eta}_{or}$.

$$\sin\left(\alpha\right) = \frac{\hat{\eta}_{or} \cdot V_R}{\left|V_R\right|} \tag{25}$$

To calculate the aerodynamic forces, the equations obtained from (Anderson, 2001) are used as:

$$L = C_l\left(\frac{1}{2}\rho\left|V_R\right|^2\right)S_A \tag{26}$$

$$D = C_d\left(\frac{1}{2}\rho\left|V_R\right|^2\right)S_A \tag{27}$$

$$M = C_m \left(\frac{1}{2} \rho \left| V_R \right|^2 \right) S_A c \tag{28}$$

Where L is the lift force, which has a perpendicular direction to the stroke plane; D is the drag force that has an opposite direction to the aircraft movement; M is the torque generated by the aerodynamic force at $c / 4$ of the wing profile, measured from the leading edge; ρ is the air density; S_A is the planimetric area of the wing, that is, the area obtained by multiplying the chord by the current wingspan of the wing under study; and C_l, C_d and C_m are the lift, drag and moment coefficients respectively, which are obtained through tables, being referenced from Selig et al. (2012) and which values depend on the wing's angle of attack. In *Figure 6* is presented a diagram of the aerodynamic forces and torques.

To calculate the aerodynamic torque $\tau_a \in \mathbb{R}^{3\times1}$ that actuate on the wing, and will complement the formulation of Equation (23), the total momentum according to (Beer et al., 2013) by the aerodynamical torque vector seen from frame $\{0\}$ is:

$$\tau_a = M_{\frac{c}{4}} + p_{\frac{c}{4}} \times F_a^3 \tag{29}$$

Where $p_{\frac{c}{4}}$ is the position vector of the Torque's point of application, located at $c / 4$ of the leading edge; and the aerodynamic force F_a^3, in relation to the system $\{3\}$, is given by $\boldsymbol{F}_a^3 = \begin{bmatrix} 0 & D & L \end{bmatrix}^T$. The torque $M_{\frac{c}{4}}$ is already in the main direction with respect to D and so it is not necessary to do axes transformation.

Figure 6. Aerodynamic forces and torques saw from the wing profile. The wing chord extends upon c_2 axis meanwhile c_3 is always perpendicular to it. The axes location is purely illustrative

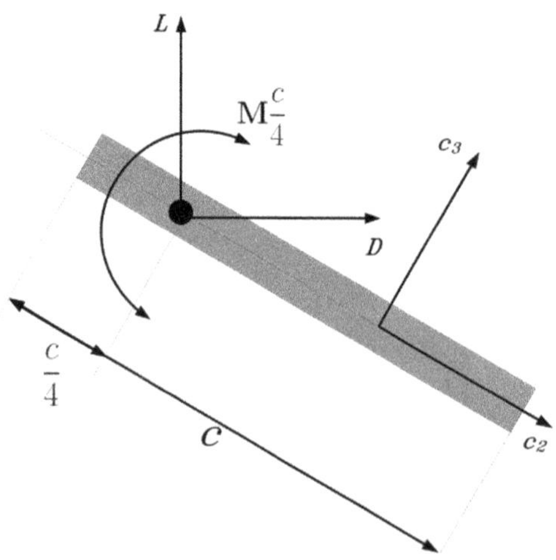

Flight Kinematic Patterns

Based on the assumption that the wing doesn't suffer deformations, the joint trajectories that the angle α_1, β_2 and γ_3 perform in time were proposed by Sane et al., (2001) (Sane et al., 2001) and Roccia et al., (2007). These trajectories were based in kinematic patterns of *Drosophila melanogaster* stroke, which characteristics approximate flying patterns described for a variety of insects and birds.

To describe the patterns performed by the insect in α_1 and β_2 angles, a set of harmonic functions were proposed. Variations of α_1 patterns could be found in the mentioned documents, while Sane et al., (2001) proposed a smooth-triangular waveform, Roccia et al. (2007) proposed a sine form, for both cases, the movement allows to maintain a translational velocity through a complete stroke. Due to both patterns perform the same effect, the simple harmonic function was chosen for this work.

To describe the pattern performed by β_2 angle two different functions were considered. One describes an oval pattern in the wingtip movement and is executed by a half-sine function, while the second performs a figure-of-eight or infinite pattern that is executed by a full sine function.

Finally, to describe the pattern in angle γ_3 a trapezoidal function was considered. This model allows to maintain the attack angle constant on each half-stroke, this function was build using Fourier Transform, considering three harmonics.

In *Figure 7* is described the kinematical patterns performed by the three angles to describe the oval (a) and infinite (b) wingtip trajectory. The movements were built considering an excitation frequency of 7 Hz, expecting the worst operational scenario (Tobalske & Dial, 1996), this allowed to ensure that the proposed control could achieve an appropriate response under normal operating conditions. The signal amplitudes were established following Roccia et al. (2007) work.

Figure 7. Wing kinematics describing the angles time course around three cycles. The graphics describe variations in the angular position (rad), velocity (rad/seg) and acceleration (rad/seg²) of the three degrees-of-freedom in accordance with the pattern wingtip performed, for circular (a), and infinite (b) trajectories

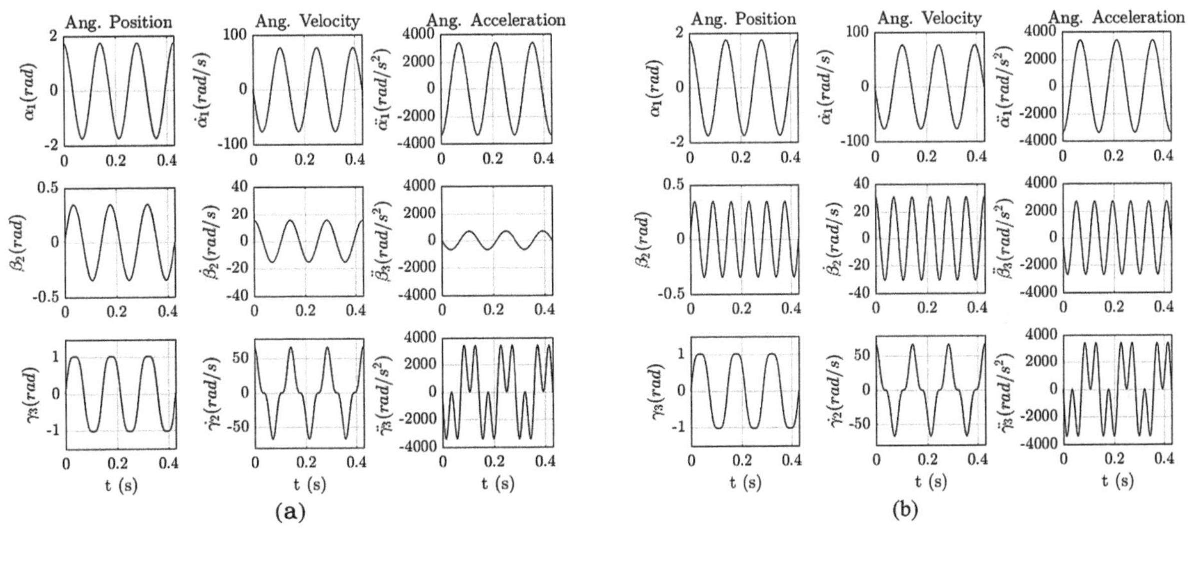

(a) (b)

Control of the System

As was mentioned in the previous section, to control the system it is considered a two-degree-of-freedom controller. To establish the feedforward part from the Equation (23), the contributions made by the respective aerodynamical torques τ_a are neglected, as they will be interpreted as uncertainties that will be controlled by the feedback part. Therefore, according to this considerations Equation (23) is rewritten in a compact form as:

$$M(q)\ddot{q} + H(q,\dot{q}) = u \tag{30}$$

Where

$$H(q,\dot{q}) = C(q,\dot{q})\dot{q} + K(q).$$

Considering Equation (31) as the control law for the feedback loop:

$$u = M(q)(\ddot{q}_d - k_v\dot{e} - k_p e) + H(q,\dot{q}) \tag{31}$$

Where $\ddot{q}_d$, $\dot{q}_d$ and q_d are the vectors that represent the desired trajectories in acceleration, velocity and position of the system angles respectively, defining $e = q - q_d$ and $\dot{e} = \dot{q} - \dot{q}_d$ as the errors in position and velocity respectively and establishing k_v and k_p as vectors that contain the proportional and derivative gains. Substituting (31) into (30) yields to:

$$\ddot{e} + k_v\dot{e} + k_p e = 0 \tag{32}$$

Thus, the error **e** will be asymptotically tending to zero when were correctly chosen the control constants k_v and k_p. According to Ogata (2010), the poles in Equation (32) can be obtained as follows:

$$p = -\xi\omega_n \pm \omega_n\sqrt{\xi^2 - 1} \tag{33}$$

Considering that *p* must take real or imaginary values, the stability of the system depends solely on the choice of the control constants. To obtain the value of k_v and k_p constants, the damping factor ξ value was set to *1* to annulate oscillations when the generalized coordinates are positioned, thus, the selection of the gains will depend on the natural frequencies $\omega_{n,i} i = 1..3$, that is:

$$k_{p,i} = \omega_{n,i} \tag{34}$$

$$k_{v,i} = 2\xi\omega_{n,i} \tag{35}$$

The selection of ω_n for Equations (34) and (35) will depend on the behavior desired to be induced into the system. Considering the description above the control system proposed to take the form described in *Figure 8*.

The vector $x \in \mathbb{R}^{6\times1}$ defines the states of the dynamic system describing the model as was presented in Equation (5), thus the dynamical model has as states both position and velocity of the three generalized coordinates described as $x_1 = \alpha_1, x_3 = \beta_2$ and $x_5 = \gamma_3$. The vector $y \in \mathbb{R}^{6\times1}$ is the output of the dynamic system, defined as $y = Cx$, in other words:

$$y = \begin{bmatrix} \begin{bmatrix} 1 & \cdots & 0 \\ \vdots & \ddots & \vdots \\ 0 & \cdots & 1 \end{bmatrix} \end{bmatrix} \begin{bmatrix} x_1 \\ x_2 \\ x_3 \\ x_4 \\ x_5 \\ x_6 \end{bmatrix} \tag{36}$$

With $C \in \mathbb{R}^{6\times6}$ being an identity matrix.

The ornithopter model is a Lagrangian dynamic system build respect to three generalized coordinates, all being actuated with their respective input u_n. Hence, the system could be considered as differentially flat, being the generalized coordinates and their derivatives the system's flat outputs, as them describes both their states and outputs, as was seen in equations (5) and (6). Thus, it is had:

Figure 8. Proposed control applied to the dynamical system of the Ornithopter

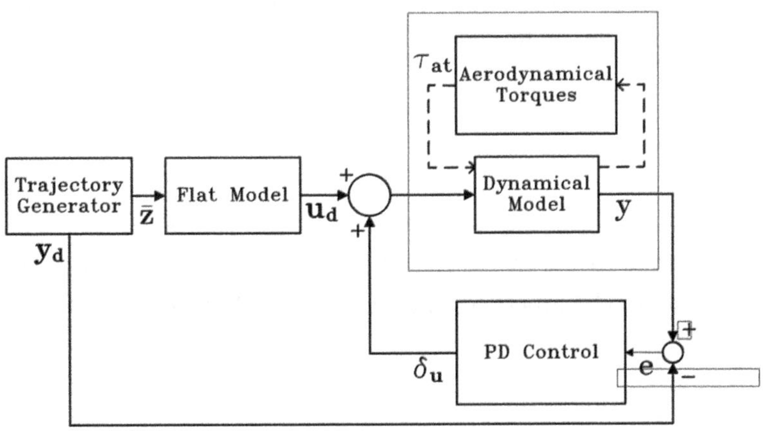

$$\bar{z} = \begin{bmatrix} z & \dot{z} & \ddot{z} \end{bmatrix} = \begin{bmatrix} q & \dot{q} & \ddot{q} \end{bmatrix} = \begin{bmatrix} \alpha_1 & \dot{\alpha}_1 & \ddot{\alpha}_1 \\ \beta_2 & \dot{\beta}_2 & \ddot{\beta}_2 \\ \gamma_3 & \dot{\gamma}_3 & \ddot{\gamma}_3 \end{bmatrix} \quad (37)$$

Therefore, the flat model that constitutes the feedforward loop is defined as:

$$u_d = M(z)\ddot{z} + C(z, \dot{z})\dot{z} + K(z) \quad (38)$$

The vector u_d gives a set of desired inputs required to drive the system to the desired output, which values are given by the trajectory generator box.

Being the desired outputs y_d, the kinematic patterns described by the harmonic functions seen in *Figure 7*, that is:

$$y_d = \begin{bmatrix} \alpha_{1d} & \beta_{2d} & \gamma_{3d} \end{bmatrix}^T \quad (39)$$

$$\dot{y}_d = \begin{bmatrix} \dot{\alpha}_{1d} & \dot{\beta}_{2d} & \dot{\gamma}_{3d} \end{bmatrix}^T \quad (40)$$

Due to the aerodynamic torques are influencing the entire system dynamic, the feedback loop estimates the compensation input $\delta_u \in \mathbb{R}^{3\times1}$ from the error vector $e \in \mathbb{R}^{3\times1}$, that is the difference between the desired output and the actual output, being, $e = y - y_d$. As the control law is based on a PD controller, the derivative of error needs to be found, thus, $\dot{e} = \dot{y} - \dot{y}_d$.

The entire system above ensures the system control, it is expected that it drives the model through the desired trajectories neglecting the effects of the aerodynamical torques. Additionally, it is expected that the control loop also guarantees the trajectory tracking considering changes in their initial conditions, as it computes the required inputs to drive the model. All these considerations will be tested in the next section, as are evaluated the system with the two wingtip patterns described previously.

RESULTS

The pigeon morphometric data, extracted from Tobalske et al. (1996) was used as reference to test the system. In *Table 1*, the selected values are described. For wing's mass calculation in relation to the total bird mass m_b, the following model proposed by Kirkpatrick (1990) is used.

$$m = 9,74 \times 10^{-2} m_b^{1,10} \quad (41)$$

Table 1. Morphometric parameters of a pigeon, values originating of (Tobalske et al., 1996)

Description	Symbol	Value	Units
Longitude of a single wing	L_1	0,279	Meters (m)
Average wing chord	c	0,112	Meters (m)
Pigeon mass	m_b	0,316	Kilograms (kg)

The angle between the body and the horizontal plane was considered equal to zero radians. The gravity value used was $g = 9,81m / s^2$; to calculate the Reynolds number (Equation (42)), were considered the constants to describe the air surrounding the wing that is extracted from Rubenstein et al. (2012), being: specific mass $\rho = 1,23kg / m^3$, and dynamic viscosity $\mu = 1,79 \times 10^{-5} N.s / m^2$.

$$\mathrm{Re} = \frac{\rho}{\mu} v_\infty c \tag{42}$$

Where v_∞ is the wing velocity and *c* is the average wing chord.

System Response With No Perturbations

In this section are presented the application's results of the two degrees-of-freedom controller on the wing. An initial experience simulating the system with the inexistence of aerodynamical torques and no changes on the initial conditions $x_0 = x_d(t_o)$ was made. In *Figure 9* is presented the system response (a) and the estimated output (b) made by the controller to follow the desired path, considering a value of $\omega_n = 25$ for angles α_1 and β_2 and $\omega_n = 3$ for the γ_3 angle. The simulation was done using a 4th

Figure 9. Trajectory path tracking (a) and input compensation (b) time courses. In (a) is presented the behavior of the system's position (dashed line) respect to the desired path (bold line)

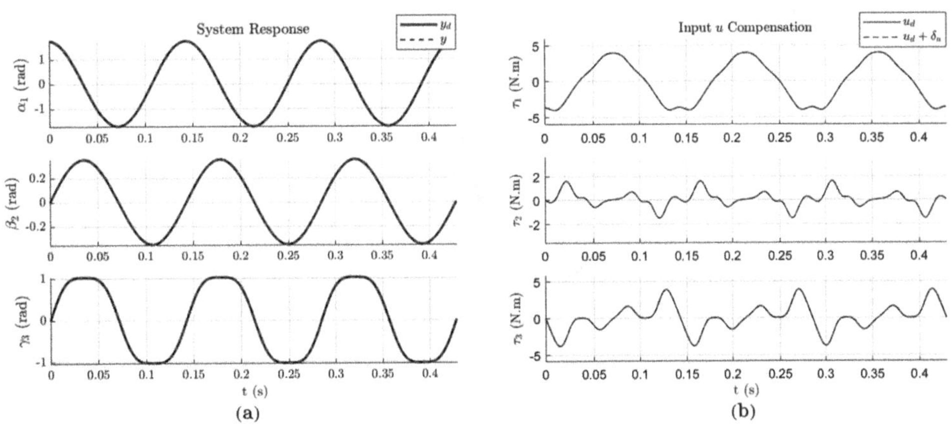

degree Runge-Kutta numerical integrator under MATLAB's environment, the entire simulation time course was made along with three cycles of the featured harmonic functions, width a bandwidth of 4 times the harmonic frequency.

For this experience, no disturbs caused by the aerodynamic forces that influenced the wing dynamics were presented, as well, there were no deviations into the initial conditions. Thus, the error in the feedback loop for all the three angles through the entire time course was zero ($e = 0; \dot{e} = 0$). In consequence, no compensations from the feedback part of the controller were made $\left(\delta_{u,n} = 0, n = [1,2,3]\right)$, and, the system was driven only by the feedforward part, which response was made by the flat model through the set of desired inputs u_d for each coordinate.

Wingtip Circular Pattern Response

Was performed a new set of simulations to test the controller, was considered the existence of a set of aerodynamical forces caused by a changing wind velocity on the wing, which profile is described in *Figure 10.*

In this case, the wind velocity v_∞ goes from a mean velocity of 14 m/s at the beginning of the run, increasing to a final mean velocity of $20\,m/s$. For simplification purposes, was considered that exits a mean Reynolds number (Re) of 120×10^3 during the entire time course, thus, the lift, drag and moment coefficients of equations (26), (27) and (28) respectively were obtained from the same table. For this, the hysteresis present along the values was neglected, and instead, was build a function between the force coefficients C_l and C_m and the angle of attack α, being:

Figure 10. Wind Velocity profile applied to the system. The perturbation's influence over the system will be increased as the air velocity raise, due to their effects on the aerodynamical forces.

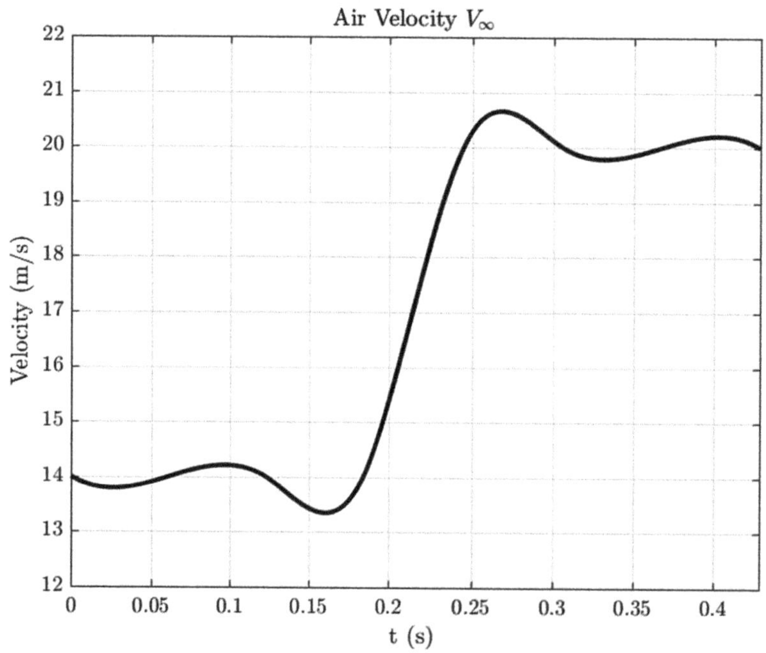

$$C_l = F\left(a_t\right) \tag{43}$$

$$C_m = G\left(a_t\right) \tag{44}$$

To simplify the values in Equation (43) and (44), the expressions were defined in function of $a_t = \sin\left(\alpha\right)$, instead of α of the originals tables. Remembering that the expression to estimate this value was presented in Equation (25). The draft coefficient C_d is 0 for this case due to was selected a flat-plate profile for the wing, as their thickness is considered negligible (Selig et al., 2012). Also, was considered that the moment coefficient C_m is zero when the angle of attack angle minor of zero $\left(\alpha < 0\right)$ (Anderson, 2001).

To do the control test two wingtip kinematics patterns were considered, in the first experience the wingtip circular pattern was used, the functions that perform that kinematic pattern was described in *Figure 7* (a).

Considering that, the wing's generalized coordinates starts with the following initial conditions

$$q = \left[\pi / 40 \quad \pi / 10 \quad \pi / 20\right]^T$$

and $\dot{q} = \left[0 \quad 0 \quad 0\right]^T$, it is presented in *Figure 11*: (a) the behavior of the system's generalized coordinates (angular positions expressed in radians and plotted in the thin dashed line) compared with their respective desired path (expressed in radians and plotted in the bold gray line); and (b) the desired input (ex-

Figure 11. Trajectory path tracking (a) and input compensation (b) time courses of the system when a circular wing pattern is performed by the wing, and the Aerodynamical Forces actuates upon the system

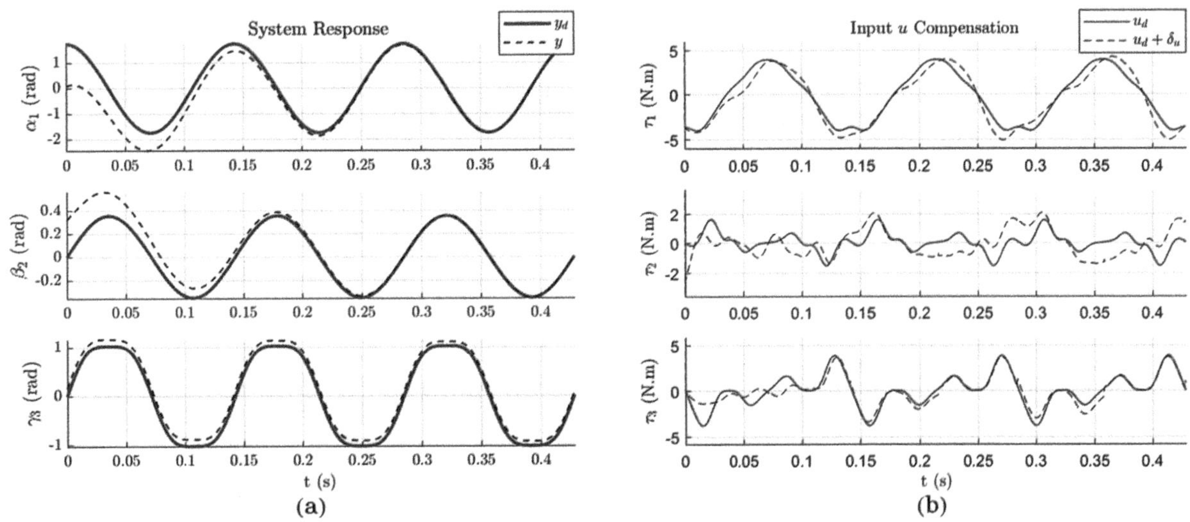

pressed in $N \cdot m$ and plotted in the thin gray line) and their changes resulting from the compensation made by the feedback (expresses in $N \cdot m$ and plotted in the dashed line) required to solve the influences generated by the aerodynamic forces on the wing. The simulation was made during a time interval equivalent to 3 cycles, the controller gains for this simulation where calculated considering the same natural frequencies of the previous experience. The control of the angle γ_3 seems to be slower in comparison with the other two, as was required more time to reduce the position difference to zero.

While in *Figure 11* was described the dynamic behavior of the proposed model, in *Figure 12* is presented the phase portraits resulting of that performance in (a) the position angle, (b) the deviation angle, and (c) the rotation angle. The phase portrait describes the evolution of the angle's position respect to their own velocity during the simulation (plotted as the gray line), the arrows describe the direction of the executed movement from the initial conditions. To compare the path tracking considering not only the desired position but also the desired velocity the graphs compare the executed movement with the expected path, illustrated as an expected phase portrait illustrated as a dashed bold line. In all the three-movement, the trajectory performed evolved to the expected movement, responding to the controller compensation.

In *Figure 13* is presented (a) the pattern performed by the wingtip (dotted pattern) seen from the Z-Y plane which is perpendicular to the stroke plane, compared with the desired pattern (crossed pattern) and (b) the mean error of the spatial position reflected through the positioning along the X, Y, and Z coordinates. As was expected, the controller execution tracked the system to the desired behavior through the simulated conditions. At the end of the time course, both trajectories were nearly coincidental as the wingtip mean position error was closer to zero.

Figure 12. Dynamics phase portrait of the three generalized coordinates. Comparing the wing's movement driven by the control's law

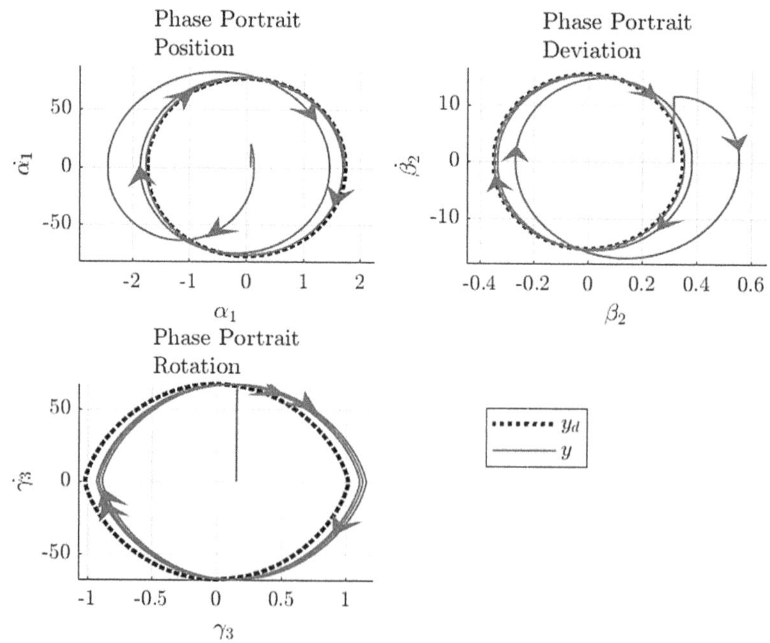

Figure 13. (a) Comparison of the wingtip pattern performed by the controlled wing (dotted pattern) with the desired path (crossed pattern) performed by the desired kinematic patterns (b)

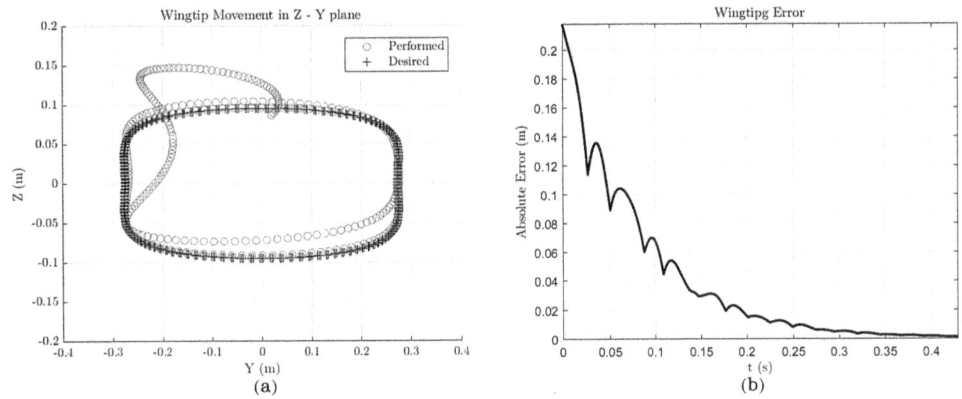

Finally, in *Figure 14* is presented the behavior of the aerodynamic torques according to the wing position. The moment patterns on the three main axes are the result in accordance to the system response during simulation, thus, consequences of the initial positions and the path tracking are presented in them. Was observed the presence of an abrupt changes of the moment on the x-axis as consequence of the angle of attack, position, and the attenuate augmentation of the moment force due to the increase of the wind velocity v_{∞}. Which is visible after $t > 0.2$ seconds in the three plots.

Figure 14. Aerodynamic torque response for the three main axes according to the wing position

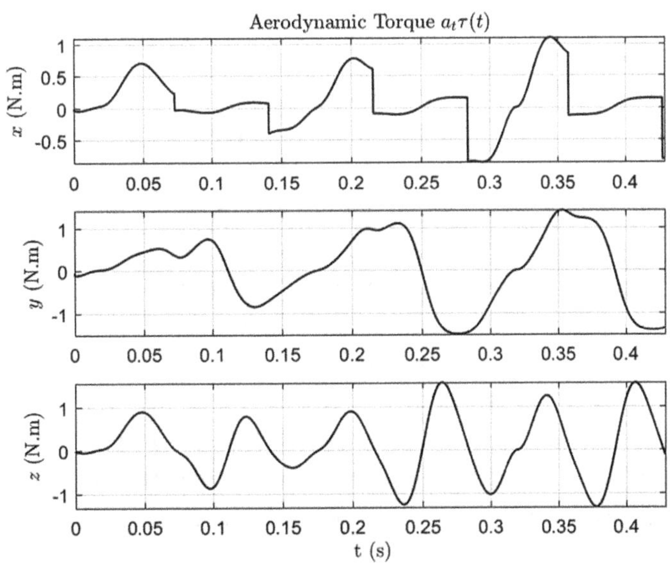

Figure 15. Trajectory path tracking (a) and input compensation (b) time courses of the system when an infinite wing pattern is performed by the wing, and the Aerodynamical Forces actuates upon the system

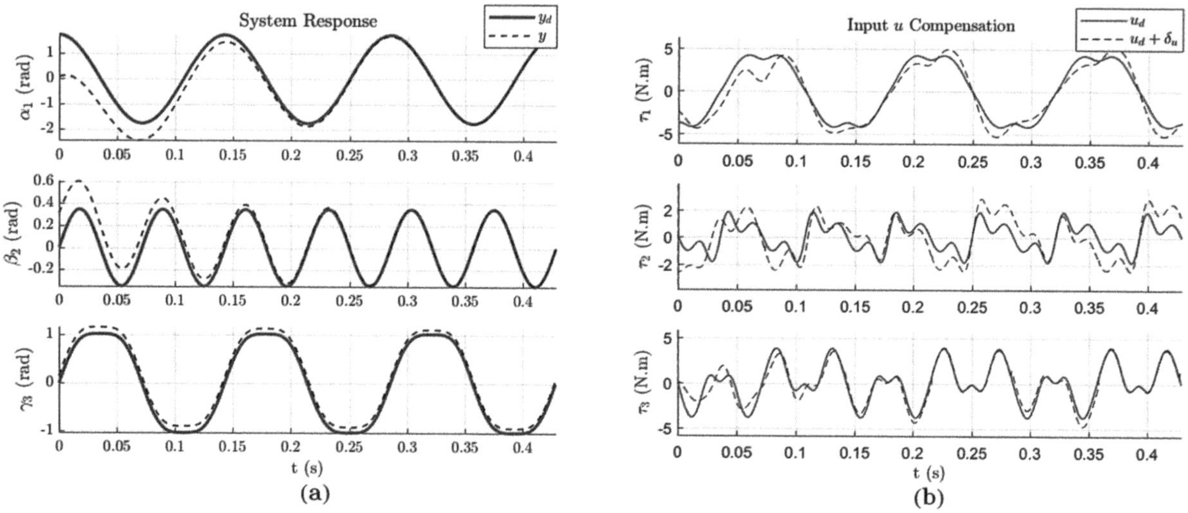

Wingtip Infinite Patter Response

The second experience corresponds to the application of the set of trajectories that performs an infinite pattern in the wingtip, which was presented in *Figure 7* (b). The wing initial conditions were

$$q = \begin{bmatrix} \pi / 40 & \pi / 10 & \pi / 20 \end{bmatrix}^T$$

and $\dot{q} = \begin{bmatrix} 0 & 0 & 0 \end{bmatrix}^T$, similar as the presented in the previous experience, also, where considered the same natural frequencies for the controller's gains.

As was mentioned, to effectuate an infinite pattern on the wingtip movement, a duplicaton in angle's β_2 oscillation frequency was required, thus, an increase in both velocity and acceleration of that coordinate movement was presented.

In *Figure 14* is presented: (a) the behavior of the system's generalized coordinates (angular positions expressed in radians and plotted in the thin dashed line) compared with their respective desired path (expressed in radians and plotted in the bold gray line); and (b) the desired input (expressed in $N \cdot m$ and plotted in the thin gray line) and their changes resulting from the compensation made by the feedback (expresses in $N \cdot m$ and plotted in the dashed line) required to solve the influences generated by the aerodynamic forces. The simulation was made during a time interval equivalent to 3 cycles. Changes in the torque applied in the *y*-axis were found, indicating a requirement of more energy because of the velocity increase of the angle movement. As was presented in the previous experience, γ_3 angle seems to have a slower position correction in comparison with the other two.

In *Figure 16* is presented the phase portraits resulting from the experiment in (a) the position angle, (b) the deviation angle, and (c) the rotation angle. Because of the changes in β_2 angle, the phase portrait in that coordinate experiment changes increasing the value along the velocity axes. Nevertheless, the movement requirement experimented variations, during all the three movements, the trajectory performed evolved to the expected movement, responding to the controller compensation.

In *Figure 17* is presented (a) the pattern performed by the wingtip (dotted pattern) seen from the Z-Y plane in comparison with the desired pattern (crossed pattern), and (b) the mean error of the spatial position reflected through the positioning along with the X, Y and Z coordinates. As was expected, the controller execution tracked the system to the desired behavior through the different external conditions. At the end of the time course, both trajectories were nearly coincidental as the wingtip mean position error goes to zero. Changes in the angle β_2 angle leaded to an alteration of the wingtip movement forming the expected infinite pattern. No additional behaviors were presented during the system response.

Finally, in *Figure 18* is presented the behavior of the aerodynamic torques according to the wing position, the moment patterns on the three main axes are the result in accordance to the system response during simulation, thus, consequences of the initial positions and the path tracking are presented in them. Similar effects corresponding to C_m value respect of the angle of attack and the increase in the wind velocity was found. However, no significative changes in the effects of the aerodynamical torques were presented, considering that the angle of attack performed for both studied patterns is the same, this result could be reasonable, thus, the increases of the β_2 angle velocity and acceleration don't affect considerably the external force's response.

Figure 16. Dynamics phase portrait of the three generalized coordinates. Comparing the wing's movement driven by the control's law

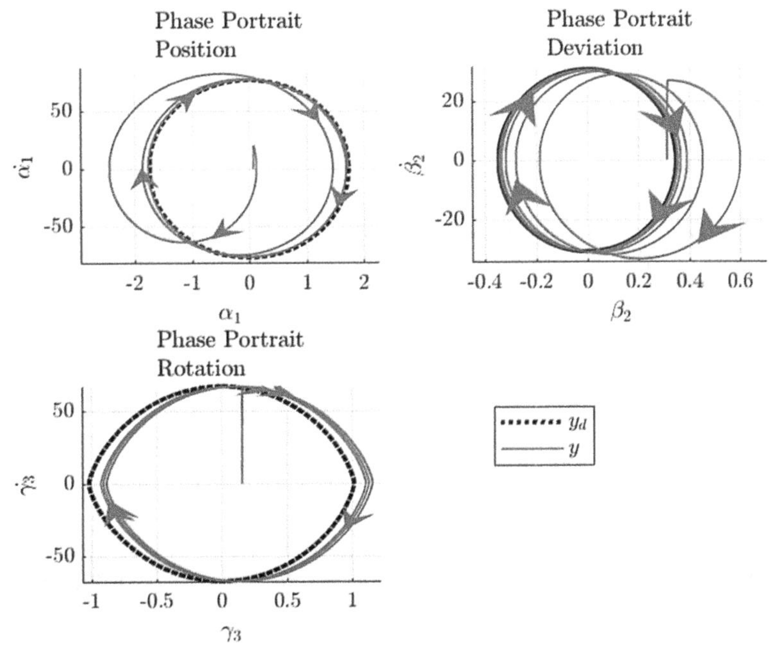

Figure 17. (a) Comparison of the wingtip pattern performed by the controlled wing (dotted pattern) with the desired path (crossed pattern) performed by the desired kinematic patterns (b)

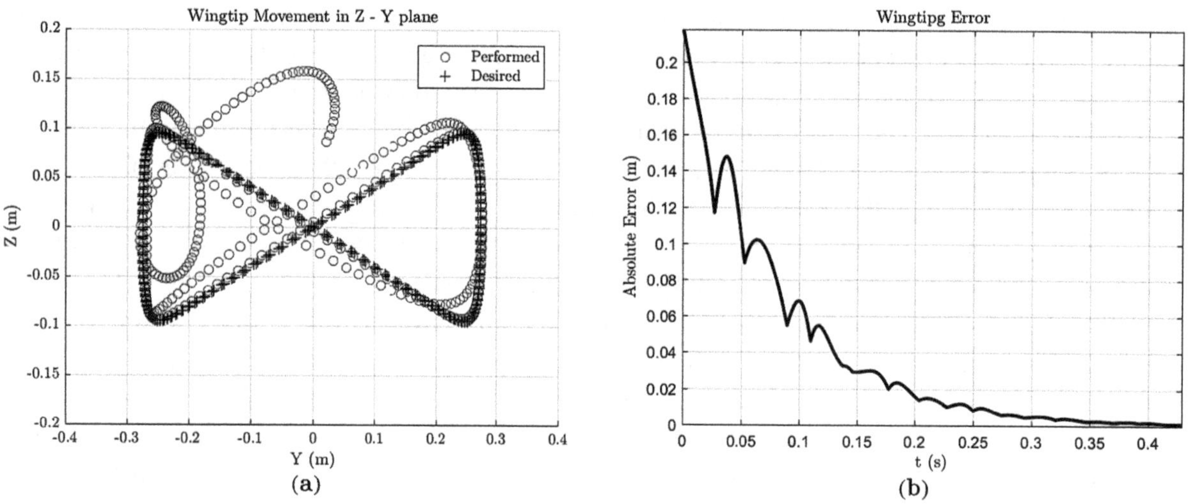

Figure 18. Aerodynamic torques response in the three main axes according to the wing position

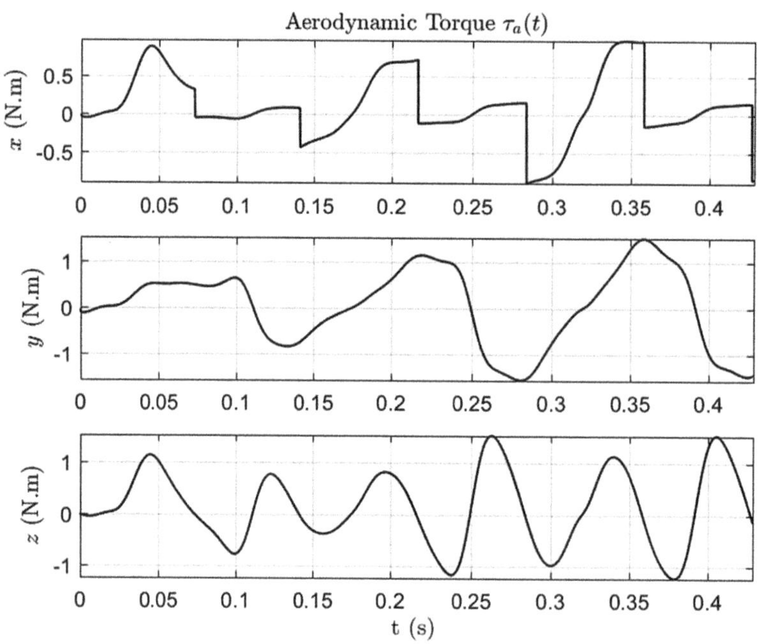

DISCUSSION AND RECOMMENDATIONS

In this chapter were presented the control of an ornithopter's wing using a two-degree-of-freedom scheme that exploits the virtue that the dynamical system is differentially flat, due to all the generalized coordinates in the system was actuated. The presented method allows the controlling of the wing considering changes in the initial conditions, and external disturbs occasioned by changing aerodynamical forces that actuate upon the wing. The control study was made around two wingtip patterns, a circular and infinite trajectory patterns. Due to changes in the angle movement to effectuate the pattern, increases of their velocity and acceleration were found, increasing therefore, the energy requirements to actuate on the coordinate. Kinematical patterns build by harmonic functions in a frequency of 7 Hz were applied to the system, this was done, expecting the worst working conditions, thus, movements with high velocity and acceleration were performed by the three studied coordinates.

The commanded patterns effectuated by the wing correspond to natural trajectories inspired by insect flying, this movement was considered as it is commonly performed by different insect and bird species. Also, morphometric data of a pigeon was used to describe the body of the structure, therefore, modeling both wing kinematics and dynamics was totally inspired in nature. No mass in the bird's body was considered as also, no other wing connected to the system, due to the authors were interested to test the control performance.

The system's structure was constituted of a three degree of freedom system model constituted by a single link, disregarding the existence of the elbow and pulse movement that birds had, this last assumption in the body structure was inspired by previous works (DeLaurier et al., 1993), that build a full-scale piloted ornithopter and proved that a device with such structure could fly. However, the design made by (DeLaurier et al., 1993) was based in a 1-DOF wing structure that executed flapping movement with a 1.1 Hz frequency. More recent works considered ornithopters' models with multiple degrees-of-freedom: as models inspired in seagulls that considered a structure with two degrees-of-freedom (Jiang et al., 2016) and (Li et al., 2017); or models that considered active tails to generate balancing control in the aircraft (Yang et al., 2018). Nevertheless, the works evaluate the structure's kinematics, dynamics and forces actuating upon the aircraft based in simulations, and no feedback control was applied to the studied systems.

In (Sane et al., 2001) and (Roccia et al., 2007) works, were presented a study of ornithopters flapping-wing that presented a three degree-of-freedom structure, which inspired the development of this chapter. Regarding, both considered works presented an extensive study of the proposed wing's kinematics, and a structure design performance, no dynamic and control test were made in both works.

In (Murphy et al., 2013) could be found a study and control of a flapping wing robot with Lagrangian modeling, as was done in this work. To control this system was used feedback linearization combined with Hard Switching Control and Soft Switching Control, considering unknown aerodynamic coefficients, and requiring a method to estimate the coefficients. In (Bogado Martínez et al., 2018), the system was modeled using Newton-Euler dynamics and was presented the control of the proposed model applying a PD control. To applicate, such controller, a linearization of the wing dynamic was required, as also was take into consideration that the aerodynamical coefficients are known, as they were taken from tables and parametrized trough functions that depend on the angle of attack.

In this work, the system was modeled applying Newton-Lagrange, and the differential flatness were introduced, this consideration, allowed to control the wing's dynamics through a two-degrees-of-freedom scheme without model linearization, and, based on a set of desired inputs estimated through direct dynamics using a set of desirable trajectories to control it. Also, as the work presented in (Bogado Martínez et al., 2018), a set of aerodynamical forces were introduced to the system, being considered as disturbs that alters the wing's dynamic by summation of their resultant moments.

The performance of the controller was tested through two wingtip patterns, in both cases, the proposed system was able to do drove the wing around the desired path, with non-significative differences on their performance. An initial test was made, considering the inexistence of the aerodynamical forces and no displacements in the initial conditions, and was found that the system was controlled in an open-loop fashion as being observed that the desired inputs were not altered by the feedback loop, in this case with no requirement of compensations in the inputs, the system was driven directly by the desired trajectory.

Comparing the behavior of the system on both wingtip patterns, the desired inputs and the external forces caused by the aerodynamic effects presented different alterations. The first, generated a continuous time course of inputs that were corresponding to the movement requirements, being, that was presented an augment of the velocity in the angle required to move the wingtip into the infinite pattern, the model respond with an increase of the energy required to manipulate the articulation on this velocity; and the second, despite this event, performed minor variations on the applied torque that actuates on the wing, indicating that changes in the movement velocity don't affect significantly their magnitudes. Instead, the effects of the external wind upon the aerodynamic forces exerted an important increase on the moment actuating upon the wing, being a factor that causes major alterations in the system's dynamics. No matter, the magnitude of the exerted movement, the applied controller law achieved to modify the desired inputs trough feedback linearization, compensating them according to the occasionally disturbs values generated during the wing movement.

The controller was able to drive the generalized coordinate into the performed trajectory in a short time. However, was noticed that the recovery time of the third degree-of-freedom (angle actuating on the x-axis) to recuperate their position is larger than the other angles. In other words, the control of the angle of attack experimented a delay in comparison with the position and deviations angles (Z and Y respectively). This issue could be addressed with an increase in the value, however, doing this led to increasing the instability of the model. Although the proposed control allowed to drive a non-linear system, the proposed methodology may cause that the internal motion to have unacceptably large transient performance, in (Nieuwstadt et al., 1997) this problem is avoided by allowing an explicit tradeoff between the trajectory tracking performance and the stability of the internal motions.

The present control scheme was able to compensate the effects caused by external factors without the requirement of internal changes on the control gains, even though the environmental conditions were changing around time. As were made a set of feasible trajectories defined by the flat model to drive the system, the controller could maintain the stability of the system around the desired path avoiding the performance of abruptly changes in their movement, as was shown in the respective phase portraits.

The authors considerate that uncertainties caused by non-modeled configurations of the original system could be handled by the feedback linearization, but, the effective response of the controller will depend on a proper selection of the control gains. In addition, the considering of a PID controller could address the issue of the slow response around the X-axis rotation, due to its capacity to address the accumulative error, also, the study of others non-linear control techniques could achieve a better performance into their control, but no other assumptions must be done until being observed their results

FUTURE RESEARCH DIRECTIONS

As was presented in the Introduction section Flapping-wing control is an research area that provides interesting approaches to be considered in future works, control strategies and performance tests are required to provide future insights for autonomous flapping-wing based robots, robust control techniques, adaptative control or fuzzy control, among other techniques of non-linear control techniques applications, could be implemented and tested. Also, more complex body structures as adding more degrees-of-freedom to the wing structure, or the implementation of another wing profiles, needs to be addressed; also, structures based in the two-wings system without tail to gain stability must be studied, however, the influence of the dynamic between and their effects upon external conditions requiring more extensive work. Results of this work will help us to address future directions to build a flapping wing physical system to test the theoretical results obtained from simulations.

CONCLUSION

In this work the wings dynamical and control test of a flapping-wing for an ornithopter composed by a three degrees-of-freedom system was studied. The control of the non-linear dynamics model was addressed using differential flatness implemented in a two-degree-of-freedom control scheme. As control law was considered a PD feedback, driving both position and velocity around a set of trajectory patterns, inspired in living species observation. The effects of the wind upon the wing, presented as aerodynamical forces were studied through the performance of the controller to drive the system through two wingtip pattern trajectories. In both cases, the controller was able to drive the system performing the movements around the desired patterns, no matter the wind variations caused an important increment of the aerodynamical moments. The control performance and system stability when disturbs were present was dependent on the selected control gains, otherwise, as no disturbs presence, the controller was able to drive the system in an open-loop fashion.

ACKNOWLEDGMENT

This research was supported by FINEP, CNPq, FAPERJ, Fundação COPPETEC, and DIPPG/CEFET-RJ.

REFERENCES

Anderson, J. D. (2001). *Fundamentals of aerodynamics* (3rd ed.).

Armanini, S. F., de Visser, C. C., de Croon, G. C. H. E., & Mulder, M. (2016). Time-Varying Model Identification of Flapping-Wing Vehicle Dynamics Using Flight Data. *Journal of Guidance, Control, and Dynamics*, *39*(3), 526–541. doi:10.2514/1.G001470

Baek, S. S., Garcia Bermudez, F. L., & Fearing, R. S. (2011). Flight control for target seeking by 13 gram ornithopter. In *2011 IEEE/RSJ International Conference on Intelligent Robots and Systems* (pp. 2674–2681). IEEE. 10.1109/IROS.2011.6094581

Beer, F. P., Johnston, E. R., Jr. ., Mazurek, D. F., León Cárdenas, J., Ríos Sánchez, M. A., & Zamora Gallardo, E. (2013). *Mecánica vectorial para ingenieros : estática*. McGraw-Hill Interamericana.

Bogado Martínez, C. F., Dutra, M. S., & Raptopoulos, L. S. C. (2018). Proposed control for wing movement, type flat plate, for ornithopter autonomous robot. *Journal of the Brazilian Society of Mechanical Sciences and Engineering*, *40*(9), 456. doi:10.100740430-018-1362-7

Chirarattananon, P., Ma, K. Y., & Wood, R. J. (2014). Adaptive control of a millimeter-scale flapping-wing robot. *Bioinspiration & Biomimetics*, *9*(2). doi:10.1088/1748-3182/9/2/025004 PMID:24855052

Colorado, J. (2012). BaTboT : A biologically inspired flapping and morphing bat robot actuated by SMA-based artificial muscles., 223. Retrieved from http://digital.csic.es/handle/10261/109918

De Croon, E., Groen, M. A., De Wagter, C., Remes, B., Ruijsink, R., & Van Oudheusden, B. W. (2012). Design, aerodynamics and autonomy of the DelFly Recent citations Mimicking Insect Wings: The Roadmap to Bioinspiration Jafar Hasan et al Design, aerodynamics and autonomy of the DelFly. *Bioinspiration & Biomimetics*, *7*, 16. doi:10.1088/1748-3182/7/2/025003

De Wagter, C., Tijmons, S., Remes, B. D. W., & de Croon, G. C. H. E. (2014). Autonomous flight of a 20-gram Flapping Wing MAV with a 4-gram onboard stereo vision system. In *2014 IEEE International Conference on Robotics and Automation (ICRA)* (pp. 4982–4987). IEEE. 10.1109/ICRA.2014.6907589

Delaurier, J. D. (1988). The Development and Testing of a Full-Scale Piloted Ornithopter. *Canadian Aeronautics and Space Journal*. Retrieved from http://www.ornithopter.net/Publications/TheDevelopmentandTestingofaFull-ScalePilotedOrnithopter.pdf

DeLaurier, J. D., & Harris, J. (1993). *A study of mechanical flapping-wing flight. Aeronautical Journal*. Cambridge University Press; doi:10.10170001924000026452

Dietl, J. M., & Garcia, E. (2008). Stability in Ornithopter Longitudinal Flight Dynamics. *Journal of Guidance, Control, and Dynamics*, *31*(4), 1157–1163. doi:10.2514/1.33561

Fliess, M., Levine, J., Martin, P., Rouchon, P., Lévine, J., Martin, P., & Rouchon, P. (1995). Flatness and defect of non-linear systems: Introductory theory and examples. *International Journal of Control*, *61*(6), 1327–1361. doi:10.1080/00207179508921959

Fry, S. N., Sayaman, R., & Dickinson, M. H. (2005). The aerodynamics of hovering flight in Drosophila. *The Journal of Experimental Biology*, *208*(Pt 12), 2303–2318. doi:10.1242/jeb.01612 PMID:15939772

Habib, M. K. (2008). Human adaptive and friendly mechatronics (HAFM). In *2008 IEEE International Conference on Mechatronics and Automation* (pp. 61–65). IEEE. 10.1109/ICMA.2008.4798726

Isermann, R. (2005). Mechatronic systems - Innovative products with embedded control. In *IFAC Proceedings Volumes (IFAC-PapersOnline)*.

Jiang, H., Zhou, C., & Xie, P. (2016). Design and kinematic analysis of seagull inspired flapping wing robot. In *2016 IEEE International Conference on Information and Automation (ICIA)* (pp. 1382–1386). IEEE. 10.1109/ICInfA.2016.7832035

Kirkpatrick, S. (1990). The Moment of Inertia of Bird Wings. *The Journal of Experimental Biology*, (January): 489–494. Retrieved from http://d.wanfangdata.com.cn/NSTLQK_NSTL_QK9557394.aspx

Lamers, K., Tijmons, S., De Wagter, C., & de Croon, G. (2016). Self-supervised monocular distance learning on a lightweight micro air vehicle. In *2016 IEEE/RSJ International Conference on Intelligent Robots and Systems (IROS)* (pp. 1779–1784). IEEE. 10.1109/IROS.2016.7759284

Levine, J. (2009). Analysis and control of nonlinear systems: A flatness-based approach. Berlin, Germany: Springer Science & Business Media. doi:10.1007/978-3-642-00839-9

Li, Y., Qian, C., Zhu, B., & Fang, Y. (2017). *Kinematic, Static and Dynamic Analyses of Flapping Wing Mechanism Based on ANSYS Workbench* (pp. 316–323). Cham, Switzerland: Springer; doi:10.1007/978-3-319-70136-3_34

Liu, T. (2006). Comparative Scaling of Flapping- and Fixed-Wing Flyers. *AIAA Journal*, *44*(1), 24–33. doi:10.2514/1.4035

Ma, K. Y., Chirarattananon, P., Fuller, S. B., & Wood, R. J. (2013). Controlled flight of a biologically inspired, insect-scale robot. *Science*, *340*(6132), 603–607. doi:10.1126cience.1231806 PMID:23641114

Ma, K. Y., Felton, S. M., & Wood, R. J. (2012). Design, fabrication, and modeling of the split actuator microrobotic bee. In *2012 IEEE/RSJ International Conference on Intelligent Robots and Systems* (pp. 1133–1140). IEEE. 10.1109/IROS.2012.6386192

Markus, E. D., Agee, J. T., & Jimoh, A. A. (2017). Flat control of industrial robotic manipulators. *Robotics and Autonomous Systems*, *87*, 226–236. doi:10.1016/j.robot.2016.10.009

Martin, Philippe, & Paden, B. (1994). A different look at output tracking : control of a VTOL aircraft * Searching for a flat output, (December), 2376–2381.

Martin, P., Murray, R. M., & Rouchon, P. (1997). Flat Systems. In *European Control Conference* (p. 54). Brussels. Retrieved from http://pierre.rouchon.chez-alice.fr/publications/PR1997/eccflat.pdf

Muijres, F. T., Bowlin, M. S., Johansson, L. C., & Hedenström, A. (2012). Vortex wake, downwash distribution, aerodynamic performance and wingbeat kinematics in slow-flying pied flycatchers. *Journal of the Royal Society, Interface*, *9*(67), 293–303. doi:10.1098/rsif.2011.0238 PMID:21676971

Murphy, I. P., Kurdila, A. J., Bayandor, J., & Burdisso, R. A. (2013). *Modeling and Control of Flapping Wing Robots*. Retrieved from https://vtechworks.lib.vt.edu/bitstream/handle/10919/19275/Murphy_IP_T_2013.pdf?sequence=1

Murray, R. M., Sluis, W., Rathinam, M., & Sluis, W. (1995). Differential flatness of mechanical control systems: A catalog of prototype systems. *Proc. of the ASME Mechanical Engineering Congress and Expo*, 1-9. doi:10.1109/INFCOM.2011.5934932

Nasr, S., Mekki, H., & Bouallegue, K. (2019). A multi-scroll chaotic system for a higher coverage path planning of a mobile robot using flatness controller. *Chaos, Solitons, and Fractals*, *118*, 366–375. doi:10.1016/j.chaos.2018.12.002

Nguyen, Q.-V., & Chan, W. L. (2018). Development and flight performance of a biologically-inspired tailless flapping-wing micro air vehicle with wing stroke plane modulation. *Bioinspiration & Biomimetics*, *14*(1). doi:10.1088/1748-3190/aaefa0 PMID:30523879

Ogata, K. (2010). Modern control engineering. Upper Saddle River, NJ: Prentice-Hall.

Park, J. H., & Yoon, K.-J. (2008). Designing a Biomimetic Ornithopter Capable of Sustained and Controlled Flight. *Journal of Bionics Engineering*, *5*(1), 39–47. doi:10.1016/S1672-6529(08)60005-0

Pfeiffer, A. T., Lee, J.-S., Han, J.-H., & Baier, H. (2010). Ornithopter flight simulation based on flexible multi-body dynamics. *Journal of Bionics Engineering*, *7*(1), 102–111. doi:10.1016/S1672-6529(09)60189-X

Phan, H. V., Kang, T., & Park, H. C. (2017). Design and stable flight of a 21 g insect-like tailless flapping wing micro air vehicle with angular rates feedback control. *Bioinspiration & Biomimetics*, *12*(3). doi:10.1088/1748-3190/aa65db PMID:28281468

Pons, J. L. (2008). *Wearable Robots: Biomechatronic Exoskeletons*. Hoboken, NJ: John Wiley & Sons. doi:10.1002/9780470987667

Rahn, C. D. (2002). Mechatronic Control of Distributed Noise and Vibration: A Lyapunov Approach. *Measurement Science & Technology*, *13*(4), 643–644. doi:10.1088/0957-0233/13/4/704

Roccia, B. A., Preidikman, S., & Massa, J. (2007). Desarrollo de un código computacional para simular y analizar la cinemática de alas batientes Diseño estructural mecánico View project. *Mecanica Computacional, XXVI*, 3224–3245. Retrieved from http://www.ing.unrc.edu.ar

Rubenstein, D. A., Yin, W., & Frame, M. D. (2012). Fundamentals of Fluid Mechanics. In Biofluid Mechanics (pp. 11–48). doi:10.1016/B978-0-12-381383-1.00002-3

Ryu, J.-C., & Agrawal, S. K. (2011). Differential flatness-based robust control of mobile robots in the presence of slip. *The International Journal of Robotics Research*, *30*(4), 463–475. doi:10.1177/0278364910385586

Sane, S. P., & Dickinson, M. H. (2001). Control of Flight force by a flapping wing. *The Journal of Experimental Biology*, *2626*(15), 2607–2626. PMID:11533111

Selig, M., Guglielmo, J. J., Broeren, A. P., & Giguere, P. (2012). *Summary of Low-Speed Airfoil Data - Vol 5* (Vol. 5). doi:10.1115/1.1793208

Tenenbaum, R. A. (2013). *Fundamentals of applied dynamics. Choice Reviews Online* (Vol. 41). USA: Springer; doi:10.5860/choice.41-6560

Tobalske, B., & Dial, K. (1996). Flight kinematics of black-billed magpies and pigeons over a wide range of speeds. *The Journal of Experimental Biology*, *199*(2). PMID:9317775

Van Nieuwstadt, M. J. (1996). *Trajectory Generation for Nonlinear Control Systems*. Pasadena, CA: California Institute of Technology.

Van Nieuwstadt, M. J., & Murray, R. M. (1997). Real Time Trajectory Generation for Differentially Flat Systems. *International Journal of Robust and Nonlinear Control*, 28.

Verboom, J. L., Tijmons, S., De Wagter, C., Remes, B., Babuska, R., & de Croon, G. C. H. E. (2015). Attitude and altitude estimation and control on board a Flapping Wing Micro Air Vehicle. In *2015 IEEE International Conference on Robotics and Automation (ICRA)* (pp. 5846–5851). IEEE. 10.1109/ICRA.2015.7140017

Veslin Diaz, E., Slama, J., Dutra, M., Lengerke, O., & Morales Tavera, M. (2011). Trajectory tracking for robot manipulators using differential flatness. *Ingenieria e Investigacion*, *31*(2), 84–90.

Yang, S., Shen, Y., Li, B., Li, Y., & Zhang, J. (2018). Modeling and Simulation of a Flapping-Wing Robot with Active Tails for Balancing Control during Wheeled Running. In *2018 IEEE International Conference on Mechatronics and Automation (ICMA)* (pp. 1806–1811). IEEE. 10.1109/ICMA.2018.8484330

ADDITIONAL READING

Anderson, J. D. (2001). *Fundamentals of aerodynamics* (3rd ed.).

Armanini, S. F., de Visser, C. C., de Croon, G. C. H. E., & Mulder, M. (2016). Time-Varying Model Identification of Flapping-Wing Vehicle Dynamics Using Flight Data. *Journal of Guidance, Control, and Dynamics*, *39*(3), 526–541. doi:10.2514/1.G001470

Baek, S. S., Garcia Bermudez, F. L., & Fearing, R. S. (2011). Flight control for target seeking by 13 gram ornithopter. In *2011 IEEE/RSJ International Conference on Intelligent Robots and Systems* (pp. 2674–2681). IEEE. 10.1109/IROS.2011.6094581

Beer, F. P., Johnston, E. R. (Jr. ., Mazurek, D. F., León Cárdenas, J., Ríos Sánchez, M. A., & Zamora Gallardo, E. (2013). *Mecánica vectorial para ingenieros : estática*. McGraw-Hill Interamericana.

Bogado Martínez, C. F., Dutra, M. S., & Raptopoulos, L. S. C. (2018). Proposed control for wing movement, type flat plate, for ornithopter autonomous robot. *Journal of the Brazilian Society of Mechanical Sciences and Engineering*, *40*(9), 456. doi:10.100740430-018-1362-7

Chirarattananon, P., Ma, K. Y., & Wood, R. J. (2014). Adaptive control of a millimeter-scale flapping-wing robot. *Bioinspiration & Biomimetics*, *9*(2), 025004. doi:10.1088/1748-3182/9/2/025004 PMID:24855052

Colorado, J. (2012). BaTboT : a biologically inspired flapping and morphing bat robot actuated by SMA-based artificial muscles., 223. Retrieved from http://digital.csic.es/handle/10261/109918

De Croon, E., Groen, M. A., De Wagter, C., Remes, B., Ruijsink, R., & Van Oudheusden, B. W. (2012). Design, aerodynamics and autonomy of the DelFly Recent citations Mimicking Insect Wings: The Roadmap to Bioinspiration Jafar Hasan et al Design, aerodynamics and autonomy of the DelFly. *BIOINSPIRATION & BIOMIMETICS. Bioinspiration & Biomimetics*, *7*, 16. doi:10.1088/1748-3182/7/2/025003

De Wagter, C., Tijmons, S., Remes, B. D. W., & de Croon, G. C. H. E. (2014). Autonomous flight of a 20-gram Flapping Wing MAV with a 4-gram onboard stereo vision system. In *2014 IEEE International Conference on Robotics and Automation (ICRA)* (pp. 4982–4987). IEEE. 10.1109/ICRA.2014.6907589

Delaurier, J. D. (1988). The Development and Testing of a Full-Scale Piloted Ornithopter. *Canadian Aeronautics and Space Journal*. Retrieved from http://www.ornithopter.net/Publications/TheDevelopmentandTestingofaFull-ScalePilotedOrnithopter.pdf

DeLaurier, J. D., & Harris, J. (1993). *A study of mechanical flapping-wing flight. Aeronautical Journal*. Cambridge University Press; doi:10.1017000192400026452

Dietl, J. M., & Garcia, E. (2008). Stability in Ornithopter Longitudinal Flight Dynamics. *Journal of Guidance, Control, and Dynamics*, *31*(4), 1157–1163. doi:10.2514/1.33561

Fliess, M., Levine, J., Martin, P., Rouchon, P., Lévine, J., Martin, P., & Rouchon, P. (1995). Flatness and defect of non-linear systems: Introductory theory and examples. *International Journal of Control, 61*(6), 1327–1361. doi:10.1080/00207179508921959

Fry, S. N., Sayaman, R., & Dickinson, M. H. (2005). The aerodynamics of hovering flight in Drosophila. *The Journal of Experimental Biology, 208*(Pt 12), 2303–2318. doi:10.1242/jeb.01612 PMID:15939772

Habib, M. K. (2008). Human adaptive and friendly mechatronics (HAFM). In *2008 IEEE International Conference on Mechatronics and Automation* (pp. 61–65). IEEE. 10.1109/ICMA.2008.4798726

Isermann, R. (2005). Mechatronic systems - Innovative products with embedded control. In *IFAC Proceedings Volumes (IFAC-PapersOnline).*

Jiang, H., Zhou, C., & Xie, P. (2016). Design and kinematic analysis of seagull inspired flapping wing robot. In *2016 IEEE International Conference on Information and Automation (ICIA)* (pp. 1382–1386). IEEE. 10.1109/ICInfA.2016.7832035

Kirkpatrick, S. (1990). The Moment of Inertia of Bird Wings. *The Journal of Experimental Biology*, (January): 489–494. Retrieved from http://d.wanfangdata.com.cn/NSTLQK_NSTL_QK9557394.aspx

Lamers, K., Tijmons, S., De Wagter, C., & de Croon, G. (2016). Self-supervised monocular distance learning on a lightweight micro air vehicle. In *2016 IEEE/RSJ International Conference on Intelligent Robots and Systems (IROS)* (pp. 1779–1784). IEEE. 10.1109/IROS.2016.7759284

Levine, J. (2009). *Analysis and control of nonlinear systems: A flatness-based approach.* Springer Science & Business Media. doi:10.1007/978-3-642-00839-9

Li, Y., Qian, C., Zhu, B., & Fang, Y. (2017). *Kinematic, Static and Dynamic Analyses of Flapping Wing Mechanism Based on ANSYS Workbench* (pp. 316–323). Cham: Springer; doi:10.1007/978-3-319-70136-3_34

Liu, T. (2006). Comparative Scaling of Flapping- and Fixed-Wing Flyers. *AIAA Journal, 44*(1), 24–33. doi:10.2514/1.4035

Ma, K. Y., Chirarattananon, P., Fuller, S. B., & Wood, R. J. (2013). Controlled flight of a biologically inspired, insect-scale robot. *Science, 340*(6132), 603–607. doi:10.1126cience.1231806 PMID:23641114

Ma, K. Y., Felton, S. M., & Wood, R. J. (2012). Design, fabrication, and modeling of the split actuator microrobotic bee. In *2012 IEEE/RSJ International Conference on Intelligent Robots and Systems* (pp. 1133–1140). IEEE. 10.1109/IROS.2012.6386192

Markus, E. D., Agee, J. T., & Jimoh, A. A. (2017). Flat control of industrial robotic manipulators. *Robotics and Autonomous Systems, 87*, 226–236. doi:10.1016/j.robot.2016.10.009

Martin, Philippe, & Paden, B. (1994). A different look at output tracking : control of a VTOL aircraft * Searching for a flat output, (December), 2376–2381.

Martin, P., Murray, R. M., & Rouchon, P. (1997). Flat Systems. In *European Control Conference* (p. 54). Brussels. Retrieved from http://pierre.rouchon.chez-alice.fr/publications/PR1997/eccflat.pdf

Muijres, F. T., Bowlin, M. S., Johansson, L. C., & Hedenström, A. (2012). Vortex wake, downwash distribution, aerodynamic performance and wingbeat kinematics in slow-flying pied flycatchers. *Journal of the Royal Society, Interface*, *9*(67), 293–303. doi:10.1098/rsif.2011.0238 PMID:21676971

Murphy, I. P., Kurdila, A. J., Bayandor, J., & Burdisso, R. A. (2013). *Modeling and Control of Flapping Wing Robots*. Retrieved from https://vtechworks.lib.vt.edu/bitstream/handle/10919/19275/Murphy_IP_T_2013.pdf?sequence=1

Murray, R. M., Sluis, W., Rathinam, M., & Sluis, W. (1995). Differential flatness of mechanical control systems: A catalog of prototype systems. *Proc. of the ASME Mechanical Engineering Congress and Expo*, 1–9. 10.1109/INFCOM.2011.5934932

Nasr, S., Mekki, H., & Bouallegue, K. (2019). A multi-scroll chaotic system for a higher coverage path planning of a mobile robot using flatness controller. *Chaos, Solitons, and Fractals*, *118*, 366–375. doi:10.1016/j.chaos.2018.12.002

Nguyen, Q.-V., & Chan, W. L. (2018). Development and flight performance of a biologically-inspired tailless flapping-wing micro air vehicle with wing stroke plane modulation. *Bioinspiration & Biomimetics*, *14*(1), 016015. doi:10.1088/1748-3190/aaefa0 PMID:30523879

Ogata, K. (2010). *Modern control engineering*. Prentice-Hall.

Park, J. H., & Yoon, K.-J. (2008). Designing a Biomimetic Ornithopter Capable of Sustained and Controlled Flight. *Journal of Bionics Engineering*, *5*(1), 39–47. doi:10.1016/S1672-6529(08)60005-0

Pfeiffer, A. T., Lee, J.-S., Han, J.-H., & Baier, H. (2010). Ornithopter flight simulation based on flexible multi-body dynamics. *Journal of Bionics Engineering*, *7*(1), 102–111. doi:10.1016/S1672-6529(09)60189-X

Phan, H. V., Kang, T., & Park, H. C. (2017). Design and stable flight of a 21 g insect-like tailless flapping wing micro air vehicle with angular rates feedback control. *Bioinspiration & Biomimetics*, *12*(3), 036006. doi:10.1088/1748-3190/aa65db PMID:28281468

Pons, J. L. (2008). *Wearable Robots: Biomechatronic Exoskeletons*. John Wiley & Sons, Ltd. doi:10.1002/9780470987667

Rahn, C. D. (2002). Mechatronic Control of Distributed Noise and Vibration: A Lyapunov Approach. *Measurement Science & Technology*, *13*(4), 643–644. doi:10.1088/0957-0233/13/4/704

Roccia, B. A., Preidikman, S., & Massa, J. (2007). Desarrollo de un código computacional para simular y analizar la cinemática de alas batientes Diseño estructural mecánico View project. *Mecanica Computacional, XXVI*, 3224–3245. Retrieved from http://www.ing.unrc.edu.ar

Rubenstein, D. A., Yin, W., & Frame, M. D. (2012). Fundamentals of Fluid Mechanics. In Biofluid Mechanics (pp. 11–48). doi:10.1016/B978-0-12-381383-1.00002-3

Ryu, J.-C., & Agrawal, S. K. (2011). Differential flatness-based robust control of mobile robots in the presence of slip. *The International Journal of Robotics Research*, *30*(4), 463–475. doi:10.1177/0278364910385586

Sane, S. P., & Dickinson, M. H. (2001). Control of Flight force by a flapping wing. *The Journal of Experimental Biology*, *2626*(15), 2607–2626. PMID:11533111

Selig, M., Guglielmo, J. J., Broeren, A. P., & Giguere, P. (2012). *Summary of Low-Speed Airfoil Data - Vol 5* (Vol. 5). doi:10.1115/1.1793208

Tenenbaum, R. A. (2013). *Fundamentals of applied dynamics. Choice Reviews Online* (Vol. 41). Springer; doi:10.5860/choice.41-6560

Tobalske, B., & Dial, K. (1996). Flight kinematics of black-billed magpies and pigeons over a wide range of speeds. *The Journal of Experimental Biology*, *199*(2). PMID:9317775

Van Nieuwstadt, M. J. (1996). *Trajectory Generation for Nonlinear Control Systems*. California Institute of Tecnhology.

Van Nieuwstadt, M. J., & Murray, R. M. (1997). Real Time Trajectory Generation for Differentially Flat Systems. *International Journal of Robust and Nonlinear Control*, 28.

Verboom, J. L., Tijmons, S., De Wagter, C., Remes, B., Babuska, R., & de Croon, G. C. H. E. (2015). Attitude and altitude estimation and control on board a Flapping Wing Micro Air Vehicle. In *2015 IEEE International Conference on Robotics and Automation (ICRA)* (pp. 5846–5851). IEEE. 10.1109/ICRA.2015.7140017

Veslin Diaz, E., Slama, J., Dutra, M., Lengerke, O., & Morales Tavera, M. (2011). Trajectory tracking for robot manipulators using differential flatness. *Ingenieria e Investigacion*, *31*(2), 84–90.

Yang, S., Shen, Y., Li, B., Li, Y., & Zhang, J. (2018). Modeling and Simulation of a Flapping-Wing Robot with Active Tails for Balancing Control during Wheeled Running. In *2018 IEEE International Conference on Mechatronics and Automation (ICMA)* (pp. 1806–1811). IEEE. 10.1109/ICMA.2018.8484330

KEY TERMS AND DEFINITIONS

Aerodynamical Forces: The aerodynamical forces are a set of forces actuating on a body produced by the air or fluid that surrounds a body, it is generated by the relative motion between the body and the fluid.

Differential Flat Systems: A specific kind of system which structure of the trajectories of the dynamics can be characterized. A system is considered flat if their states and inputs could be determined by a set outputs and their derivatives.

Kinematic Patterns: A set of time course functions that describes the wing movements during flying in reference to a relative set of degree-of-freedom. Their objective is to replicate the kinematics of free flying in birds and insects.

Lagrangian Mechanics: The Lagrangian mechanics is a reformulation of the classical mechanical theory. In the Lagrangian mechanic the trajectory made by an object is obtained through the finding of the trajectory that minimizes the action, that is, the integral of the Lagrangian in time. The Lagrangian is obtained by the difference between kinetical energy and the potential energy.

Ornithopter: An ornithopter is defined as an aircraft that generates flies by the flapping of their wings. It structures and movement is inspired by the flight of birds and insects.

PD Control: A PD control stands for Proportional-Derivative control, it is a control loop feedback strategy used in industrial control to minimize the error over time by the adjust of a control variable or input, by means of calculating the error value as the difference between the desired setpoint and the measured state. The P term stands for the proportional part of the controller that generates a corrective input proportionally to the existing error. The D term acts on the derivative rate of change of the control error providing a fast response.

Two Degree-Of-Freedom Controllers: A non-linear control technique composed by a feedforward compensator block that modifies the input of the feedback compensator block which is used to correct any errors due to the noise of uncertainties in the plant.

Chapter 10
Locomotion Interfaces for Legged Robots:
Design Inspiration From Natural Locomotion Interfaces

Hisham A. Abdel-Aal
Drexel University, USA

ABSTRACT

This chapter presents a comparative study of the topographical structure of three common biological robotic inspirations: human, canine, and feline feet. It is shown that the metrological roughness of each of the examined feet is customized for the specific locomotion demands of the species. The textural parameters manifest close correlation to the pressure distribution experienced in movement and gait. This correlation enhances the durability and structural integrity of the bio-analogue. It is also shown that the metrological function of the human (plantigrade) feet pads combine that of the back and the front feet pads of the digitigrade mammals examined. It is argued that integrating the targeted engineering of roughness within the design process of robotic feet can enhance the function of walking robots. Further, it offers elegant solutions to some of the current problems encountered in design of humanoids and other bio-inspired walking robots.

INTRODUCTION

Autonomous robotic systems comprise two main groups. The first is termed as manipulation robotics, whereas the second is referred to as mobile robotics. Mobile robotics are conventionally categorized in terms of their locomotion configuration. The first group is referred to as wheeled, legged, or crawling machines. Each of these locomotion configurations offer certain advantages that renders the particular robot class suitable for special applications. Each of the robotic classes have limitations as well. One limitation faced by wheeled and crawling robots is mobility in natural terrains. This disadvantage, however, is offset in legged robots.

DOI: 10.4018/978-1-7998-0137-5.ch010

Legged robots offer superior mobility and they allow traversing difficult and inaccessible terrains. This is because their mode of locomotion, which uses legs, is advantageous in irregular terrains. Through varying the configuration of the legs, legged robots can selectively establish contact with the ground and adapt to surface irregularities. Legs also offer advantages in soft terrains (e.g. sand). The ability of selectively contacting the ground reflects positively on the energy consumption of the machine, improves stability, and widens the scope of applications. It is no surprise, therefore, that successful construction of legged robots have revolutionized application and utilization of robots in general.

Legged robots often mimic a biological model. Humanoids (which are biped machines) for example, mimic human locomotion (Hirai et al., 1998; Kaneko et al., 2008; Park et al., 2005; Radford et al., 2015; Kojima et al., 2015). Other machines are inspired by dinosaurs (Fukuoka and Akama, 2014), apes (Leong and Johnston, 2016), dogs (Tan et al., n.d; Sheba et al., 2018; Hunt, Szczecinski, and Quinn, 2017; He et al., 2019), cats (Kwon et al., 2015; Hayashi, Kato and Chobonyan, 2015; Liu et al., 2017; Dupeyroux et al., 2017), insects (Koh et al., 2015; Szczecinski et al., 2015; Cardinaels et al., 2017), and reptiles (Gao et al., 2014; Yu et al., 2018; Luo et al., 2014; Vespignani et al., 2015; Tanaka and Tanaka, 2015; Liu et al., 2016), among other analogs. The field is growing rapidly and many biological examples are being studied for inspiration.

Mammalian locomotion stances may be classified in terms of the part of the foot engaging the ground while moving. Accordingly, three locomotion styles are typically mentioned in literature. The first is sole walking, termed as plantigrade, as that practiced by humans, bears and some species of apes. The second is the so-called digitigrade, which stands for toe walking (practiced by quadrupeds such as cats and dogs). The third class of locomotion is known as unguligrade, which denotes walking on nails or hoofs as in the case of horses and cattle.

The anatomy of mammals manifest differences that accommodate the diversity of their locomotion stances. Digitigrade and unguligrade animals differ than plantigrade animals in length and position of the structure comprising the foot-ankle assembly. The diversity of mammalian anatomy combined to their form of locomotion provide vast design possibilities for legged robots. A crucial step in the design of these machines is that of designing feet. Similar to their biological analog, efficiency of locomotion for the machines depends on optimizing the interaction between feet and the supporting ground. This interaction is influential to the gait of the machine and thereby the energetic cost of motion. It also affects posture, balance, and stability of the machine. The interaction between the feet the ground is twofold. First there is the biomechanical aspects of the motion of the foot, how the transfer must effort, relative motion of its parts, and generation forces while. The second concerns the interaction of the underfoot (i.e. the soles or the pads) with the ground. That is the actual interfacial contact with the ground and the mechanistic steps involved in force transmission.

Foot pads are an intrinsic component of the locomotor assemblies in mammals. They display a myriad of configurations, and form, all while reflecting universal composition. Regardless of the mammal, footpads are all composed of thick epidermal skin and subdermal adipose tissue permeated with fibrous septa (Raibert et al., 2008). This adipose tissue, which is essentially fat, is specialized. It does not respond to the physiological state of the animal, and instead of acting as a reservoir for energy, it serves structural purposes. Therefore, it is often termed as "structural adipose tissue" (Tenreiro and Silva, 2006). It follows that footpads are parts of the skin and they share many of the mechanical, thermal, and tactile its functions. From a locomotion perspective, in particular, they form the initial contact interface between a moving mammal and the ground. Thus, mechanically they accommodate locomotion forces generated by the animal, support forces exerted by bodyweight, provide traction, and accommodate reactions from the contacting ground. In addition, they accommodate friction forces that arise from the contact between the surface and the ground.

Footpads in biology and soles in robotic feet share several mechanical functions. They cushion impact forces resulting from stepping. This may include dissipating part of the impact energy, they transmit locomotion forces during the stance. In addition, depending on the material used in a robotic foot, they store strain energy for elastic recoil Zielinska, 2004. Footpads also have a stabilization role why the feet out on the ground (Tummala, 2002). Most of these functions, point at the functional role for the metrology of the surface of the footpad (both on the level form and the level of roughness) in biology. This role is foreseen as a complimentary factor that optimizes or tunes the function of the footpad material in a mammal. Such a role is important, and perhaps crucial, to the durability and structural integrity of the animal. While robotic feet may undergo periodic maintenance to mend or replace damage to their material, a mammal does not have such a luxury. Animal feet have to last throughout their entire lifespan, and a self-sustaining durability optimizing factor has to exist to minimize their injury. Here it is important to note that an injury of the footpad in the biological world may be crippling and therefore can threaten the survivability of the species. Bearing this in mind, an argument for the benefit of "designing" surfaces for the sole of robotic feet may be advanced.

The advantage of the metrological design of the soles, along the principles deduced from the biological world may extend the structural lifespan of the machine. This is especially beneficial for machines that serve in hostile environments (especially those that interact with rough terrains) where periodic maintenance is not a possibility.

Despite the perceived importance of the role of metrology of the pad surfaces most of the efforts concerning design of robotic feet are based on biomechanical or structural issues. In particular, the vast majority of work concerns force transmission through the feet (Sellaouti et al., 2006; Daltorio et al., 2006; Narioka, Tsugawa, and Hosoda, 2009), evolution of reaction forces (Nabulsi et al., 2009; Chenkun et al., 2015; Kalamdani et al., 2007; Wisse et al., 2006; Eckert and Ijspeert, 2016), alignment and posture (Colasanto et al, 2015; Muscolo et al, 2017; Ninomiya et al, 2011; Harada et al, 2007; Venaman et al, 2007), and joint-force interaction (Burlayenko et al, 2010; Cheung et al, 2005; Chi et al, 2005). The generation of tractions during motion as relating to surface structure of the sole (or biological footpads) are not encountered in literature if investigated at all. Accommodating grip, generation of tractions, or prevention of slip is typically approached in walking machines from a mechanics point of view. That is, the design process stems from the study of forces, reactions, and pressure distribution resulting from loading. The consequence, is that the current art of designing feet for robots does not include the metrological dimension. Indeed, common practice is to apply a compliant material comprising periodic structures to the soul (Muhlberger et al, 2015; Abdel-Aal et al, 2014). The process of applying this material, again, does not consider the metrological structure of the biological analog. Rather, it only considers a measured coefficient of friction (or adhesion) of the applied material. Given the rarity of the deal studies of surface morphology and metrological structure of footpads in biology it is seems to conclude that this particular information is lacking. Indeed, when considering the interplay between metrology of soles, and footpads, and forces experienced in walking designers are incomplete vacuum. It follows that customization of feet-ground interfaces for particular machines, or to accommodate operation in a specific terrain, does not exist.

Previous studies for the metrology of ventral snakeskin (Greiner et al, 2015) showed that friction optimization in legless-reptiles depends on the surface morphology of the species in relation to its habitat. Further, it was shown that certain areal surface parameters (e.g. kurtosis and skewness) combine to optimize the locomotion mode of the reptile and to aid in saving muscular effort. Studies on human tactile perception (Gil et al, 2003b; Dario et al, 1987; Kummer et al, 1992; Johansson et al, 1979; Brown et al, 1973) also have emphasized the importance of the metrological structure of fingers for tactile perception. To this end, one would expect that the metrological layout of footpads and human foot soles, and other robot bio-analogues, are related in some manner to forces generated in motion, reactions, and savings in the energetic cost of motion. This work, therefore, aims at exploring this relationship, and to verify whether the metrology of footpads in different species has a specialized role in maintaining mobility and promoting motion specialization.

Morphology and Structure of Footpads

A distinctive feature of mammals is that their body is supported by their limbs. The orientation of the limbs with respect to the body closely relates to the manner of locomotion. Limbs of plantigrade mammals, like that of humans, are parallel to the longitudinal body axis (LBA) (figure-1-a). Quadrupeds, such as dogs and cats, manifests a body that is supported by limbs oriented normal (or perpendicular) to the LBA (figure-1-b). Limbs in case of quadrupeds are located closer beneath the body so that the elbows are pointing backwards whereas the knee joint points forward. The elbow and knee joint orientation is also shared by humans despite the orientation of the limbs in their case. The orientation of the limbs

Figure 1. General layout of body supporting limbs in Quadruped-a ans biped locomotors

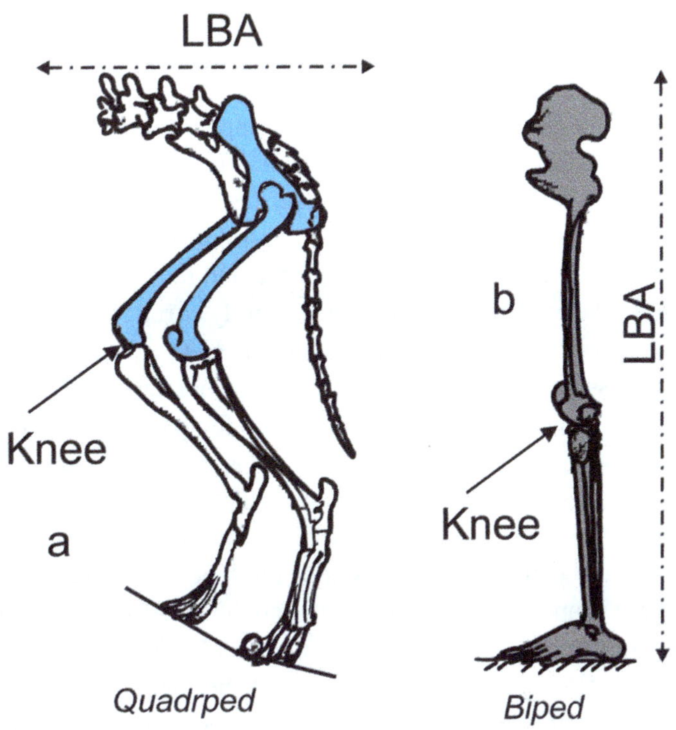

with respect to the LBA reflects some functional adaptation that aims at achieving speedy locomotion. In doing so, it advances the prospects of survivability through facilitating the chase of prey and avoiding potential predators by outrunning them.

Locomotion is a major energy expense for animals. It is common in the biological world to encounter structural and form adaptations that aim to facilitate optimized energy consumption. The orientation of limbs in mammals, therefore, may be understood in this context. The orientation of joints observed in mammals allows propulsion through swinging the joint in a manner similar to the motion of a pendulum.

Mimicking oscillations of a pendulum ensures that the energy invested in propulsion is directed towards overcoming the inertia of the limbs. Thus, at the end of each "swing" the limb is stopped and then accelerated in the opposite direction. This form of propulsion, as explained by Gray (2013) constitutes using the limbs as extensible struts. Swinging, Gray argues, minimizes the energy invested in moving the body. However, a consequence of implementing pendulum mechanics is a "bouncy" path for the center of gravity of the body. Due to the swinging of the limbs the center of gravity will oscillate up and down in proportion to the period and frequency of limb swinging. The oscillations are energized by the work done by the muscles i.e. it is an additional energy expense. An additional consequence of swinging relates to the inertia of the limbs. In particular, since mass factors in the magnitude of inertia created upon motion, heavier animals will spend more energy to stop and then reverse direction. To optimize locomotion energy, heavier animals typically cover longer distances per stride so that the energy consumed per unit distance is optimized.

The transfer of energy, it's consumption in braking, stoppage, and reversing the direction, start at the interface between the animal and the ground. That is, any optimization of the energetics of locomotion, energy consumption for propulsion, and dissipation of energy to overcome resistance to motion start at the level of the footpads. This, in principle, defines the functional role of footpads in locomotion and thereby should reflect on their morphology and structure.

Structure of Footpads

The overall structure, and properties of footpad material compliments their functional demands. Planter surfaces, which primarily are frictional surfaces, are cornified at different degrees. Cornification is greatest where most pressure is experienced. Thus, the dermis layer of the pad skin is thicker at the high-pressure areas. The increased thickness provides cushions through the dense connective tissue comprising the local structure. The connective tissue serves to support, and reinforce, deeper skeletal elements and protect against dislocation. The area of a footpad is not entirely cornified since cornification can hinder efficient motion. As such, cornified skin is present within discrete regions within the footpad.

Footpads manifest species-specific structural modifications which closely correlate to foot functions and form of locomotion. The basic structure of a plantigrade footpad comprises three sets of pads. Near the tip of each digit there is a digital pad (also known as an apical pad). The so-called interdigital pads are located near the tip of each digit. Near the wrist or ankle there exists a pair of pads with the inner known as the "thenar" and the lateral as the "hypothenar". Interdigital pads may also be called plantar pads (Hicks et al, 1954). These pads bear the weight of the foot when placed on the ground. Once the foot starts lifting, the weight support progressively shifts on the forward pads in a manner that originates the final thrust from the digital pads. There are some functional differences that pertain to the different parts of the footpads. These closely relate to the stance of the mammal. For plantigrade locomotion, the weight is borne by the digital, interdigital, thenar, and hypothenar pads. The weight in case of the digitigrade locomotion is borne on the digital and interdigital pads. The thenar and hypo-thenar pads are therefore reduced. In the case of unguligrade locomotion, the weight is carried by the terminal phalanges throughout their claws, hooves, and or digital pads.

In humans, the footpad i.e. the sole of the foot, his hairless. It contains a higher density of sweat pores (similar to the rest of the skin and the palms of the hand). Compatible with other mammalian footpads, the sole of a human foot (figure-3-a) contains the thickest layers of skin in order to meet its loadbearing function. Being covered with skin, it performs a tactile function as well. For analysis, the human foot may be divided into four main regions (figure-3-b). These are the heel, the midfoot, the metatarsal head region, and the toe region. These may further be divided into the nine sub regions shown in the figure. Paws, i.e. footpads for dogs and cats, comprise thin pigmented and keratinized hairless epidermis (like the human sole). The epidermis covers a subcutaneous, collagenous, adipose tissue. The adipose tissue comprises the pads. Such a composition meets the loadbearing function of the pads.

Paws of dogs and cats manifest similar constructional traits. They comprise five basic parts (figure-4-a and b). These are the claw, the digital pads, and the metacarpal pads. There are also the dewclaw and the carpal pad (which are not shown in the figure). The digital metacarpal, on front paws, and metatarsal pads primarily bear several forms of loads (impact and frictional). In addition they perform braking upon the acceleration or stoppage.

Figure 2. Different anatomical parts of the human foot-foot pad complex

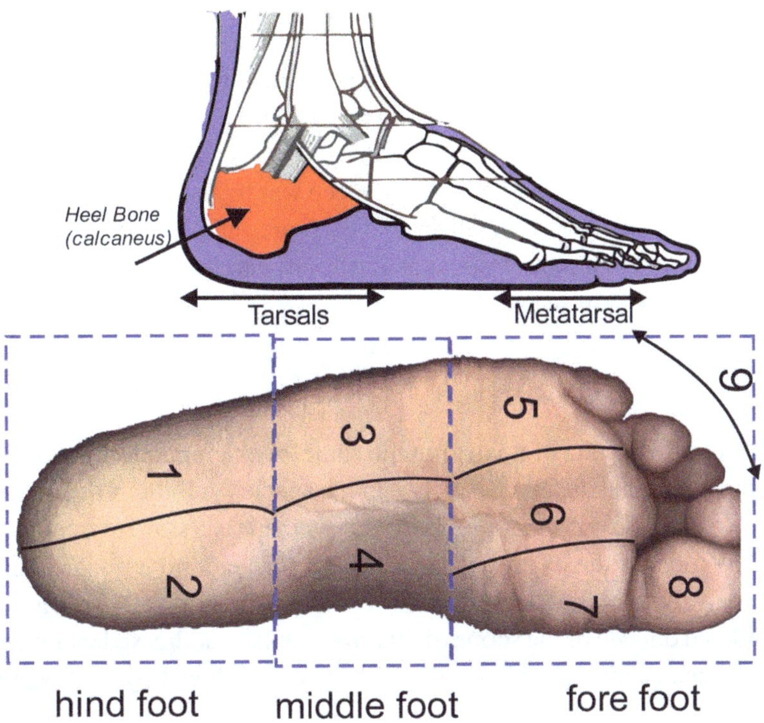

Figure 3. Areas of foot comprising foot pads in digitigrades

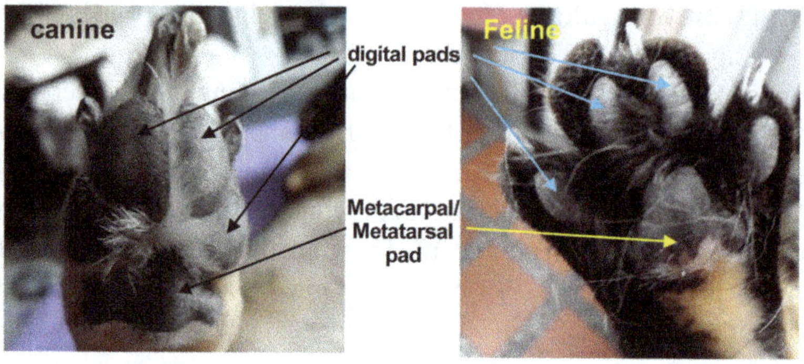

MATERIALS AND METHODS

To explore the surface structure of footpads we chose three examples. These examples represent popular inspirations for robots. The first is human (representing biped plantigrade locomotion). In addition, we chose to examine canine footpads and feline (domestic cat). Both of these pads represent quadruped-digitigrade animals.

The examination of human feet pads included making replicas of the soles for two groups of men and women. Each group contained five individuals. The choice of the group subjects did not consider race or age. This is because at this stage of the study the focus is on exploring the general traits of the surface metrology of the footpads. Similarly, for each of the quadruped animals we examined groups of five individual animals. Thus information was collected for a group of five mixed breed dogs and a group of five domestic cats.

Examination of the pads is a two-step process. The initial step of the examination was to obtain replicas of selected parts of the footpads using a polymer resin. Following replication, white light interferograms were obtained for the replicas. In conjunction to replication, high resolution scanning of the animal and human footpads took place. The following step was to analyze the white light interferometer information to obtain the depth of the skin ridges and each preselected spot. The obtained depth values were subsequently substituted in a metrological analysis software package (Mountains 7.0) to perform in-depth image analysis of the skin scans.

Results

In this section we present the results of analyzing the samples examined in this study. In presenting the results, we observe the biological differences between the species in light of the objectives of this study. Given the lack of information about the topic of the study, there is a need to explore the general differences between the structures of the examined surfaces. To this end, no emphasis in the presentation is placed on in-depth analysis of individual values for roughness parameters (profile or aerial). In-depth analysis, while considered outside of the scope of this work, will be addressed when comparison between members of the same species, along with starting differences induced by gender, breeds and race are started. As such, the scope of the current presentation concerns comparative presentation of three aspects of surface structure. Namely, we focus on the loadbearing parameters (R_K family), textured anisotropy, and the kurtosis – skewness combination. Choice of these parameters was based on experience gained from previous studies of other species (Meyer et al, 1990) as well as the reported functions (mechanical and tribological) discussed earlier.

The order of presentation is similat for each of the footpads examined in the work. The presentation starts with the general topographical maps of the particular footpads. This is followed by presenting data for the load bearing curves, loadbearing functional parameters, and texture anisotropy. Finally, we present distribution maps of the kurtosis – skewness combination. This order is observed for all footpads examined including human feet soles.

Feline – Domestic Cat Feet

This section presents results of examining pads of domestic cats. Figure 4 presents topographical maps for a representative sample. The map displays the topography layout of the left front pad of the cat. Numbers in the figure denote the various anatomical portions of the footpad. Thus, numbers 1 through 4 denote the digital pads whereas the pad labeled 5 denotes the metacarpal pad for this particular sample.

It is noted that the pads are not flat rather they display curved surfaces, with the peaks of the surface approximately located in the middle section of each pad. The topographical maps also show that the area was the peak height of the surface is relatively large which indicates that the slopes of the pads are sharper along the boundaries.

An initial query of the analysis concerns the degree of symmetry between the topography of front and back paws of the animal. To explore this point, first we introduce figure 5, which plots the distribution of the arithmetic mean height parameter, S_a, for the front and hind paws of the right side of the representative cat sample of figure 4.

Figure 5-a presents the distribution of the S_a parameter within the pads of the front paw, whereas figure 5-b presents the distribution within the pads of the back paw. Although detailed measurements are not exhaustively presented, the data plotted in both figures reflect some degree of similarity. The values of S_a for the medial digital pads of the front and back paws, reflect only a slight variation ($17\ \mu m < S_a < 20\ \mu m$). The same remark applies within the data of the lateral and the carpal/tarsal pads. This points at possible symmetry in the value of the parameters along the lateral and the longitudinal axis of the animal. As such, in what follows we will base presentation of additional results on such an assumption for. Exploring to what degree the assumption of symmetry is valid and whether it is breed-specific or a common feature within the species is considered out of the scope of this presentation. Detailed investigation of such a question is, however, a subject of ongoing work.

Figure 6, presents detailed plots of the aerial material ratio curves for the left foot pad shown in figure 4. Labels of curves follow the same order as that used in figure 4. Namely, the curves labeled POS: 1 through POS: 4 pertain to the digital pads identified by the labels 1 through 4 and figure 4. The curve labeled POS: 5 pertains to the metacarpal pad labeled 5 in figure 4. The curves also identify the areal family of functional parameters (S_p, S_{pk}, and S_{vk}).

Initially, it is noted that the material ratio curve is asymmetrical for any pad. Noted also that the curve shows the same trend for all pads (although that the values of the individual functional parameters vary by location). The value of the reduced peak height, S_{pk}, is visibly dominant within the curve. It's magnitude is considerably larger than that of the reduced Valley depth parameter S_{vk}. The values of the

Figure 4. Topographical maps for representative front left paw of a domestic cat

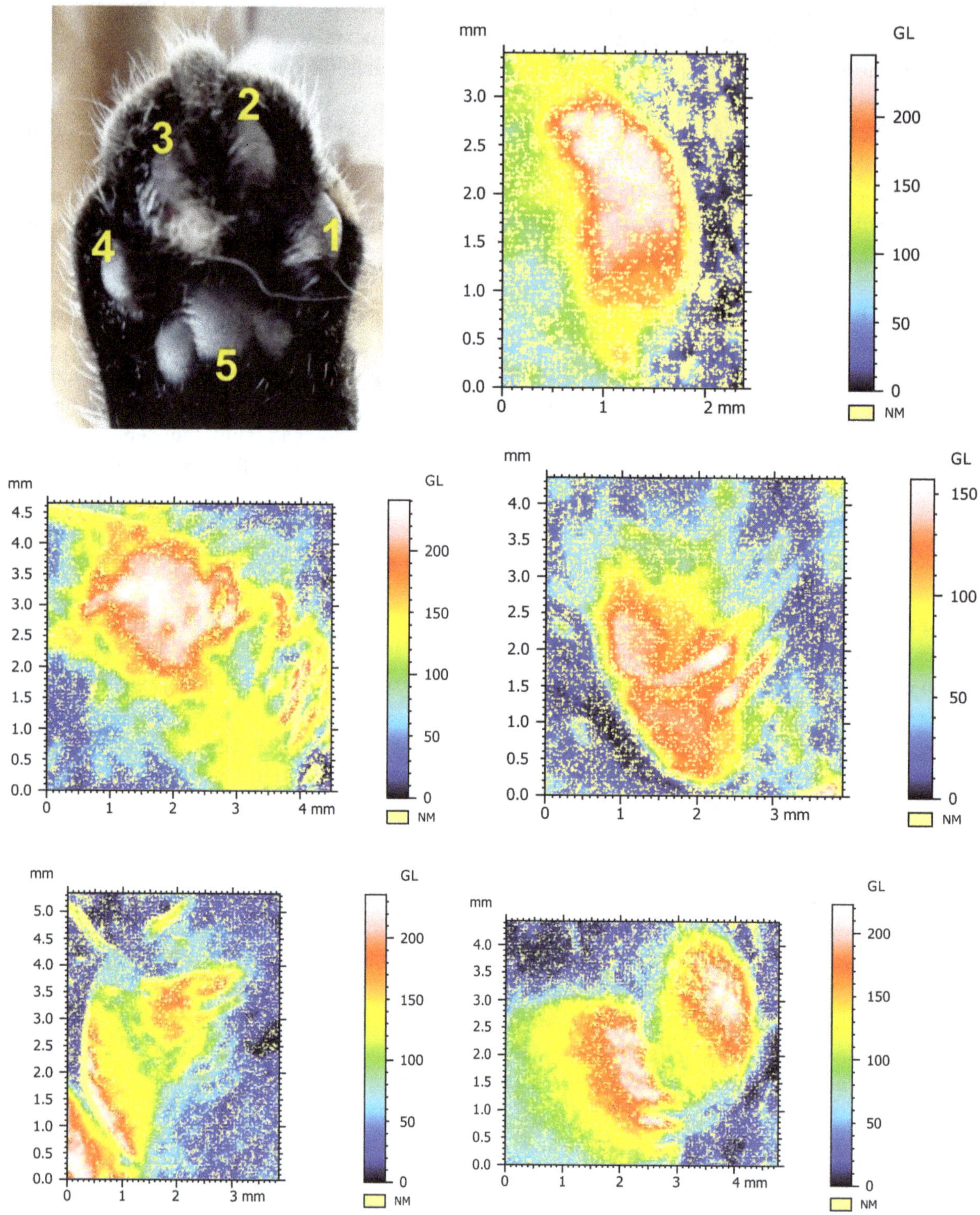

Figure 5. The arithmetic mean height, Sa, for the front and hind paws of the right side of a representative cat. Figure 5– a presents the distribution of the Sa parameter for the front paw, and figure 5-b presents the distribution for the back paw

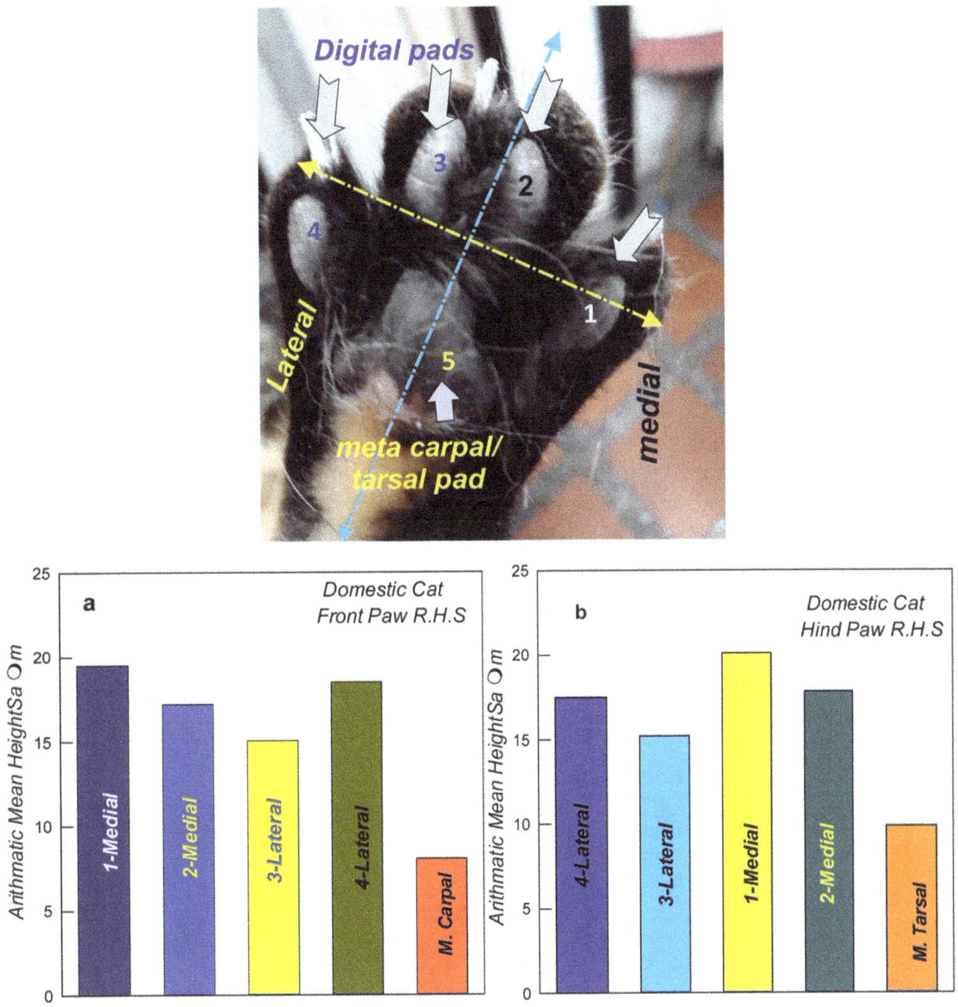

S_{pk} parameter, however, are not equal among all pads. Rather the values differ by position. The lateral digital pads, position 3 and 4, manifest the largest, and the smallest S_{pk} and S_{vk} values respectively. The rest of the pads manifest smaller S_{pk}. Their S_{vk} values, however, are almost within the same order of magnitude of their S_{pk} values. The metatarsal pad however manifests a unique trend. The S_{pk} value for this pads is almost of the same order of magnitude as that of the medial digital pads, whereas the S_{vk} value is the smallest among all pads.

It is useful to express the distribution of the material functional parameters as a ratio of the surface core roughness depth, S_k. This yields an insight into topographical build-up of the surfaces especially when considering whether or not there exists considerable differences in the S_a values of the different surfaces. Such data is presented in figures 7 and 8. The figures compare two ratios: S_{pk}/S_k and S_{vk}/S_k for the front and the back paws. The plots pair information by body side location of the digits. For example,

Figure 6. Aerial material ratio curves for the left foot pad shown in figure 4. Labels of curves follow the same order as that used in figure 4. Curves labeled POS: 1 through POS: 4 pertain to the digital pads identified by the labels 1 through 4 and figure 4. The curve labeled POS: 5 pertains to the metacarpal pad labeled 5 in figure 4.

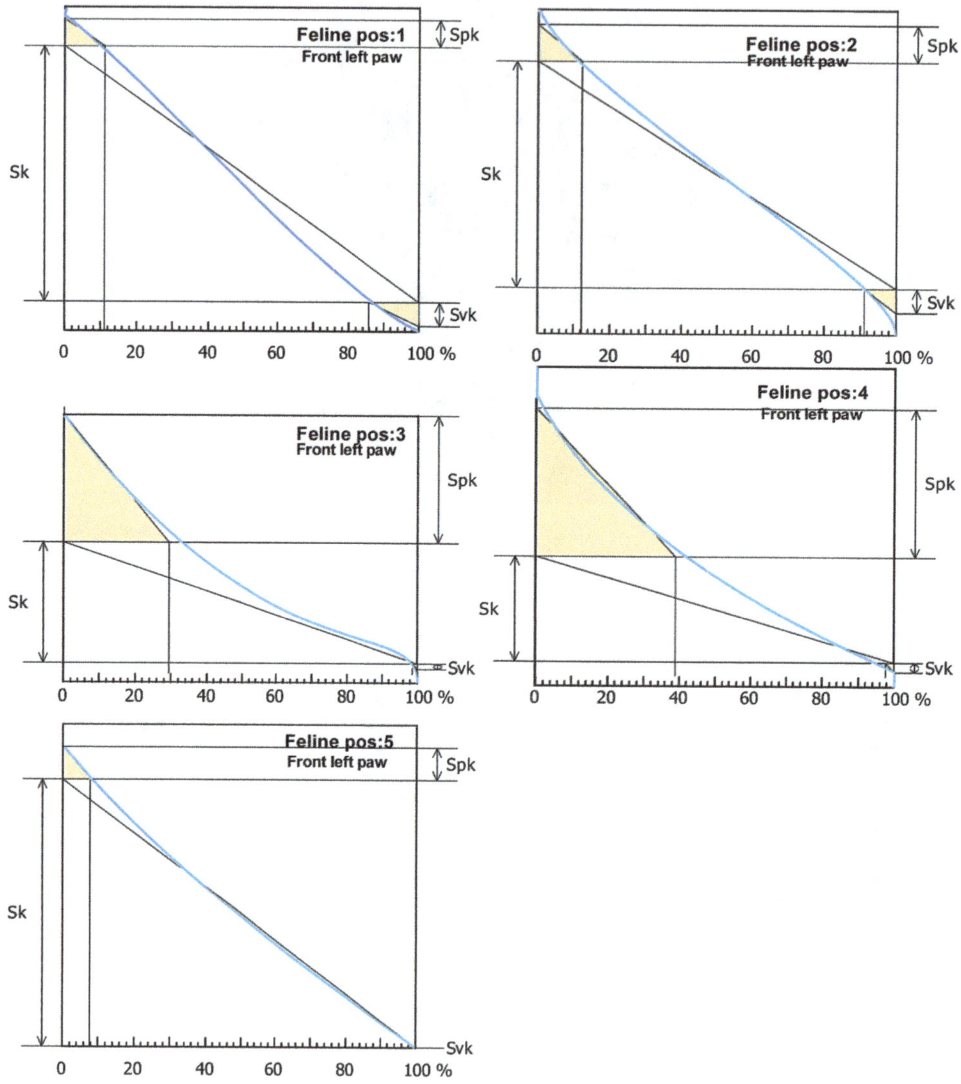

figure 7-a, presents the ratios S_{pk}/S_k and S_{vk}/S_k for the front medial pad, labeled 1, for the left and the right front paws. Pairing the data in this manner provides direct means to investigate if symmetry of functional parameters takes place along the longitudinal or the lateral body axes. The plots in figure 7-b, which depicts the functional ratios for the front second medial pad, reflects a small difference between values of left and right pads. The same remark applies to figure 8 which compares left on the right hand side back pads (see for example figure 8-d which depicts the ratios for the second lateral pad). Another form of parameter ratio consistency is noted upon comparing the relative values for the two functional ratios among all pads within the front and back paws. Comparison of all plots within figure 7 (a through

Figure 7. Ratio of the material functional parameters S_{pk} and S_{vk} compared to the surface core roughness depth, S_k for the front paws of the domestic cat

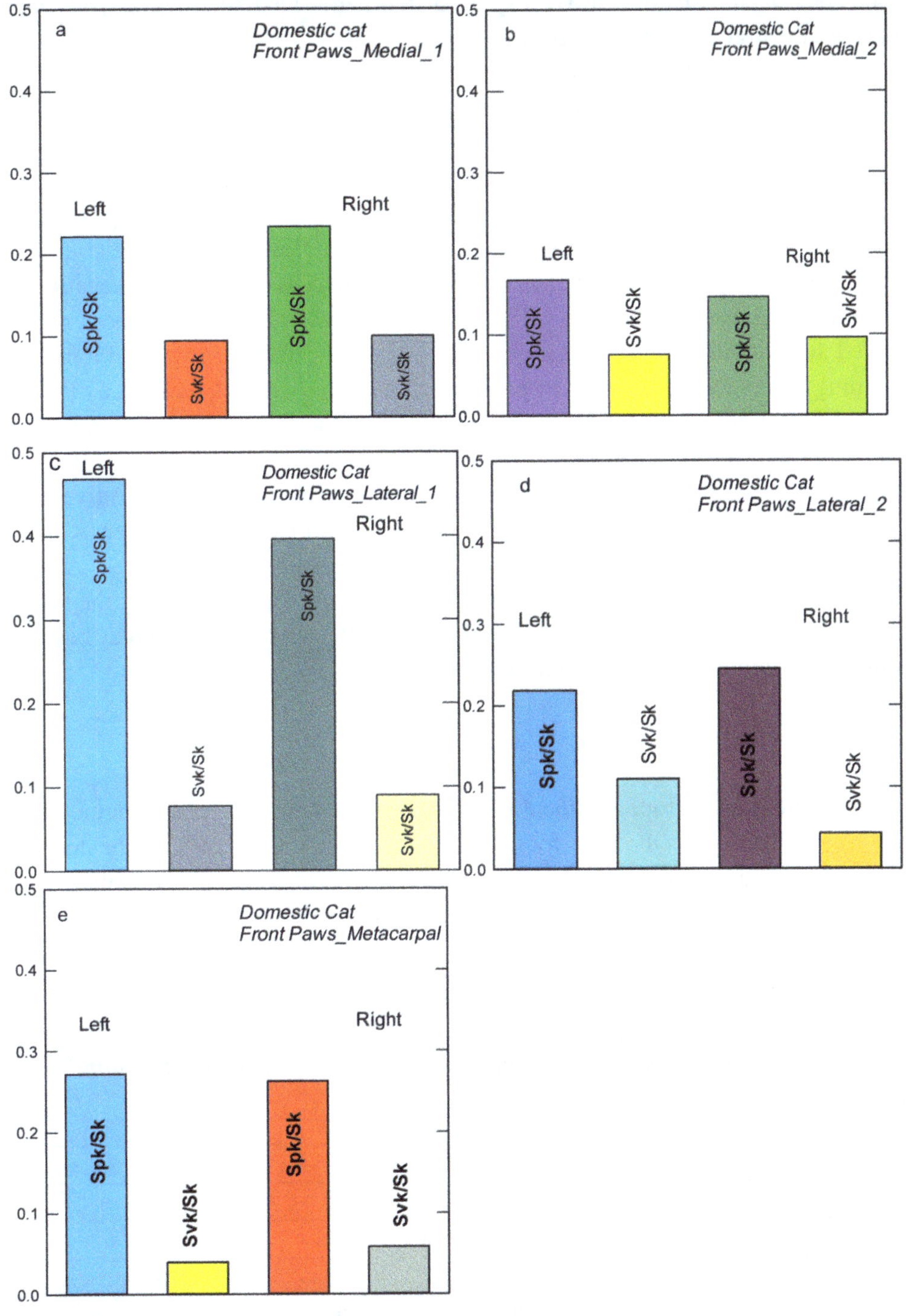

e) show that the first lateral digital pad has the highest value of the ratio S_{pk}/S_k. This is also valid for the left and right hand side pads and for the back paws. In all, comparing values presented in figures 8 (a through e), one notes, again, that the ratio S_{pk}/S_k for the left hand side lateral digital pads and that of the right hand side are the highest among all pads.

Value of the ratio S_{vk}/S_k for the digital pads, in the figure, manifest are higher than that of the metacarpal pads. Among the back paws, the metatarsal pad (both on the left and right feet) has the highest value of the ratio S_{vk}/S_k.

To complete analysis we examined the surface kurtosis, S_{ku}, and skewness, S_{sk}, of the pads. Figure 9, along with table 1, present the extracted data. The figure shows the left hand side paw, front and back, of a representative cat. The arrows shown on each of the pads indicate whether the combination of the S_{sk} and S_{ku} parameters are positive or whether the S_{ku} is positive while the S_{sk} is negative. When both arrows are pointing at the same direction it indicates that S_{ku} as well as S_{sk} are both positive. When, however, the red arrow points upwards, it indicates a positive S_{ku} (blue arrow) and a negative S_{sk} (red arrow). Examining the combined signs of the parameters revealed that each paw has a dominant positive -positive combination indicating a tendency for load bearing. On the one digital pad in the front paw, medial pad-1 manifests a positive-negative S_{ku}-S_{sk} combination.

On the back paw, lateral pad-4 which is on the counter opposite end shows a positive-negative combination. The metacarpal pad (pad- 5) in the front paw shows in loadbearing tendency, positive S_{ku}-S_{sk} combination. On the contrary. The metatarsal pad manifests a positive-negative combination. This implies a tendency for traction. It is interesting to note, however, that considering the positions of the traction pads indicates folding symmetry between the front and back paws which may have implications for stability in locomotion or during climbing activities.

Canine-Dog Feet Pads

Figure 10 (a and b) presents the topographical maps of the right-hand side paws. Figure 10-a shows the front pad, and figure 10-b shows the back footpads. Again, similar to the case of the domestic cat, dog pads are not flat. They manifest curved surfaces, with the slopes of the surfaces relatively steep around the boundaries of the pads and almost flat around the central section of individual pads.

Table 1. Values of surface kurtosis, S_{ku}, and skewness, S_{sk}, parameters for the front and back pads of a representative domestic cat (left hand side)

Cat					
	M-1	**M-2**	**L-3**	**L-4**	**5**
Left-Front					
Ssk	-0.037	0.109	0.661	0.689	0.485
Sku	1.87	1.96	2.08	2.27	2.22
Back-Left					
Ssk	0.479	0.307	0.399	-0.171	-0.168
Sku	2.573	1.934	1.89	1.737	2.246

Figure 8. Ratio of the material functional parameters S_{pk} and S_{vk} compared to the surface core roughness depth, S_k for the front paws of the domestic cat

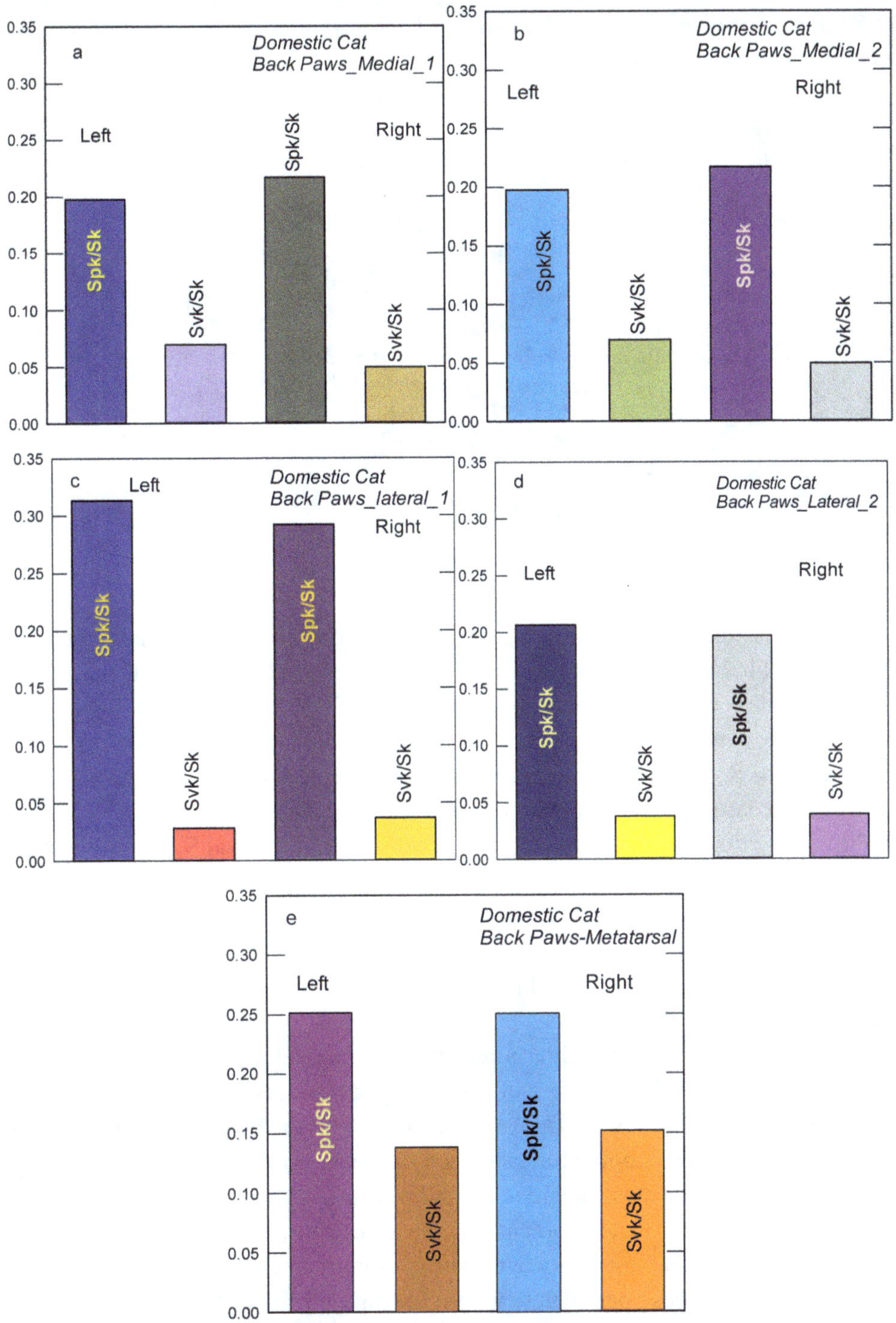

Figure 9. Mapping of the surface kurtosis, S_{ku}, and skewness, S_{sk}, parameters for the front and back pads of a representative domestic cat (left hand side). Arrows shown on the pads indicate whether the combination of the S_{sk} and S_{ku} parameters are positive or negative. Arrows pointing at the same direction indicate that S_{ku} and S_{sk} are both positive. When the red arrow points upwards, it indicates a positive S_{ku} (blue arrow) and a negative S_{sk} (red arrow)

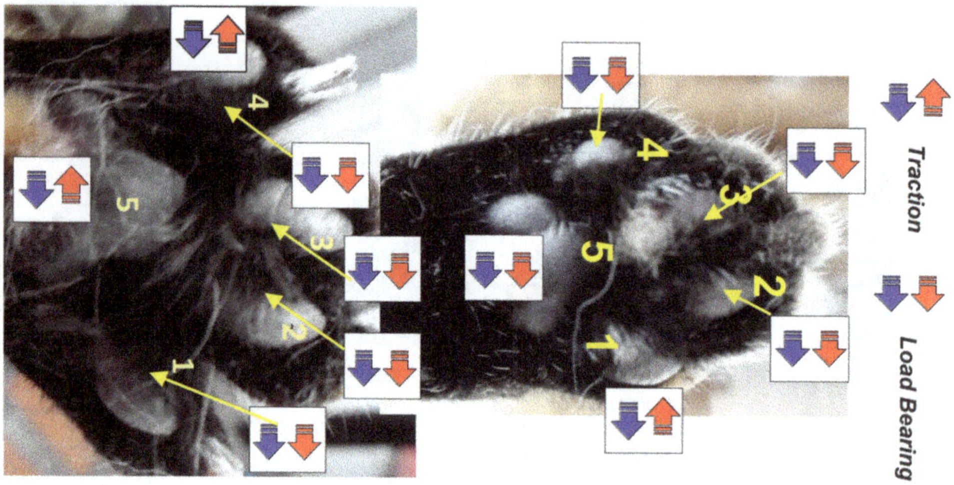

Distribution of the arithmetic mean height parameter S_a, among the digital and meta pads of the feet is presented in figure 11. Individual values of the S_a are rather random distribution. At definite order that correlates the location of a particular pad and the value of S_a is not apparent.

Figure 12 plots selective load bearing curves for the dog front right paw similar to the case of cats, the curves show that the S_{PK} portion of the curves are not balanced by the S_{vk} portion. The curve itself is not symmetrical. The largest S_{pk} is located on the outer lateral digital pad. For the rest of the pads, the S_{pk} seems slightly larger than the S_{vk}. Detailed plots of the ratios S_{pk}/S_k and S_{vk}/S_k for the front and back right paws are given as figure 13. The presentation within the figure is consistent with that followed in case of the cat. The figure comprises five plots that pair like digits. In general, the ratio S_{pk}/S_k is the highest for all pads (except for the second lateral pad shown in figure 13-d). The ratio S_{vk}/S_k meanwhile is almost the same order of magnitude for all pads (around$0.2 \leq S_{vk}/S_k \leq 0.7$). The largest Spk/Sk value is found on the metacarpal and the metatarsal pads. Proportion symmetry between front and back pads, similar to the case of the cat, is also noted (compare for example values in figure 13-a, and 13-e).

Figure 14 (a and b) presents the mapping of the S_{ku}-S_{sk} parameters within the front and back paws. The mapping of the front paw is shown as figure 13-a, and that for the back paw is shown as figure 14-b. Consistent with the presentation adopted for cats when the small arrows, representing S_{sk} and S_{ku}, point toward the same direction it indicates a positive S_{sk} and positive S_{ku}. This is taken to indicate suitability for load bearing. When, however, that is point at opposite directions, it implies positive S_{ku} and negative S_{sk}. This, again, is taken to indicate suitability of the surface for tractions and accommodation of frictional reactions. The distribution of these parameters within the digital pads of the front paw shows tendency

Figure 10. Topographical maps for representative front left paw of front (10-a) and back(10-b) foot pads of a dog

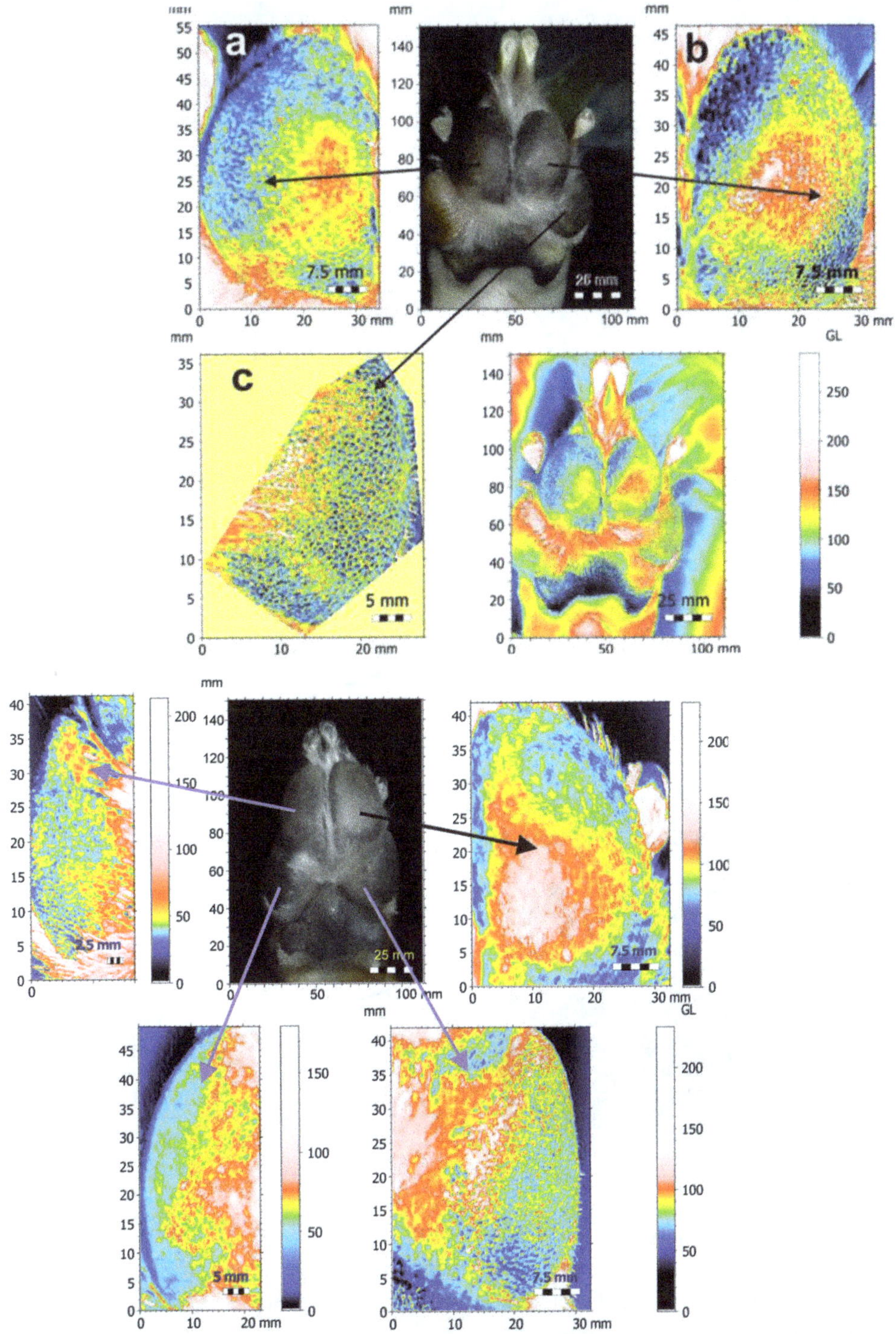

Figure 11. The arithmetic mean height, Sa, for the front and hind paws of the right side of a representative Dog. Figure 11– a presents the distribution of the Sa parameter for the front paw, and figure 5-b presents the distribution for the back paw

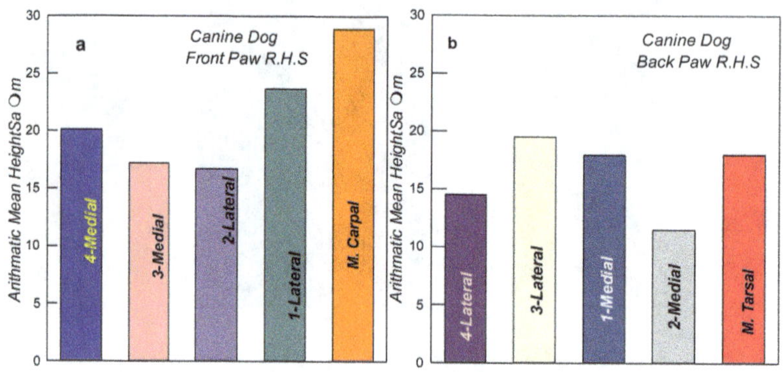

Figure 12. Aerial material ratio curves for the front paw shown in figure 10-a. Labels of curves follow the same order as that used in figure 10-a

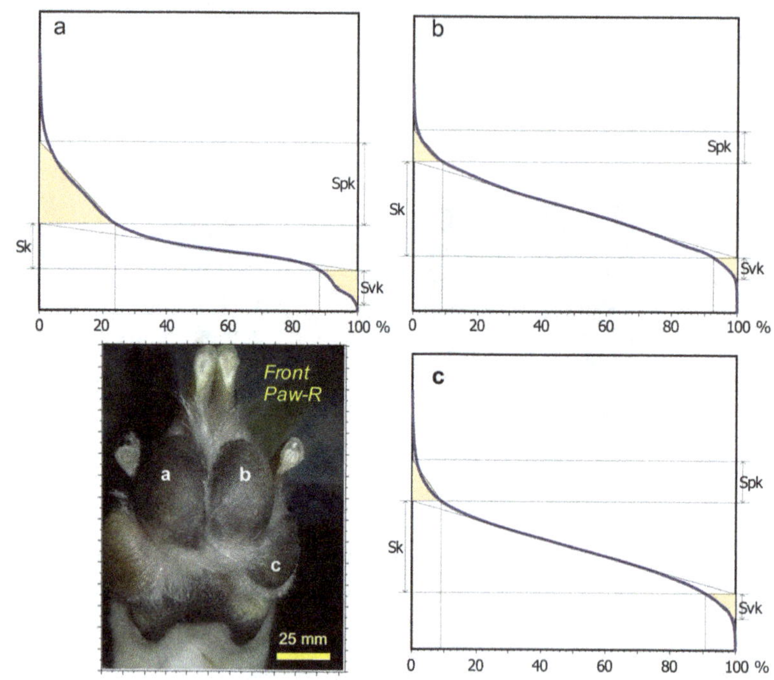

for load bearing. On the back paw, however, the digital pads manifest suitability for traction (note the opposite direction of the arrows). The outer lateral digital pad however manifests loadbearing tendency (both arrows point toward the same direction). Both meta pads (carpal and tarsal) show loadbearing tendency (i.e. positive kurtosis and skewness).

Figure 13. Ratio of the material functional parameters S_{pk} and S_{vk} compared to the surface core roughness depth, S_k for the front and back R. H. S paws of a dog

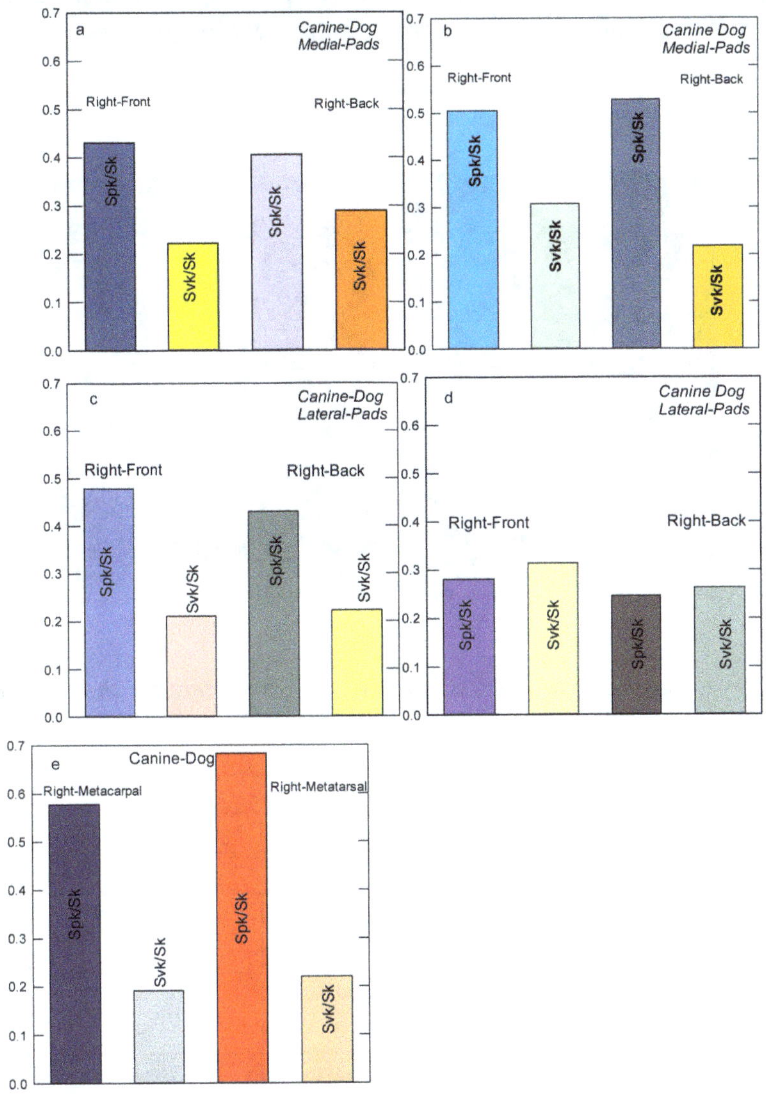

Biped Human Feet Pads

In this section we introduce the results of analyzing the surface structure of human feet pads. The digitigrade human footpad, as shown in figure 2, has a different layout than the feet of quadruped animals. These differences impact the style of presenting the data. In essence, the same set of parameters, as those evaluated for canid and felid feet, are present. However in presenting these data we emphasize three different parts. These are the heel (hind foot), the fore foot, and the toes. In line with the general objectives of the current work no detailed analysis of differences between genders or effective age would be given. Such details are considered beyond the scope of this presentation and are the subject of on going work. Consistent with data presented for quadrupeds we start by presenting the distribution of the S_a parameter.

Figure 14. Mapping of the surface kurtosis, S_{ku}, and skewness, S_{sk}, parameters for the front and back pads of a representative dog (left hand side). Arrows shown on the pads indicate whether the combination of the S_{sk} and S_{ku} parameters are positive or negative. Arrows pointing at the same direction indicate that S_{ku} and S_{sk} are both positive. When the yellow arrow points right to left, it indicates a positive S_{ku} (blue arrow) and a negative S_{sk} (yello arrow)

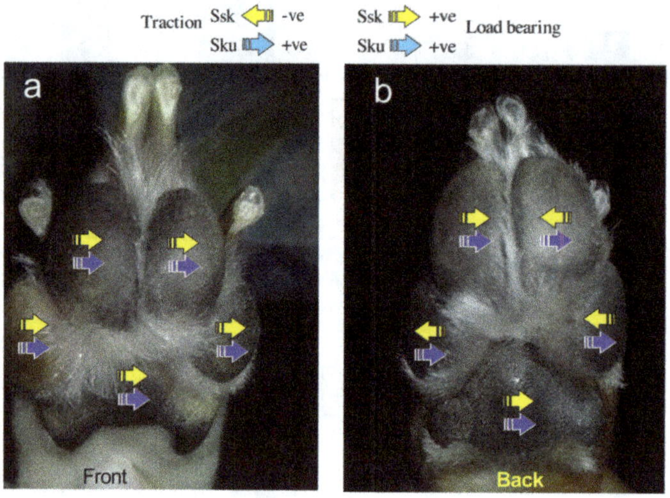

Table 2. Values of surface kurtosis, S_{ku}, and skewness, S_{sk}, parameters for the front and back pads of a representative dog

Canine-Dog				
Position	**I**	**II**	**III**	**IV**
Front Digital Pads				
Ssk	0.72	0.215	0.351	0.289
Sku	3.67	2.77	3.5	3.8
Back Digital Pads				
Ssk	1.29	-0.3	-0.61	-0.6
Sku	5.2	3.52	3.96	3.67

The figures present the distribution of the S_a parameter for male toes, figure 15, and for a female subject, figure 16. It is noted in figure 16 that the second toe in both feet has the highest value of S_a, the big toe, however has the lowest value. This trend is not reflected in figure 15 which depicts the S_a for a male. Topography of roughness of the human feet pads are influenced by many factors other than age and gender. Mainly the frequency of walking barefoot, the interaction of the skin with the roughness of the foreground, and the roughness of the stepping grounds itself. Another factor is the friction between the footpad skin and socks or shoe paddings if any. Bearing this in mind it is seen that any statement about the origin of the difference in values of the S_a would be rather subjective if the walking habits of the particular individual are not considered. As such, no objective discussion of differences in S_a between individual toes will be given in this work.

Figure 15. Distribution of the arithmetic mean height parameter S_a for male toes

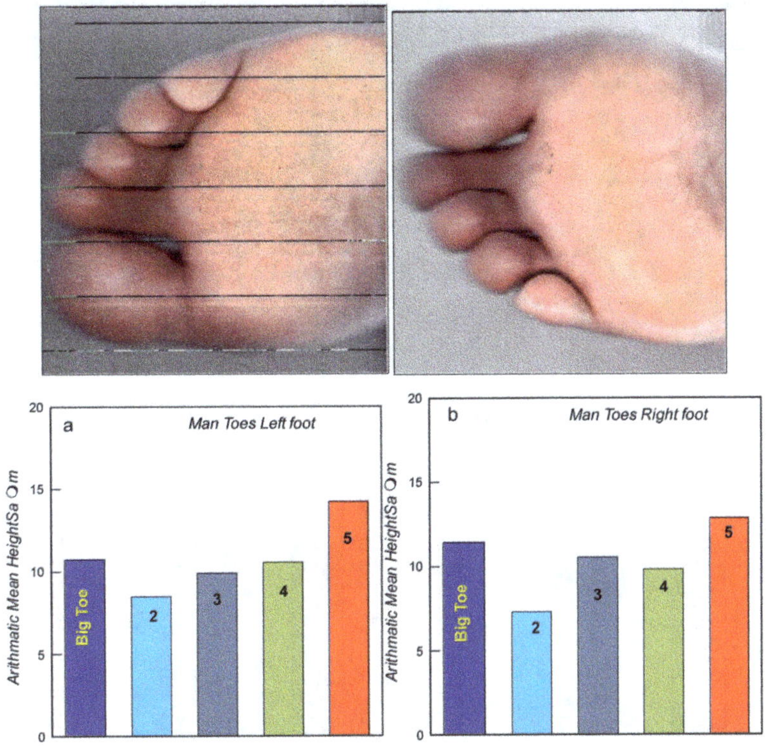

A sample distribution of the loading curve, and the functional material aerial parameters, is shown in figure 17. The foot shown in the figure pertains to an 80 years old female. Data are shown for the toes of the foot. Consistent with the case of the canine and that of the feline the curve is asymmetric. However opposite to the case of the feline pads, the S_{vk} parameter is apparently larger than the S_{pk} parameter. This remark is fairly consistent among both male and female subjects. Data presented in table-3 confirms that remark. Table-3 lists data for the ratios S_{pk}/S_k and S_{vk}/S_k four female feet and for the sample male foot. The data show, consistently and regardless of gender, that the ratio S_{vk}/S_k is larger than the ratio S_{pk}/S_k for all toes. It is also noted that the Svk for the toes of the male is slightly larger than S_{pk}. For the female toes however the S_{vk} is almost twice the S_{pk}.

Sample load bearing curves of selected spots within the foot are shown as figure 18 (a and b). The curve pertains to a male (50-60 years old). Figure 18-a shows the load bearing curves, and distribution of the height parameters, for three spots on the heel. Figure 18-b, meanwhile, depicts similar information as figure 18-a except that the figure considers the fore foot of the same male subject. The distribution of the bearing curves reveals opposing trends.

Points located on the heel show mostly asymmetric curves. The asymmetry reflects an S_{pk}/S_k, for point 1, located on the outer perimeter of the feet and for point 3, located on the lateral side of the heel. The degree of these asymmetry is not equal (as indicated from data presented in table-4). Point 2, however, reflects a quasi-symmetric load bearing curve (note that the ratios S_{pk}/S_k and S_{vk}/S_k table 4 are almost equal). The curse pertain to points 4 and 5, located on the forefoot again reflect curve asymmetry bite opposite trends. Point 4, located on the lateral side of the foot shows an S_{pk}/S_k that is almost twice the

Figure 16. Distribution of the arithmetic mean height parameter S_a for female toes

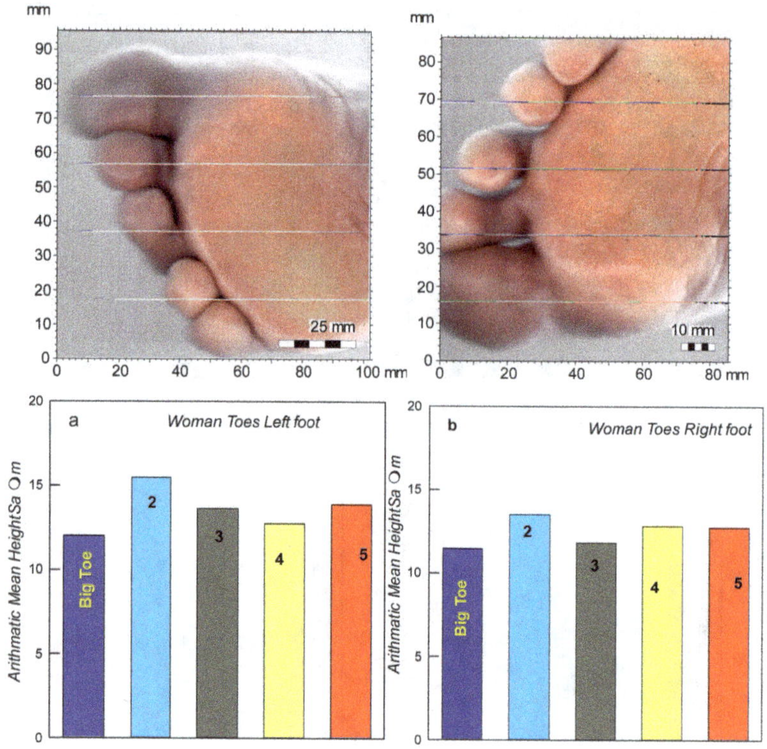

value of the ratio S_{vk}/S_k. Incidentally, such a trend is consistent with that of point 3, again located on the lateral side of the heel. The load curve for point 5, located on the medial side of the forefoot shows an S_{vk}/S_k that is larger than S_{pk}/S_k.

The change in the proportions of the functional aerial parameters pointed at the possibility of a correlation between the order of magnitude of the parameter and location on the footpad. To investigate this possibility we identified 12 areas on the perimeter of the foot for further analysis. These areas, shown in figure 19, are approximately 3 mm^2 each. As shown in the figure, the yellow squares present areas within the lateral side of the foot, whereas the same gray present locations on the medial side. Data from the analysis may also indicate differences, if any, between the heel and the forefoot regions as indicated by the dashed squares within the figure.

The location of the twelve areas shown in the figure roughly coincide with the lperimeter of the so-called sub-calcaneal fat pad comprising the complex fibrous fatty structure intertwined with the dermis layer of the footpad. The heel pad, as mentioned in section 2, is a specialized structure that facilitates transmission of body weight to the ground during locomotion cycles (Alexander et al, 1986) (among other mechanical and loadbearing functions). The analysis of the areas entailed evaluation of the area functional parameters (S_k-family) for each of the pre-chosen locations (data are given as table-5). Organization of the data within the table follow the location of each cluster of areas. Thus data for the ratios S_{pk}/S_k, S_{vk}/S_k, and S_{pk}/S_{vk} are organized by location on the side of the foot pad (i.e. lateral or medial). The last two rows within the table present the ratio S_{pk}/S_{vk} organized by location with respect to the distal-proximal axis of the foot. Points located on the heel are highlighted within the shaded cells of the table.

Figure 17. Aerial material ratio curves for the female human toes

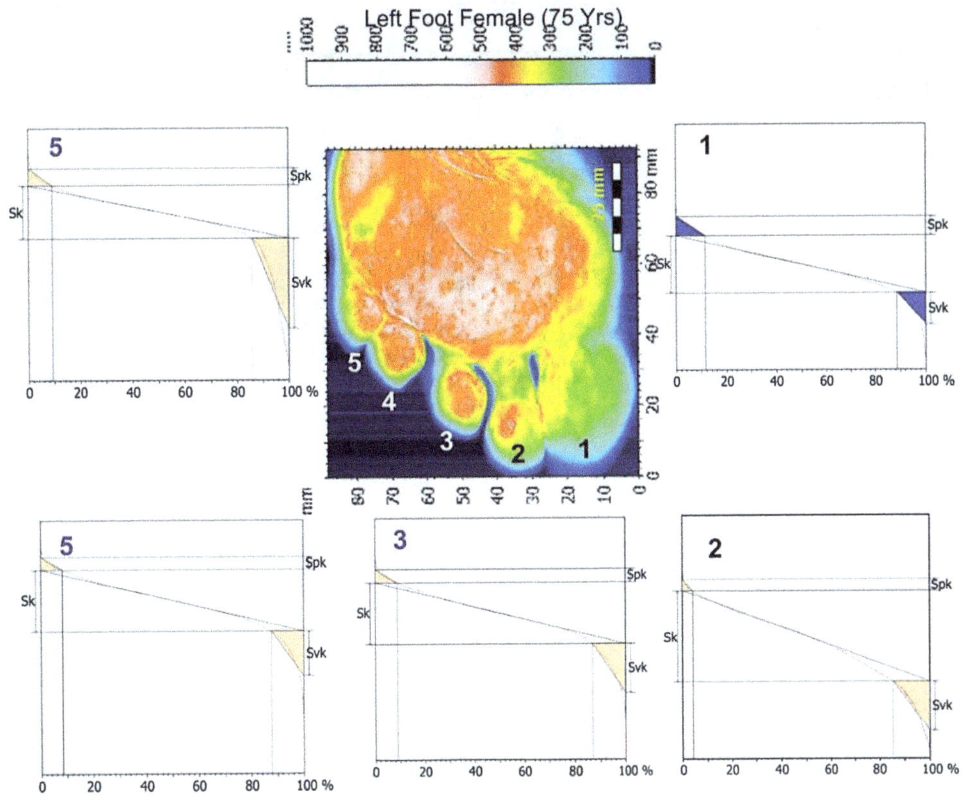

Table 3. Sample values of Aerial material ratio parameters for toes of a male and female subjects.

	Big Toe	2	3	4	5
Woman					
Left Foot					
S_{pk}/S_k	0.339189	0.127485	0.217959	0.202966	0.322111
S_{vk}/S_k	0.567568	0.546784	0.791837	0.720339	1.693467
Right Foot					
S_{pk}/S_k	0.503185	0.276804	0.169737	0.195279	0.532857
S_{vk}/S_k	0.614013	0.484536	0.894737	1.124464	0.370714
Man					
Left Foot					
S_{pk}/S_k	0.227481	0.339419	0.295588	0.20101	0.257551
S_{vk}/S_k	0.608142	0.275104	0.145221	0.172054	0.306939

Figure 18. (a Aerial material ratio curves for selected positions on a human-male heel (b, Analyses on Engineering Mechanics of Robotic Arm for Sorting Multi-Materials

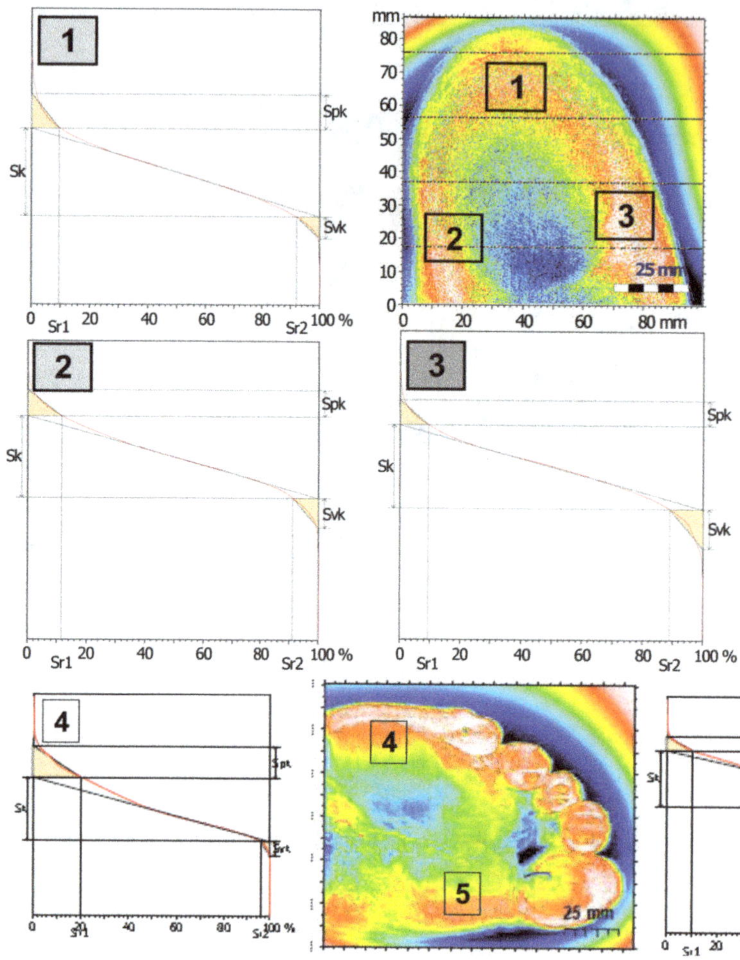

Table 4. Values Aerial material ratio curves for the human-male subject of figure 20

Position	1	2	3	4	5
spk/Sk	0.391	0.326	0.6	0.499	0.249
Svk/Sk	0.248	0.365	0.385	0.225	0.326

The last two rows present data pertaining to the heel, and those pertaining to locations on the forefoot. It is seen that in general, the ratios between the various Sk parameters is not constant. On the lateral side of the foot, the S_{pk} is larger than the S_{vk} ($S_{pk}/S_{vk} > 1$). At the very distal part of the foot areas 1 and 2, S_{pk} is almost twice that of S_{vk}. This coincides with the thickest part of the sub calcaneal patent. Past this location the ratio S_{pk}/S_{vk} drops in the proximal direction. However, such a drop is consistent with the general trend observed for each side. Namely, the S_{pk} remains smaller than S_{vk} within the lateral side

Figure 19. Location of the additional areas examined on human foot pads to determine the distribution of aerial parameters

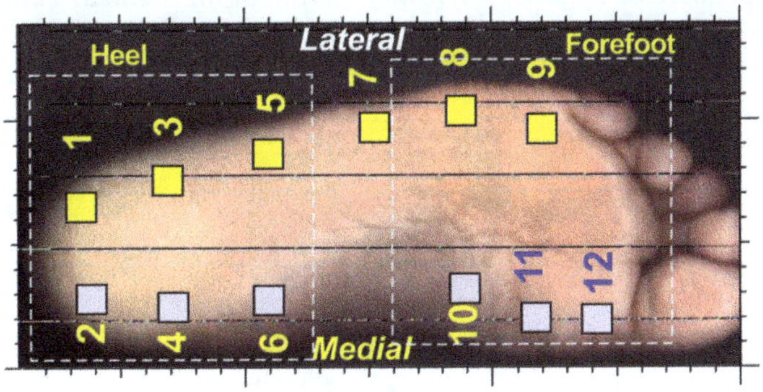

Table 5. Distribution of the ratios of the aerial parameter for the areas examined in figure 19

Lateral						
	1	3	5	7	8	9
Spk/Sk	0.462	0.413	0.410	0.437	0.408	0.395
Svk/Sk	0.279	0.356	0.278	0.272	0.276	0.339
Spk/Svk	1.658	1.158	1.473	1.605	1.477	1.164
Medial						
	2	4	6	10	11	12
Spk/Sk	0.441	0.360	0.396	0.135	0.265	0.378
Svk/Sk	0.178	0.552	0.436	0.280	0.344	0.751
Spk/Svk	2.482	0.652	0.909	0.483	0.771	0.503
Heel						
Spk/Svk	1.658	1.158	1.473	2.482	0.652	0.909
Forefoot						
Spk/Svk	1.605	1.477	1.164	0.483	0.771	0.503

although that the difference diminishes going forward. Within the medial side, however, although the Spk remains smaller than the S_{vk}, the difference fluctuates by location.

Figure 20 depicts the map of the S_{sk} and S_{ku} parameters within the forefoot. Point I, an area about 4 mm^2 represents the medial side of the forefoot and point II represents the lateral side of the footpad. The rest of the points, areas of the same magnitude, present the same distribution for the toes. The toes in the figure are numbered 1 through 5 starting with the big toe. Similar to previous presentations in case of the cat and dog, the red arrows represent the skewness parameter S_{sk}. The blue arrow meanwhile represents the kurtosis parameter S_{ku}. Again when both arrows point at the same direction it indicates a positive S_{sk}-S_{ku} combination. When, however, the red arrow points opposite to the blue arrow it indicates a positive S_{ku} and negative S_{sk}.

Figure 20. Mapping of the surface kurtosis, S_{ku}, and skewness, S_{sk}, parameters for a representative toe-forefoot complex (male 50-55). Arrows shown on the pads indicate whether the combination of the S_{sk} and S_{ku} parameters are positive or negative. Arrows pointing at the same direction indicate that S_{ku} and S_{sk} are both positive.

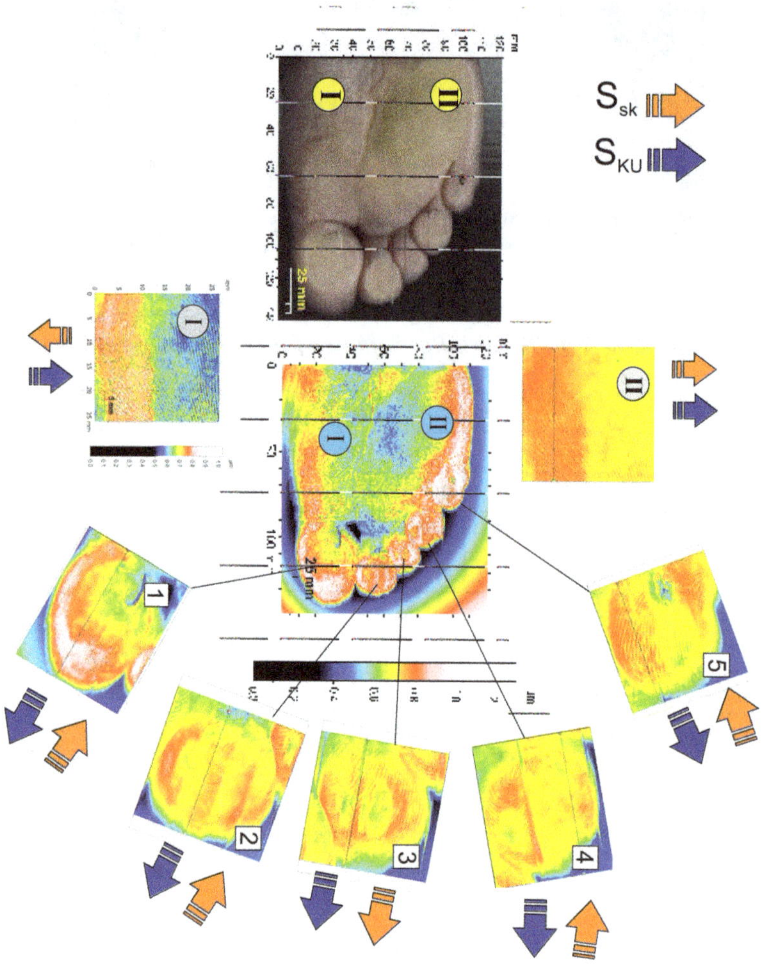

Examining the mapping within the figure reveals that the lateral side of the forefoot shown has negative skewness and positive kurtosis. This indicates that this part of the foot is suitable for generation, and accommodation, of tractions. The other extreme of the forefoot on the other hand shows positive skewness-kurtosis combination which is interpreted as suitability for load bearing. The distribution of the signs of the toes manifest suitability for generation and accommodation of tractions. The opposite sign combination found on the extreme edges of the forefoot imply that there is a transition area on the foot where tendency for load bearing ends and accommodation of friction start. Alternatively, this may indicate nested contours of friction and loadbearing tendencies within the foot area. To investigate which of these assumptions, if any, are valid we mapped an additional array of areas on the footpads. Results are shown as figure 21. The figure depicts the distribution of the array of areas examined on that particular foot (belonging to a male subject 45-50 years old). The white squares indicate areas were

Figure 21. Complete mapping of the surface kurtosis, S_{ku}, and skewness, S_{sk}, parameters for the toe-forefoot complex for a human male (45-50). Dark squares indicate friction accommodation zones whereas white squares are load bearing zones

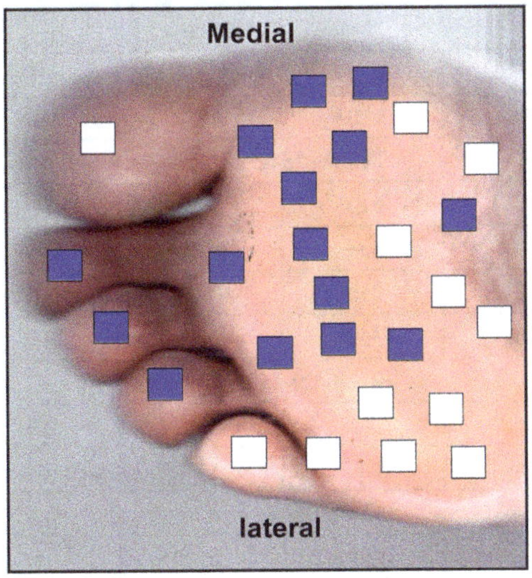

roughness favors loadbearing function and areas shown as blue squares present locations where fraction accommodation function is favored. The distribution of the two zones confirm the transition questioned earlier. However, a different pattern of transition is not clear. Load bearing clusters are nested within the tractions zones. Curious within the new scans is that the big toe for this particular individual shows tendency for load bearing. The same is noticed for the small toe. This is in contrast to what was found for other samples for which data is shown in table 6. Whether such finding is an anomaly or related to the gait of that individual is not known thus far. The answer awaits detailed analysis of data that is currently undergoing. Data shown in table 6 manifest some dependence on age especially when the skewness of the roughness is examined. For example, the skewness for females between 20-25 years old is higher than that for the male subjects between 75-85 years old.

To complete the analysis we investigated the metrological layout of heel pads. Representative sample of one of the analyzed heels, belonging to a male 55-60 years old, shown in figure 24. The analysis of the heels proceeded based on the lessons learned from analyzing fore feet. In particular, we divided the heel into contours that follow the outer perimeter of the. The different controls (numbered 1 through 6) where divided along the longitudinal distal-proximal axis of the foot. The individual topography maps of each of individual controls are shown to the sides of the figure. Ten small areas (around 2 mm^2) within each contour were examined for the S_{sk}-S_{ku} sign combination. In addition, four areas within the midfoot, labeled I through IV, were chosen. Again ten small areas within each region were analyzed to determine the dominance of load bearing or traction adaptation. The results, indicated by the particular arrow combinations, revealing nested contour structure in which accommodation of load bearing and friction tractions alternate within the heel. The outer rim of the heel, contours 1 and 4, show loadbearing tendency. The following contour however, zones 2 and 5, manifests contrasting tendencies. Contour 2 which is located toward the lateral part of the heel, displays loadbearing tendency. That adjacent contour, labeled 5 in the figure, which is located toward the medial side, display traction tendency. In the inner

Table 6. Comparative values of the skewness-kurtosis values for different human subjects examined in this work.

Position	BigToe	2	3	4	5	I	II
Male 50-55 Left Foot							
Ssk	-0.20425	-0.226	-0.157	-0.139	-0.059	-0.25	0.125
Sku	3.67	2.985	3.203	3.087	2.988	3.515	3.183
Female 75-85 Left Foot							
Ssk	-0.444	-0.649	-0.837	-0.85	-1.94		
Sku	3.95	2.93	3.92	4.25	7.9		
Ssk	-0.904	-1.28	-1.17	-0.459	-0.105		
Sku	4.15	5.01	3.97	3.05	3.81		
Female 20-25 Right Foot							
Ssk	-1.3	-0.915	-1.5	-1.78	-2.34		
Sku	4.65	3.6	3.17	4.77	8.24		
Female 20-25 Left Foot							
Ssk	-1.68	-1.127	-1.585	-1.679	-1.898		
Sku	3.47	3.046	2.544	3.207	4.351		
Male Heel							
Position	**1**	**2**	**3**	**4**	**5**	**6**	
Ssk	1.29	0.694	-0.297	-0.066	-0.059	-0.173	
Sku	5.62	4.5	2.84	2.55	2.9875	2.8	
Position	**I**	**II**	**III**	**IV**			
Ssk	-0.259	0.0172	-0.172	0.36			
Sku	3.07	2.76	2.74	3.43			

parts, contours 3 and 6, manifest the roughness that favors accommodation of frictional tractions. The representative zones, I through IV, show a similar behavior. Zone I, which is close to the inner core of the heel, favors traction accommodation, whereas those on the outer perimeter favor loadbearing. Noted, however, is that while the lateral side of the heel favors loadbearing, the lateral side of the forefoot favors tractions. The same also applies for the medial outer perimeter (compare figures 20-22).

DISCUSSION

This work examined two classes of locomotors. The first is the so-called plantigrade represented by the human feet pads. The second is the so-called the digitigrade locomotors-represented by canids, dogs, and felids, cats. The analysis reveal differences and similarities in both surface and dimensional metrology of the footpads in both locomotor classes. Differences, and how they relate to locomotion, are the subject of this discussion. Findings are best understood within the context of the distinct locomotion features of the examined species.

Figure 22. Mapping of the surface kurtosis, S_{ku}, and skewness, S_{sk}, parameters for a representative human heel complex (male 50-55). Arrows shown on the pads indicate whether the combination of the S_{sk} and S_{ku} parameters are positive or negative. Arrows pointing at the same direction indicate that S_{ku} and S_{sk} are both positive.

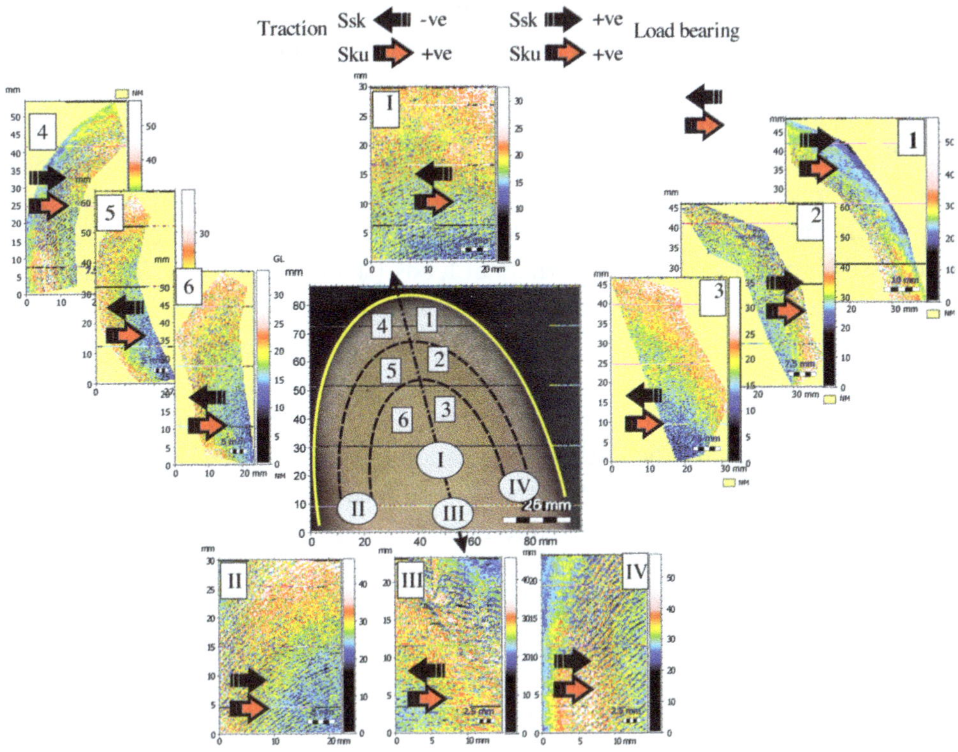

Plantigrade mammals, such as humans, utilize the whole sole of the feet for walking. Digitigrade mammals, such as dogs and cats, use only part of their feet in contacting the ground while moving. This is due to the elevated carpal (equivalent to the wrist in humans). Due to this construction, only the phalanges contact the ground while moving.

Configuration during walking, use of part or whole area of the foot, in addition to the construction of the foot-footpad complex present species-specific functional challenges. The challenges, in general, relate to optimization of energy spent in locomotion. This energy may be considered as a cumulative sum of two components. The first is the energy spent in initiation, and maintaining, motion. The second represents the energy spent in dissipating the ground actions resulting from moving on a surface or through a medium. The second component is closely connected to accommodation of ground actions and dissipation of shear tractions (including frictional tractions) during ground contact. Given that humans use their whole foot (the sole) in walking, whereas cats and dogs used the digital pads, one would expect that both form and textural roughness of the parts used in walking (or ground contact in general) contribute toward optimization of the second component of energy cost.

Kerr (1999) observed that digitigrade footpads and plantigrade footpads have the same composition (fatty tissue pads and adipose tissue permeated with a fibrous membrane composed of collagen fibers). As such, in the greater scheme of behavior, the trend of mechanical property variation, deformation, and dissipation of energetic loads in both locomotion classes should reflect similarity. Naturally, there are differences in the numerical values of the properties involved, and in their proportion to body mass. Similarity, or resemblance in mechanical behavior, advances the argument concerning the contribution of form and surface construction in meeting the specific mechano-functional challenges of a species. In other words, the overall shape of the foot sole in humans (including the heel), and the digital-metapad complex in cats and dogs, along with the surface construction, assume a principal role in meeting the functional demands of ground contact. This is not surprising since pattern is often an important factor in meeting the locomotion constraints in other species (Abdel-Aal et al, 2016).

Essentially, the construction of pads (heel, digital, and metacarpal) offer hydrostatic support being mainly composed of incompressible adipocytes (essentially rheological fluids). These fluids are confined by tensile membranes (Natali et al, 2010; Ninomiya et al, 2011; Bennett et al, 1990). The liquid, mostly incompressible, offers support for horizontal loads (static and impact). The fibrous structures (the collagen fibers) act as cushioning rings that allow the control dissipation of lateral deformation and for accommodating shear deformation. Accommodation of vertical deformation due to normal forces is customized. Vertical deformation causes the fatty fluid to deform at a rate that depends on the nature and duration of the normal force. To preserve the integrity of the fluid, and thereby the integrity of the footpads, there is a need to control the rate of deformation of the pad material and the stress acting on the different layers of the pad. It is here where the general form of the footpads play an important role in regulating the rate of stressing the material through the control of geometry.

Implications of Height Parameters

The stress caused by a force acting on an area is well known to be the result of dividing the magnitude of the force over the magnitude of the area. If a force acts in the same direction through layers of the material than the stress caused by that force on each of the layers would change in relation to the change in the area of the layer that supports the load. The deformation of the different layers will also be proportional to the change in the land areas in the direction of the force (through relation to stress). As such, if the gradient of the area (i.e., the changes from layer to layer in the direction of the force) is properly controlled, the stress gradient and deformation gradients on the material may be also controlled. In all, the rate of deformation due to normal force may be controlled through controlling the geometry along the direction of the force. In this case, the force acts from the ground contact plane two words the bulk of the foot pad. As such, a gradual increase in the cross sectional area from the point of the contact toward the bulk of the footpad can control the rate of dissipation of stress acting in the ground contact phase. Another requirement for controlling the formation is to confine the action of the normal force at the plane of initial contact to a very small area. Confining the initial contact area to a minimum contributes towards controlling the dissipation as well. Those two requirements essentially determine the shape of the pads and renders them almost hemispherical, in cross section, to a first approximation which indeed is observed in figures 4 and 10 for cats and dogs respectively. The figures show that the shape of individual pads is approximately that of the hemi spheroid with its boundaries surface convex toward the ground.

The point of maximum height on that hemi spheroid is roughly located on the center. The geometry of the pads, and the location of the point of maximum height, causes the normal ground forces and impact forces to be initially concentrated within a very small area approximately at the center. Interesting in that respect is that some studies consider the location of the maximum contact pressure at the center of the pad as a sign of a healthy animal which does not suffer deformity or limbness (Lascelles et al, 2006;

Qian et al, 2014; Budsberg et al, 1988). A gradual increase in the cross-sectional area of the pad, towards the foot, allows for the reduction of the gradient of stress in the normal direction (layer by layer). Since deformation is tied to stress, it follows that the layer by layer deformation, both in the normal and the lateral directions will also be reduced. The rate of this reduction is proportional to the rate of increase in the cross-sectional area. Mitigation of lateral deformation causes the collagen fibers surrounding the pad to stiffen upon impact (Budsberg et al, 1987). The mechanism of that stiffening is beyond the scope of this work. However, in all, the rate of mitigation of deformation, both lateral at normal, are rooted in the control of the geometry of the pads (i.e. the form) being complementary to the mechanical properties of the pad materials. That is form assists in compliance of the material of the footpad. Tuning of the deformation mitigation process, is assisted by the roughness of the pad. This is manifested in the roughness of the pads of both dogs and cats being mostly suitable for load bearing (negative S_{sk} and positive S_{ku}). The role of the roughness and form in compliance and optimization of the function of the pads may be appreciated in light of the absence of muscles within the footpads of dogs and cats. Absence of muscles render the performance of the pet functions dependent on geometry and mechanical properties of the pad materials. Therefore, any tuning or refinement of performance or compliance to maintain the structural integrity of the pad has to initiate at the roughness-texture geometry level.

Friction accommodation within the pads, figure 10 for cats and 15 for dogs, is not dominant as loadbearing accommodation. In particular, for both species, the front pads were found to be loadbearing oriented, whereas the hind pads included some accommodation of friction. Namely, in both species, only to pads within the hind paws manifested accommodation of friction. This may be correlated to the nature of foot function in quadruped locomotion. Several studies (Lieberman et al, 2004; Cavagnaa et al, 1977; Biewener et al, 1983; Golubisky et al, 1999) concluded that in digitigrade quadruped locomotion, the fore limbs participate in braking while the hind limbs have mostly propulsion duties. This arrangement requires the dissipation of shear forces during braking for the forward limbs and generation of tractions with some bearing of the weight of the animal for the hind limbs. Again, considering figures 4 and 15 the distribution of the load bearing texture zones within the pads are almost identical for cats and dogs. Cats however, manifests its action friction zone on the fore paw. For dogs, all the forepads are load bearing suitable. This finding, considering the dominance of the braking function requirement within texture of the forelimbs and the absence of shear accommodation zones, seems contradicting. The initial step of braking the motion, however, is to absorb the impact forces resulting from touching the ground. This obviously requires loadbearing texture. Shear loading, however, is absorbed in the linkage above the paw (i.e. the so-called bar-linkage arrangement). The hind pad S_{sk}-S_{ku} sign combination indicates an arrangement for partial traction accommodation. This could satisfy the locomotion traction requirement through generating enough grip to initiate motion and maintain balance. The load bearing pads would thus work to accommodate the impact force experienced at landing. The additional friction-pad found in the fore paw of the cat could relate to client being assistance. Cats are climbers and therefore the need to generate enough grip and force to support their body weight against gravity thus reducing muscular effort. Most of the clinging is achieved through penetration by nails. However, due to the small contact area of the nails, high stresses can take place. This may threaten the structural integrity of the nails. To alleviate distress and minimize shield loading on the nails additional grip is needed. This requirement is met by the hind paw-pads and, for stability, that one pad on the front paw manifests frictional accommodation. Note that on the hind paw, the metatarsal pad (the one with the highest area) accommodates friction. The dog breed examined in this work did not reflect the exact arrangement of the cat. This can be explained by the fact that the requirement for climbing does not apply in case of dogs.

Human feet pads examined in this work manifested several traits common with canid and felid feet pads specially when meeting the complex demands of stress relief within the contact phase of locomotion. The interference between the loadbearing and the traction accommodation zones is consistent with the inner composition of the heel pads. The alternating zones follow the thickness of the sub calcaneal fat pad. The overall picture however shows compatibility with the function of the pads during locomotion. The heel undergoes ground reaction forces during gait cycles. As such, the heel manifests complex interaction between three fundamental tasks: reduction of shock and impact (Ker et al, 1996), shock absorption, and protection against excessive local stress (Miller-Young et al, 2002). Managing this complex interaction requires unique mechanical properties that are complemented by compatible surface properties. An interesting role that starts at the surface, whence depends initially on the roughness functionality, is energy dissipation during gait cycles. The human heel pad manifests anisotropic stiffness. It is stiff and proximal-distal shear while being compliant in distal-proximal shear (Ledoux et al, 2007). It is also resilient in compression while being damped in shear. In all, it is an anisotropic entity was the respect to composition and to mechanical properties. This anisotropy facilitates performing the multitude of functions necessary for gait optimization. Compliance in distal-proximal shear aims at transmitting tractions for propulsion. The heel, therefore, dissipates friction energy in shear to stabilize the foot while emerging from backward motion. It also stores elastic energy in compression to help lifting the foot upward at the end of each step. In summary the foot functional demands accommodation of friction while optimizing impact loads and reaction forces. From a metrological point of view such a complex task requires an interference between traction- accommodating features and textural features that enhance loadbearing. This is manifested in the interfering clusters where the skewness-kurtosis show different signs.

The structure of the toe-fore foot complex and human footpad bears striking resemblance to the construction of the foot-footpad complex in digitigrade animals. This reflects on the form and the dextral function of the toes and the fore foot. The analysis of the S_{sk}-S_{ku} sign combination for that those, illustrated in figure 23 and table-6, indicated that toes in human feet mainly accommodate tractions. Such a function appears to be complementary to shear accommodation through the joints in the fore foot-toe complex. Such an observation implies that the pointed roughness dictated by the sign of the S_{sk}-S_{ku} combination enhances the compliance of toe pads. The surface within the toes is mainly suitable for gripping as indicated by the positive kurtosis and negative skewness. Locally, this implies that the surface is dominated by peaks. This in turn indicates that at the roughness level load transmission mechanism within the ground contact phase is similar, in principle, to that taking place within digital pads of cats and dogs. Namely, regulation of the rate of transition of stress layer-by-layer and streamlining the associated deformation have the same structural origins both in human toes and digital pads of animals. Here it is worth pointing at the striking resemblance between the shape of toe pads influence and the shape of the digital pads in cats and dogs. Both, toes and digital pads, manifest hemi spheroidal form. In addition, both structures contain only fatty adipose tissue and are free of muscles.

Loading Curve Implications

Pads of dog feet comprise three layers: an external epithelium, and intermediate dermis layer, and the subcutaneous layer. The external epithelium layer is composed of the hardest material and is in direct contact with the ground. This stratified epithelium layer, along with the outermost stratum corneum layer, is subject to considerable pad-ground wear, friction, and impact during locomotion (Meyer et al, 1990). The bottom surface of the paw, that which establishes contact with the ground (i.e. the epithelium layer), is covered by a layer of spike -like stratum corneum (Alexander et al, 1986). The innermost layer, the subcutaneous layer is the softest and the foremost energy absorber of footpads (Ker et al, 1999; Weissenbgruber et al, 2006). A similar arrangement is also present in cats and human feet pads. The hardness of the very outer layer enforces complicated mechanical constraints on the nature of the surface texture on the very contact layer of the pad. The response to these constraints manifests in the loadbearing curves examined in this work.

The contacting layer, being hard and stiff, will establish a smaller real area of contact under the same road compared to a softer layer. This results in a lower friction force since the force of friction is proportional to the real area of contact. On the other hand, the high stiffness, generally associated with heart materials would cause a higher wear rate. Further, due to the high stiffness the contact forces when transfer faster to the bulk of the pad additional dissipation mechanisms take place. Accommodating the potential higher rate of wear of the contacting layer should reflect on the loadbearing curve of the pad roughness. In addition, since the surface is subject to a high frequency of loading-unloading cycles that should be some provision for reducing fatigue-induced wear and material failure.

In general, where reduction, especially within the running-in period is accommodated through the functional loading parameters (the so-called S_K-family of functional parameters). The surface of which loading curve reflects a large S_{pk} value manifests high peaks that provide for a small initial real area of contact. Establishing a small area of contact leads to higher contact stress and thereby a higher wear rate. The larger the S_{pk} parameter the higher the probability that the surface want to reach catastrophic wear during running-in. The running in in case of mammals is not as continuous as in machines, it could be taken as periods of transient contact. Transient contact is contact. A very short duration typically taking place during walking and for shorter periods during running.

The hardness of the outer layer of the pads, as mentioned earlier, lead to high stress upon contact and therefore is likely to undergo a higher wear rate. To ensure the integrity of the surface, there is a need to increase the height of the peaks to enhance the structural lifespan with respect wear. Consistent with this requirement, a noticeably high S_{pk} is manifested in all of the bearing curves of canid and felid subjects tested (see figures 6,7, 12, and 13). Such an arrangement allows for enhancing the way life of the pads especially that impact wear may also take place at higher speeds. It is also noted that the S_{vk} parameter is either negligible or considerably small in all animal pads examined. So that, the loading curve for animals is considerably asymmetric. The asymmetry of the loadbearing curve in cats and dogs imply small void volume parameter V_{vv}. Although this parameter was not evaluated for test subjects of the study, but the shape of the curves confirm that the value of this parameter is indeed very small. The void volume parameter relates to the tip radius at the root of the valleys of the elements of roughness. This, in turn, reduces stress concentration and enhances fatigue performance (Ker et al, 1999).

Functional parameters of human footpads, however, reflect two trends. Within the heel, and the medial side of the foot sole, the S_{pk} parameter, consistent with animals, is almost twice the value of the S_{vk}. Such a feature is compatible with the trend noticed in cats and dogs. Therefore for those lesions wear resistance and fatigue requirements are dominant and satisfied. The order of magnitude of S_{pk} relative to S_{vk} is less than that reflected in animals however. A possible explanation could be rooted in the speed of locomotion in humans and the height of the foot in the swing phase compared to the same parameters in animals. In animals speed and height of the foot are greater than those observed in human locomotion. Consequently the developed impact forces upon establishing ground contact would be higher in animals. In addition the frequency of loading-unloading cycles would increase due to the higher speed which leads to a higher fatigue failure probability. In humans since speed and hide the feet are less than those observed in animals the demands for where protection and fatigue failure resistance may not be equal among all parts of the foot. Thus the proportions of the loading curves are different. Values for the toes, table-3, reflect the consistent trend. For men and women S_{pk} is less than the S_{vk} (i.e. reverse asymmetry compared to the heel and the medial side of the footpad). A small S_{pk} enhances stability of friction especially in the presence of a lubricant. The smaller S_{pk} also indicates a lower average coefficient of friction and a shorter transition to a steady-state friction condition (Bunningham et al, 2010). This enhances tractions generation in walking especially in wet conditions. Enhancing the grip for the toes also enhances stability of locomotion.

The examined footpads reflect design objective differences. On a fundamental biomechanical level there is a distinction of the essence of that configuration and conception. Humans are hypothesized to be specialized, and physiological and anatomical sense, for endurance running. This Specialization manifests into the long-distance stamina that is surpassed by very few mammals. Such an advantage is argued to be more visible in warm and hot environments.

Considerations of the anatomical aspects of human feet is paradoxical. A human heel-down plantigrade foot posture is rare among mammals and is considered incompatible with adaptation for economy of effort and running. Fast running mammalian species, including those notorious for speed and energy economy, fall within the digitigrade or unguligrade classes. The heels in these locomotion classes, are held elevated above the ground so that the animal would move on the balls of the feet or that those respective. This design promotes locomotor economy through an increased step length. In contrast, foot posture for plantigrades manifests in a heel that contacts the substrate, and thereby dissipates energy up on the start of a running step. Thus, the design, in theory, paradoxically seems to demote storage and recovery of elastic energy.

Recent studies (Sedlacek et al, 2012), however, analyze the mechanics of plantigrade versus low-digitigrade walking. The results suggested that plantigrade posture contributes to lowering the energetic cost of transport due to lower correlational losses of the body, increased recovery of kinetic and potential energy, and lower ground reaction force moments especially at the ankle joint. These results point at a possible strong correlation between the metrological structure of human footpads and accommodation of contact forces. In particular, the nature of roughness features, and their topological layout contributes to stabilizing and minimizing of ground frictional and normal reaction forces. The same concept, in principal, may be extended for link feed design and cats and dogs to their locomotion performance and energetic cost of transportation. The analysis such a topic in detail is a direction for future work especially with accommodating age and gender differences (or alternatively when differences in breed and habitat for animals) are invoked.

Implications for Legged Robots

There are two approaches that allow legged robots to traverse difficult terrains. The first entails application of general control approaches whereas the second comprises implementing special robot structures and materials with specific control strategies. The second approach, which is of recent origin and is still evolving, has come to be known as *morphological computation* (Pfeifer et al, 2006; Pfeifer et al,

2007). The aim is to increase friction either by using a bio-inspired methodology to generate adhesive or tractions; or to increase "gripping" by applying a high friction material (Kim et al, 2008; Baltorio et al, 2006; Voigt et al, 2012; Spenko et al, 2008; Palmer et al, 2010; Bretl et al, 2006). Design of locomotion-contact surfaces for moving robots therefore, is more demanding than merely producing a structure that enhances friction or "grip".

The ground contact phase of natural locomotion depends on an intricate relationship between surface geometry and mechanical compliance. Terrestrial mammals establish contact with a substrate through the footpads. Footpads therefore, have to fulfill multiple functional roles. These roles are sometimes competing while in others they may involve contradicting mechanical and structural demands. Pads must be compliant to cushion impact effects, yet their stiffness is to be high enough for force transmission. There is also the requirement for a damping to enhance stability of contact while being resilient for harvesting elastic energy.

In this study we compared plantigrade human feet pads to canid and felid pads. Previous studies revealed different species-specific mechanical attributes that facilitate meeting the functional demands of locomotion subject to particular species constraints. This work focused on the morphology and textural engineering of pad surfaces across the species. The analysis revealed that while the mechanistic steps of meeting functional demands may seem different, the logic of the metrological layout of footpad surfaces is almost universal. In fact, the result demonstrated that a graphical layout of the pads induces mechanical compliance and replaces muscle and tendon actuation. Thus the data pointed at the tight interplay between the topography of surfaces, four of the pads, and the mechanical properties of pad materials.

The overall form over the pads facilitate the streamlining of force transmission and reduces stress gradients and associated deformation. The controlled reduction of stress gradients allows for managing external forces without sacrificing the structural integrity of the pad materials. This reduction also reduces the stresses and deformations to the level needed for efficient energy dissipation and elastic energy recovery. One interesting aspect is that the pad materials that contain muscles. As such, reduction of stress at the roughness contact level is also driven by geometry. The geometry itself is hierarchical (micro geometry is similar to the macro geometry).

In humans, hierarchy is partially responsible for ensuring the flexibility of the heel to facilitate shock absorbance during heel rocker. Flexibility of the heel in humanoids is important and ensuring that all human-like properties are included in a single-foot solution is not currently possible due to the technical complexity of the human foot. Consequently, most humanoids still implement basic stiff ankle actuation with no compliant properties. Surface engineering solutions may be a powerful solution in this respect. An example is the heel rocker problem which requires controlled rotation about a small area of support represented by the heel. A crucial step in the rotation is the initiation of rotation when a very small and narrow strip of the ankle initially establishes contact with the ground. This narrow strip becomes the fulcrum about which the foot and tibia move (in humans). The bony segment between this fulcrum and the center of the ankle rolls toward the ground as body weight is dropped onto the stance foot, preserving the momentum of forward progression. For stability and prevention of skid there is a need for load bearing and traction. In humans part of these requirements are met through the alternating contours of load bearing and friction accommodation shown in figure 22 (zones 1, 4, and 5). Thus the customized roughness in the human heel is what meets the local functional demand. Such a solution has not been implemented in robotic feet so far. In fact, surface engineering of robotic feet is completely bypassed despite its potential.

CONCLUSION

This work examined the metrological layout of two classes of natural locomotors (humans, cats and dogs). The difference is in the use of the foot. Humans are plantigrade mammals and use the full foot for walking. Canids and felids use both front and back pads for walking and motion in general.

The general makeup of the texture of all examined species reflected the generalized theme of roughness complimenting the locomotion function of the footpads. It was found that the generalized form, along with the roughness layout, are essential for controlling the stress gradients developed upon contact with the ground. The hemi-spheroid for of animal pads and of human toes assist in regulating stress gradients developed in the ground contact phase.

The human footpad was found equivalent in function to the back and front pads in digitigrade mammals. This is accomplished through an intricate nesting of load bearing and friction accommodation zones in the heel and forefoot.

The main lesson for design of robotic feet is that including the engineering of roughness can enhance the function of walking robots. It can offer elegant solutions to some of the current problems encountered in design humanoids and other bio-inspired walking robots.

REFERENCES

Abdel-Aal, H. A. (2016, November 3). Functional surfaces for tribological applications: Inspiration and design. *Surface Topography: Metrology and Properties.*, *4*(4), 043001.

Abdel-Aal, H. A., & El Mansori, M. (2014, October 23). Characterization of load bearing metrological parameters in reptilian exuviae in comparison to precision finished cylinder liner surfaces. *Surface Topography: Metrology and Properties.*, *2*(4), 045002.

Abdel-Aal, H. A., El Mansori, M., & Mezghani, S. (2010, March 1). Multi-scale investigation of surface topography of ball python (Python regius) shed skin in comparison to human skin. *Tribology Letters*, *37*(3), 517–527. doi:10.100711249-009-9547-y

Alexander, R., Bennett, M. B., & Ker, R. F. (1986, July 1). Mechanical properties and function of the paw pads of some mammals. *Journal of Zoology*, *209*(3), 405–419. doi:10.1111/j.1469-7998.1986.tb03601.x

Alexander, R. M. N., Bennett, M. B., & Ker, R. F. (1986). Mechanical properties and function of the paw pads of some mammals. *J. Zool.*, *209*, 405-419. .1986.tb03601.x doi:10.1111/j.1469-7998

Baum, M. J., Kovalev, A. E., Michels, J., & Gorb, S. N. (2014, May 1). Anisotropic friction of the ventral scales in the snake Lampropeltis getula californiae. *Tribology Letters*, *54*(2), 139–150. doi:10.100711249-014-0319-y

Bennett, M. B., & Ker, R. F. (1990, August). The mechanical properties of the human subcalcaneal fat pad in compression. *Journal of Anatomy*, *171*, 131. PMID:2081699

Biewener, A. A. (1983, July 1). Allometry of quadrupedal locomotion: The scaling of duty factor, bone curvature and limb orientation to body size. *The Journal of Experimental Biology*, *105*(1), 147–171. PMID:6619724

Biewener, A. A. (1990). Biomechanics of mammalian terrestrial locomotion. *Science, 250*(4984), 1097–1103. doi:10.1126cience.2251499 PMID:2251499

Bretl, T. (2006). Motion planning of multi-limbed robots subject to equilibrium constraints: The free-climbing robot problem. *The International Journal of Robotics Research, 25*(4), 317–342. doi:10.1177/0278364906063979

Brown, J. C., & Yalden, D. W. (1973, December). The description ofmammals– 2 Limbs and locomotion of terrestial mammals. *Mammal Review, 3*(4), 107–134. doi:10.1111/j.1365-2907.1973.tb00178.x

Budsberg, S.C., Verstraete, M.C., & Soutas-Little, R.W. (1987). Force plate analysis of the walking gait in healthy dogs. *American Journal of Veterinary Research, 48*(6).

Budsberg, S. C., Verstraete, M. C., Soutas-Little, R. W., Flo, G. L., & Probst, C. W. (1988, September). Force plate analyses before and after stabilization of canine stifles for cruciate injury. *American Journal of Veterinary Research, 49*(9), 1522–1524. PMID:3223659

Burlayenko, V. N., & Sadowski, T. (2010). Effective elastic properties of foam-filled honeycomb cores of sandwich panels. *Composite Structures, 92*(12), 2890–2900. doi:10.1016/j.compstruct.2010.04.015

Cardinaels, R., Maes, J., Deseure, J., Saldien, J., Aelterman, K., Verstockt, S., & Wyffels, F. (2017). GECO: the development of a wall-climbing robot. *20th International Conference on Climbing and Walking Robots and the Support Technologies for Mobile Machines (CLAWAR 2017).*

Cavagna, G. A., Heglund, N. C., & Taylor, C. R. (1977, November 1). Mechanical work in terrestrial locomotion: Two basic mechanisms for minimizing energy expenditure. *American Journal of Physiology. Regulatory, Integrative and Comparative Physiology, 233*(5), R243–R261. doi:10.1152/ajpregu.1977.233.5.R243 PMID:411381

Chenkun, Q., Feng, G., Qiao, S., Xianbao, C., Yilin, X., & Xianchao, Z. (2015). A foot force sensing approach for a legged walking robot using the motor current. In *Robotics and Biomimetics (ROBIO), 2015 IEEE International Conference on 2015 Dec 6* (pp. 1078-1083). IEEE. 10.1109/ROBIO.2015.7418915

Cheung, J. T.-M., Zhang, M., Leung, A. K.-L., & Fan, Y.-B. (2005). Three-dimensional finite element analysis of the foot during standing—A material sensitivity study. *Journal of Biomechanics, 38*(5), 1045–1054. doi:10.1016/j.jbiomech.2004.05.035 PMID:15797586

Chi, K. J. (2005). *Functional morphology and biomechanics of mammalian footpads* (PhD thesis). Duke University, Durham, NC.

Colasanto, L., Van der Noot, N., & Ijspeert, A. J. (2015). Bio-inspired walking for humanoid robots using feet with human-like compliance and neuromuscular control. In Humanoids 2015 Nov 3 (pp. 26-32). Academic Press. doi:10.1109/HUMANOIDS.2015.7363518

Cunningham, C. B., Schilling, N., Anders, C., & Carrier, D. R. (2010). The influence of foot posture on the cost of transport in humans. *JEB, 213*(5), 790–797. doi:10.1242/jeb.038984 PMID:20154195

Dahiya, R. S., Metta, G., Valle, M., & Sandini, G. (2010, February). Tactile sensing—From humans to humanoids. *IEEE Transactions on Robotics, 26*(1), 1–20. doi:10.1109/TRO.2009.2033627

Daltorio, K. A., Gorb, S., Peressadko, A., Horchler, A. D., Ritzmann, R. E., & Quinn, R. D. (2006). A robot that climbs walls using micro-structured polymer feet. In *Climbing and walking robots* (pp. 131–138). Berlin: Springer. doi:10.1007/3-540-26415-9_15

Daltorio, K. A., Gorb, S., Peressadko, A., Horchler, A. D., Ritzmann, R. E., & Quinn, R. D. (2006). A robot that climbs walls using micro-structured polymer feet. In *Climbing and walking robots* (pp. 131–138). Berlin: Springer. doi:10.1007/3-540-26415-9_15

Dario, P., & Buttazzo, G. (1987, September). An anthropomorphic robot finger for investigating artificial tactile perception. *The International Journal of Robotics Research*, *6*(3), 25–48. doi:10.1177/027836498700600302

Dupeyroux, J., Viollet, S., & Serres, J. (2017). Providing an autonomous hexapod walking robot with the ability to reorientate: application of a novel ant-inspired celestial compass. *Journée des Jeunes Chercheurs en Robotique*.

Eckert, P., & Ijspeert, A. (2016). *Adaptive foot design for small quadruped robots*. Academic Press.

Fukuoka, Y., & Akama, J. (2014, October). Dynamic bipedal walking of a dinosaur-like robot with an extant vertebrate's nervous system. *Robotica*, *32*(6), 851–865.

Gao, X. S., Shao, J., Dai, F. Q., Zong, C. G., & Guo, W. Z. (2014). A Gecko-Inspired Robot for Wind Power Tower Inspection. Applied Mechanics and Materials, 461, 831-837.

Gill, H. S., & O'Connor, J. J. (2003b). Heelstrike and the pathomechanics of osteoarthrosis: A simulation study. *Journal of Biomechanics*, *36*(11), 1617–1624. doi:10.1016/S0021-9290(03)00190-8 PMID:14522202

Golubitsky, M., Stewart, I., Buono, P. L., & Collins, J. J. (1999, October). Symmetry in locomotor central pattern generators and animal gaits. *Nature*, *401*(6754), 693–695. doi:10.1038/44416 PMID:10537106

Gray, J. (2013). *How animals move*. Cambridge University Press.

Greiner, C., & Schäfer, M. (2015, June 30). Bio-inspired scale-like surface textures and their tribological properties. *Bioinspiration & Biomimetics*, *10*(4), 044001. doi:10.1088/1748-3190/10/4/044001 PMID:26125522

Harada, K., Kajita, S., Kanehiro, F., Fujiwara, K., Kaneko, K., Yokoi, K., & Hirukawa, H. (2007, February). Realtime planning of humanoid robot's gait for force-controlled manipulation. *IEEE/ASME Transactions on Mechatronics*, *12*(1), 53–62. doi:10.1109/TMECH.2006.886254

Hayashi, R., & Kato, S. (2015). Chobonyan: soft-stuffed robot in palmtop size and light weight for therapy. In *Consumer Electronics (GCCE), 2015 IEEE 4th Global Conference on 2015 Oct 27* (pp. 112-113). IEEE. 10.1109/GCCE.2015.7398537

He, Z., Meng, F., Liu, H., Wang, L., Zhu, X., Fan, X., . . . Huang, Q. (2019). Method of Design Optimization and Trajectory Implementation on a Small Cat-Like Robot. In ROMANSY 22–Robot Design, Dynamics and Control 2019 (pp. 323-330). Springer. doi:10.1007/978-3-319-78963-7_41

Hicks, J. H. (1954, January). The mechanics of the foot: II. The plantar aponeurosis and the arch. *Journal of Anatomy*, *88*(Pt 1), 25. PMID:13129168

Hirai, K., Hirose, M., Haikawa, Y., & Takenaka, T. (1998). The development of Honda humanoid robot. In *Robotics and Automation, 1998. Proceedings. 1998 IEEE International Conference on 1998 May 16* (Vol. 2, pp. 1321-1326). IEEE. 10.1109/ROBOT.1998.677288

Hunt, A., Szczecinski, N., & Quinn, R. (2017, April 4). Development and training of a neural controller for hind leg walking in a dog robot. *Frontiers in Neurorobotics*, *11*, 18. doi:10.3389/fnbot.2017.00018 PMID:28420977

Johansson, R. S., & Vallbo, A. B. (1979, January 1). Tactile sensibility in the human hand: Relative and absolute densities of four types of mechanoreceptive units in glabrous skin. *The Journal of Physiology*, *286*(1), 283–300. doi:10.1113/jphysiol.1979.sp012619 PMID:439026

Kalamdani, A., Messom, C., & Siegel, M. (2007, October). Robots with sensitive feet. *IEEE Instrumentation & Measurement Magazine*, *10*(5), 46–53. doi:10.1109/MIM.2007.4343567

Kaneko, K., Harada, K., Kanehiro, F., Miyamori, G., & Akachi, K. (2008). Humanoid robot HRP-3. In *Intelligent Robots and Systems, 2008. IROS 2008. IEEE/RSJ International Conference on 2008 Sep 22* (pp. 2471-2478). IEEE.

Ker, R. F. (1996, July 1). The time-dependent mechanical properties of the human heel pad in the context of locomotion. *The Journal of Experimental Biology*, *199*(7), 1501–1508. PMID:8699155

Ker, R. F. (1999, December 1). The design of soft collagenous load-bearing tissues. *The Journal of Experimental Biology*, *202*(23), 3315–3324. PMID:10562514

Ker, R. F. (1999). The design of soft collagenous load-bearing tissues. *The Journal of Experimental Biology*, *202*, 3315–3324. PMID:10562514

Kim, S., Asbeck, A., Cutkosky, M., & Provancher, W. (2008). Spinybot II: Climbing hard walls with compliant microspines. In Proc. of the Kim, S. et al. Smooth vertical surface climbing with directional adhesion. IEEE Trans. Robot., 24, 65–74.

Koh, J. S., Yang, E., Jung, G. P., Jung, S. P., Son, J. H., Lee, S. I., ... Cho, K. J. (2015, July 31). Jumping on water: Surface tension–dominated jumping of water striders and robotic insects. *Science*, *349*(6247), 517–521. doi:10.1126cience.aab1637 PMID:26228144

Kojima, K., Karasawa, T., Kozuki, T., Kuroiwa, E., Yukizaki, S., Iwaishi, S., . . . Nozawa, S. (2015). Development of life-sized high-power humanoid robot jaxon for real-world use. In *Humanoid Robots (Humanoids), 2015 IEEE-RAS 15th International Conference on 2015 Nov 3* (pp. 838-843). IEEE. 10.1109/HUMANOIDS.2015.7363459

Kummer, F., & Michelson, J. D. (1992). Investigations into the fat pads of the sole of the foot: heel pressure studies. *Foot Ankle Int.*, *13*, 227-232. doi:10.1177/107110079201300501

Kwon, J. Y., Lee, B. H., Mun, K. H., & Jung, J. S. (2015, September 1). A study on the effects of using an Eco-friendly cat robot to treat children with autism spectrum disorder. *Indian Journal of Science and Technology*, *8*(23), 1.

Lascelles, B. D., Roe, S. C., Smith, E., Reynolds, L., Markham, J., Marcellin-Little, D., ... Budsberg, S. C. (2006, February). Evaluation of a pressure walkway system for measurement of vertical limb forces in clinically normal dogs. *American Journal of Veterinary Research, 67*(2), 277–282. doi:10.2460/ajvr.67.2.277 PMID:16454633

Ledoux, W. R., & Blevins, J. J. (2007). The compressive material properties of the plantar soft tissue. *Journal of Biomechanics, 40*(13), 2975–2981. doi:10.1016/j.jbiomech.2007.02.009 PMID:17433335

Ledoux, W. R., & Blevins, J. J. (2007, January 1). The compressive material properties of the plantar soft tissue. *Journal of Biomechanics, 40*(13), 2975–2981. doi:10.1016/j.jbiomech.2007.02.009 PMID:17433335

Leong, T. W., & Johnston, B. (2016). Co-design and robots: a case study of a robot dog for aging people. In *International Conference on Social Robotics 2016 Nov 1* (pp. 702-711). Springer. 10.1007/978-3-319-47437-3_69

Lieberman, D. E., Polk, J. D., & Demes, B. (2004, February). Predicting long bone loading from cross-sectional geometry. *American Journal of Physical Anthropology: The Official Publication of the American Association of Physical Anthropologists., 123*(2), 156–171. doi:10.1002/ajpa.10316 PMID:14730649

Liu, G., Liu, Y., Wang, X., Wu, X., & Mei, T. (2016). Design and experiment of a bioinspired wall climbing robot using spiny grippers. In *Mechatronics and Automation (ICMA), 2016 IEEE International Conference on 2016 Aug 7* (pp. 665-670). IEEE. 10.1109/ICMA.2016.7558642

Liu, Q., Sa, M., Lee, H., Lee, M., Kim, S.W., & Eune, J. (2017). Peekabot: Robot that Helps Children's Cognitive/Physical Development. In *Proceedings of the 2017 CHI Conference Extended Abstracts on Human Factors in Computing Systems 2017 May 6* (pp. 468-468). ACM.

Lu, Z., Aoyama, T., Sekiyama, K., Hasegawa, Y., & Fukuda, T. (n.d.). Motion transfer control from walking to brachiation through vertical ladder climbing for a multi-locomotion robot. IEEE/ASME Transactions on Mechatronics, 19(5), 1681–1693.

Luo, Y., Liu, J., Gao, Y., & Lu, Z. (2014). Smartphone-controlled robot snake for urban search and rescue. *In International Conference on Intelligent Robotics and Applications* 2014 Dec 17 (pp. 352-363). Springer. 10.1007/978-3-319-13966-1_35

Meyer, W., Bartels, T., Tsukise, A., & Neurand, K. (1990). Histochemical aspects of stratum corneum function in the feline foot pad. *Archives of Dermatological Research, 281*(8), 541–543. doi:10.1007/BF00412742 PMID:2322014

Meyer, W., Bartels, T., Tsukise, A., & Neurand, K. (1990). Histochemical aspects of stratum corneum function in the feline foot pad. *Archives of Dermatological Research, 281*(8), 541–543. doi:10.1007/BF00412742 PMID:2322014

Miller-Young, J. E., Duncan, N. A., & Baroud, G. (2002, December 1). Material properties of the human calcaneal fat pad in compression: Experiment and theory. *Journal of Biomechanics, 35*(12), 1523–1531. doi:10.1016/S0021-9290(02)00090-8 PMID:12445605

Mühlberger, M., Rohn, M., Danzberger, J., Sonntag, E., Rank, A., Schumm, L., ... Trappl, E. (2015, June 15). UV-NIL fabricated bio-inspired inlays for injection molding to influence the friction behavior of ceramic surfaces. *Microelectronic Engineering*, *141*, 140–144. doi:10.1016/j.mee.2015.02.051

Muscolo, G. G., Caldwell, D., & Cannella, F. (2017). Biomechanics of human locomotion with constraints to design flexible-wheeled biped robots. In *Advanced Intelligent Mechatronics (AIM), 2017 IEEE International Conference on 2017 Jul 3* (pp. 1273-1278). IEEE. 10.1109/AIM.2017.8014193

Nabulsi, S., Sarria, J. F., Montes, H., & Armada, M. A. (2009, October). High-resolution indirect feet–ground interaction measurement for hydraulic-legged robots. *IEEE Transactions on Instrumentation and Measurement*, *58*(10), 3396–3404. doi:10.1109/TIM.2009.2017650

Narioka, K., Tsugawa, S., & Hosoda, K. (2009). 3D limit cycle walking of musculoskeletal humanoid robot with flat feet. In *Intelligent Robots and Systems, 2009. IROS 2009. IEEE/RSJ International Conference on 2009 Oct 10* (pp. 4676-4681). IEEE.

Natali, A. N., Fontanella, C. G., & Carniel, E. L. (2010). Constitutive formulation and analysis of heel pad tissues mechanics. *Medical Engineering & Physics*, *32*(5), 516–522. doi:10.1016/j.medengphy.2010.02.018 PMID:20304698

Ninomiya, H., Akiyama, E., Simazaki, K., Oguri, A., Jitsumoto, M., & Fukuyama, T. (2011, December). Functional anatomy of the footpad vasculature of dogs: Scanning electron microscopy of vascular corrosion casts. *Veterinary Dermatology*, *22*(6), 475–481. doi:10.1111/j.1365-3164.2011.00976.x PMID:21438930

Ninomiya, H., Akiyama, E., Simazaki, K., Oguri, A., Jitsumoto, M., & Fukuyama, T. (2011). Functional anatomy of the footpad vasculature of dogs: Scanning electron microscopy of vascular corrosion casts. *Veterinary Dermatology*, *22*(6), 475–481. doi:10.1111/j.1365-3164.2011.00976.x PMID:21438930

Palmer, L. R., Diller, E. D., & Quinn, R. D. (2010). Toward a rapid and robust attachment strategy for vertical climbing. *Proc. of the IEEE Int. Conf. on Robotics and Automation*, 2810-2815. 10.1109/ROBOT.2010.5509680

Park, I. W., Kim, J. Y., Lee, J., & Oh, J. H. (2005). Mechanical design of humanoid robot platform KHR-3 (KAIST humanoid robot 3: HUBO). In *Humanoid Robots, 2005 5th IEEE-RAS International Conference on 2005 Dec 5* (pp. 321-326). IEEE.

Pfeifer, R., Iida, F., & Gomez, G. (2006). Morphological computation for adaptive behavior and cognition. *International Congress Series*, *1291*, 22–29. doi:10.1016/j.ics.2005.12.080

Pfeifer, R., Lungarella, M., & Iida, F. (2007). Self-organization, embodiment, and biologically inspired robotics. *Science*, *318*(5853), 1088–1093. doi:10.1126cience.1145803 PMID:18006736

Qian, Z., Miao, H., Shang, Z., & Ren, L. (2014). The foot-ground contact analysis of german shepherd dog based on normal walking, trotting and jumping gait. *J. Jilin Univ.*, *44*, 1692–1697.

Radford, N. A., Strawser, P., Hambuchen, K., Mehling, J. S., Verdeyen, W. K., Donnan, A. S., ... Berka, R. (2015, May). Valkyrie: Nasa's first bipedal humanoid robot. *Journal of Field Robotics*, *32*(3), 397–419. doi:10.1002/rob.21560

Raibert, M., Blankespoor, K., Nelson, G., & Playter, R. (2008). Bigdog, the rough-terrain quadruped robot. *Proc. 17th World Congress, The Int. Federation of Automatic Control*, 10822–5.

Sedlaček, M., Podgornik, B., & Vižintin, J. (2012, April 1). Correlation between standard roughness parameters skewness and kurtosis and tribological behavior of contact surfaces. *Tribology International*, *48*, 102–112. doi:10.1016/j.triboint.2011.11.008

Sellaouti, R., Stasse, O., Kajita, S., Yokoi, K., & Kheddar, A. (2006). Faster and smoother walking of humanoid HRP-2 with passive toe joints. In *Intelligent Robots and Systems, 2006 IEEE/RSJ International Conference on 2006 Oct 9* (pp. 4909-4914). IEEE.

Sheba, J. K., Salman, A. A., Kumar, S., Elara, M. R., & Martínez-García, E. (2018, January 1). Development of Rehabilitative Multimodal Interactive Pet Robot for Elderly Residents. *Procedia Computer Science*, *133*, 401–408. doi:10.1016/j.procs.2018.07.049

Spenko, M. J., Haynes, G. C., Saunders, J. A., Cutkosky, M. R., Rizzi, A. A., Full, R. J., & Koditschek, D. E. (2008, April). Biologically inspired climbing with a hexapedal robot. *Journal of Field Robotics*, *25*(4-5), 223–242. doi:10.1002/rob.20238

Stout, K. J., Sullivan, P. J., Dong, W. P., Mainsah, E., Luo, N., Mathia, T., & Zahouani, H. (1994), The development of methods for the characterization of roughness on three dimensions. Publication no. EUR 15178 EN of the Commission of the European Communities, Luxembourg.

Szczecinski, N. S., Chrzanowski, D. M., Cofer, D. W., Terrasi, A. S., Moore, D. R., Martin, J. P., . . . Quinn, R. D. (2015). Introducing mantisbot: hexapod robot controlled by a highfidelity, real-time neural simulation. In *Intelligent Robots and Systems (IROS), 2015 IEEE/RSJ International Conference on 2015 Sep 28* (pp. 3875-3881). IEEE. 10.1109/IROS.2015.7353922

Tan, Y.G., Li, Z., Chen, Y.H., & Wang, H. (n.d.). *Bionic mechanism and kinetic characteristic for quadruped robot dog*. Academic Press.

Tanaka, M., & Tanaka, K. (2015, April). Control of a snake robot for ascending and descending steps. *IEEE Transactions on Robotics*, *31*(2), 511–520. doi:10.1109/TRO.2015.2400655

Tenreiro, J. A., & Silva, M. F. (2006). An Overview of Legged Robots. *Proceedings of the International Symposium on Mathematical Methods in Engineering*, 27-29.

Tummala, R. L., Mukherjee, R., Xi, N., Aslam, D., Dulimarta, H., Xiao, J., ... Dang, G. (2002, December). Climbing the walls. *IEEE Robotics & Automation Magazine*, *9*(4), 10–19. doi:10.1109/MRA.2002.1160067

Veneman, J. F., Kruidhof, R., Hekman, E. E., Ekkelenkamp, R., Van Asseldonk, E. H., & Van Der Kooij, H. (2007, September). Design and evaluation of the LOPES exoskeleton robot for interactive gait rehabilitation. *IEEE Transactions on Neural Systems and Rehabilitation Engineering*, *15*(3), 379–386. doi:10.1109/TNSRE.2007.903919 PMID:17894270

Vespignani, M., Melo, K., Mutlu, M., & Ijspeert, A. J. (2015). Compliant snake robot locomotion on horizontal pipes. In *Safety, Security, and Rescue Robotics (SSRR), 2015 IEE International Symposium on 2015 Oct 18* (pp. 1-8). IEEE. 10.1109/SSRR.2015.7442941

Voigt, D., Karguth, A., & Gorb, S. (2012). Shoe soles for the gripping robot: Searching for polymer based materials maximizing friction. *Robotics and Autonomous Systems*, *60*(8), 1046–1055. doi:10.1016/j.robot.2012.05.012

Weissengruber, G. E., Egger, G. F., Hutchinson, J. R., Groenewald, H. B., Elsässer, L., Famini, D., & Forstenpointner, G. (2006). The structure of the cushions in the feet of African elephants (Loxodonta africana). *Journal of Anatomy*, *209*(6), 781–792. doi:10.1111/j.1469-7580.2006.00648.x PMID:17118065

Wisse, M., Hobbelen, D. G., Rotteveel, R. J., Anderson, S. O., & Zeglin, G. J. (2006). Ankle springs instead of arc-shaped feet for passive dynamic walkers. In *Humanoid Robots, 2006 6th IEEE-RAS International Conference on 2006 Dec 4* (pp. 110-116). IEEE. 10.1109/ICHR.2006.321371

Yu, Z., Shi, Y., Xie, J., Yang, S. X., & Dai, Z. (2018). Design and analysis of a bionic adhesive foot for gecko robot climbing the ceiling. *International Journal of Robotics and Automation*, *33*(4). doi:10.2316/Journal.206.2018.4.206-5412

Zielinska, T. (2004). Development of walking machines; historical perspective. In *International Symposium on History of Machines and Mechanisms 2004* (pp. 357-370). Springer.

Chapter 11
A Modular Structured Architecture Using Smart Devices for Socially-Embedded Robot Partners

Jinseok Woo
Tokyo University of Technology, Japan

Naoyuki Kubota
Tokyo Metropolitan University, Japan

ABSTRACT

Recently, robot architectures with various structures have been developed to improve the human quality of life. Such a robot needs various capabilities such as learning, inference, and prediction for human interaction, and such capabilities are interconnected with each other as a whole system. In the development of a socially-embedded robot partner, human-robot interaction plays an important role. Therefore, in order to develop a socially embedded robot partner, we must consider human communication system. Human Cognition, Emotion, and Behavior should be considered in the development process of the robot partner, and if these factors are fully reflected in the robot partner, then the robot can be used as a socially-friendly robot partner. This book chapter is organized as follows: First, we describe the hardware and software structures. Next, we discuss the cognitive model of the robot partners. Third, we discuss interaction content design for various services. Finally, we discuss the contents of society implementation, and discuss the applicability of robots for social utilization.

DOI: 10.4018/978-1-7998-0137-5.ch011

INTRODUCTION

In recent years, the number of elderly people living alone has been increasing, and various social problems have arisen. For example, elderly people who are living alone without medical or social support have difficulties in providing care for themselves. In fact, the problem of lonely death is a social problem in Japan . In order to solve these social problems, the introduction of coexisting human-friendly robot partners is one of the possible solutions to realize robot elderly care support. In addition to support for the elderly people, robot partners are being developed for various parts of daily life based on the background of personal robot industry development. Besides that, the remarkable growth of mechatronics in accordance to that has also contributed to the improvement of human life quality (Alfred et al., 1990; Takahiro et al., 2017; Renteria-Bilbao et al., 2019; Broadbent et al., 2009; Tang et al., 2015; Sheridan et al., 2016; Habib et al., 2008).

In robotics, in order to improve the quality of life of humans, various robots are developed by using social robot technology (Esposito et al., 2016; Woo et al., 2012). Social robots can communicate with people by combining network technologies such as artificial intelligence, Internet of things technology, and cloud computing, which are becoming technologically advanced recently. Therefore, in order for these robots to be put to practical use, social behaviors and interactions should be studied so that they work well in human society (Lemaigan et al., 2017).

Conventional robots moved by using sensors to detect obstacles in the environment. However, in the case of social robots, it is necessary to interact not only with the obstacles present in the environment, but also with human conditions (So et al., 2018). In order to take full account of these factors, there are many researches on cognitive models of robots that imitate human cognitive models (Breazeal et al., 2002). Cognitive model is composed of cognitive architecture and knowledge to take a behavior (Howes et al., 1997). Among them, a typical cognitive architecture is the "Subsumption Architecture" by Rodney Brooks (Brooks, 1986).

In the case of human interaction, Piaget's theory of cognitive development suggests that human cognitive development is an adaptation process that is made by interaction with the environment, and that it develops through several steps (Piaget et al., 1964). Robots living with humans also need to learn and develop by interacting with the environment in accordance with human learning rules. Therefore, various cognitive models of the robot have been discussed (Khamassi et al., 2011; Burghart et al., 2005). Here, we have developing robot partner system based on a modular cognitive model (Woo et al., 2017).

In this paper, we describe from the hardware modules to the software modules of the robot, explain the cognitive model for constructing the social robot, and explain the system integration according to modular structure of robot partner.

This paper is organized as follows: Section 2 explains the development of robot partners. This section describes the robot's hardware and software modules. Furthermore, we explain the system integration using each module. Section 3 explains interaction design for social implementation. Here, we describe the interaction contents design of the robot that is configured to make it easy to create, maintain, and manage the contents necessary for the human-robot interaction. Section 4 describes several examples of the proposed methods to social implementation using robot partner system. Section 5 concludes this paper, and discusses the applicability of the proposed method and future direction of this research.

THE DEVELOPMENT OF ROBOT PARTNERS

In our research, we tried to develop a robot partner platform at a level that could be used for human care supporting at home. Among the many devices, we considered using smart devices for development of robot partners. The smart device basically has functions that enables voice recognition and voice synthesis by using the telephone function, and it can sense with the environment by using inner sensors such as camera, touch screens, acceleration and GPS sensors (Woo et al., 2014). Moreover, it can be applied to artificial intelligence, Internet of Things (IoT), cloud technology based on the increment of hardware specifications of smart device. Furthermore, the miniaturization and low-cost microcontrollers and wireless modules are making it easier to develop robots in combination with smart devices. Therefore, taking this advantage into account, we have used smart devices as a robot platform. Figure 1 shows an example of robot development using smart devices. Regarding personal preferences, robot design can be made differently according to personal preferences and usage environment.

There are many considerations that needs to be taken when developing a robot. Especially, the guidelines of system design are important to consider achieving specification of robot application. In our robot partner, design guidelines of our robot system, the hardware is of a freely customizable design by using 3D printer. These 3D models could link to open source services such as "GitHub" (n.d.), they can develop better their robotic structure by sharing their models for improved development as shown in Figure 2. By exchanging each module, a new function can be used by robots. Eight motors are used to control the robot, and motor ID are assigned to each part (Robotis et al., n.d.). Therefore, they can be controlled when the ID has same setting even if the 3D model design is different. And one microcontroller is used for controlling the motor (Figure 3(a)). Bluetooth 4.0 BLE (Bluetooth Low Energy) is used for communication between smart device and microcontroller as shown in Figure 3(b). As a concept of the robot, the robot mode is changed while charging at home, and when the user goes out of the house, the

Figure 1a. The example of robot partners based on smart devices

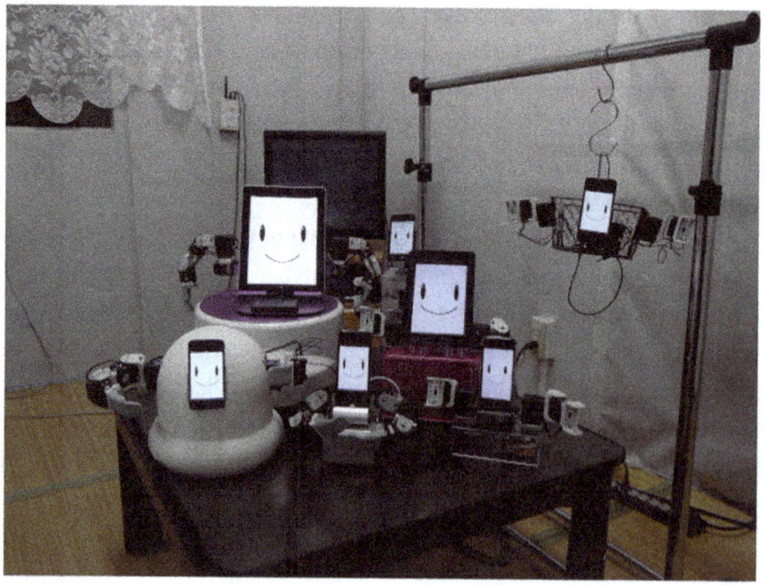

Figure 1b. The example of robot partners based on smart devices

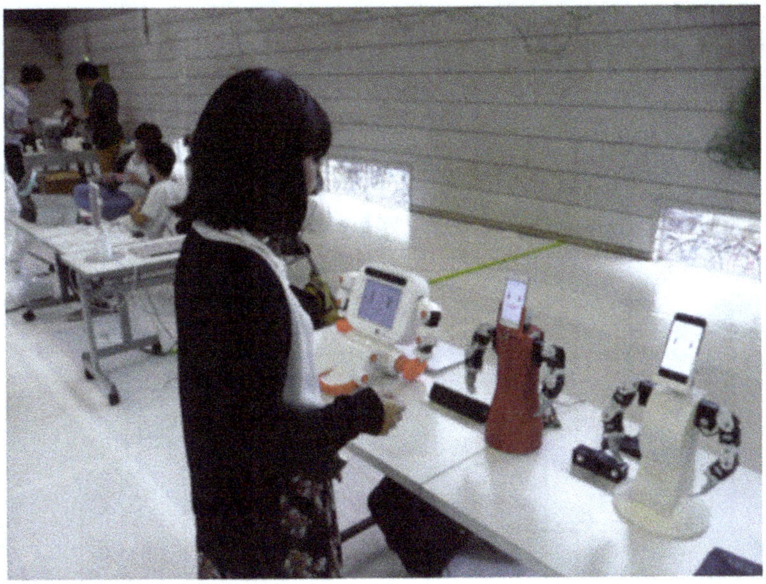

Figure 2. Modularization of Robot Partner: (a) Basic modules (b) 3D CAD design (c) Various robot design

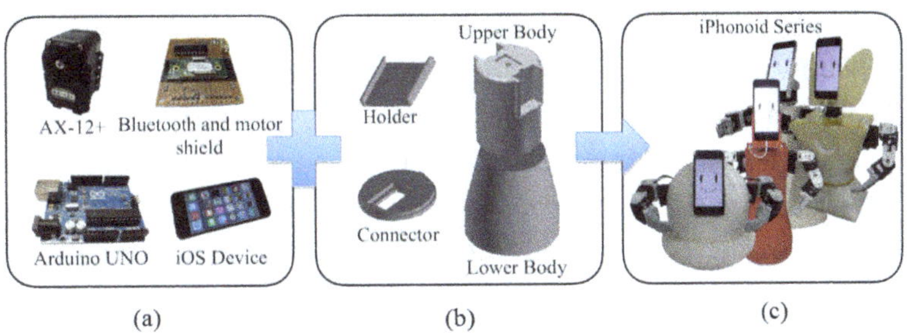

Figure 3a. Configuration of robot motors: Motor module of robot

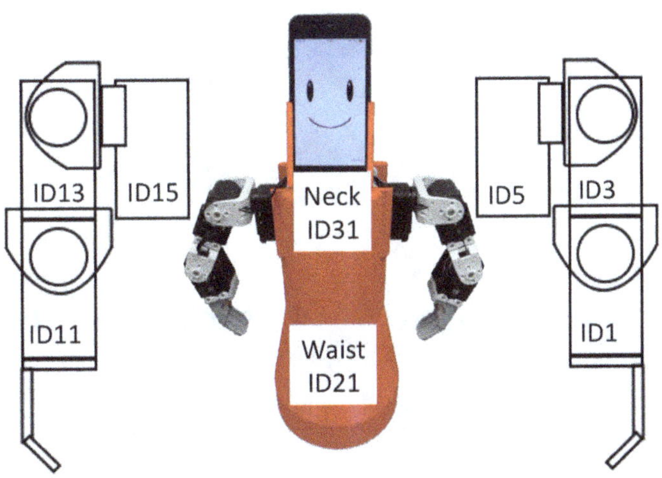

Figure 3b. Configuration of robot motors: Motor control module

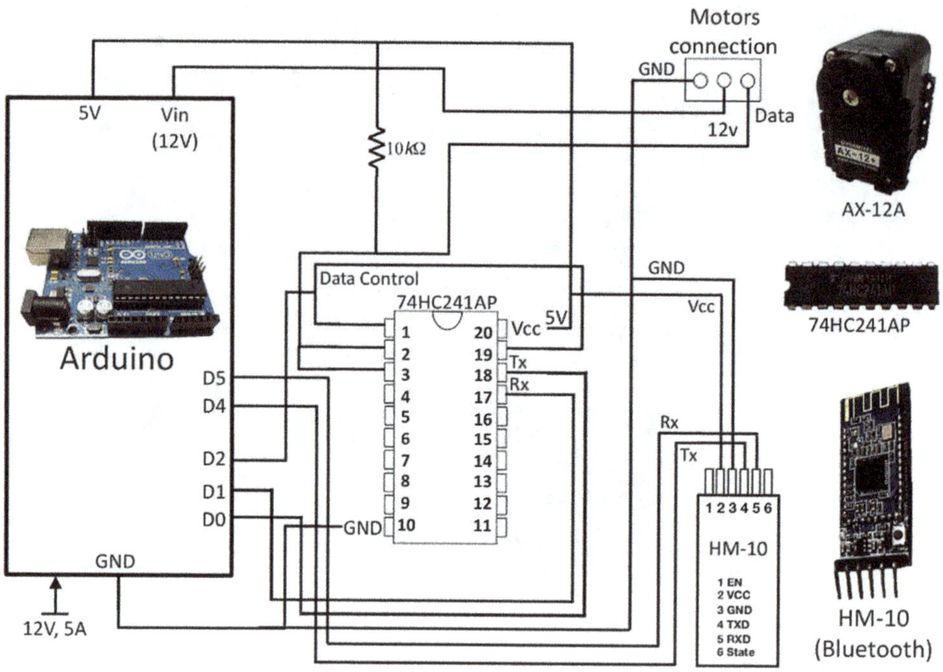

robot is disconnected from the body (Figure 4(a)). In order to set the robot mode, the mode is changed by the presence or absence of the charging base as shown in Figure 4(b).

The software is designed to achieve the realization of system integration based on the modular structural systems. In order to develop a social robot, we have to consider how to design the robot partner to become widely used in the world. A robot with minimum functionality is required in order to develop the robot for business market with low price. Recently, various AI elements have been developed, so that different robot systems can be developed depending on how these elements are integrated.

Figure 4a. Changes in robot partner mode: Daily life activity dependency

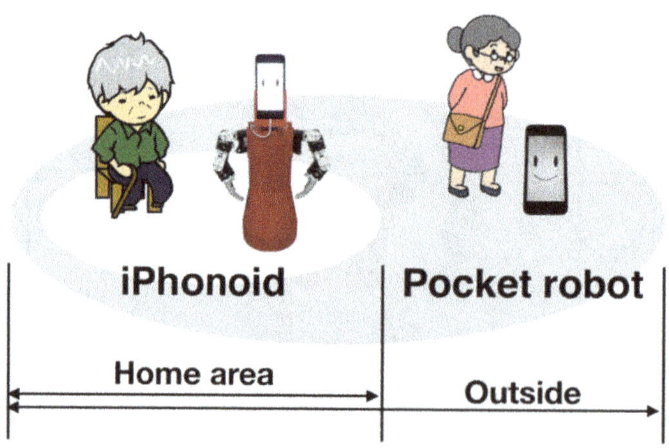

Figure 4b. Changes in robot partner mode: Robot partner mode transition

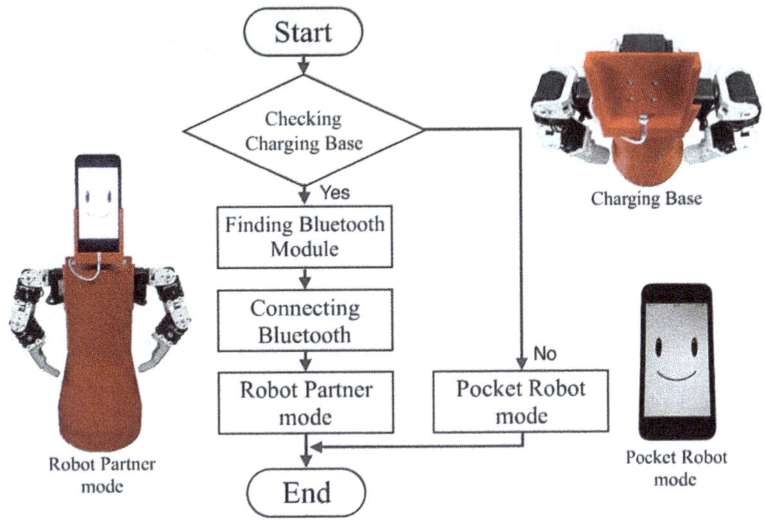

Accordingly, our "iPhonoid" system has modular structure for cognitive model. Each robot system module consists of four layers to combine the robot's basic function: 1st-Hardware Layer, 2nd- Library Layer, 3rd-Component Layer, 4th-Application Layer (Figure 5) (Woo et al., 2017). The hardware layer consists of the basic modules to realize the robot structure. For example, the smart device has display, touch sensor, microphone, camera and other sensors. The charging stand (robot body) has microcontroller, actuators, supporting body structure and Bluetooth module. This hardware layer is the most basic layer to define a robot partner system for service. The robot partner design is also selected in this layer. The second layer is the library layer to control the hardware part. Based on this structure developers can consider what libraries and components are needed for the application layer. In the component layer, we proposed conversation modules to interact with human. The component layer is a fundamental framework for creating robotic partner applications.

INTERACTION DESIGN FOR SOCIAL IMPLEMENTATION

Here, we describe the structure of the robot's cognitive model for social implementation. The explanation of robot system is divided into the cognitive part, the conversation part, and the contents designing part, which are necessary for social implementation.

Development of Robot's Cognitive Model for Human-Friendly Robot Partner Design

In the development of a socially embedded robot partner, human-robot interaction plays an important role. If the robot can socially interact with the human, the robot can provide the human with many services such as information support, health promotion, and rehabilitation. Therefore, in order to develop a socially embedded robot partner, we have to consider the human communication system.

Figure 5. An example of a basic module of robot system that can be referred to during robot designing

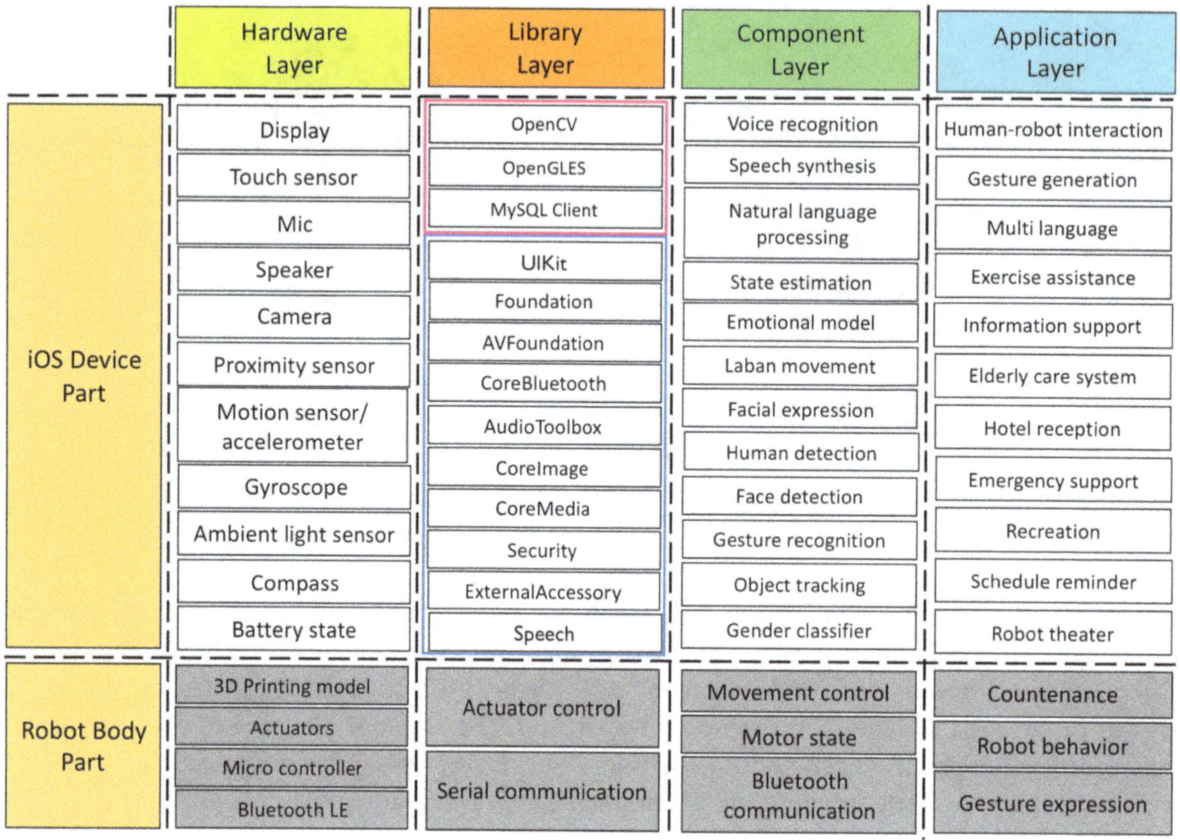

In order to study communication with robots, it is necessary to understand human cognitive ability, emotion, and behavior. Human's cognition, emotion, and behavior have a close relationship and they flexibly react to each other. Each outcome of the system will be different based on the environmental structure (Salter et al., 2006) This is interpreted as the individuality of robot, which means, that the robot could have different individuality (Coeckelbergh et al., 2011). As well as models for individuals, social relations are also of great importance. Therefore, a robot partner requires a cognitive model like human being to conduct social interaction. Various types of cognitive models have been proposed in psychology and cognitive science (Lee et al., 2018; Hollander et al., 2016). For example, a simple cognitive model based on subsumption architecture has been successfully applied to easy tasks, but it is difficult to apply those kinds of cognitive models practically to social applications based on human interaction.

The cognitive model of our robot partner is illustrated in Figure 6. In order for a robot to live with human, it is necessary to express robot behavior according to information received from the outside. Therefore, our research proposed a conversation system with emotional empathy. The robot interacts with humans by using verbal and nonverbal communication. The detailed description of this model has been discussed in our previous papers (Woo et al., 2018; Woo et al., 2017; Woo et al., 2014). In the cognitive model, emotional part is an important part of the cognitive structure. Each person has various experiences about emotions during his/her own life. Therefore, we propose emotions in the robot's cognitive model in order to realize a human-like robot partner system.

Figure 6. Cognitive flows of robot partner for human-robot interaction

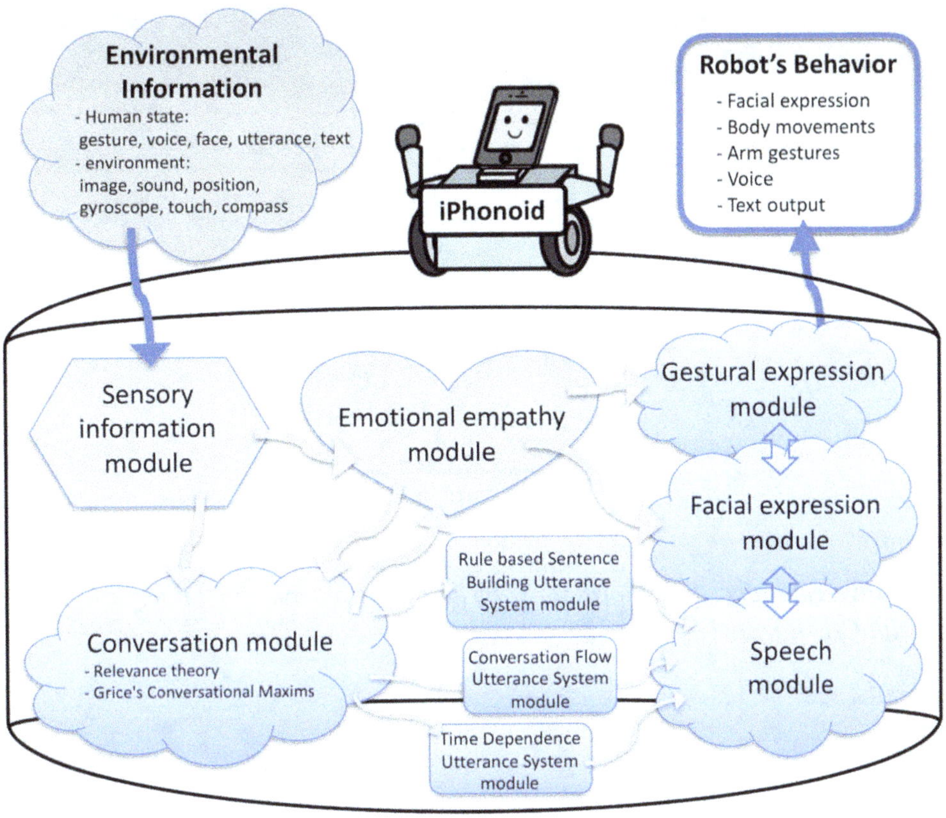

The cognitive-emotional-behavioral model is able to give individuality to the robot partner based on the control of the C-E-B model in order to adapt to the user's interaction style. For example, among the people, there are types that are easily affected by emotions, and some people think deeply to act. Therefore, we develop a system that assigns cognitive priority, emotional priority, and behavioral preference to a robot partner system to make a difference in robot interaction style. For the robot system structure of C-E-B model, we are getting inspiration from "Social systems theory", that the structure will incorporate with other structure. Here, each cognitive, emotion and behavior system have a strong relationship through communication (Seidl et al., 2004; Luhmanne et al., 1993) .Our robot partners compete with each other for these three systems based on the C-E-B model (Figure 7(a)), and the system with the strongest factor appears on the surface for human-robot interaction as shown in equation (1).

$$O = \max\left(C, E, B\right) \tag{1}$$

The C-E-B model is used to interact with humans, where each module has a weight, and a module with a strong weight determines the interaction style of the robot. For example, when "E" parameter is strong, emotion factor tends to be a priority (Sentimental reaction), and when "C" is strong, the dependence on utterance is high (Rational reaction). And if "B" is strong, the use dependence of non-verbal elements increases (Instinctive reaction) (Woo et al., 2017). In Figure 7(b), based on the results of the emotion

Figure 7a. Cognitive-emotional-behavioral (C-E-B) model for human-robot interaction: Coupling relations of C-E-B model

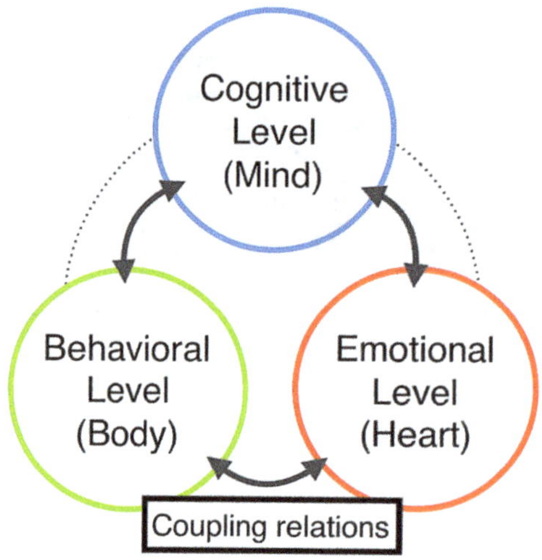

Figure 7b. Cognitive-emotional-behavioral (C-E-B) model for human-robot interaction: Facial expression of robot with C-E-B model

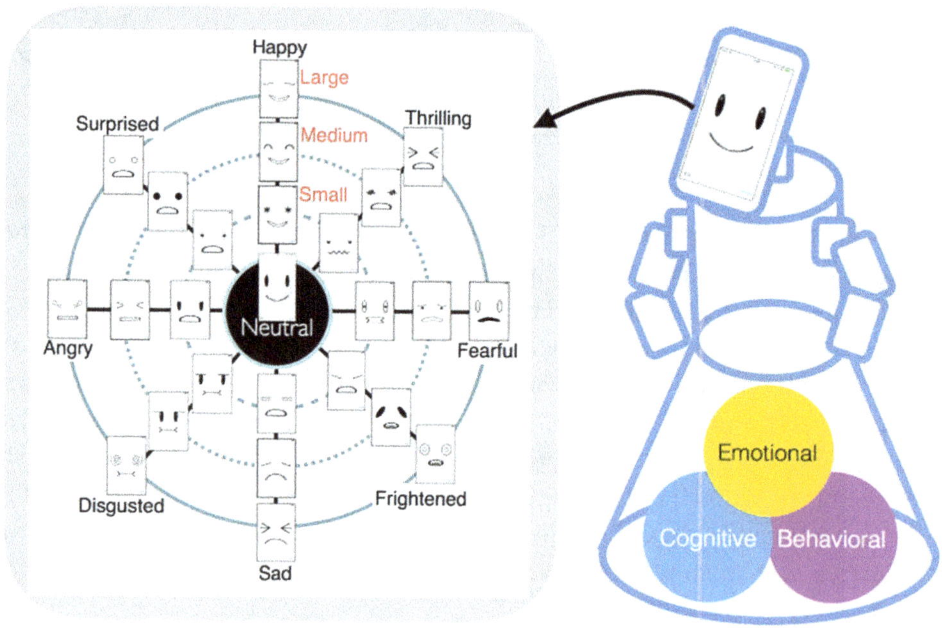

model, the robot partner has 25 facial expression designs that are expressed through the touch screen. The emotional expression of the robot is determined through the emotion model. In human interactions, interactions based on mutual emotional states play an important role in social activities. Therefore, it is necessary to interact with each other in consideration of their feelings. Our previous researches proposed emotional empathy model by considering human emotional state as shown in Figure 8.

Figure 8. Emotional empathy model of robot's for human-robot interaction

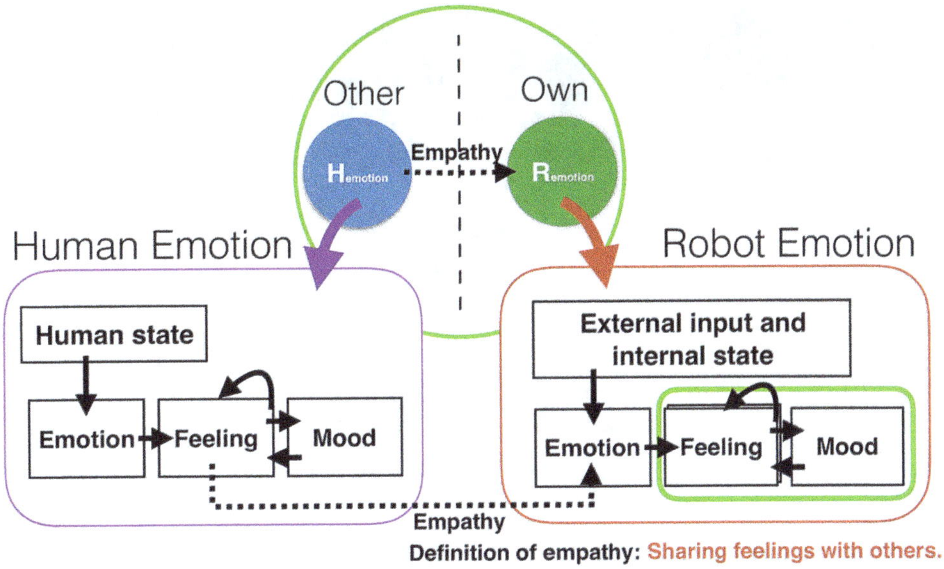

And it can realize robot movements with 8 degrees of freedom like neck, waist arm based on its specification. In particular, the degree of freedom for the neck and the waist can be used to change directions for camera photography as shown in Figure 9 ("Apple Inc.", n.d.).

Conversation System of Robot for Human-Robot Interaction

There are various ways of implementing robot's conversation systems when the robot partner interacts with human. In our study, the conversation of robot partner has three modes such as conversation flow utterance mode, sentence building utterance mode and time dependent utterance mode as shown in Figure 10 (Woo et al., 2017).

Firstly, the conversation flow utterance mode represents the robot utterance content using a data log in which the user has spoken to prevent a monotonous situation in a user's daily life. The user's utterance sentences are stored in a database for analysis of communication patterns. The database saves the spoken utterance sentences input by user in order. Thereafter, the relevance of the keywords used in the stored data is compared with the currently input utterance sentence, and the most relevant sentence is selected and used for robot utterance. Accordingly, it is possible to interact with the same style as the user. This utterance mode is implemented to the daily conversation mode and the content conversation mode for robot conversation.

Secondly, sentence building utterance mode is used based on the predefined sentence generation rule (Woo et al., 2015). Simple sentences composed of subject, verb and objects are used for applying sentence building mode. In this case, since the number of corresponding utterances is limited, the information-assisted service system of the robot is configured mainly using time-dependent utterance mode of the event driven type. In the time-dependent mode, each utterance content has a temporal property, and the utterance pattern of the robot changes based on time parameter. The time attribute is mainly used for schedule management and time related greetings. As shown in Figure 11, in the time-

Figure 9. Example of panoramic photography using a waist joint

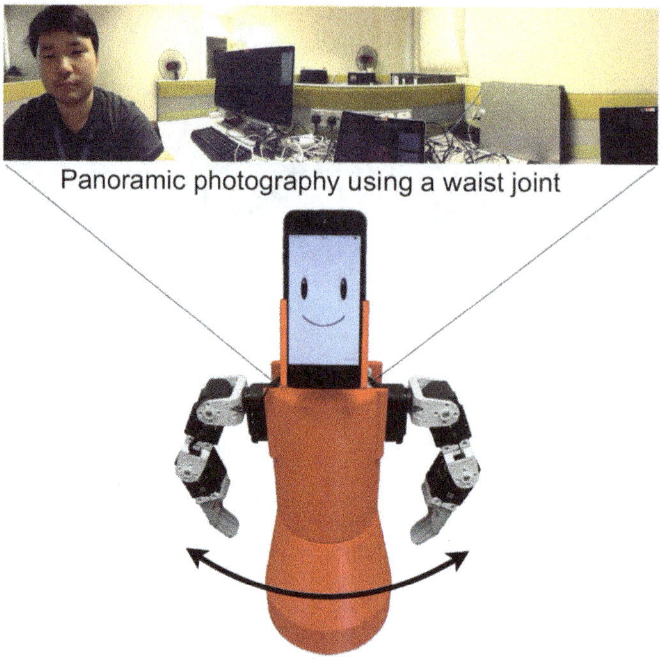

Figure 10. Structure of Conversation System

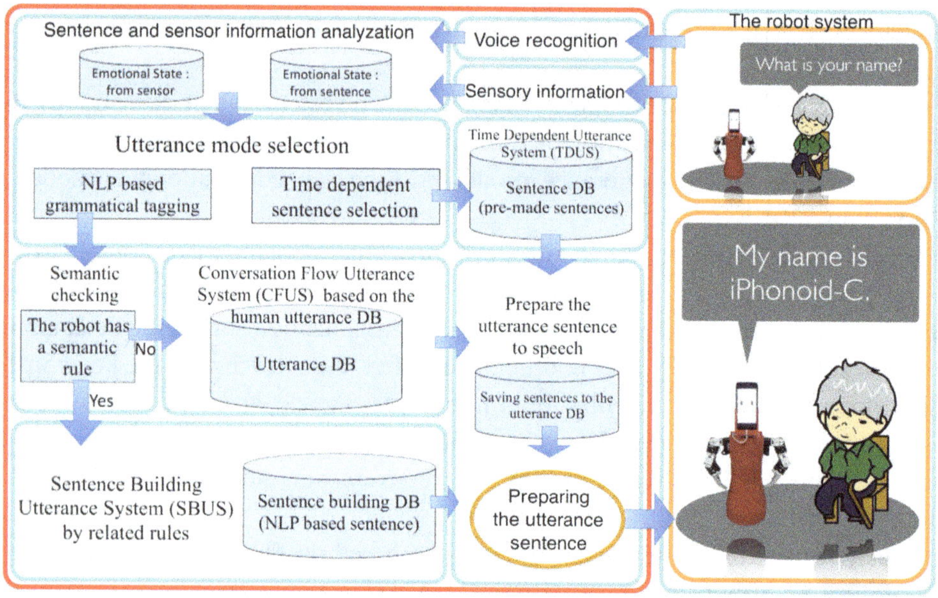

Figure 11. Conversation modes for robot partners

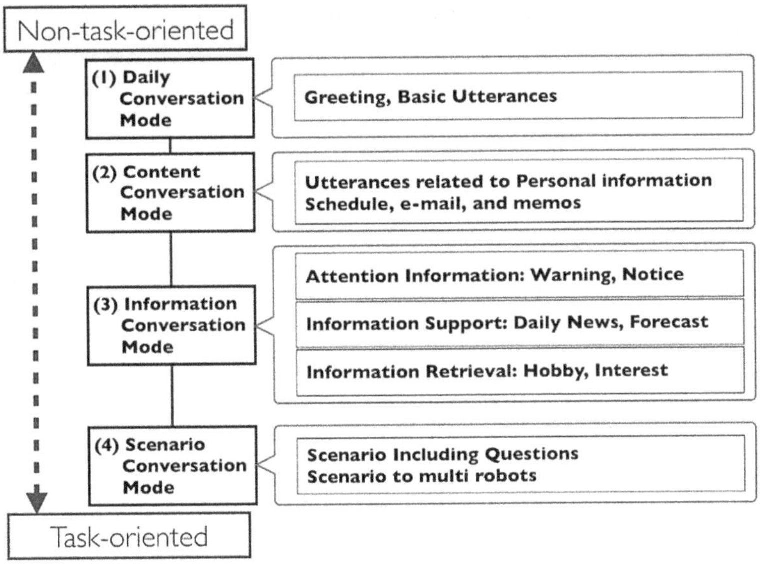

dependent utterance mode, there are four conversation modes consisting of daily conversation mode, content conversation mode, information conversation mode, and scenario conversation mode (Woo et al., 2017). A conversation system is classified into two parts of (1) task-oriented conversation and (2) non-task-oriented conversation.

In the daily conversation mode, conversation is performed based on the lifestyle extracted from the daily life log. In the content conversation mode, the robot partner utters using inner information. This is because smart devices have a lot of useful information such as calendar, music and email application information. Therefore, the robot can refer to personal preference of user for information supporting.

In the information conversation mode, not only the utterance such as the news and the weather forecast but also the information gathering is performed on which topic the elderly person might be interested in. In the scenario conversation mode, a large amount of collected information is interactively provided to the elderly. The robot conversation package is prepared according to robot partner's purpose, such as "Health improvement support conversation package", "Elderly care conversation package", and "Support to encourage going out conversation package".

Furthermore, we develop a robot utterance system using a human persona that takes into account the difference in the place where the user lives ("Infitechm", 2017) "Persona" is a fictitious person that represents various types of users within a target population that are likely to use the product or service. Robot becomes more familiar as partner towards the user when these attributes are considered.

Conversation Contents Design Template for Robot Utterance

Robot partners usually have an utterance database, many of which are downloaded from the network such as internet. As a result, robot utterance tends to depend on external or internal databases. Therefore, using a smart device as a basic module of a robot partner facilitates connection to the database.

Table 1. Interaction design template for robot interaction

robot_id	scenario_id	context_id	face	gesture	time	language	sentences
robot_A	0	1	1	2	0	en-US	Are you sure you want to use English?
robot_A	0	1	1	2	0	ko-KR	日本語でよろしいですか・
robot_A	0	1	1	2	0	ja-JP	한국어를 사용하시겠습니까?
…	…	…	…	…	…	…	…

The robot partner's utterance contents are necessary to provide update for sustainable service. However, it takes a lot of money and time to develop and use content. Accordingly, even if a robot is introduced, it becomes impossible to utilize it properly. Therefore, we have developed a design template for utterance content so that anyone can easily design robot utterance contents. As a basic concept of the design template, we have designed a system that can update the spoken contents by text editing. With this system, users who have difficulty in computer programming can easily update the utterance contents. In order to update the utterance content from the user, we propose a database structure as shown in Table 1 to support interaction content design. The user can manage utterance contents such as speech sentences, language codes, time intervals, emotional elements, and behavioral elements according to a database format (Woo et al., 2017). Currently, the database structure of robot utterance system is designed by a series of categories based on robot ID, scenario ID, context ID, emotional element, action element, time interval, language code, speech content, and external inputs. The external input is used for event driven speech according to the touch interface and voice input. Emotional factors (face expression ID) and behavioral factors (gesture ID) are used to represent the state of emotion parameter from the emotion model. Even if the user does not understand the programming language of the robot system, we are promoting the development of the robot system so that the interaction design of the robot can be performed by adding and deleting the utterance contents. Such a conversation contents design template can be used at the reception counter of a hotel or as a guiding role of a restaurant. We also have used these designed contents in exhibitions.

SOCIAL IMPLEMENTATION OF ROBOT PARTNER SYSTEM

According to recent robot development trends, the areas in which social robots can be used include public support systems and personal life support. Here, we describe some examples of social implementation of robot partner.

Public Support System Using Robot Partners

As a general way of using social robots, we can consider the usage of social robot for public support such as educational systems, health promotion systems, airports information supporting and robot theater. First, we describe the system for the robot partner used at the airport (Quan et al., 2018; Quan et al., 2019). There are many people who are worried about getting lost at the airport. Therefore, we design the function and interface of the airport guidance robot with high usability. When designing the system, it is possible to create new services by integrating the module for the smart device and the robot body

Figure 12a. Concept of airport guide robot

Figure 12b. Concept of airport guide robot

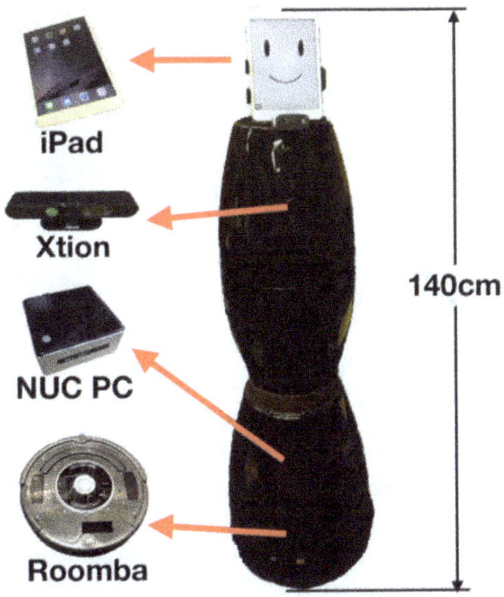

to be used at the airport. This robot is composed of Apple iPad for robot partner system, Asus Xtion for human detection, Intel NUC PC for database and control system and iRobot Roomba for moving. This robot can support airport information by touch screen and voice information. The robot moves to the detected human after performing human detection using the Xtion sensor, and provides guidance of the place using the touch screen and utterance information.

Furthermore, here we introduce social implementation examples of human-robot interactions for users such as utterance content design for robot partners as shown in Figure 13. Users developing robot content can demonstrate the interactions from multiple robots by editing the interaction contents parameters

Figure 13a. Application of multi-robot partner: Basic system of multi-robot partner

Figure 13b. Application of multi-robot partner: Robot theater

shown in Table 1. For example, in order to maximize the effects of multi-robots, we developed a system that can interact with humans through robot theater. Here, the user can create and edit the interaction contents without any special programming knowledge for the robot, thereby making it possible to produce a variety of robot plays. For example, during the utterance content design for each robot ID, the server system controls the overall interaction by transmitting the information to the robot in accordance with the play timing (Woo et al., 2017).

To assess the usefulness of the actual system, we developed a guest participatory interactive robot theater and collected opinions from the audience. The contents of the story of theater used "Momotaro", which is the story of Japan as shown in figure 13 (Umetsu et al., 2019). As a result of interviews with the audience, the robot theater was enough to be enjoyed by the audience. Therefore, based on such a system, contents such as event promotion can be easily created even those who are not proficient in complex programming as shown in Figure 14.

Information Support of Personal Life Using Robot Partner

In order to provide a wide range of services in an environment unknown to the robot, it is necessary to confirm the environmental information to which the user belongs. If the measured environment information can be processed and configured so that the robot can use it, the robot can perform service more efficiently (Szeles et al., 2017; Woo et al., 2012). The main role of sensor network used for robots is measurement of environmental data, environmental monitoring, human tracking, and object detection. The sensor data thus measured is sent to the database server (Woo et al., 2018). Environments with humans

Figure 14a. Application example of robot partners in exhibition for promotional purposes

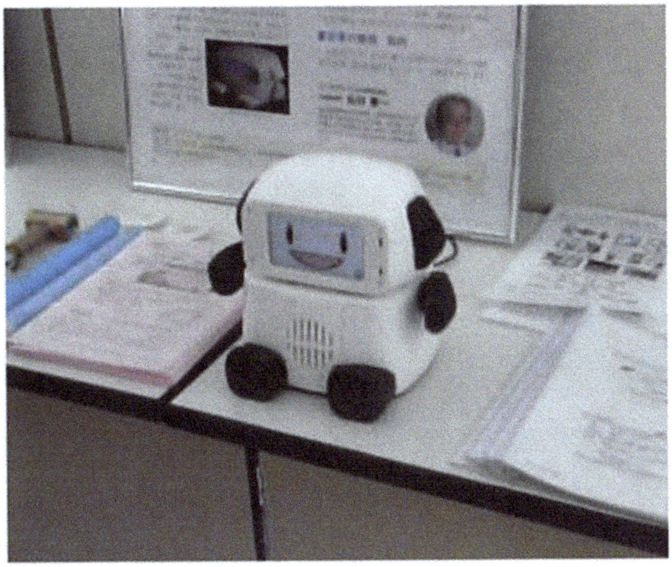

Figure 14b. Application example of robot partners in exhibition for promotional purposes

and robots require a structured platform for collecting, storing, transforming, and providing information as shown in Figure 15. The user's behaviors are changed as follows; (a) nobody, (b) come in the room, (c) walking (d) sitting, (e) walking, (f) on the bed. Based on such information, the interaction mode of the robot was changed according to the user's behavior.

Here, we call this environment as Informationally Structure Space (ISS). In this space, we use various technologies and libraries to estimate the human condition to obtain necessary information. For example, to measure human motion with a typical camera, OpenPose is used as shown in Figure 16. OpenPose is the library of realtime multi-person 2D poses estimation (Cao et al., 2018). By using these technologies and internal sensor information using smart devices, it is possible to estimate a more accurate user state. The most important task in Informationally structure space is estimation of human position, state and behavior for natural communication.

In order to construct an interaction support system in the environment in which the user lives, we are researching and developing a robot partner system that supports watching over using information in the environment (Woo et al., 2018). Here, we use the information support mode for robot interaction. This information support mode includes contents such as local event information, information on the user's hobbies, and information support for the area of interest based on the residence of the user. As shown in Figure 16, we show a case where a robot provides information to a person living in an informationally

Figure 15. System components of informationally structured space for robot partners

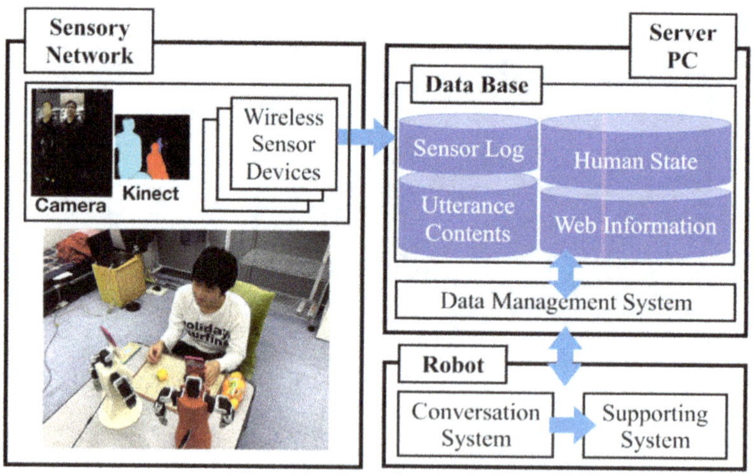

Figure 16. Analysis of user's condition in living space by OpenPose

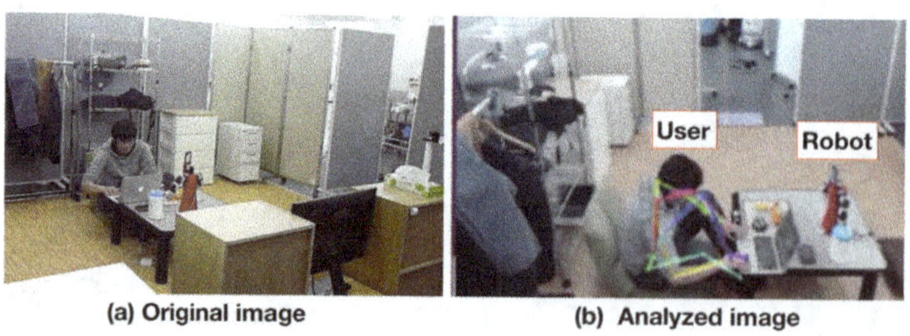

structured space. As in this example, the robot system analyzes human behavior patterns and confirms that user's safe and in secure state.

And for the language acquisition of the robot partner in this environment, we are developing nonverbal communication system to realize natural human-robot interaction (Tanaka et al., 2019). Thus, the use of information structured space enables the realization of verbal and nonverbal communication.

CONCLUSION AND FUTURE WORKS

This paper introduces several technologies and examples of robot partner development aiming at social implementation, and explains modularization technology and service application examples of robot systems for various needs of users. Through such applications, we can consider that a robot system that can provide a variety of contents is needed in order for future robots to become more familiar with people. Therefore, we have developed various types of robots to realize the interaction and information support for people.

Our robot partner system is designed based on a modularized architecture of robot partners using smart device to realize flexible update based on the usability of hardware and software modules. The robot body can be designed according to users' knowledge, but the robot modules can be combined based on the purpose of usage. The robot software can also be modularized based on well-developed libraries from the existing market, e.g., the speech recognition module and voice synthesis module. Based on the versatility of the configuration modules, this chapter proposed a basic structure that facilitates easy content design for verbal and nonverbal communication.

In future research, in order to develop a robot system that can adapt to various social situations, we will consider the modular structure of the system that enables the design of more detailed interaction content of the association between modules.

ACKNOWLEDGMENT

The authors would like to thank the students of the Kubota Lab for their research and experiment supports. And the authors would like to thank Noel Tay Nuo Wi for his proofreading.

REFERENCES

Alfred, D., Kolody, B., & Wood, P. (1990). Effects of social support from various sources on depression in elderly persons. *Journal of Health and Social Behavior*, 148-161. PMID:2102494

Apple Inc. (2019). panoramaPhoto – INPhotoAttributeOptions. Retrieved from https://developer.apple.com/documentation/sirikit/inphotoattributeoptions/1902458-panoramaphoto

Breazeal, C., & Scassellati, B. (2002). Robots that imitate humans. *Trends in Cognitive Sciences*, *6*(11), 481-487. doi:10.1016/S1364-6613(02)02016-8 PMID:12457900

Broadbent, E., Stafford, R., & MacDonald, B. (2009). Acceptance of healthcare robots for the older population: Review and future directions. *International Journal of Social Robotics*, *1*(4), 319-330. doi:10.100712369-009-0030-6

Brooks, R. (1986). A robust layered control system for a mobile robot. *IEEE Journal on Robotics and Automation*, *2*(1), 14-23. doi:10.1109/JRA.1986.1087032

Burghart, C., Mikut, R., Stiefelhagen, R., Asfour, T., Holzapfel, H., Steinhaus, P., & Dillmann, R. (2005). A cognitive architecture for a humanoid robot: A first approach. In *5th IEEE-RAS International Conference on Humanoid Robots*. 10.1109/ICHR.2005.1573593

Cao, Z., Hidalgo, G., Simon, T., Wei, S.-E., & Sheikh, Y. (2018). OpenPose: Realtime Multi-Person 2D Pose Estimation using Part Affinity Fields. In *arXiv preprint arXiv:1812.08008.*

Coeckelbergh, M. (2011). Humans, animals, and robots: A phenomenological approach to human-robot relations. *International Journal of Social Robotics*, *3*(2), 197-204. doi:10.100712369-010-0075-6

de Hollander, G., Forstmann, B. U., & Brown, S. D. (2016). Different ways of linking behavioral and neural data via computational cognitive models. *Biological Psychiatry: Cognitive Neuroscience and Neuroimaging*, *1*(2), 101-109. PMID:29560872

Dolan, R. J. (2002). Emotion, cognition, and behavior. science 298(5596), pp. 1191-1194. doi:10.1126cience.1076358

Esposito, R., Fiorini, L., Limosani, R., Bonaccorsi, M., Manzi, A., Cavallo, F., & Dario, P. (2016). "Supporting active and healthy aging with advanced robotics integrated in smart environment." Optimizing assistive technologies for aging populations (pp. 46–77). Hershey, PA: IGI Global.

GitHub, Inc. (2019). Retrieved from http://github.com/.

Habib, M. K. (2008). Human Adaptive and Friendly Mechatronics (HAFM), IEEE International Conference on Mechatronics and Automation 'ICMA2008', Takamatsu-Kagawa, Japan, pp. 61-65. 10.1109/ICMA.2008.4798726

Howes, A., & Young, R. M. (1997). The role of cognitive architecture in modeling the user: Soar's learning mechanism. *Human-Computer Interaction*, *12*(4), 311-343. doi:10.120715327051hci1204_2

Infitechm Co. Ltd. (2017). Robot-type watching system. Available at http://infitechm.com/en/index.html

Kato, T. A., Shinfuku, N., Sartorius, N., & Kanba, S. (2017). *Loneliness and single-person households: issues of Kodoku-Shi and Hikikomori in Japan* (pp. 1-15). Mental Health and Illness in the City.

Khamassi, M., Lallée, S., Enel, P., Procyk, E., & Dominey, P. F. (2011). Robot cognitive control with a neurophysiologically inspired reinforcement learning model. *Frontiers in Neurorobotics*, *5*, 1. doi:10.3389/fnbot.2011.00001 PMID:21808619

Langley, P., Laird, J. E., & Rogers, S. (2009). Cognitive architectures: Research issues and challenges. *Cognitive Systems Research*, *10*(2), 141-160. doi:10.1016/j.cogsys.2006.07.004

Lee, M. D., & Vanpaemel, W. (2018). Determining informative priors for cognitive models. *Psychonomic Bulletin & Review*, *25*(1), 114-127. doi:10.375813423-017-1238-3 PMID:28194721

Lemaignan, S., Warnier, M., Sisbot, E. A., Clodic, A., & Alami, R. (2017). Artificial cognition for social human–robot interaction: An implementation. *Artificial Intelligence*, *247*, 45-69. doi:10.1016/j.artint.2016.07.002

Luhmann, N. (1993). *Communication and Social Order: Risk: A Sociological Theory*. Piscataway, NJ: Transaction Publishers.

Piaget, J. (1964). Part I: Cognitive development in children: Piaget development and learning. *Journal of Research in Science Teaching*, *2*(3), 176-186. doi:10.1002/tea.3660020306

Quan, W., Toda, Y., Woo, J., & Kubota, N. (2018). Integrated Robotic Control System for Public Nursing. World Automation Congress. *16th International Symposium on Robotics and Applications (ISORA 2018)*, Washington, DC. 10.23919/WAC.2018.8430398

Quan, W., Woo, J., Toda, Y., & Kubota, N. (2019). Human Posture Recognition for Estimation of Human Body Condition. [JACIII]. *Journal of Advanced Computational Intelligence and Intelligent Informatics*, *23*(3), 519-527. doi:10.20965/jaciii.2019.p0519

Renteria-Bilbao, A. (2019). Ethical Concerns when Robots Assist Eldery People. *Dilemata*, *30*, 43-49.

ROBOTIS. (2017). DYNAMIXEL. Retrieved from http://www.robotis.us/dynamixel/

Salter, T., Dautenhahn, K., & te Boekhorst, R. (2006). Learning about natural human-robot interaction styles. *Robotics and Autonomous Systems*, *54*(2), 127-134. doi:10.1016/j.robot.2005.09.022

Seidl, D. (2004). *Luhmann's theory of autopoietic social systems* (pp. 36-37). Ludwig-Maximilians-Universität München-Munich School of Management.

Sheridan, T. B. (2016). Human–robot interaction: Status and challenges. *Human Factors*, *58*(4), 525-532. doi:10.1177/0018720816644364 PMID:27098262

So, W.-C., Wong, M. K.-Y., Lam, C. K.-Y., Lam, W.-Y., Chui, A. T.-F., Lee, T.-L., ... Fok, D. C.-W. (2018). Using a social robot to teach gestural recognition and production in children with autism spectrum disorders. *Disability and Rehabilitation. Assistive Technology*, *13*(6), 527-539. doi:10.1080/17483107.2017.1344886 PMID:28673117

Szeles, J., Kubota, N., & Woo, J. (2017). Weather Forecast Support System Implemented into Robot Partner for Supporting Elderly People Using Fuzzy Logic. *Joint 17th World Congress of International Fuzzy Systems Association and 9th International Conference on Soft Computing and Intelligent Systems (IFSA-SCIS 2017), Otsu, Japan.* 10.1109/IFSA-SCIS.2017.8023309

Tanaka, R., Woo, J., & Kubota, N. (2019). Nonverbal Communication based on Instructed Learning for Socially Embedded Robot Partners. [JACIII]. *Journal of Advanced Computational Intelligence and Intelligent Informatics*, *23*(3), 584-591. doi:10.20965/jaciii.2019.p0584

Tang, D., Yusuf, B., Botzheim, J., Kubota, N., & Chan, C. S. (2015). A novel multimodal communication framework using robot partner for aging population. *Expert Systems with Applications, 42*(9), 4540–4555. doi:10.1016/j.eswa.2015.01.016

Umetsu, K., Woo, J., & Kubota, N. (2019). Audience Attention in Robot Interactive Theater. In *Proceedings of the 2019 JSME Conference on Robotics and Mechatronics (ROBOMECH2019)*, Hiroshima, Japan.

Woo, J., János, B., & Kubota, N. (2017). A modular cognitive model of socially embedded robot partners for information support. *ROBOMECH Journal 4*(1), 10.

Woo, J., Botzheim, J., & Kubota, N. (2014). Facial and Gestural Expression Generation for Robot Partners. In *25th 2014 International Symposium on Micro-NanoMechatronics and Human Science (MHS2014)*, Nagoya, Japan, 218-223.

Woo, J., Botzheim, J., & Kubota, N. (2017). Emotional empathy model for robot partners using recurrent spiking neural network model with Hebbian-LMS learning. *Malaysian Journal of Computer Science, 30*(4), 258-285. doi:10.22452/mjcs.vol30no4.1

Woo, J., Botzheim, J., & Kubota, N. (2018). System Integration for Cognitive Model of a Robot Partner. [AUTOSOFT]. *Intelligent Automation & Soft Computing, 24*(4), 829-842.

Woo, J., Botzheim, J., & Kubota, N. (2018). A Socially Interactive Robot Partner Using Content-based Conversation System for Information Support. [JACIII]. *Journal of Advanced Computational Intelligence and Intelligent Informatics, 22*(6), 989-997. doi:10.20965/jaciii.2018.p0989

Woo, J., Botzheim, J., & Kubota, N. (2015). Verbal conversation system for a socially embedded robot partner using emotional model. In *Proceedings of the 24th IEEE International Symposium on Robot and Human Interactive Communication (RO-MAN2015)*, Kobe, Japan, pp. 37-42. 10.1109/RO-MAN.2015.7333685

Woo, J., & Kubota, N. (2017). Interaction Content Design for Information Support based on Robot Partner, The 10th International Conference on Human System Interactions (HSI 2017), Ulsan, Republic of Korea, 35-40.

Woo, J., & Kubota, N. (2017). Interaction Content Design for Information Support based on Robot Partner, *The 10th International Conference on Human System Interactions (HSI 2017), Ulsan, Republic of Korea,* pp. 35-40. 10.1109/HSI.2017.8005019

Woo, J., Shimazaki, J., Matsuo, Y., Masuta, H., Hun-ok, L., & Kubota, N. (2013). Communication based on Frankl's Psychology for Humanoid Robot Partners using Emotional Model. In *2013 IEEE International Conference on Fuzzy Systems (FUZZ-IEEE 2013),* Hyderabad, India.

Woo, J., Tanaka, R., & Kubota, N. (2017). *A Robot Partner Interaction System based on Persona with Modular Cognitive Model.* The 18th International Symposium on Advanced Intelligent Systems, Daegu, Republic of Korea, pp. 51-58.

Woo, J., Wada, K., & Kubota, N. (2012). Robot Partner System for elderly people care by using sensor network. In *4th IEEE RAS & EMBS International Conference on Biomedical Robotics and Biomechatronics (BioRob2012),* Rome, Italy, 1329-1334.

Woo, J., Wada, K., & Kubota, N. (2012). Robot Partner System for elderly people care by using sensor network. In *Proceedings of 4th IEEE RAS & EMBS International Conference on Biomedical Robotics and Biomechatronics (BioRob2012),* Rome, Italy, pp. 1329-1334. 10.1109/BioRob.2012.6290933

Woo, J., Yamamoto, K., & Kubota, N. (2018). Interaction Design Using Multi-Robot System for Information Support. *2018 International Conference on Fuzzy Theory and Its Applications (iFuzzy 2018),* Daegu, Korea.

Chapter 12
Incorporation of Human Facial Expression Into Robot Control

Naoya Hasegawa
Kanagawa Institute of Technology, Japan

Yoshihiko Takahashi
Kanagawa Institute of Technology, Japan

ABSTRACT

An amusement robot that can recognize and respond to human facial expressions is proposed. The integration of human facial recognition into the control systems of robots is necessary to improve their entertainment value. We are proposing a new technology that uses a camera to read and respond to human facial expressions. As the preliminary research, a portable bubble ejection robot, which produces soap bubbles like a fountain, has been fabricated. The ejected soap bubbles are illuminated by LEDs and the direction of the ejection nozzle and the amount of the ejected bubbles are controlled by using an Android tablet. Subsequently, human facial expressions are read by a facial expression detection system. This paper will explain the design of the bubble ejection robot, its control system, which reads human facial expressions, and the experimental results of human reaction to the bubble ejection. Consequently, the relationship between the soap bubble ejection and the human emotions is revealed.

INTRODUCTION

We propose a technology in which human facial expressions are incorporated into the control systems of amusement robots. The purpose is to create fun amusement events that every participant will enjoy. Therefore, it is important to utilize human facial expressions when designing a robot control system to develop intelligent amusement systems.

The degree of entertainment value and the success of the amusement events are commonly evaluated by the number of attendees, the return rate of customers, the result of questionnaires, and so on. As gathering and evaluating the enormous amount of event data is a problem. We are proposing a new

DOI: 10.4018/978-1-7998-0137-5.ch012

technology using human facial expressions, which are instantly read and analyzed by a camera. This new technology can, thereafter, be utilized with more traditional methods to provide better design schemes for the entertainment industries.

As an application example, a simple face robot that helps elderly people with their medications was developed (Takahashi, Goto, & Shigemoto, 2006) and its expressions were evaluated. The robot schedules their medication, reminding them with an angry face if the elderly person forgets to take them.

It is important that human facial expressions can be accurately synthesized and interpreted. Human emotions have been detected by using a galvanic skin response (GSR) sensor, and the characteristics have been studied in many reports (e.g. Darrow & Gullickson, 1970). Although this equipment can detect sudden changes of emotions, it is unable to classify or identify them. Such ability has become necessary to improve interactions between humans and robots.

Soap bubbles are loved by everyone; from the young to the elderly (e.g. Yokota, Sato, & Fujimori, 1963). In this work, therefore, soap bubble machines are selected to analyze human emotions. Our findings may mark an important step in the field of human-robot interaction.

We are proposing a portable bubble ejection robot (Hasegawa & Takahashi, 2017; Hasegawa & Takahashi, 2019) which produces soap bubbles illuminated by LEDs. The direction of the ejection nozzle and the amount of ejected bubbles are controlled by using an Android tablet. A detection system of human facial expressions was installed into the control system, which is able to actuate the robot depending on human emotional reactions. Field experiments were, afterward, conducted using two female subjects. From now on, we refer to the detection system of human facial expressions as a face sensor.

This paper will explain the conceptual, mechanical, and control system design in which the face sensor is incorporated as well as the analysis of the experimental results. The experimental results revealed the relationship between soap bubble ejection and the human emotions.

BACKGROUND

At various exhibitions, amusement robots are able to interact with humans. Questionnaires and interviews are used to evaluate how a robot performs and how humans feel (Odashima, Hata, & Goto, 2014). There are examples of human facial expression studies (Kitamura, Takemura, Iwai, & Sato, 2014) or human reactions to robot facial expressions (Takahashi, Goto, & Shigemoto, 2006) employing this strategy. However, the long periods of time required for feedback make short-term or real-time feedback.

On the other hand, GSR is often used as a method to measure human emotion; several basic research examples exist from a long time ago (Darrow & Gullickson, 1970; Yokota, Takahashi, Kondo, & Fujimori, 1959; Fujimori & Yokota, 1962; Iwama & Abe, 1953; Umehara & Sekido, 1976; Yokota, Sato, & Fujimori, 1963). Furthermore, GSR is applied in various fields, such as music (Zimny & Weidenfeller, 1963), medicine (Lorens, Jr., & Darrow, 1962; Volavka, Matousek, & Roubicek, 1967), education (Watanabe & Amano, 1994; Inoshita, Ogata, Tokunaga, Bando, Yamada, & Marumoto, 1993; Kai, 1962), and robotic nursing (Takahashi, Hasegawa, Takahashi, & Hatakeyama, 2001; Takahashi, Hasegawa, Takahashi, & Hatakeyama, 2002; Shirai, Takahashi, 2018). Although GSR can detect if a human is upset, it cannot measure emotions such as joyfulness or sadness.

There are many research examples (Takahashi, Manabe, Takahashi, & Hatakeyama, 2002) on human measurement using an EMG (electromyography). However, to the best of our knowledge, there have been no studies applied the EMG to the evaluation of human emotions.

As many children are attracted to soap bubbles, many studies have tried to comprehend the emotions of children by using soap bubbles (Yamada, 2017; Kato, 2013; Matsumura & Shiono, 2016; Yabe, 2012). In addition, because many people are interested in soap bubbles, there are also events using soap bubbles (Nakamura, Inaba, Tamaoki, Hoshino, & Shiratori, 2006; Hasegawa & Takahashi, 2017).

In this research, we read human emotions from facial expressions and perform control that incorporates data into a robot controller. By doing this, we aim to achieve better communication between people and robots.

Bubble Ejection Robot as Amusement System

When managing an exhibition, the presentation time and the ease of installation and removal are the major issues to consider. Although fountains can be directed for a long time, the scale of the equipment is large, the installation and remove time are long, and the cost is high. In an exhibition that shoots balloons and pigeons, time is spent to collect and clean them. Although rice or flower showers are very easy to remove, the presentation time is very short and people are reluctant to treat food and plants as expendable for exhibitions.

For children, we are developing portable soap bubble ejection robots that can be easily utilized at exhibitions. To do that, the bubble ejection robot has a built-in battery, allowing long time presentations to be conducted, and soap liquid; it can also be used with a power supply to remove the restrictions on usage time. To make the children delighted, the ejection amount is controlled and the soap bubbles are irradiated with LED lights to color them, therefore, allowing the solution to work even in indoor environments where low ambient light intensity conditions exists, making it more effective to color soap bubbles with LEDs.

Figure 1 shows an example of a bubble ejection robot at a children's exhibition. A great number of bubble ejection robots are synchronized and eject soap bubbles at various angles. By observing the photographs, it can be concluded that (A) quiet children and (B) active children are present. Quiet children do not approach the robot even though they are attracted to soap bubbles. On the other hand, active children move closer to the robot, the source of soap bubbles, and interact with it. The purpose of our soap bubble ejection robot is to protect the equipment from group B while pleasing group A. Therefore, the robot looks for a quiet child and only ejects soap bubbles for them.

To perform such control, the robot searches for a face expressing Indifference, evaluates it with a value, and ejects a soap bubble according it.

Experimental Setup

Experiments were conducted incorporating human facial expressions into robot control system while humans' reaction to bubble ejection was investigated. The robot used had undergone various changes in development, such as the insertion of a face sensor (OMRON: HVC-P2).

The control system of the bubble ejection robot is shown in Figure 2, where (a) is the control command transfer system, and (b) is the mechanism control system. Meanwhile, Figure 3 depicts the movement (a) and mechanism (b) of the bubble ejection robot. A tray, located bellow the robot, contains soap bubble liquid. A soap bubble ejection frame, which blows air with the liquid by using a fan, generates bubbles and their ejection velocity as well as wind speed are measured.

Figure 1. Children's exhibition using soap bubble ejection robots. The children of (A) in the photograph are quiet. Meanwhile, the children of (B) are too energetic and interact with the robot. The robot stops the soap bubble ejection and calms the children. The children's emotions are read using a facial expression sensor, and the soap bubbles are ejected based on the level of emotions. The robot searches for differences among multiple faces and ejects soap bubbles based on the difference.

Figure 2. Control command transfer system and mechanism control system of bubble ejection robot. (a) shows a system for transmitting control commands. The control commands are sent from an Android tablet to the robot using IEEE802.15.4; 128 robots are controlled at one time. (b) shows a mechanism control system of the robot. The horizontal and vertical movements of a bubble ejection nozzle, coloring the bubble using LEDs, the fan rotation for bubble ejection, and other such parameters are controlled.

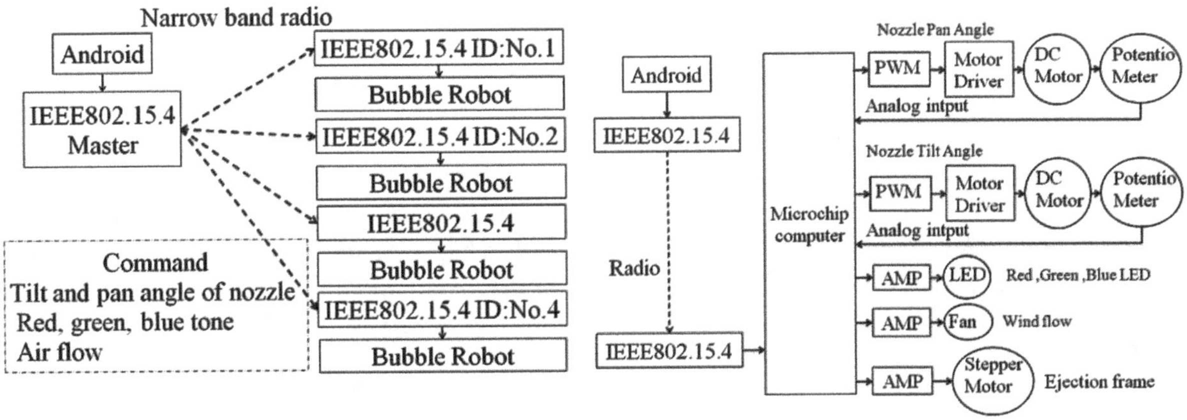

An operator commands the robots from the screen of an Android tablet which is able to communicate with up to 128 robots simultaneously using IEEE802 15.4. Brightness and colors of the LEDs, direction of the ejection nozzle, wind speed of the fan, and rotation speed of the bubble ejection frame are variables that can be controlled by the operator. The direction of the nozzle is also determined by the feedback control system, with the aid of a DC motor and a potentiometer, while the other variables are set by an open loop control system.

To quantify human emotions, five values, corresponding to the intensity of each emotion (Surprise, Joy, Anger, Sorrow, and Indifference) ranged from 0% to 100%, are outputted. As the main objective of this work is to employ robots to cheer up quiet children, the Indifference value is the most critical.

The fabricated control system using a microcomputer detects the data from the face sensor, and controls the bubble ejection robot. Figure 3(b) shows the photograph of the bubble ejection robot and the face sensor, which may be installed on top of the robot. Figure 4 shows an example of the bubble ejection operation, proving that the equipment is able to eject soap bubbles in any direction.

EXPERIMENTAL RESULTS

In this section, the experiments on human emotional reaction to the bubble ejection phenomena are analyzed. The subjects were two females in their 20s and 30s. The bubble ejection robot was placed in front of the subjects at a live event while the face sensor was installed on or in front of the robot. Furthermore, bubbles were ejected manually by an operator.

The experimental results are shown in Figs. 5 and 6, which plot the level of human emotion and the fan voltage based on the experiment time. A high voltage corresponds to high fan speed. The time periods that the sensor was not able to identify the subject's face is also indicated by the solid vertical lines.

Figure 5 shows the results of the female subject in her 20s, and the following observations could be made based on the time periods. In approximately:

1. 15 s to 28 s, 36 s to 44 s, 53 s to 61 s, 65 s to 73 s, the fan voltage (adjusted manually) rises and soap bubbles are ejected.

Figure 3. Movement and photograph of bubble ejection robot. The soap bubble generating nozzle is moved in the horizontal and vertical directions; a tray containing the soap liquid is placed on the lower side, and a bubble ejection frame scoops the soap liquid. Wind is blown from the fan to the ejection frame and the soap bubbles are ejected. The bubble ejection robot has a plastic cylindrical cover attached. The face sensor that reads the facial expressions can be attached to the robot body or may be installed away from the robot. The LEDs are arranged in a ring around the nozzle, and the bubbles are colored.

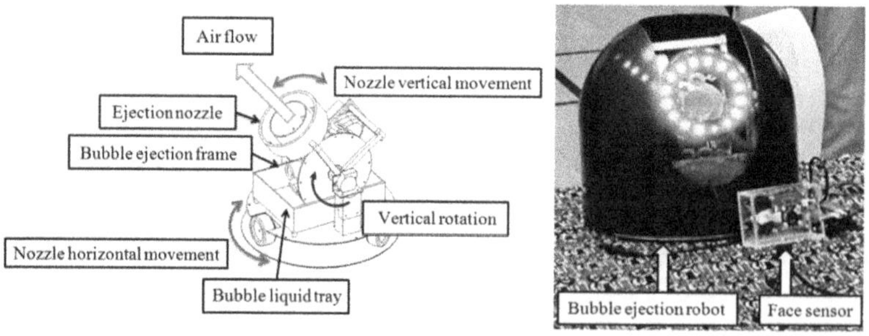

Figure 4. Bubble ejection. Bubbles are ejected in any direction between 0 and 90 degrees. The ejection speed of bubbles is also changed.

2. 0 s to 16 s, the subject expressed the emotions of Indifference and Surprise before the robot produced bubbles.
3. 16 s to 69 s, the Joy values were high, as the subject was delighted owing to the soap bubbles ejection.
4. 20 s, 35 s, 42 s, the subject expressed Indifference.
5. 20 s, 40 s, the subject expressed Surprise.

From the results, it is clear that the human emotions could be correctly detected and the expected result that bubble ejection increases subject joyfulness could be observed.

Figure 6 indicates the results of the female subject in her 30s. At approximately:

Figure 5. Experimental results of bubble ejection for a woman in her 20s. The horizontal axis denotes time and the vertical axis denotes the emotion level. The five emotions of Indifference, Surprise, Sorrow, Joy, and Anger, and the voltage of the fan were measured. In the figure, the vertical lines of the two pairs represent the duration when the face cannot be measured as the subject was not looking directly into the camera. The numbers above the two lines represent the time intervals that could not be measured. Before ejecting the soap bubbles, the Indifference and Surprise values are high; however, after the soap bubbles are ejected, the Joy value is high. These results indicate that human emotions can be measured; and that they are changed by soap bubbles.

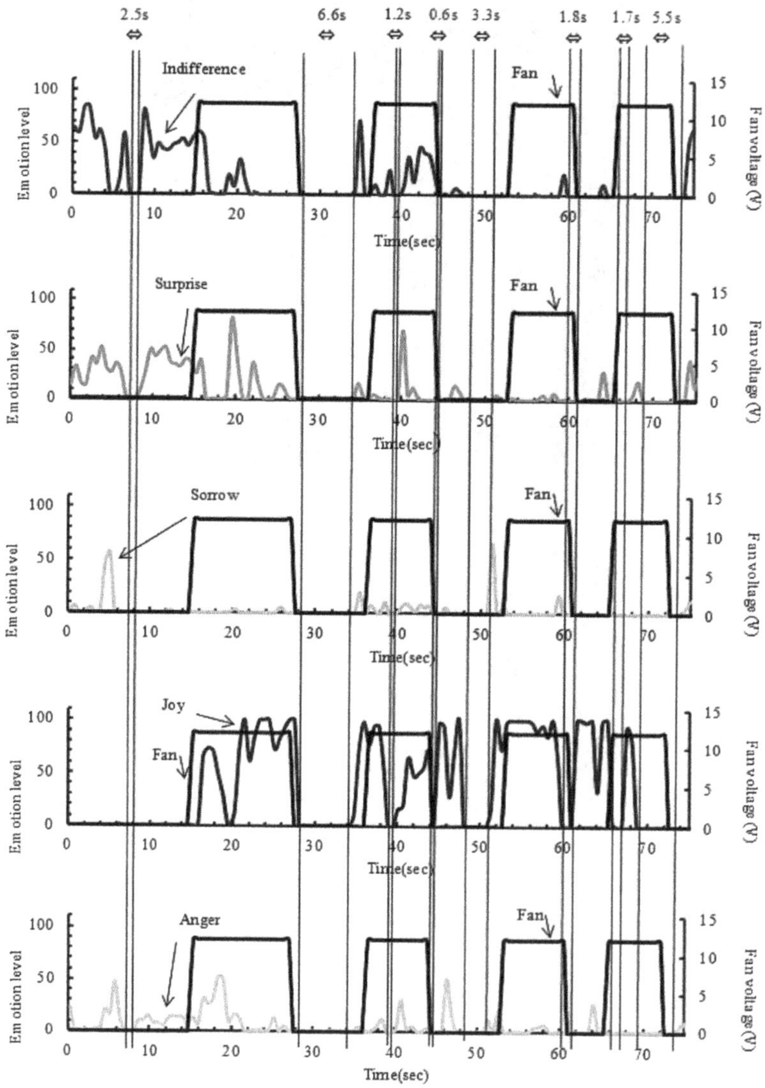

1. 9 s to 20 s, 26 s to 38 s, 42 s to 51 s, bubbles were ejected.
2. 0 s to 14 s, the subject expressed Sorrow and Anger before the robot produced bubbles. In the case of this subject, Sadness also could be observed.
3. 16 s to 50 s, the Joy values were high, indicating delight because soap bubbles were ejected.

Figure 6. Experimental results of bubble ejection with a woman in her 30s. The horizontal axis denotes time and the vertical axis denotes the emotion level. The five emotions, Indifference, Surprise, Sorrow, Joy, and Anger and the voltage of the fan were measured. Before ejecting the soap bubbles, the value Sorrow value is high; however, after the soap bubbles are ejected, the Joy value is high. These results indicate that human emotions can be measured; and that they are changed by soap bubbles.

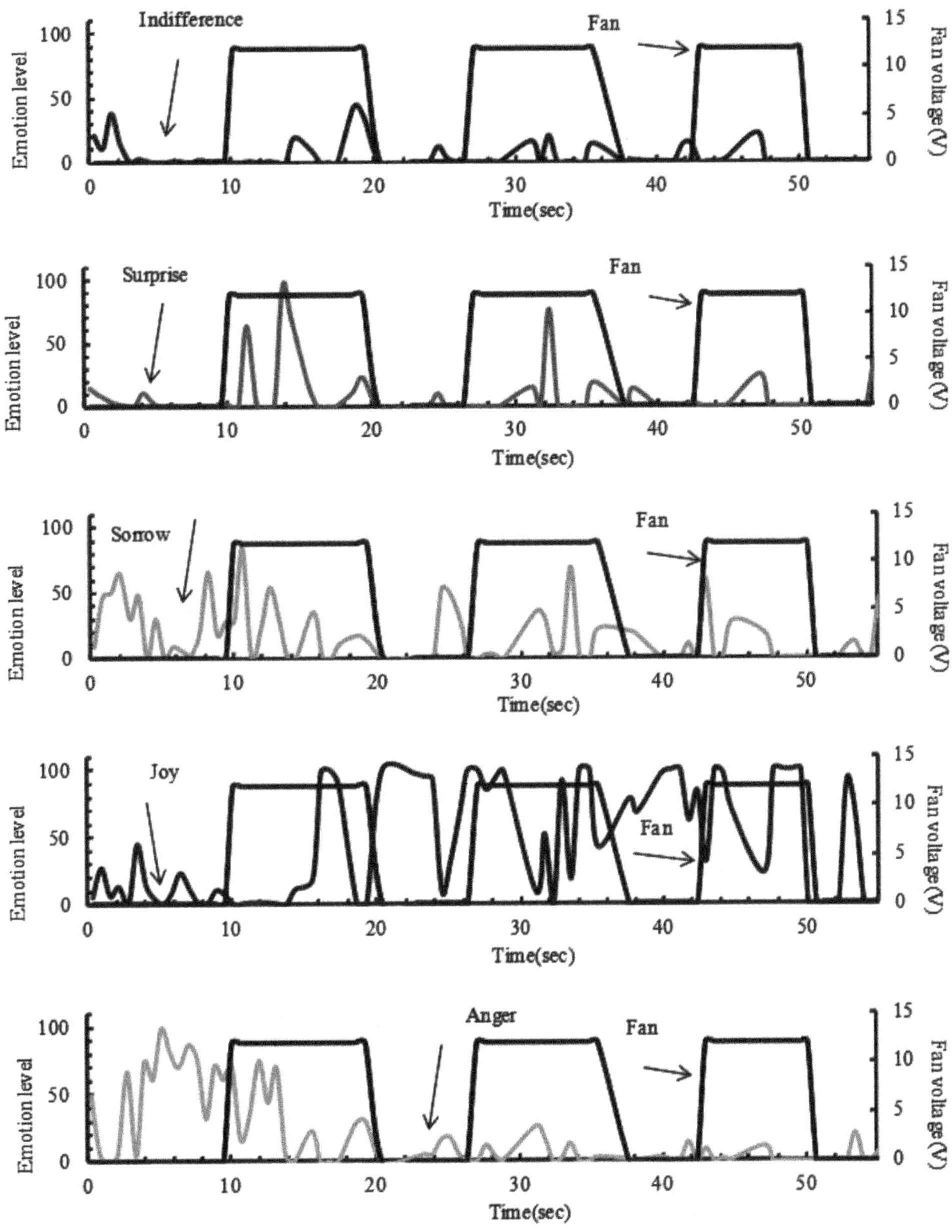

4. 12 s, 14 s, 32 s, the subject expressed Surprise.
5. 11 s, 13 s, 23 s, 34s, 43 s, the subject expressed Sorrow.

These results indicate that the experimental conclusion for the subject in her 20s could also be replicated for the subject in her 30s.

In the present work, bubbles were manually ejected by an operator; however, in future research, real time automatic response to human emotions is desirable.

FUTURE TRENDS

It was found that, by using face measurements, it is possible to quantify human emotions, and identify their change when soap bubbles are ejected. In future works, robot control based on human emotion data obtained by face sensor readings will be studied.

Experiments will first be performed with one subject. A study will be conducted on what emotion data should be used and how to eject soap bubbles based on it. The Indifference value will be first evaluated, and soap bubbles will be ejected if it reaches a pre-determined maximum value.

In children's events, it is possible to apply the experiments with multiple subjects. To identify children with high Indifference values a horizontal swing movement of the robot, with the face sensor attached to it, is required. After this identification, the ejection control of the soap bubbles needs to follow the same procedure described in the one subject case.

Figure 7 shows the program flowchart applied for finding a face that expresses Indifference and ejecting bubbles according to emotions. A program that searches for faces which exhibit high Indifference value and ejects bubbles according to the emotion level has been, therefore, developed. The program procedure is as follows.

1. Determine if the subjects are on the camera attached to the robot.
2. If the camera does not show the subject, swing the robot until only one subject is found.
3. Determine if the subject is one.
4. If there is only one subject, eject soap bubbles according to the Indifference value.
5. When there are two or more subjects.
6. Compare the Indifference values.
7. Direct the robot to the person with high Indifference value, and proceed the same procedure.

Figure 8 depicts the algorithm for ejecting bubbles in accordance with the level of Indifference. The program procedure is as follows.

1. A face sensor captures a subject face as explained in Figure 7.
2. Ejects soap bubbles according to the Indifference value.
3. Soap bubbles are ejected, and subject expression changes to another emotion such as Surprise or Joy, and the Indifference value decreases.
4. Decreases the fan output when the Indifference value decreases.
5. If the Indifference value rises again, the bubbles are ejected according to (2).

Figure 7. Program flowchart for finding the face of Indifference and ejecting bubbles according to the emotion level. The program finds a face showing Indifference from among the faces of multiple people, and then, it ejects bubbles according to the emotion level.

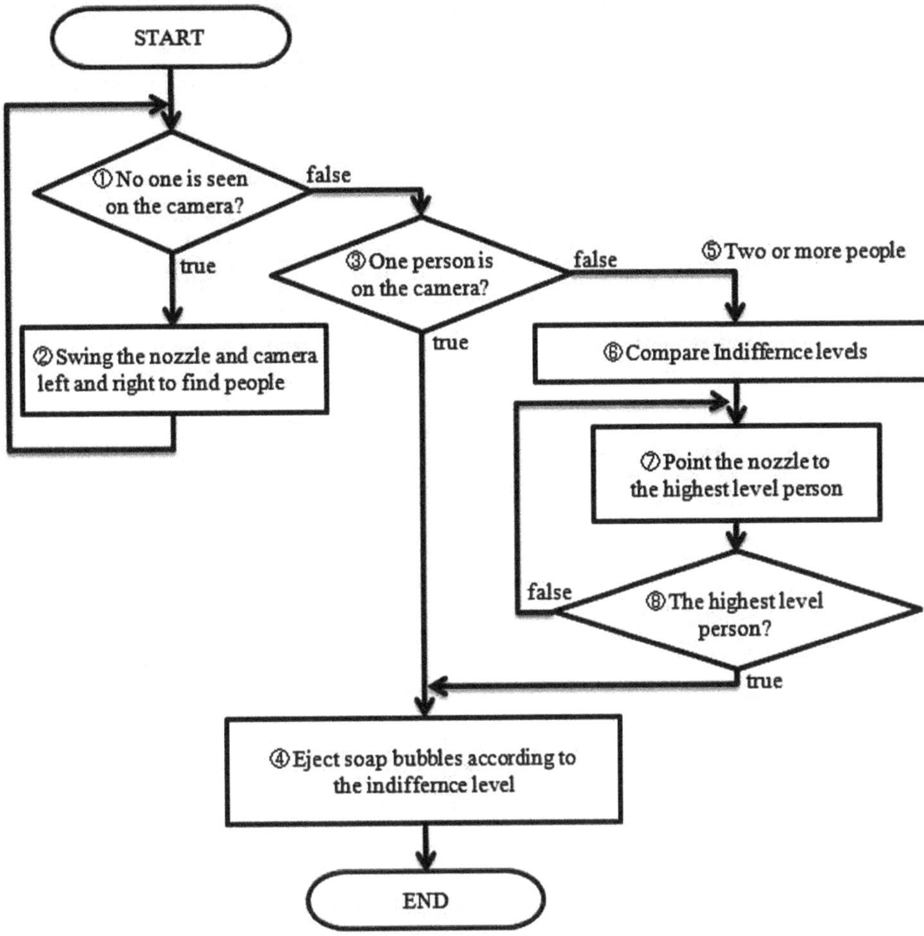

CONCLUSION

A bubble ejection robot with a face sensor was built to investigate the interaction between humans and robots. The bubble ejection robot is able to produce soap bubbles in any direction depending on the control system signal, which is based on the face sensor data. The face sensor is able to detect the human emotions of Surprise, Sorrow, Joy, Anger, and Indifference. Experiments on the interaction between a human and the bubble ejection robot were conducted; the results indicated that it was possible to read human emotion and that bubble ejection increased the Joy value of the subject.

Bubbles were ejected manually by a robot operator; however, real time automatic bubble ejection will be conducted in future research. Moreover, the face sensor was sometimes unable to read human faces if the subject looked away from the robot. Face detection techniques, such as multiple sensors, will also be the subject of future research to avoid this problem.

Figure 8. Illustration of algorithm for ejecting bubbles in accordance with the level of Indifference. The horizontal axis denotes time, and the vertical axis denotes the emotion level. (1) A face sensor captures the face of the subject. (2) If it is the face showing Indifference, soap bubbles are ejected based on the emotion level. (3) Soap bubbles are ejected, and the subject expression will change to Surprise or Joy, and the Indifference value decreases. (4) The fan output decreases when the value of Indifference decreases. (5) If the Indifference value rises again, the bubbles are ejected based on the level of the expression.

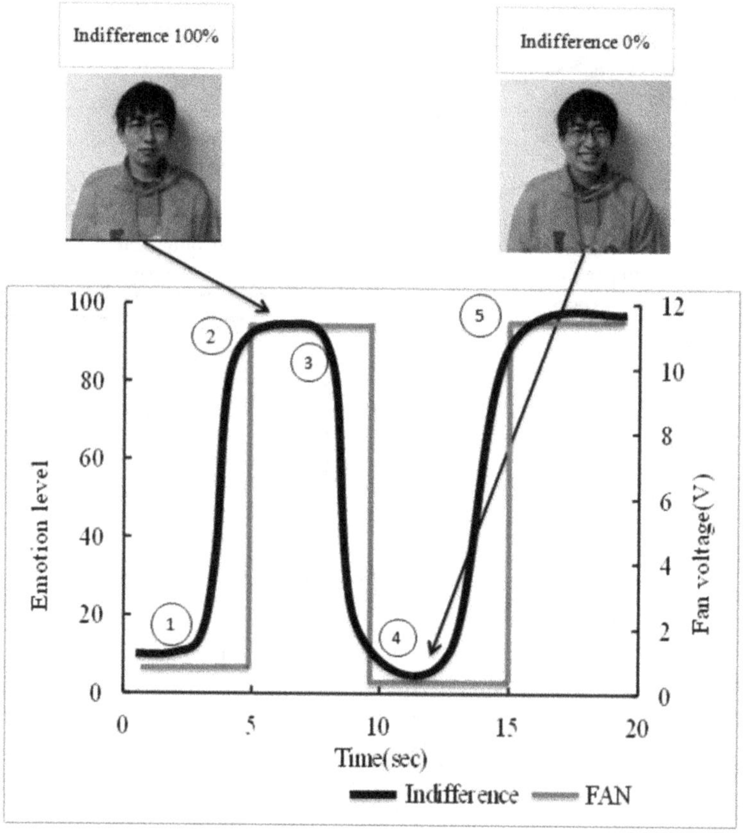

This is the invited paper from the 20th International Conference on Research and Education in Mechatronics (REM 2019).

REFERENCES

Darrow, C. W., & Gullickson, G. R. (1970). The Peripheral mechanism of the galvanic skin response. *Psychophysiology*, *6*(5), 597–600. doi:10.1111/j.1469-8986.1970.tb02248.x PMID:5485543

Fujimori, B., & Yokota, T. (1962). Central and peripheral mechanisms of galvanic skin reflex. *Acta Neurovegetativa*, *24*(1-4), 241–249. doi:10.1007/BF01233000

Hasegawa, N., & Takahashi, Y. (2017). Preliminary Research into Control System using Quantified Human Emotions. In *Proceedings of International Conference on Control, Automation and Systems*, pp. 1519-1524. 10.23919/ICCAS.2017.8204230

Hasegawa, N., & Takahashi, Y. (2019). How recognition of human facial expression can be incorporated in robot control. *International Conference on Research and Education in Mechatronics.*

Inoshita, H., Ogata, M., Tokunaga, A., Bando, M., Yamada, Y., & Marumoto, K. (1993). Studies on Galvanic skin response (GSR) in relation to psychological condition test for the students. *Bull. Sch. Health Sci. Okayama Univ.*, *4*, 99–103, (in Japanese).

Iwama, K., & Abe, M. (1953). Conditioned galvanic skin reflex and electroencephalogram. *The Tohoku Journal of Experimental Medicine*, *57*(4), 327–335. doi:10.1620/tjem.57.327 PMID:13090218

Kai, N. (1962). An Experimental Study of Emotional Stability in Galvanic Skin Reflection (GSR), Memoirs of the Faculty of Education, Kumamoto University, *Humanistic science, 10,* pp.109-121 (in Japanese).

Kato, N. (2013). Incipient Scientific Thinking in Five-year-old Children's Activities: Through Sandbox Play, Soap Bubble Play, Playing in the Water Color and Mud Ball Play, *Cross-cultural business and cultural studies, 18*(1), pp. 17-28 (in Japanese).

Kitamura, K., Takemura, N., Iwai, Y., & Sato, K. (2014). Detection of Unconscious Facial Reactions to Uncomfortable Illumination, IEEJ Transactions on Electronics. *Information Systems*, *134*(2), 218–224 (in Japanese).

Lorens, S. A. Jr, & Darrow, C. W. (1962). Eye movements, EEG, GSR and EKG during mental multiplication. *Electroencephalography and Clinical Neurophysiology*, *14*(5), 739–746. doi:10.1016/0013-4694(62)90088-3

Matsumura, K., & Shiono, M. (2016). Interference Experiments of Soap-Bubble Films Using a Flashlight as a Light Source and Its Applications to Teaching Material of Science Educations, Studies in human sciences, Seinan gakuin Univ, 12(1), pp. 83-100 (in Japanese).

Nakamura, M., Inaba, G., Tamaoki, J., Hoshino, J., & Shiratori, K. (2006). Mounting and Application of Soap Bubble Display Method. *The Virtual Reality Society of Japan*, *11*(2), 339–346 (in Japanese).

Odashima, K., Hata, A., & Goto, Y. (2014). How to satisfactorily enjoy an amusement park, Kokyuroku, Research Institute for Mathematical Sciences. *Kyoto University*, *1887*, 28–35 (in Japanese).

Shirai, S., & Takahashi, Y. (2018). Personal Mobility Vehicle User's Psychological Reaction to Surrounding People. In *Proceedings of International Conference on Control, Automation and Systems*, pp. 1730-1735.

Takahashi, Y., Goto, H., & Shigemoto, H. (2006). Facial expressions of simplified robot face for health care system. *Transactions of the Institute of Measurement and Control*, *28*(2), 163–176. doi:10.1191/0142331206tm144oa

Takahashi, Y., Hasegawa, H., Takahashi, K., & Hatakeyama, T. (2001). Human Interface Using PC Display with Head pointing Device for Eating Assist Robot and Emotional Evaluation by GSR Sensor. In *Proceedings of International Conference on Robotics and Automation*, pp. 3674-3679. 10.1109/ROBOT.2001.933189

Takahashi, Y., Hasegawa, H., Takahashi, K., & Hatakeyama, T. (2002). Experimental Evaluations of Human Interface Using Head Space Pointer for Eating Assist Robot and Human's Emotional Response to Robot Movement, *Journal of Japan Society for Design Engineering*, *37*(3), 122–128 (in Japanese).

Takahashi, Y., Manabe, G., Takahashi, K., & Hatakeyama, T. (2002). Experiments on Burden Measurement of Human and Transfer Aid System for Developing Self-Transfer Aid System. *The transactions of the Institute of Electrical Engineers of Japan. C, 68*(674), pp. 2977-2983 (in Japanese).

Umehara, H., & Sekido, T. (1976). Measurement of Human Emotion by GSR. *The SICE Journal of Control, Measurement, and System Integration*, *15*(9), 753–757 (in Japanese).

Volavka, J., Matousek, M., & Roubicek, J. (1967). Mental arithmetic and eye opening. An EEG frequency analysis and GSR study. *Electroencephalography and Clinical Neurophysiology*, *22*(2), 174–176. doi:10.1016/0013-4694(67)90158-7 PMID:4163688

Watanabe, K., & Amano, A. (1994). A Study on Vital Reactions of School Children during the Class - Fluctuations of GSR, Heart Rate, and Subcutaneous Temperature -, *Japanese journal of school health, 36*(7), pp. 470-478 (in Japanese).

Yabe, T. (2012). Development Process of Musical Expression under Varying Environmental Configurations: Practical study of bubble play in four-year-old preschool class. *Journal of the Study of School Music Education Practice*, *16*, 49–56 (in Japanese).

Yamada, S. (2017). A Study on Environmental Setting to Expression Activity (1): The Case of Two Soap Bubble Arts, Shukutoku University Junior College bulletin, 57, pp. 73-87 (in Japanese).

Yokota, T., Sato, A., & Fujimori, B. (1963). Analysis of inhibitory influence of bulbar reticular formation upon sudomotor activity. *Japanese Journal of Physiology*, *15*(13), 145–154. doi:10.2170/jjphysiol.13.145 PMID:14002614

Yokota, T., Takahashi, T., Kondo, M., & Fujimori, B. (1959). Studies on the diphasic wave form of the galvanic skin reflex. *Electroencephalography and Clinical Neurophysiology*, *11*(4), 687–696. doi:10.1016/0013-4694(59)90109-9 PMID:13846558

Zimny, G. H., & Weidenfeller, E. W. (1963). Effects of Music upon GSR and Heart-Rate. *The American Journal of Psychology*, *76*(2), 311–314. doi:10.2307/1419170 PMID:14003728

Chapter 13
Design and Implementation of a Wireless Robot for Image Processing

Md. Kamaruzzaman
Aliah University, India

Rafiqul Haque
Aliah University, India

ABSTRACT

Robotics has developed into a solid discipline of study that incorporates the background, knowledge, and creativity of mechanical, electronics, electrical, computer, industrial, and manufacturing engineering. The versatility of a robot can be translated into increased productivity, improved product quality, and decreased production costs in numerous ways. The robot can be wired or wireless, and can have a controller device. Radio frequency control system is one of the most efficient controls for a robot. Other than control, image processing is one of the important parts of robotics. Moreover, it also requires hardware for interfacing with the system for radio frequency control. It is one of the key technologies which can be used in defense, safety, intelligent transport system, and efficient management traffic. In recent years, there has been an increased scope for image processing in robotics. The main goal of this work is to show a system that solves the practical problem of identification of real scenes. All the simulations have used MAT LAB software.

INTRODUCTION

Robotics is an interdisciplinary subject which drawing ideas and tools from mathematics, physics, engineering and computing. The field of robotics has its origin in science fiction. The word "robot" is based on the Czech word for slave, and was introduced into our culture in the early 1921 in a play by Czech playwright, novelist; and essayist Karel Capek about mechanical men that rebel against their human masters. Capek dreamt of a situation where bioprocess could create human-like machines,

DOI: 10.4018/978-1-7998-0137-5.ch013

devoid of emotions and souls, who were strong, obeyed their masters, and could be produced quickly and cheaply. Czech word robota means "forced laborer" or "slave laborer". The word "robotics" was coined by the renowned science fiction writer, Issac Asimov, in 1942 science fiction story, "Runabout". A robot is a reprogrammable multifunctional manipulator designed to move material, parts, tools, or specialized devices through variable programmed motions for the performance of a variety of tasks. The versatility of a robot can be translated into increased productivity, improved product quality and decreased production costs in several ways. In the recent history we see a close relationship between the state of industry, the revolution in numeric and computer control of machinery, space exploration of imagination of creative people. A robot consists of various elements such as Manipulator, Sensory Devices, Controller and Power Conversion Unit. Actually, the robotic technology deals with the design, manufacture, operation and application of robots. The robotic technology uses computer system control, sensory feedback, actuator action and manipulation to execute the whole operation. The robotic technology deals with automatic technology that can be instructed by an operator to perform critical task in hazardous environment. Also, the robotic system created a new cluster of robotics and flexible robotics (Habib et al., 2007; "How to Make", 2014). The mechanical system technique was introduced through the manufacturing era as there are various practical applications in industry such as automated machines, mobile-control robotic system and long distance remote-control. Now a day's Image processing is one of the emerging fields of robotics. Advances in vision technology for robotics are expected to broaden the capabilities of robotic vision systems to allow for vision-based guidance of the robot arm, complex inspection for close dimensional tolerances and improved recognition and part location capabilities. The typical vision system consists of the camera and digitizing hardware, a digital computer and hardware & software necessary to interface them. This interface hardware and software is often referred to as a preprocessor. In the introduced work background subtraction technique is used and image processing is done by MATLAB software and robotic car is controlled by radio frequency. The software is used for real-time computer vision. A 14-16 mega pixel camera is placed over the robotic car to process image using MATLAB and battery-operated robot car is controlled by RF. To use the system wirelessly, all the applications of image processing are wrapped in robots. The robots can take their self-decision in any industry or else, whether man or machine works properly or not, sending information to supervisor by using digital camera placed at the top of the robotic car for image processing (Viola et al., 2004). The system has many applications in pattern recognition and machine vision for a complex security system to common areas and for parking admission to urban traffic control.

Image sensing functions involve the input of vision data by means of a vision camera focused on the scene of interest. The image consists of relative light intensities corresponding to the various portions of the scene. These light intensities are continuous analog values which must be sampled and converted into digital form. Special lighting techniques are needed to obtain an image of sufficient contrast. Various lighting techniques are there for good illumination of the scene. The image viewed by the camera is typically digitized by analog to digital converter (ADC) and stored in computer memory. The digital image is called a frame of vision data & is frequently captured by a hardware device called frame grabber and these devices are capable of digitizing images at the rate of 30 frames/second. ADC is either a part of a digital video camera or the front end of frame grabber. The choice is dependent on the type of hardware of the system. The digitized image matrix for each frame is stored and then subjected to image processing and analysis functions for data reduction and interpretation of the image. These steps are required in order to permit the real time application of vision analysis required in robotic application. Typically, an image frame will be thresholded to produce a binary image and then various feature measurements will further reduce the data representation of the image. This data reduction can change the representation of a frame from several hundred thousand bytes of raw image data to several hundred bytes of feature value data. The resultant feature data can be analyzed for action by the robot system. The vision system frequently must be trained to accomplish image processing and analysis. In training information is obtained on prototype objects and stored as models. The current applications of machine vision in robotics include inspection, part identification, location, and orientation. Vision systems can be classified in a number of ways. One classification is whether the system deals with a 2D or 3D model of the scene. Another way of classifying vision systems is according to the number of gray levels (light intensity levels) used to characterize the image. In a binary image the gray level values are divided into either two categories, black or white. Other systems permit the classification of each pixel's gray level into various levels, the range of which is called a gray scale.

In charge coupled devices (CCD) the image is estimated by a video camera on to CCD which senses the supplies and reads out the stored charge which is produced by the light on each part of the image. It is essential to change the analog signal for each pixel into digital signal. Analog to digital converter consist of three steps: Sampling, Quantizing and Encoding. λ is the central length of the lens because of the camera is in focus for separate objects. The assumption is that camera co-ordinate system is associated with the world coordinate system (X, Y, Z).

Let (X, Y, Z) be the world co-ordinate of any point in a 3D scene. Let that $Z > \lambda$, i.e. all position are in front of the lens. There is a relationship which is obtained, That provides the coordinates (x,y) of the projection of the position (X,Y,Z) on to the image plane.

By using similar triangles

$$x/\lambda = (-X)/(Z-\lambda) = X/(\lambda-Z)$$

and

$$y/\lambda = (-Y)/(Z-\lambda) = Y/(\lambda-Z)$$

Where the –ve sign indicates the points are actually inverted.

From the above two equations it can be written that

$$x=\lambda X/((\lambda-Z)\ y) = \lambda Y/(\lambda-Z).$$

By obtaining the camera constraints using recognized world co-ordinates is said to as Camera Calibration.

Wireless communication is one of the emerging fields in communication area. There are two important parts in wireless communication, transmitting and receiving. The transmitting part is responsible for transmitting the signal whereas the receiving part is responsible to receive the transmitted signal. In wireless communication, the signal is transmitted over a distance without the help of wires or any other electrical conductors. The main objective of wireless communication is to communicate between two or more devices by using wireless signal through wireless communication technology.

Today's scenario is that the robots are straight forwardly interacting with people, so the task is to find easier interface. Some of the robotic systems are used with user friendly interfaces that possess the ability to control the robotic system by some wireless way. The directed signals are generated from wireless technique by using the radio frequency switches. Then the directed signals are passed to the robotic system to navigate the specified directions which is needed for operation. The wireless robot utilizes the radio frequency module which contains transmitter and receiver to make a wireless monitoring system that could be used to operate an output from a remote place. Radio frequency element uses broadcasting frequency to send the signals for execution. The radio frequency signals transmit at a particular frequency and a baud rate. The receiving part is able to receive this signal when it is monitored for that particular frequency. Beside that, a four-channel, encoder/decoder pair has been used in the wireless robotic arrangement. The input signals have been taken via four switches at the transmitter side while the output signals have been controlled on a set of four LEDs analogous to each input switch. For scheming the remote equipment control system this circuit can be introduced. The receiver output signal can actuate correspondent relays which can be linked to any domestic appliance. The other important need of this system is that the thought system can be used for installing of robot system in industries. The robotic systems can save the manpower, improved the product quality, efficiency and capability to work in any critical conditions.

RESEARCH BACKGROUND

Since last few decades there are various research have been in the field of design and development of robotic system. So, video tracking, surveillance systems with robotic platforms are fields that have been well studied in the last few years. Now a day's surveillance monitoring and video tracking monitoring is dominant area in the field of robotics in which the incorporated sensors are stationary. If the object goes away from this range, the stationary systems need the desired object to stay within the observation range of the system. Design and control of by various means such as gesture control, Bluetooth control, pc control and mobile control methods are the emerging research area in the field of robotics since last few years. Since last few years, an extensive work has been done in the field of design and control of robot by various means like gesture control, Bluetooth control, pc control and mobile control methods. Many researchers also have been done in the field of environment monitoring where human and robot can work concurrently.

Some of the robotic systems have been used with the help of user-friendly interfaces which can control the robot by natural means. To make possible a practical explanation to this necessity, they have been developed a system through which the operator can give instruction to a wireless robot using gesture (Habib, 2008; Otsu et al., 1975; Wang, 2015). Most of the communication systems have been done through mobile which is very common in robotics. The wireless connectivity that makes the robotic system hassle free and the knowledge of monitoring a robot by using a mobile phone has highlighted (Patwari et al., 2010; Tse et al., 2005). The development of robotic system completes their communication by using transmitter and receiver the signals. Whole the controls have been done through a mobile phone. A sensor has been fed into the robotic system for gaining all the information. The sensor actually detects the object and gains some information and transfer the information to the control unit. The control unit execute whole the operation for desired result (Ahmed et al., 2000; Patra et al., 2007).

Robot actuation is determined according as per the dual tone multiple frequencies (DTMF) technique which is determined by mobile phone in wireless mode. A complex programmable logic devices (CPLD) is introduced for interfacing issues in the robotic system with the help of various sensors, actuating system, power source and communicating devices. The drawing of flexibility of mobile controlled robot and various interfacing issues linked with a variety of means with CPLD methods are introduced (Kamaruzzaman et al., 2017; Wang, 2015). Many of the domestic appliances are controlled by a remote for better lifestyle and many of the communication process is done through a mobile and a GSM module of the scheme host is placed in computer TC35i which is based on the DTMF technology. The introduced technology has lot of improvement, such as easy monitoring, good viability and large appliance assortment etc (Rampa et al., 2015; Jain et al., 2014). To process the image, an image frame has been taken as an input and processed data. After that the processed image is worn about to take out the gesture program (Otsu et al., 1975; Patra et al., 2007). Autonomous underwater vehicle (AUV) operators works on underwater acoustic telemetry for monitoring and control during mission. This requires operators to maintain acoustic communications by radio frequency (RF) communications range (Bocca et al., 2004; Habib, 2008). The smooth movement function of a robotic system is very much important for a doctor, patient and robot itself, so as to responsible for the modification of the safety security for human being and immovability of the robot (Nicholas et al., 2011; Patra et al., 2007; Viola et al., 2004). A number of various viewing methodologies have been introduced which is responsible for allowing the snapped images to find out in 3D structure, throughout the utilize of particular glasses, filters, or journey of one's eyes (Wang et al., 2015) .

HISTORY OF WORK

The purpose of the work is to build a cost efficient and efficient robotic system with video streaming for industrial, medical and many other applications with wire and wireless live video streaming. To provide a platform for many industrial applications that requires many inspections and surveillance which is difficult to reach at hazardous area? Therefore, it acts as a substitute for human in hazardous environment for image detection and recognition. To design and develop a system that will work on factory floor to detect and recognize the workers and machines. The workers and machines are working efficiently or not. From the control room, a supervisor can able to visualize the staffs of all the sections simultaneously. Wireless robot is designed by using motors, motor driver IC, RF communication module which consists of encoder and decoder. Two breadboards are used. One breadboard is used for implementing encoder circuit and another is used for decoder circuit. Encoder circuit breadboard is placed over robotic car. The radio frequency works on 434 MHz. Designing of circuit is done on dip trace software. There are numerous countries in the world who have tailored this method of image processing in various areas like defense, meteorology, vehicle detection systems, etc.

Block Diagram of Wireless Robot

The block diagram of wireless robot consists of all basic components used to design wireless robot. The main components used to design the wireless robot are voltage regulator IC, motor driver IC, radio frequency module. Battery is used to supply power all components. Voltage regulator called 7805 gives 5V supply to both the circuit as both circuit requires only 5V. Motor driver IC called L293D is dual in line package IC. The switches transfer parallel signals to encoder. The encoder converts it into serial form and transfers these signals to radio frequency transmitter. The transmitter again, transfer these serial signals to receiver wirelessly using radio frequency of 434 MHz. At the receiver end, the decoder decodes the serial data into parallel form. The indicator led glows at the encoder circuit indicating the transfer of signal and the robot moves as per the instruction given by the switch. The decoder circuit is also connected to another power supply battery with another voltage regulator IC. In the experiment, a image is taken using digital camera placed over the RF controlled robotic car and the image is processed using MATLAB. The input colour image is converted to grayscale for simplicity, then after converting it to binary image, the image is dilated using sobel edge detection technique. Then image is segmented and recognized. The unwanted quantity of an image can be removed by using dilation methods. The distinction of gray value in between adjacent pixels at the edge of an object can also be enhanced by formative the edges sharper which can be facilitated in edges recognition. A camera is positioned in front of the robotic car. The USB is used to attach the camera with the robotic car and computer is responsible to detect and competition the object online. In recent sciences and technologies, image processing is one of the emerging field. There are various application of image processing in the field of science and engineering such biomedical, robotic system etc. Beside that wireless communication is one of the emerging fields in communication area which is used robotic system for communication. There are two important parts in wireless communication, transmitting and receiving. The transmitting part is responsible for transmitting the signal whereas the receiving part is responsible to receive the transmitted signal.

In the introduced work a full-featured robotic car is designed for detection, tracking the skeleton, particularly designed to work on video footage. This difficulty can be overcome by using the tools of some software platform like Mat lab, Lab View, and Open CV. In this scheme MATLAB platform is used to execute the operation. MATLAB platform is used for image processing in the module using RF control, which track and match the objects. Whole the operations have been done by using a mobile phone through RF technology. The robotic car is controlled with the help of HC-05 Bluetooth technology. In this process, the operator needs to turn on the Bluetooth in the mobile. The wireless communication techniques used to control the robot is Bluetooth technology. According to users command, the movement function of the robotic car will be examined. The receiver unit of HC-05 Bluetooth takes the commands and gives it to the controller circuit to control the motors. The controller then transmits the signal to the motor driver to stimulate the motors.

Development of a Wireless Robot

The robotic model consists of a mechanical chassis with four 100 rpm motor and a wheel is connected to each of them. The chassis is used for mechanical support for the robot and in the chassis two motor is screwed of 100rpm in which both wheels are attached to each of them for movement. This D.C. motor is connected to 5V supply via motor driver IC i.e. L293D which is 16 pin IC. The system is monitored by a remote from transmitter side. HT12D and HT12E are used as decoder and encoder IC for signal manipulation. L293D is used as motor driver IC for amplification the accepted signals. A 16-megapixel camera is used for image processing which is fixed at the top of the robotic car.

A software platform is used to detect and identify the object. The D.C motor is connected with 5V supply via motor driver IC i.e. L293D which is 16 pin IC. Out of two bread board, one is placed over robotic system and another is with the operator in which ground and supply voltage is given by 5V supply from 7805 IC, which takes 18V input supply. The output from voltage controller is supplied to the whole circuit for execution viz. transmitter Tx02. the design model consist of the RF unit which includes transmitter part as well as receiver part that could be used to operate an output signal from a remote place.

Figure 1. Block Diagram of Wireless Robot

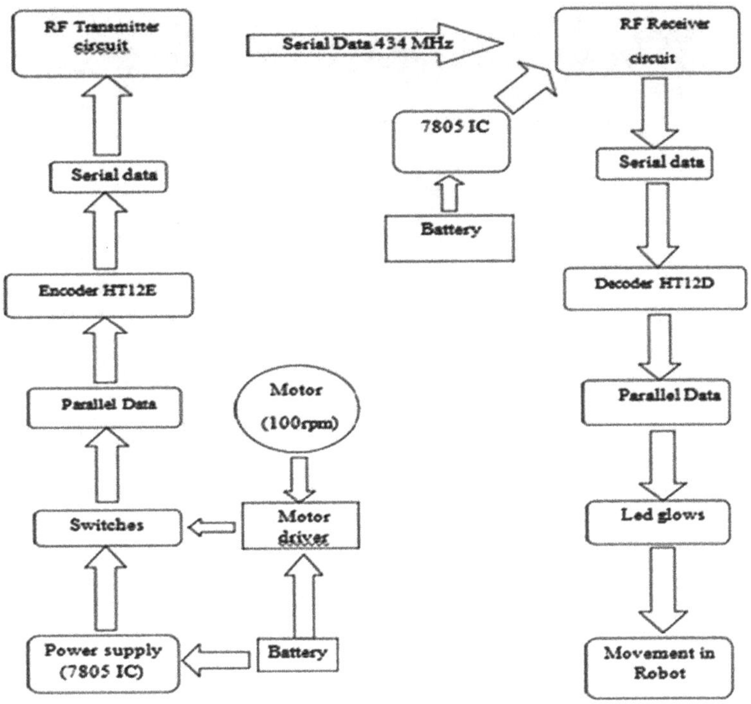

The development of robotic system completes their communication by using transmitter and receiver the signals. Transmitter is responsible to transmit the signal whereas the receiver is responsible to receive the signal to that particular frequency 434 MHz frequency. The input signals, at the transmitter side, are taken from other breadboard .Encoder IC HT12E receives parallel data in the form of address bits and control bits. The control signals from remote switches along with 8 address bits constitute a set of 12 parallel signals. The encoder HT12E encodes these parallel signals into serial bits. Transmission is enabled by providing ground to pin14 which is active low. The control signals are given at pins 10-13 of HT12E. The serial data is fed to the RF transmitter through pin17 of HT12E. Transmitter, upon receiving serial data from encoder IC HT12E, transmits it wirelessly to the RF receiver. The receiver, upon receiving these signals, sends them to the decoder IC HT12D through pin2. The serial data is received at the data pin DIN, pin14 of HT12D. The decoder then retrieves the original parallel format from the received serial data. Whole the controls have been done through a mobile phone. A sensor has been fed into the robotic system for gaining all the information. The sensor actually detects the object and gains some information and transfer the information to the control unit. The control unit execute whole the operation for desired result. A four-channel encoder/decoder pair has also been used in this system i.e. HT12D and HT12E. The following figure shows the designed structure of wireless robot.

Some of the components used in the design are as follows:

1. **Radio Frequency Module:** The RF element consists of transmitter and receiver. The transmitter is responsible to send the continuous signal to receiver part at 434 MHz. Actually, the receiver part

Figure 2. Wireless robot in standstill situation

is responsible to receive the signals only if it is organized to the exact frequency. The two different circuits are used for transmitter and receiver which are shown in following figure.

2. **DC Motor:** At the time of designing 100 rpm dc motor is used for actuation purpose. To convert direct current electrical power into mechanical power, Any DC motor at any class can be used. The motor is tightening with the framework and is a wheel attached to each motor shaft. 18 volt battery is used for motor actuation. Below figure shows the used DC motor.

3. **Encoder and Decoder IC:** For design, IC HT12E is used as encoder and IC HT12D is used as decoder. Encoder is used to encode the signal for transmitter and decoder is used to decode the signal from receiver. Encoder and Decoder are used in remote applications.

Figure 3. Radio frequency module

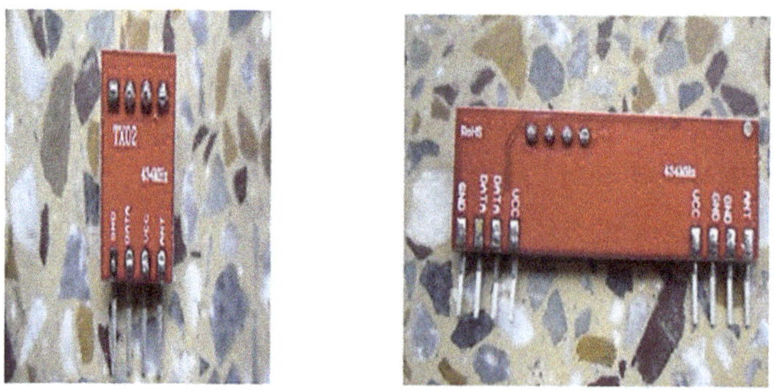

Figure 4. DC motor

Wireless Robot in Image Processing

Image processing is one of the emerging field of robotics. To convert an image into the digital form, Image processing method is used. The performances have been done for some execution on it to get an improved image. Usually image processing methods consist of indulgence images as 2-D signals when a bunch of signal processing methods have been applied on it. Mainly image Processing associated with three phases introducing the image with optical scanner, scrutinizing and manipulating that includes data firmness and image improvement etc. The output is based on the image analysis. Advances in vision technology for robotics are expected to broaden the capabilities of robotic vision systems to allow for vision based guidance of the robot arm, complex inspection for close dimensional tolerances and improved recognition and part location capabilities.

The typical vision system consists of the camera and digitizing hardware, a digital computer and hardware & software necessary to interface them. The idea of digital image processing is visualization which is used to monitor the objects that are imperceptible, reinstatement etc. The main purpose is to create a better image, image repossession, pattern events for many objects in an image. DIP methods help in organizing the digital images by using computers. There are various types of applications of wireless robot in image processing and digital image processing such as smart transport system, moving object tracking, remote sensing technology, defense surveillance, biomedical imaging, cancer imaging, brain imaging etc.

Figure 5. Encoder and decoder IC unit

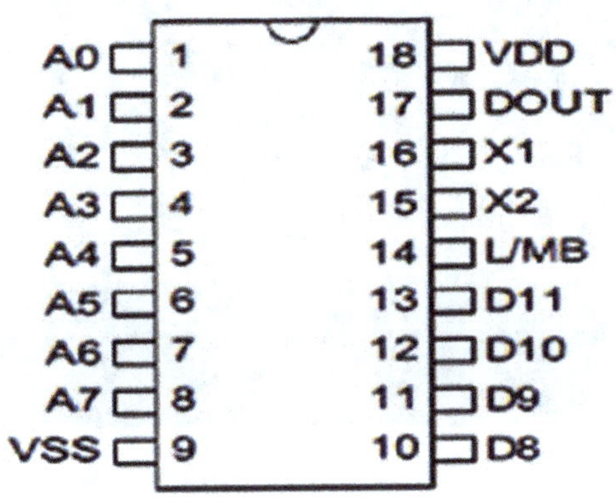

Table 1. Movement of robotic car by the action of switches

Serial No.	Action of S1	Action S2	Movement in Robotic Car
1	Press upward	Press upward	Forward motion
2	Press upward	Press downward	Right turn
3	Press downward	Press upward	Left turn
4	Press downward	Press downward	Backward motion

Movement of the Robot

The movement function of robot can be executed by transmitter circuit where two DPDT (double pole double throw) switches are allied. There are two switches in the circuit for operation. Both motors ML1,ML2 of the left side of the robot connected parallel and controlled by switch 1 S1 and the motors MR1,MR2 of the right side of the robotic system are connected in parallel which is controlled by switch 2 (S2).

CIRCUIT DIAGRAM OF THE SYSTEM

Below diagram shows the snapshot of the circuit, designed in dip trace Platform. Dip Trace is a CAD platform for which is used to create the schematic structure and responsible for printing the circuit boards. There are four modules in dip trace platform such as Schematic Capture Editor, PCB Layout Editor, 3D preview and export.

Figure 6. Schematic circuit diagram of wireless robot, controlled by RF

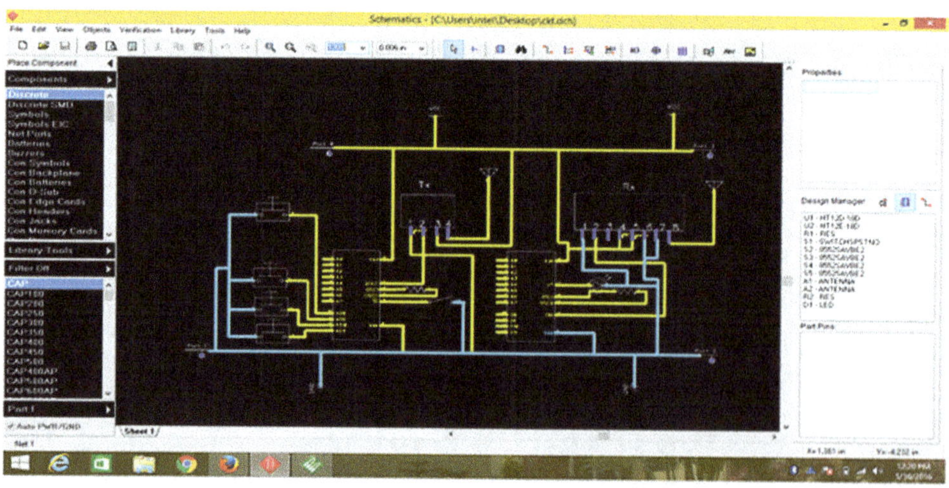

Figure 7. Line circuit diagram of the wireless robot

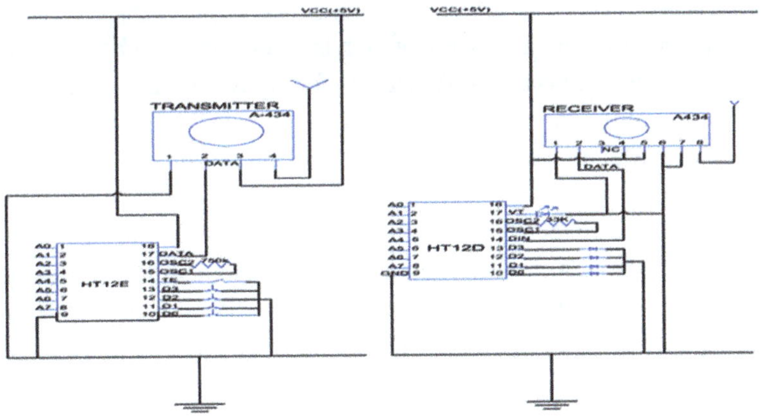

FLOW CHART OF THE SYSTEM

The overall working principle of the robot are depicted in the below flow chart for video or image processing. In the projected model, the robot control has been done by RF technology. The processing of video or image has been done by using MATLAB program. The image is captured by camera of 16 megapixels, which is positioned over robotic car. The robot is responsible to detect and track object while moving.

Figure 8. Flow chart of the wireless robot

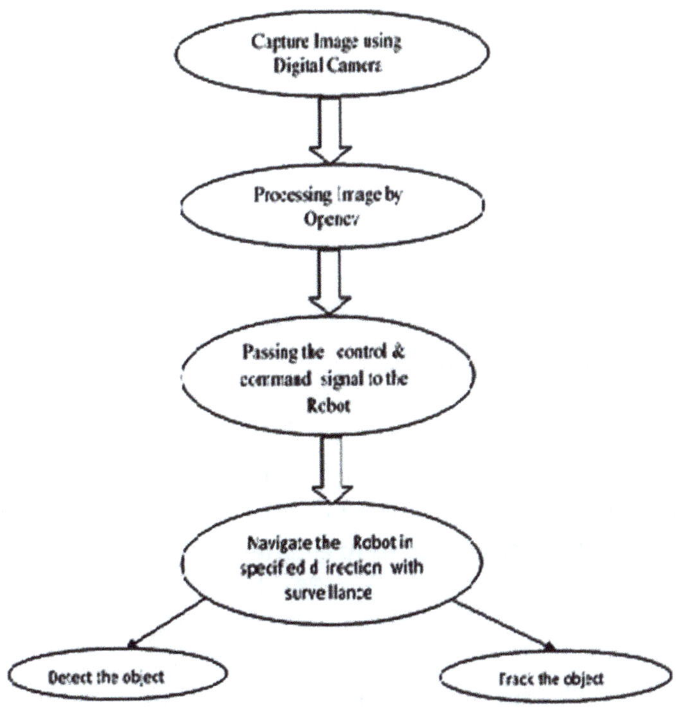

METHODOLOGY OF THE SYSTEM

A stationary camera is placed to monitor a scene which is a very natural case for a supervision scheme. To identify the interfering substance is a significant step for analyzing the scene and successful arrangement for movement function of the forefront object which comes for starting the conditions to ensure the object classification, object recognition, detecting, and function examine. The main objective of moving objective is to take out the significant information about the dynamic object from the video sequences. The result provides the expediency of object-based demonstration and exploitation of video information. The objected work is a necessary step for numerous computer vision applications: video compression, video recovery, video supervision, and pattern recognition. Traditional approach is responsible for movement function of object arrangement which includes with a structural variation, background subtraction and optical flow technology. The first phase of various surveillance systems is forefront judgment and proper arrangement. The forefront areas are noticeable (e.g., facade image) for dispensation in the successive steps. In automation system, there are two major different methodologies for estimating the forefront that uses both well-built assumption to fulfill with the aforesaid definition. To gather the information on the subject of the scene surroundings of a video sequence. Then the designed model is compared with the recent structure for identifying the differences. It is reported that the camera is at a standstill. The thought provides itself well for computer execution but leads to troubles with slow traffic. Any moving object can be considered as a forefront but static objects are missed due to be the short of motion.

SIMPLE BACKGROUND SUBTRACTION

In simple conditions subtraction method is the total difference which is taken from each current image A(x, y) and the reference conditions of image B(x, y) is used to get the motion recognition mask D(x, y) The reference image of the background is generally the first framework of a video.

$D\,(x,y) = \{1\ if\ A\,(x,y) - B(x,\,y) \geq \tau$, $at\ Oerwise\ 0$

Where τ is a threshold, which make a decision whether the pixel is front position or background. If the total difference is more than or equal to τ, the pixel is classified as forefront; otherwise the pixel is classified as background.

SIMPLE IMAGE FORMATION MODEL

In the continuous case, a planar image is signified by a two-dimensional function (x, y) f(x,y). It is determined by the source of the image that the value of ‘f’ at the spatial coordinates (x,y) is positive. The intensity values are proportional to energy emitted by a physical source when the image is produced from a physical process. So, f(x,y) must be nonzero and finite:

$0 < f(x; y) < 1$

The image-function ‘f’ may be distinguished by two mechanisms:

Figure 9. Flow chart of object recognition

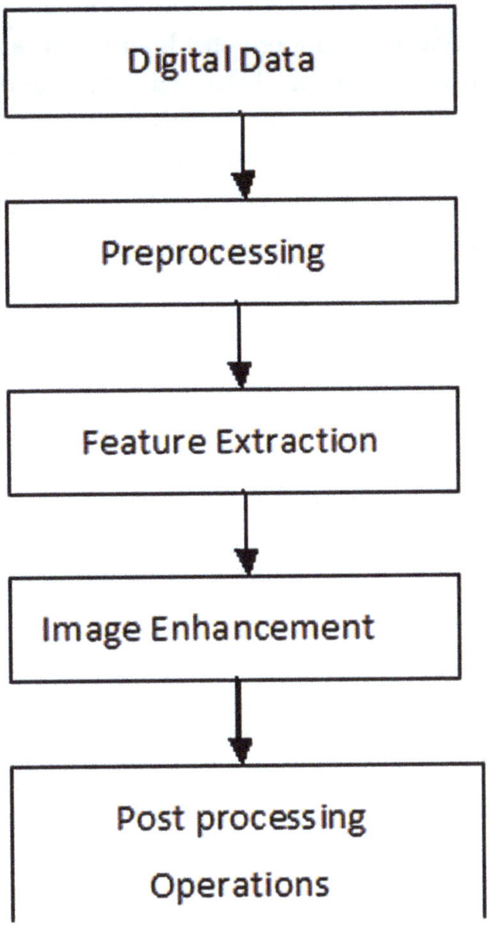

1. The amount of source illumination incident on the scene (illumination) i(x, y)
2. The amount of illumination reflected by the objects (reflectance) r(x, y)

We have,

$f(x,y) = i(x,y),r(x,y)$

where, $0 < i(x, y) < 1$ and $0 < r(x, y) < 1$

Object Recognition

First of all, the camera which is placed over robotic car is connected to computer in which MATLAB program is installed through USB cable. The input color image is converted to grayscale for easy execution, and then image is expanded by using sobel edge recognition method. Then image is segmented and recognized. If there is any character to distinguish in image then using OCR it is possible. Flowchart of the image processing is given below.

Pre Processing

The Sobel operator or Sobel–Feldman operator or Sobel filter have been used in image processing and computer vision. This method is mostly used for edge recognition algorithms in image processing where it creates an image highlighted edges. The two 3×3 kernels are used in this operator.

If we define A as the image source and G_x and G_y are two images. Then these kernels are intended to react maximally to edges running vertically and horizontally comparative to the pixel grid. The kernels can be functionalized individually to the input image to create individual measurements of the gradient. Then these can be united together to get the absolute magnitude of the gradient at each point. The gradient magnitude is specified by:

$$\left|G_x\right| = \sqrt{G_x^2 + G_y^2}$$

Usually, an estimated magnitude is computed using:

$$\left|G\right| = G_x + G_y$$

This is much faster for calculation. The angle of direction of the edge giving increase to the spatial gradient which is specified by:

$$\theta = \arctan\left(\frac{G_y}{G_x}\right)$$

The gradient is nothing but a vector by which the components calculate how rapid pixel value are altering with distance in the x and y direction. G_x and G_y are as follows.

+1 +2 +1
0 0 0
-1 -2 -1
-1 0 +1
-2 0 +2
-1 0 +1

So the following estimation may be responsible to find out the components of gradient:

$$\frac{df\left(x,y\right)}{dx} = d_x = f\left(x + d_x\right) - \frac{f\left(x,y\right)}{dx}$$

Figure 10. Movement of robot: Left: No movement in robot, white led not glows; Right: Robot moves as white led glows

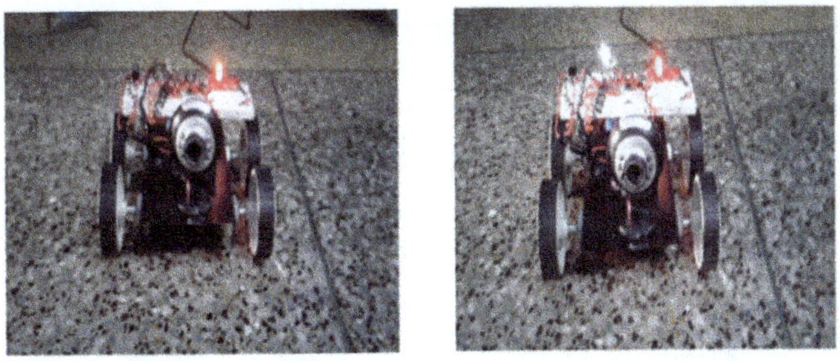

$$\frac{df(x,y)}{dx} = d_y = f(x.y + d_y) - \frac{f(x,y)}{dy}$$

Here d_x and d_y is responsible to measure the distance along x and y directions correspondingly. In discrete images, d_x and d_y can consider in terms of numbers of pixels in between two pints. $d_x = d_y = 1$ is the point at which pixel coordinates are (i, j). So:

$$d_x = f(i+1, j) - f(i, j)$$

$$d_y = f(i, j+1) - f(i, j)$$

In order to identify the occurrence of a gradient discontinuity, the change in the gradient should be calculated at (i, j). It can be done by using the following formula.

$$M^2 = d_x^2 + d_y^2$$

The gradient direction is expressed as $\tan^{-1}\frac{d_y}{d_x}$. The Intensity conversions are applied to improve the quality of the image for next step.

The below conversions are applied on the color JPEG image:

- RGB image into a Gray-scale Intensity image Conversion.
- Intensity image into a Binary image Conversion.
- Gray-scale Image into an Edge Detected image Conversion.

Figure 11. RGB to Grayscale Intensity Image Conversion: Left: RGB Image; Right: Grayscale Intensity Image

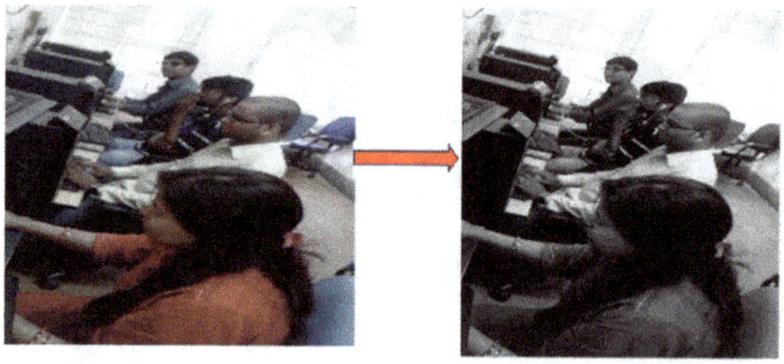

RESULTS AND DISCUSSION

In the wireless robot the control action is performed by the radio frequency switch. The robot follow the path as per switching action of the radio frequency switch. In the circuit red led shows the indication of power supply. While white led glows when RF command is given to the robot to control the movement of robot at that instant.

In proposed work, a snapshot/image is taken using digital camera placed over the RF controlled robotic car and the image is processed using MATLAB. The input colour image is converted to grayscale for simplicity, then after converting it to binary image, the image is dilated using sobel edge detection technique. Then image is segmented and recognized. The unwanted quantity of an image can be removed by using dilation methods.

The distinction of gray value in between adjacent pixels at the edge of an object can also be enhanced by formative the edges sharper which can be facilitated in edges recognition. In face detection, the capture of face may not always have the similar intensity and sunglasses. Some significant elements like variation in colour and lighter edges of image can be lost throughout in image alteration from RGB to gray scale. So to nullify the losses in image conversion dilation method can be used. There is a minimum threshold reading which is considered as a set up value. By using the set up value input image structure can be converted from threshold to binary. The set up values have some limitations on the background. So to overcome the limitations the shady background is used. The colour based thresholding can be implemented to keep away from this limitation. To improve the superiority of the image for further processing, Intensity transformations have been applied. The conversions have been introduced on the colour JPG images are depicted in following:

- RGB image into a gray-scale intensity image conversion.
- Intensity image into a binary image conversion.
- Gray-scale Image into an edge detected image conversion.

Figure 12. Horizontal Edge Processing Histogram

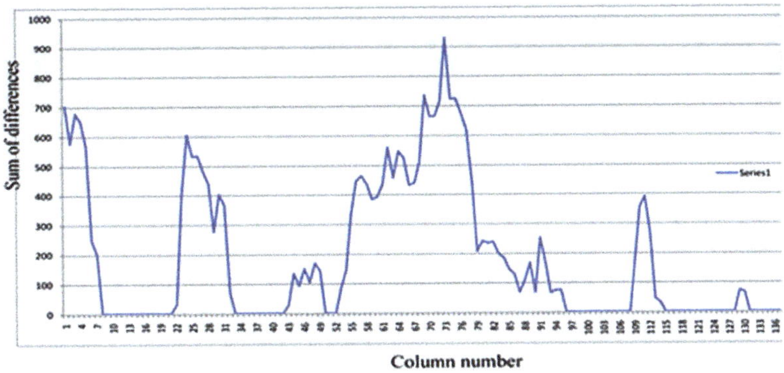

Figure 13. Vertical Edge Processing Histogram

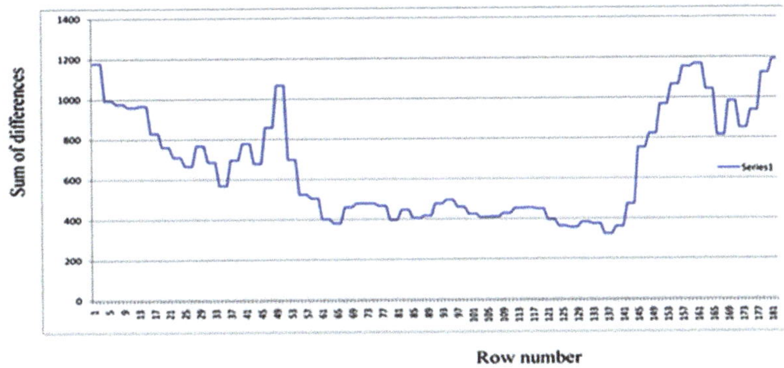

Histograms Passing Through a LPDF

There is analysis of the histogram values of the image which varies significantly in between successive columns and rows that are depicted in the below Fig. So to minimize the loss of the significant information in next step, there have been proposed a suitable step which is used for settle down in the histogram process such as radical changes in values of the histogram. To make smooth of the histogram, a low pass digital filter is used which passes the histogram via LPDF step by step. Each histogram value is averaged out to consider the values of left hand and right hand side which is responsible for performing this footstep. Both the histogram i.e. horizontal and vertical histogram is performed in this step. Before passing the histogram through a LPDF and after passing the histogram through a LPDF are depicted in the following figure14 and figure15.

Unwanted Regions Filtering Out of Image

The main function of a filter is to remove the unwanted quantity of an image. Here filter is used to eliminate the unwanted areas of an image when the histogram process is done throughout a low pass digital filter. In the pursuing case, the unwanted areas are the rows and columns which are with minimum histogram

Figure 14. Horizontal histogram after passing through LPDF

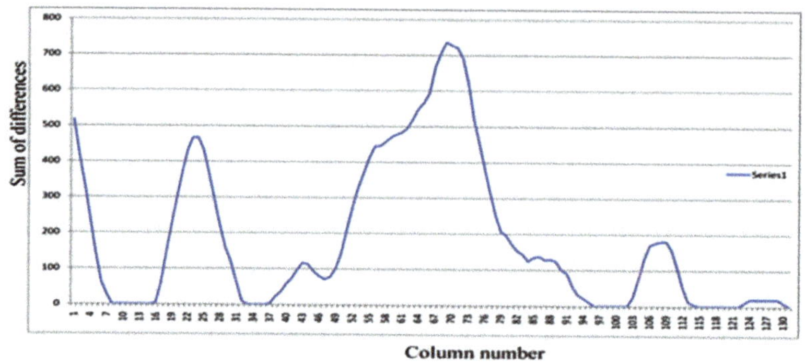

Figure 15. Vertical histogram after passing through LPDF

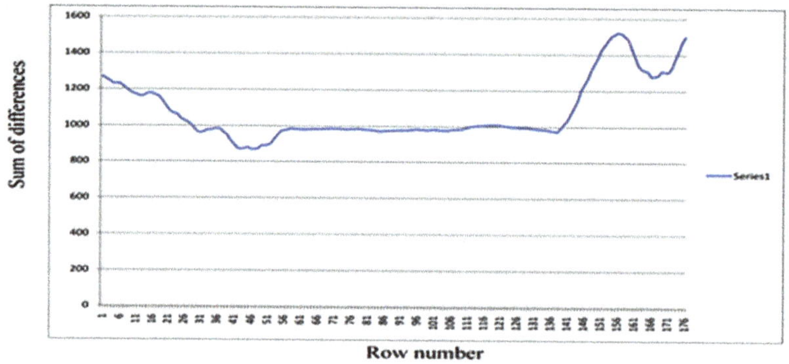

values. The element of image have very little amount of deviation among the surrounding pixels which have been shown by a less histogram rate.

The distinction in between the contiguous pixels, particularly at the edges of face and body will be very elevated because of the feasible region of the image which have a simple background with some substance in the working region. The outcome of this approach gives high histogram values for the particular region of an image. Consequently, the area with a feasible face and body has a elevated horizontal and vertical histogram value. The area with less value does not require any longer. By applying a dynamic threshold, the unwanted regions are eliminated from an image. The values of the dynamic threshold are the average value of a histogram which is reported in this algorithm. Calculations have been done on

Figure 16. Horizontal histogram after dynamic thresholding

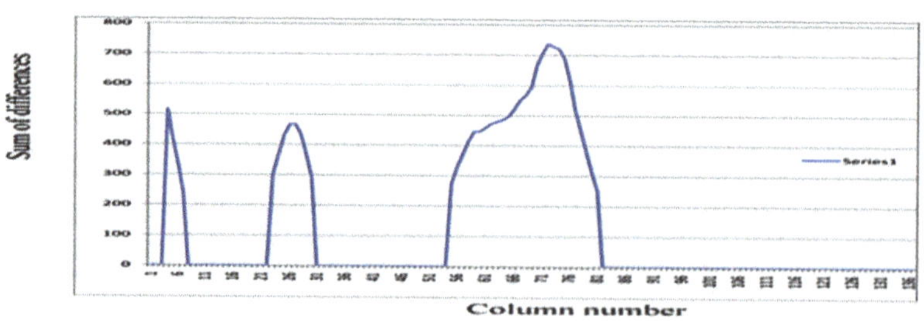

Figure 17. Vertical Histogram after dynamic thresholding

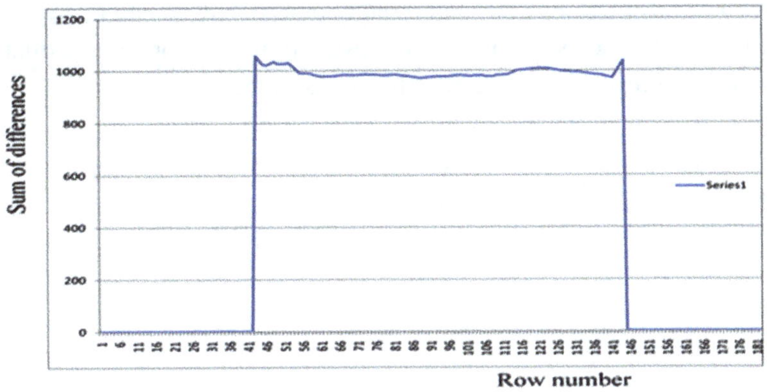

horizontal and vertical histogram when the signals passed through a LPF with the dynamic threshold value. Histogram is the outcomes of the proposed process which shows the region include with high probable of containing features. The following fig shows the histogram.

DISCUSSION OF THE PROPOSED WORK

The work focuses on the development and integration for real time location tracking of a human worker. The method is proposed to jointly track operator and robot movements. The input colour image is converted to grayscale for simplicity, then after converting it to binary image, the image is dilated using sobel edge detection technique. The value of input image structure is thresholded into binary form by fixed up the minimum threshold value. The set up values have some limitations on the background. So to overcome the limitations the shady background is used. Also it can use the colour based thresholding for avoiding this problems.

Figure 18. Results of Proposed histogram approach

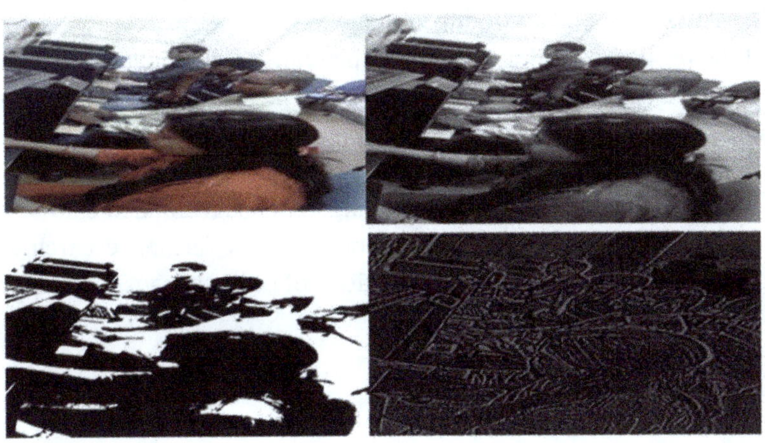

In face detection, the capture of face may not always have the similar intensity and sunglasses. Some significant elements like variation in colour and lighter edges of image can be lost throughout in image alteration from RGB to gray scale. So to nullify the losses in image conversion dilation method can be used. There is a minimum threshold reading which is considered as a set up value.

By using the setup value input image structure can be converted from threshold to binary. The deviation of gray value in between adjacent pixels can also be modified by formative the edges sharper that would be helpful in edges recognition. In next step the image is segmented and recognized for next step. Also, the unwanted quantity of the image can be removed by using dilation method. The experimental case study focuses on the development and integration of the image processing technologies for real-time location tracking of a human worker which may sharing the workspace with a robot manipulator.

CONCLUSION

The different applications of image processing have been discussed and the work on motion detection carried out. Motion detection has been done using background subtraction technique. In face detection histogram approach is used to identify vertical and horizontal edges and studied about face detection using MATLAB. In future work focus on to use some other motion detection techniques and face detection techniques, some of are mentioned in the literature. It removes the requirement of substantial existence of the user at a operational region.

To make wireless remote, the design model consists of the RF unit which includes transmitter part as well as receiver part that could be used to operate an output signal from a remote place. The development of robotic system completes their communication by using transmitter and receiver the signals. Transmitter is responsible to transmit the signal whereas the receiver is responsible to receive the signal to that particular frequency. Whole the controls have been done through a mobile phone. A sensor has been fed into the robotic system for gaining all the information. The sensor actually detects the object and gain some information and transfer the information to the control unit. The control unit execute whole the operation for desired result.

There are many applications of the proposed system in the field of image processing such as pattern identification and machine vision etc. The ranges of the system are from multifaceted security systems to common regions and from parking region to built-up traffic control. To overcome the wander effect such as rain, fog, shadows, rough lighting situations, occlusion, changeable distances etc. there are many multifaceted characteristics have been introduced in the system.

The main goal of the proposed work is to resolve the realistic difficulty of recognition for real prospects. Background subtraction techniques are used to solve these problems. To use it wirelessly, all the applications of image processing are wrapped in Robotics, controlled by radio frequency. MATLAB software is used for image processing. A digital camera is used to place over robotic car and surveillance can be done from distant place using image processing. This surveillance can be used in defense, to supervise workers and machines in industry production area. For further research work the introduced technology can be endorsed with the various applications of computer vision technology.

REFERENCES

Ahmed, E., Harwood, D., & Davis, H. (2000). Non-parametric model for background subtraction. In Computer Vision—ECCV, pp. 751-767. Berlin, Germany: Springer.

Bocca, M., Kaltiokallio, O., Patwari, N., & Venkatasubramanian, S. (2014). Multiple target tracking with RF sensor networks. *IEEE Transactions on Mobile Computing*, *13*(18), 1787–1800. doi:10.1109/TMC.2013.92

Kaura, H. K., Honrao, V., Patil, S., & Shetty, P. (2013). Gesture controlled robot using image processing. *International Journal of Advanced Research in Artificial Intelligence*, *2*(5), 69–76.

Habib, M. K. (2007). Mechatronics: A Unifying Interdisciplinary and Intelligent Engineering Paradigm. *IEEE Industrial Electronics Magazine*, *1*(2), 12–24. doi:10.1109/MIE.2007.901480

Habib, M. K. (2008, August). Human Adaptive and Friendly Mechatronics (HAFM). In *IEEE International Conference on Mechatronics and Automation 'ICMA',* Takamatsu-Kagawa, Japan, pp. 61–65.

How to make Robot. (2014). Retrieved from http://hi-robotics.blogspot.in

Hu, C., Wang, X., Mandal, M. K., Meng, M., & Li, D. (2003, May). Efficient face and gesture recognition techniques for robot control. In *CCECE 2003-Canadian Conference on Electrical and Computer Engineering. Toward a Caring and Humane Technology, 3,* (pp. 1757-1762). IEEE.

Hu, W., Tan, T., Wang, T., & Maybank, S. (2004). A survey on visual surveillance of object motion and behaviors. *Systems, Man, and Cybernetics, Part C: Applications and Reviews, IEEE Transactions on 34,* 3, pp. 334-352.

Jain, P., Firke, P. N., Kapadnis, K. N., Patil, T. S., & Rode, S. S. (2014). RF based Spy robot. *International Journal of Engineering Research and Applications, 4*(4), version 2, pp. 06-09.

Jha, A., Singh, A., Turna, R., & Chauhan, S. (2015). War Field Spying Robot with Night Vision Camera. *Journal of Network Communications and Emerging Technologies (JNCET), 1*(2).

Kamaruzzaman, M., Haque, R., & … . (2017). *Enhancement of Railway Safety Systems using Anti Collision Device, ICSMART.* Durgapur, India: CMERI.

Nicholas, A., Iphigenia, K., & Kiranoudis, C. T. (2011). A background subtraction algorithm for detecting and tracking vehicles. *Elsevier Expert Systems with Applications*, *38*(3), 1619–1631. doi:10.1016/j.eswa.2010.07.083

Otsu, P. (1975). Threshold selection method from gray-level histogram. automatic, 11, page 285-296.

Patra, A. K., & Ray, S. J. (2007). Guiding robots using mobile phone International Symposium on automation and robotics in construction. In *ISARC,* Canada.

Patwari, N., & Wilson, A. (2010). RF sensor networks for device-free localization, measurement, models and algorithms. IEEE 98(11), pp. 1961-1973. doi:10.1109/JPROC.2010.2052010

Rampa, S., Savazzi, M. N., & Amico, M. D. (2015). Physical modeling and performance bounds for device-free localization systems. IEEE Signal Process, 22(11), pp. 1864-1868.

Tse, D., & Viswanath, P. (2005). *Fundamentals of Wireless Communication*. Cambridge, UK: Cambridge University Press. doi:10.1017/CBO9780511807213

Viola, P., & Jones, M. J. (2004). Robust real-time face detection. International Journal of Computer Vision, 57(2), 137-154.

Wang, L. (2015). Collaborative robot monitoring and control for enhanced sustainability. *International Journal of Advanced Manufacturing Technology*, *81*(9), 1433–1445. doi:10.100700170-013-4864-6

Chapter 14
Detection of Traffic Signs and Road Users From a Moving Vehicle

Stefan Müller-Schneiders
Bochum University of Applied Sciences, Germany

Rajen Wiemers
Bochum University of Applied Sciences, Germany

ABSTRACT

The topic of this chapter is the implementation and analysis of vision algorithms for the detection of static and dynamic objects in videos. These algorithms are typically components of the visual perception module of modern driver-assistance systems or even autonomous cars. Whereas the vast majority of today's papers from the vision community use convolutional deep neural networks (CNNs), this chapter explores the more traditional approaches, namely HOG (Histogram of Oriented Gradients) as well as SVM (Support Vector Machine). The static and dynamic objects to be recognized are traffic signs and motorcycles, respectively. These two object classes have been chosen since traffic signs are relatively easy to detect and motorcycles state a much more complex task. Thus, this chapter tackles differently difficult tasks with a single set of algorithms.

INTRODUCTION

The visual perception of the vehicle surrounding has made significant progress in the last decades. Major improvements have been achieved by introducing HOG-features (Histogram of Oriented Gradients) and SVM-classifier (Support Vector Machine) as detectors for all kinds of visual objects (see Dalal & Triggs 2005). Recent advances are the result of the introduction of convolutional neural networks, which have been presented by LeCun et al. (1998). Besides other differences, a major difference between the approaches (Dalal & Triggs, 2005) and (LeCun et al., 1998) is that the latter approach no longer utilizes a separate feature extraction step. Instead the feature extraction became in (LeCun et al., 1998) part of the

DOI: 10.4018/978-1-7998-0137-5.ch014

neural network itself. By using a separate feature extraction step, which operates in a predefined way, i.e. it is not trained on data, this step can be optimized with respect to computational demand. Thus, the resulting detection solution will run with a higher computational efficiency. Therefore, this paper analyses the usage of (Dalal & Triggs, 2005) for the recognition of motorcycles as well as traffic signs.

The task of recognizing motorcycles is more challenging than the recognition of e.g. vehicles or lane markings, since the variation within this class is much higher. Additionally, the drivers of motorcycles can lean into a curve and therefore the recognition of motorcycles needs to take this change of position also into account.

Finding a dedicated feature descriptor for motorcycles combined with a suitable classifier is a challenging task (see e.g. Messelodi et al., 2007). The problem with the motorcycle class is increased by different motorcycle models and brands and thus various motorcycle shapes. Additionally, the driver himself increases the variation of this class. Therefore, a small moped with an overweight driver belongs to this class just as a chopper with a small driver riding it. This leads to a high-class variability. For example, Figure 1 shows a couple of examples for motorcycle models and drivers.

Using a stationary camera, it is quite easy to remove the background and use only the contour of the moving object. This is not possible with a moving camera and a moving target as this is the case for automotive applications, i.e. for driver assistance systems and autonomous driving subsystems. Another constraint for the examined approaches is

the real-time capability. Especially for autonomous driving, the individual modules should run as fast as possible, in order to use the computing resources efficiently.

In the following related work to the domain of motorcycle detection will be reviewed. Many publications concerning traffic object recognition tackle vehicles, pedestrians or traffic sign recognition (see e.g. Bensrhair et al., 2001). For vehicle detection Sun et al. 2006 present a good overview of the approaches which have been used by vision researchers.

Research in the area of pedestrian detection is a very important topic for our task. This is the case because the shape of pedestrians is very similar to the shape of motorcycles. Especially it is more similar than the shape of cars, vans or trucks. Research in pedestrian detection is highly relevant for surveillance

Figure 1. Sample images demonstrating the variations within the motorcycle class
(see also (Müller-Schneiders, 2019)

applications or for driver assistance systems. An algorithm for human body detection has been introduced by Wu et al. (2005). In their work they use socalled edgelets for classification. These edgelets define e.g. a line for the shoulder, an arc for the head-part, a symmetric pair of lines for the neck etc. Furthermore, they divide the human body into three parts (head, torso, legs) to define more suitable edgelet features. A detection of a human body is the combination of the responses of the body part detectors. In order to manage the entire edgelets in a body part, the AdaBoost algorithm was used to select the edgelets.

A highly popular image descriptor has been invented by Dalal and Triggs (2005) to build a pedestrian detector. The process of calculating the so-called HOG descriptor starts with the computation of the image gradients and follows by creating the histograms of the gradients for individual image cells. In the next step cells are grouped to form larger blocks and accumulate the cell values. This detection algorithm has been improved by Zhu et al. (2006) by using a variable size of blocks instead of a fixed size of blocks. In an AdaBoost feature selection process they identified the salient features. Furthermore, they contribute to speed up the detection by using the integral histogram.

A motorcycle detection and tracking system utilizing a stationary camera is proposed by Chung-Cheng et al. (2007). The fixed position of the camera allows for a simple moving object detection approach by subtracting the background image from an input image. Once the moving objects are identified the next task is to find a bounding box around the objects. The classification of the obtained objects is done by exploring the visual length, visual width and the pixel ratio. In order to distinguish motorcycles, an a priori threshold for the mentioned parameters is used. So, for example, if the visual length and visual width is approximately 6m and 2m, respectively, this corresponds to a vehicle, whereas values around 3m and 1.2m correspond to a motorcycle. Due to the fact that all motorcyclists have to wear helmets a further verification step is to search for the helmet of the motorcycle rider. In order to do this the region of the helmet is extracted and the distance to the edges from the center of the assumed helmet is accumulated.

A trained motorcycle recognition classifier is tested by Wen at al. 2007. The underlying image data set which is imbalanced leads to poor performance. In this approach they divide the negative samples which is a lot more than the motorcycle samples in k partitions. The size of these partitions depends on the positive data set. So after that, they get k subsets of the negative data sets combined with the one positive data set. Afterwards all combinations are trained with the SVM individually. The classification of the motorcycles is received by aggregate all SVM results to the final classification.

Besides moving objects, such as e.g. motorcycles, pedestrians and cars, static road infrastructure elements such as lane markings and traffic signs are important entities for environmental perception models. In this paper, besides the recognition of motorcycles, the recognition of traffic signs is also tackled. Traffic signs have been chosen, in order to analyse a second object class which can be detected and recognized by using the same set of algorithms, namely HOG features in combination with SVM classifier. In general, traffic sign recognition techniques have been widely studied und a large number of survey articles exist. Examples for surveys include (Swathi & Suresh, 2017; Fu & Huang, 2010; Mogelmose et al., 2012, Müller-Schneiders et al., 2008). This paper will tackle the recognition of motorcycles and traffic signs with the same set of algorithms. It is expected that using these techniques, it should be feasible to recognize in addition many more relevant road users and infrastructure elements. A major advantage of the usage of a separate feature extraction step is the re-use of this step for the classification of various classes. This will lead to an improved computational efficiency.

The remainder of this paper is organized as follows: Following this introduction the feature extraction implementation is explained in section II. The section describes the implemented features, namely the Histogram of Oriented Gradients (HOG). Section III explains the chosen classification method, whereas section IV deals with the experiments. Finally, a summary and conclusions are presented in the final section V.

FEATURE EXTRACTION

In this section the implementation of a feature extraction component for motorcycle classification and traffic sign recognition is described. Since the feature extraction step is preceded by an image preprocessing step, preprocessing algorithms are implemented first. A resizing method, which is based on the integral image has been used along with histogram transformation techniques as preprocessing steps. The Histogram of Oriented Gradients descriptor has been selected as implementation of the feature extraction step.

In the following, the approach presented in Pettersson et al. (2008) for calculating efficiently the orientation of the gradient is extended to improve the computation time of 16 bins. The common voting approach is also extended by a linear and gaussian weighting approach.

Histogram of Oriented Gradients

A highly successful Histogram of Oriented Gradients (HOG) image descriptor for human detection has been proposed in Dalal and Triggs (2005). They show that their approach improves the detection of pedestrians with respect to other feature sets. Basically this approach assumes that the local object shape can be described by the amount of the local intensity gradients. Especially the high variability concerning illumination and the robustness for small shape-shifts makes this description feature very valuable as well for this task. In the next paragraphs the implementation of this feature extraction procedure is explained which is also shown in Figure 2. The process gets an input image with a predefined resolution. First the calculation of the HOG matrix which hosts the cell values is performed. For that a gradient computation and a voting in bins is executed. The magnitudes of gradients of a certain size of area are summed up and create the cells. So once the cell computation is done the integral HOG (IHOG) is calculated. By means of this IHOG the sum of the histogram of a position x,y and width w and height h can be computed efficiently. These blocks are normalized finally and these normalized values are used to form the feature vector. In a first step, the gradients are calculated for each pixel of the input image. By using a simple 1-D derivation mask $g_x = (-1,0,1)$ and $g_y = (-1,0,1)^T$ the original image size of 32×48 shrinks to 30×46 by leaving out the border pixels. Other gradient computation masks like 3 × 3 Sobel have been tested by Dalal and Triggs (2005) without improving the performance. The next step is the calculation of the cells.

A single cell represents the histogram of gradient directions calculated over the underlying pixels. All pixels vote for one of the predefined bins. A cell represents a rectangular pixel area and accumulates the gradient information of the underlying pixels to the bins. In Figure 3 the HOG feature extraction process is depicted to get a better understanding of the relation between bins, cells and blocks.

Figure 2. The block diagram of the implemented histogram of oriented gradients approach

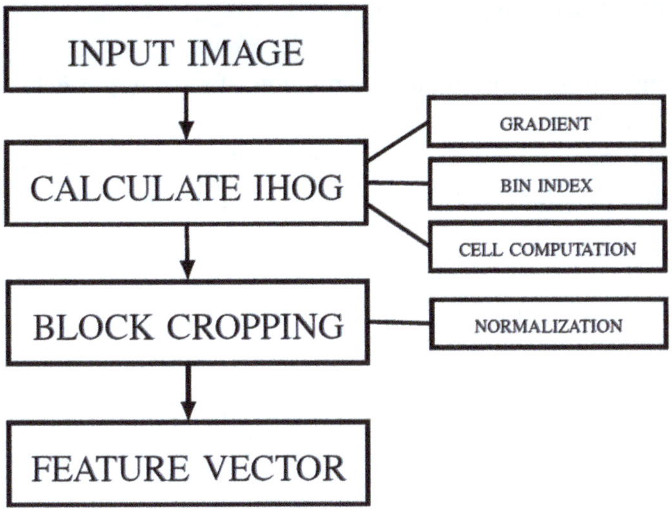

Figure 3. The Histogram of Oriented Gradients extraction process

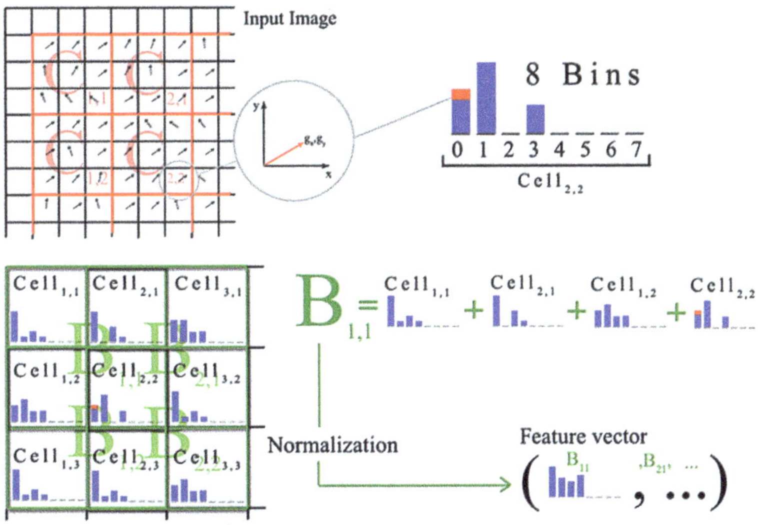

The histogram size of the cells depends on the number of bins which represent the quantization of a full circle (i.e. 360°). For example if the number of bins is set to 12, the first bin includes all gradients with angles α ranging from 0° to 30°. A very common strategy when building the histogram of gradients for a rectangular cell is to add the gradient magnitude θ to the bin. The choice of the number of the bins is a very important decision. An advantage of using more bins is an improved precision, whereas this also leads to a higher number of feature and thus to a higher computational cost.

Once the gradient orientation is given by g_x and g_y it is common to compute the orientation by $\arctan\left(g_y \cdot g_x^{-1}\right)$.

This operation is very time consuming for a real time application as this operation is done for $30\times46\times d_i$ where d_i is the number of sub windows in a frame i.

1. *8 Bins:* A more efficient approach to compute the binning b fast is proposed by Pettersson et al. (2008). They divide the orientation space into eight bins and represent the orientation with a number ranging from zero to seven. Figure 4 depicts the partitioning of the angular space and the assigned index number, respectively. The illustrated approach can be implemented in a very efficient way, namely:

$$b_{g_x,g_y} = \left[g_y < 0\right] \ll 2 + \left[g_x < 0\right] \ll 1 + \left[\left|g_y\right| > \left|g_x\right|\right] \tag{1}$$

where e.g. "$\ll 2$" is a shift operation which is equivalent to the multiplication by 4 and [...] denotes the so-called Iverson bracket. The Iverson bracket, named after K. E. Iverson is a generalisation of the Kronecker Delta operator. It converts the logical expression inside the bracket into a number that is 1 if the logical term is true and 0 otherwise (see Knuth, 1992). An example is shown in the following with the parameters $g_x = -7$ and $g_y = -2$ which has the exact angle $\arctan\left(\frac{-2}{-7}\right) = 195.95°$

Figure 4. Each pixel points to one of the 8 bins
(see also Pettersson et al. 2008)

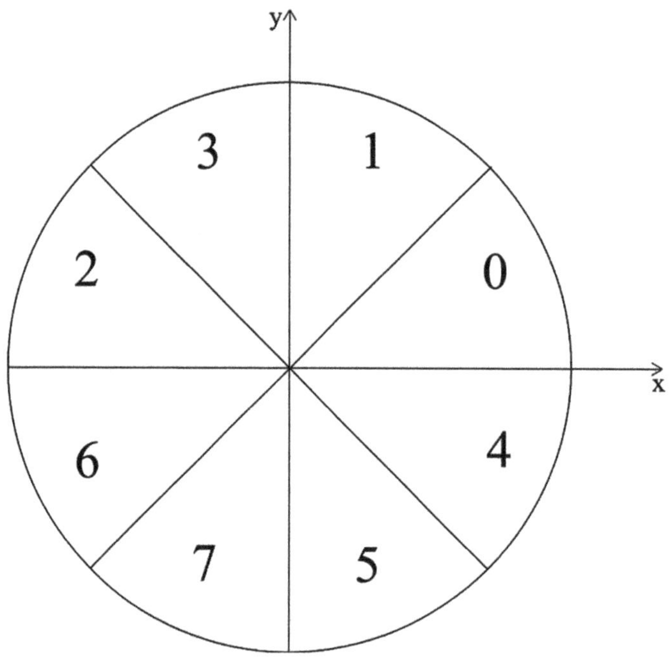

$$\left[g_y < 0\right] = \left[-2 < 0\right] = 1 \ll 2 = 100$$
$$\left[g_x < 0\right] = \left[-7 < 0\right] = 1 \ll 1 = 10$$
$$\left[\left|g_y\right| > \left|g_x\right|\right] = \left[\left|-2\right| > \left|-7\right|\right] = 0$$
$$100 + 10 + 0 = 110 = 6$$

Thus, the gradient $(-2,-7)^T$ is assigned to bin number 6, which would also be the chosen segment based on the calculation of the angle, namely 195.95°.

2. 2. **16 Bins:** This paper extends the proposed approach to 16 bins using a similar operation. The previous partition scheme is using the borders [0°, 45°, 90°,...] hence in order to get 16 bins, each bin needs to be divided once more. This results in using the following borders: [0°, 22.5°, 45°, 67.5°,...]. The idea is to include the gradients of the additional sub-divisions at 22.5° and 67.5° as an additional segmentation step which is also depicted in Figure 5. As we can set the gradient of 22.5° or 67.5° by

$$\tan(\alpha) = \frac{\sin(\alpha)}{\cos(\alpha)}; \alpha = 22.5^\circ = \sqrt{2} - 1; \alpha = 67.5^\circ = \sqrt{2} + 1 \tag{2}$$

Consequently, the binning formula is modified according to

$$b_{g_x, g_y} = \left[g_y < 0\right] \ll 3 + \left[g_x < 0\right] \ll 2 + \left[\left|g_y\right| > \left|g_x\right|\right] \ll 1 + \left[\left|g_y\right| < \left(\sqrt{2} - 1\right)\left|g_x\right|\right] + \left[\left|g_y\right| > \left(\sqrt{2} - 1\right)\left|g_x\right|\right] \tag{3}$$

Figure 5. The extension divides all eight bins once again and delivers a finer subdivision

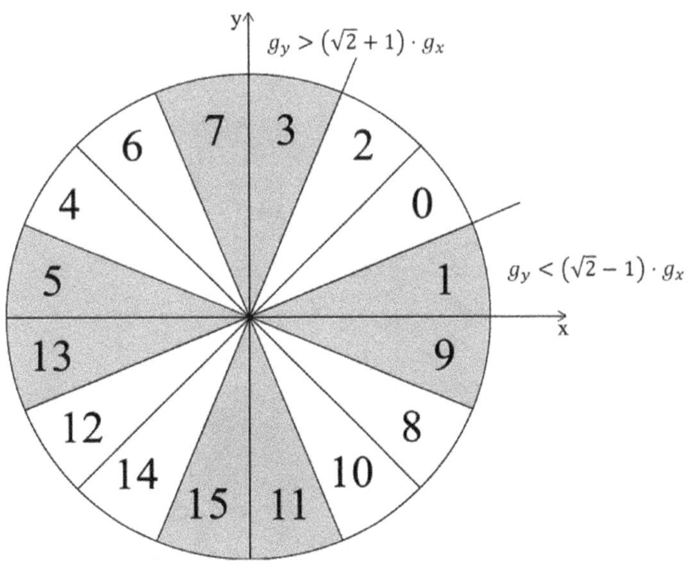

Thus, the finer division is achieved after pointing to one of the eight bins for example $g_x = 4$ and $g_y = 5$:

$$b_{4,5} = [5 < 0] \ll 3 + [4 < 0] \ll 2 + [|5| > |4|] \ll 1 + [|5| < 1.23] + [|5| > 9.64] = 2 \tag{4}$$

3. **4 Bins:** Many approaches concerning object recognition based on edge detection decide either to consider only the half circle for the gradient directions or to consider the full circle. The basic thought for this strategy is that the gradient at a certain position would be reversed by a motorcycle driver with a dark riding suit and sunshine in opposite to a motorcycle driver with a bright riding suit and a cloudy weather. As for this situation the gradient should give us the same information a common procedure is to map the opposite gradient to the same direction as shown in Figure 6. This is done by a further extension of the Petterson approach

([15]):

$$b_{g_x,g_y} = [g_y < 0] \ll 1 + [g_x < 0] \ll 1 + [|g_y| > |g_x|] - [[g_y < 0] \wedge [g_x < 0]] \ll 2 \tag{5}$$

Furthermore, the exact orientation computation is implemented to compare the effectiveness of the implemented methods and evaluate weighted accumulation by means of linear weighting and Gaussian weighting. The best result reported in Dalal and Triggs (2005) is 9 bins over the range 0° – 180°. Their results are based on images sized 64×128 pixels with 1805 challenging pedestrian samples.

Figure 6. All gradients are mapped to the opposite direction with a further extension of 4 bin

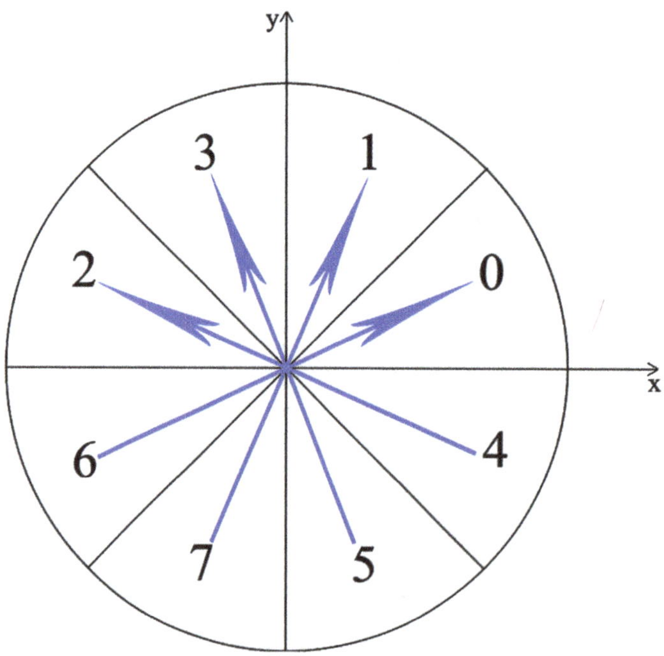

Figure 7. Linear weighting line to balance the magnitude equally and a Gaussian weighting curve

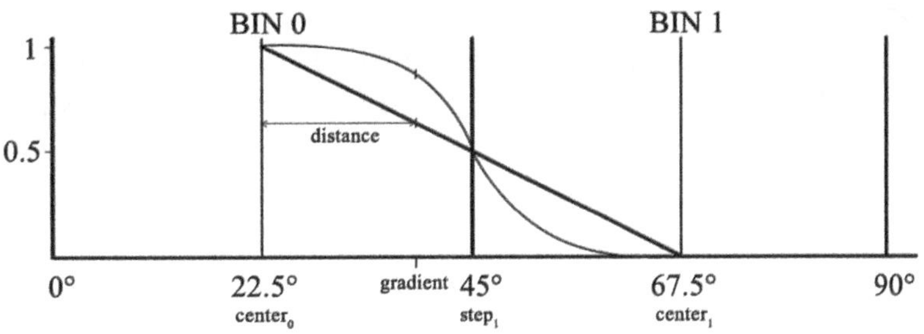

Weighted Binning Approaches

Without a weighted binning all gradients in an interval will increase that certain bin by the same amount. Hence, when looking at the example above with 16 bins a gradient with the angle of 22° is assigned to the first bin whereby a similar angle with 23° is assigned to the second bin. To compensate this problem weighting methods are added to balance the magnitude of a gradient and improve therefore the components of the feature vector.

1. *Linear weighting:* A linear weighting algorithm firstly computes exactly the gradient orientation by using $\arctan\left(g_y \cdot g_x^{-1}\right)$. The next task is to get the angle steps with a given number of bins $s_{\text{bins}} =$ 360°/bins and calculate the center c of the considered bin b_i with $c = b_i \cdot s_{bins} + \frac{s_{bins}}{2}$. With the help of the center the nearest bin is obtained to compute the exact weighted magnitude for the bin i

$$w_0 = \left(1 - \frac{d}{s_{bins}}\right)\sqrt{g_x^2 + g_y^2}$$
$$w_1 = \left(1 - \frac{s_{bins} - d}{s_{bins}}\right)\sqrt{g_x^2 + g_y^2}$$

where w_i denotes the weighted magnitude for the bin i and d denotes the distance.

2. 2. *Gaussian weighting:* Another weighting scheme, called Gaussian weighting is also considered in this paper. Other than the linear weighting, Gaussian weighting increases the weight of the bin which the gradient angle is assigned to. The weighting values are saved in a look-up table *LUT* to avoid the computation of the Gaussian function for all pixels. After the distance d is computed the Gaussian weighted magnitude is received by looking at the element at index d with

$$w_0 = LUT[d]\ (g_x^2 + g_y^2)^{(1/2)}$$

$$w_1 = (1 - LUT[d])(g_x^2 + g_y^2)^{(1/2)}$$

Descriptor Blocks

In order to compensate for illumination variations and variable foreground-background contrast it is necessary to group the cells into larger entities, so-called blocks. It is then possible to perform a local contrast normalization over the gradient magnitude which is also shown in Figure 3. A descriptor is constructed from the set of normalized histograms of the cells involved. The overlapping blocks are also used by Dalal and Triggs (2005) whereby a cell contributes to multiple blocks in the calculations. In order to compute these blocks efficiently Zhu et al. (2006) suggest an integral Histogram of Oriented Gradients *IHOG*. According to Porikli (2005) a method is proposed to compute quickly the histogram of a rectangular image region to speed up the computation time. A similar technique has been very popular based on the approach proposed by Viola and Jones (2001). The approach suggested by Porikli (2005) is used in the following. All cells are represented inside this *IHOG*.

The size of this array depends on the cell size c_x and c_y and the bin quantity b. So given the target width w and height h the border pixels are not considered and hence width and height are reduced by 2 pixels.

$$IHOG_{\text{width}} = (w - 2)/c_x$$

$$IHOG_{\text{height}} = (h - 2)/c_y$$

$$IHOG\text{size} = IHOG_{\text{width}} IHOG_{\text{height}} b$$

First all cells are computed separately and saved at the certain position following by the integral computation to create *IHOG*. So once this is computed all block-descriptors can be obtained by referencing only $4 \cdot b$ elements in the array. The result is a vector which contains all accumulated magnitudes of the underlying cells.

After the block of a certain region is computed the next step is to normalize these values.

CLASSIFICATION WITH SVM

A Support Vector Machine (SVM) divides a set of features into classes such that there is a large range to the margin of these classes. In this paper the main issue is to identify an object as a motorcycle or traffic sign. Thus, we have to distinguish between motorcycle and traffic sign or the absence of these objects. The use of a SVM classifier is a common approach in object recognition. In order to build a SVM we need a set of training data with associated labels that indicate the assigned class. The fundamental idea of a support vector classification is to separate all input objects in two classes by a hyperplane. The distance of the nearest vectors of both classes to the hyperplane will be maximized. It is not necessary to consider all vectors in order to calculate the hyperplane. Vectors which have a larger distance to the hyperplane do not affect the calculation of the hyperplane. The separation is only depended on the nearest vectors and only they are needed to describe the hyperplane mathematically. These vectors are called support vectors. A perfect linear separation is achieved for rather artificial classification problems but this is not the case in common applications as the variability of the objects in a certain class is too high. In cases where a linear separation is not possible, the so-called kernel trick might be applied. This paper deals primarily with linear SVMs so this topic is not tackled in further detail.

The evaluation in this paper is based on the linear kernel. Polynomial- or Gaussian-kernels are not used in the next experiments. The reason for this is that when using linear SVMs only the hyperplane w and the bias parameter *b* are considered for the computation. So the calculated support vectors are neglected. By using the kernel trick, it is necessary to consider all determined support vectors in the computation which increases the runtime drastically.

The evaluation with the Support Vector Machines (SVM) is done very detailedly due to the fact that Dalal and Triggs (2005) and Zhu et al. (2006) based their results on a linear SVM classifier too and reported very promising results. Another strong reason is the computation time of this classifier with a linear kernel.

EXPERIMENTS

In order to carry out classification experiments, 230 sequences have been recorded and labelled. For the evaluation of the classification modules explained in the previous sections, 6415 motorbike images have been extracted. The nonmotorcycle class is extracted by randomly sampling images from the set of 230 sequences. In order to construct a well-balanced dataset, the same number of images is used in the non-motorbike class, namely 6415 samples.

The following tests are based on the five-fold cross validation method. The whole data set is divided into five subsets. One of the five subsets is used as the testing set whereas the other four sets form the training set. This procedure is carried out with all five subsets and thus finally we get five error rates which is computed as the ratio between the false classifications and the overall amount of samples in the test set. Finally, the mean measure of the error rate is denoted as E.

Table 1 gives detailed information about the recognition rate. The check marks indicate the type of gradient computation. Rows in the table with checked marks indicate a calculation with the arctan whereas unchecked rows indicate a computation with the Pettersson approach. The column headed with n_f in Table 1 indicates the number of features used in the particular configuration. Please note that Dalal and Triggs (2005) report different results and a different *winning* configuration. They note that 9 bins over a range of 0-180° results best for human shape detection. Here it is possible that there is not a general solution but rather it is bound to the detection problem.

In subsection II-B the different weighting approaches are described. In Table 2 the results are given for each weighting approach. Note that the linear weighting approach performed best.

The same combination of feature extraction and SVM based classification has been applied to the recognition of traffic signs. In this paper, three different sign classes have been chosen, namely the stop sign, the yield sign and a 50km/h speed limit sign. The aforementioned signs have been chosen in order to demonstrate the interaction between the traffic signs and an autonomously driving model car. The experiments described here are an extension of the work presented in Müller-Schneiders (2014), especially the video database has been heavily extended and the step towards multi-class processing has been taken.

The model car in this case is an AADC model car which is shown in Figure 8. AADC is an abbreviation for Audi Autonomous Driving Cup and describes a competition for model cars carried out by Audi since 2016 (see e.g. Kuhnt et al., 2016; Zovka et al., 2016).

In a similar manner as described in the beginning of chapter 4, a database of videos has been recorded and labelled by hand. In this case, a total number of 27 sequences have been recorded. Figure 9 shows the track plan, which has been used for recording the sequences. Since this number of sequences is not

Table 1. Experimental results for the motorcycle classification task

Bins	Angular Range	arctan	n_f	E[%]
4	0°– 180°		160	2.98
8	0°– 180° 0°– 360°	X	320	2.22 2.14
9	0°– 180° 0°– 360°	X X	360	2.49 2.06
12	0°– 180° 0°– 360°	X X	480	2.88 1.87
16	0°– 180° 0°– 360°	X	640	2.93 2.06

Table 2. Experimental results for different weighting schemes

Weighting Approach	E[%]
No Weighting	2.14
Linear Weighting	1.70
Gaussian Weighting	3.28

Figure 8. Audi Autonomous Driving Cup model car, used in the experiments with traffic signs

sufficient to train a well performing classifier, additionally, publicly available data has been utilized for training. The Bochum traffic sign recognition database (Stahlkamp et al., 2012) contains more than 50000 labelled traffic signs which have been used for training the three chosen sign classes. Again, the rejecting sign class has been created by sampling random images from recorded sequences not containing the target signs.

Figure 9. Track plan used in the experiments

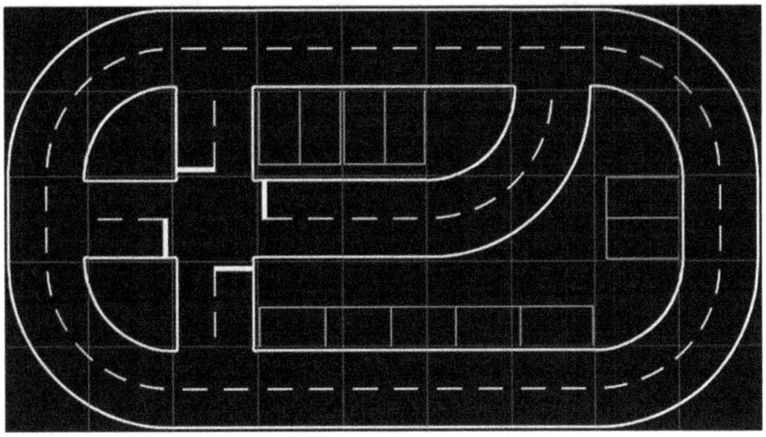

Table 3 summarizes the recognition rates for the stop sign task, evaluated on 10 recorded sequences. In the experiments, the SVM kernels have been varied, whereas the remaining parameters of HOG and SVM have been held constant. In Table 3 "Linear" denotes a linear SVM, "RBF" is a Radial Basis Function SVM, "Poly" denotes a polynomial SVM and finally "Inter" is a histogram intersection SVM.

Compared to the motorcycle task, one can see that the error is now larger, even though the stop sign task can be considered less challenging. This can be explained by the training method, where a positive training set has been used which was recorded in a different environmental setting. Please recall that for the stop signs, data from the Bochum TSR database has been used.

Luckily the best recognition rate has been achieved with the linear SVM. This is an advantage, since the processing time for this type of classifier is by far the smallest and thus, we get the best results along with the most efficient algorithm. Figure 10 shows sample detections from the sequence database. The images show the detection and correct classification of a stop sign, a yield sign as well as a 50km/h speed limit sign. The images have been taken with the color camera of the model car. Figure 11 illustrates the HOG features

SUMMARY AND CONCLUSION

This paper describes and analyses vision algorithms for the recognition of static and dynamic road elements. Histogram of Oriented Gradients (HOG) has been used as feature extraction in combination

Table 3. Experimental results for the stop-sign task

SVM Kernel	*E*[%]
Linear	8.3
RBF	10.0
Poly	9.2
Inter	9.0

Figure 10. Sample detections taken from the video database. calculated on three different traffic signs.

Figure 11. Illustration of HOG features for three different traffic sign classes.

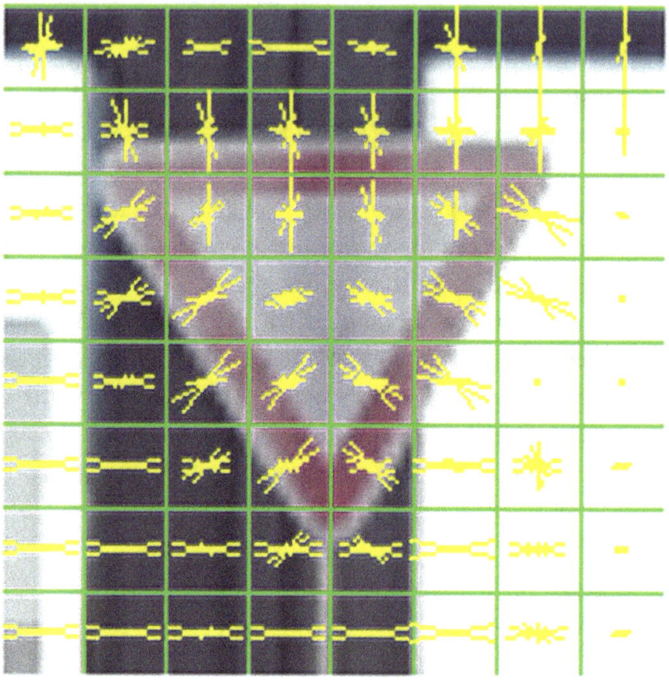

with support vector machines (SVM) as classifier. The experiments showed that using the same set of algorithms, it was possible to recognize motorcycles as well as stop signs in video images. Experiments identified the best performing parameters and analyzed the efficiency of the proposed algorithms. Overall the HOG, SVM combination is an attractive alternative to the convolutional neural networks, since the separate feature extraction step can be multiplexed between various classes and thus the computational efficiency can be improved. Future experiments will extend the number of classes which can be recognized.

REFERENCES

Bensrhair, A., Bertozzi, M., Broggi, A., Mich'e, P., Mousset, S., & Toulminet, G. (2001). A Cooperative Approach to Vision-based Vehicle Detection, *IEEE Intelligent Transportation Systems Conference*. 10.1109/ITSC.2001.948657

Chung-Cheng, C., Min-Yu, K., & Hung-Tsung, C. (2007). Motorcycle Detection and Tracking System with Occlusion Segmentation, *Eighth International Workshop on Image Analysis for Multimedia Interactive Services*, pp. 23–32.

Dalal, N., & Triggs, B. (2005). Histograms of Oriented Gradients for Human Detection, *IEEE Computer Society Conference on Computer Vision and Pattern Recognition*. 10.1109/CVPR.2005.177

Fu, M.-Y., & Huang, Y.-S. (2010). *A Survey of Traffic Sign Recognition. IEEE Intern. Conf. on Wavelet Analysis and Pattern Recognition*. doi:10.1109/ICWAPR.2010.5576425

Knuth, D. (1992). Two Notes on Notation. *The American Mathematical Monthly*, *99*(5), 403–422. doi: 10.1080/00029890.1992.11995869

Kuhnt, F., Pfeier, M., Zimmer, P., & Zöllner, J. M. (2016). *Robust Environment Perception for the Audi Autonomous Driving Cup. IEEE Intern. Conf. on Intelligent Transportation Systems*. doi:10.1109/ITSC.2016.7795744

LeCun, Y., Bottou, L., Bengio, Y., & Haffner, P. (1998). Gradient-Based Learning Applied to Document Recognition. *Proceedings of the IEEE*, *86*(11), 2278–2324. doi:10.1109/5.726791

Messelodi, S., Modena, C. M., & Cattoni, G. (2007). Vision-based Bicycle/Motorcycle Classification. *Pattern Recognition Letters*, *28*(13), 1719–1726. doi:10.1016/j.patrec.2007.04.014

Mogelmose, A., Trivedi, M. M., & Moeslung, T. B. (2012). Traffic Sign Detection and Analysis: Recent Studies and Emerging Trends. Anchorage, AK: In 15th International IEEE Conference on Intelligent Transportation Systems.

Mukhtar, A., & Tang, T. B. (2015, October). Vision based motorcycle detection using HOG features. In *2015 IEEE International Conference on Signal and Image Processing Applications (ICSIPA)*, (pp. 452-456). IEEE.

Müller-Schneiders, S. (2019). Vision-based Detection of Road Users and Infrastructure Elements for Automated Driving, *IEEE Intern. Conference on Research and Education in Mechatronics (REM 2019)*. 10.1109/REM.2019.8744089

Müller-Schneiders, S., Nunn, C., & Meuter, M. (2008). Performance Evaluation of a Real Time Traffic Sign Recognition System, *IEEE Intelligent Vehicle Symposium*. 10.1109/IVS.2008.4621164

Pettersson, N., Petersson, L., & Andersson, L. (2008). The histogram feature - a resource-efficient Weak Classifier, *IEEE Intelligent Vehicles Symposium*. 10.1109/IVS.2008.4621174

Porikli, F. (2005). Integral histogram: a fast way to extract histograms in Cartesian spaces, *IEEE Computer Society Conference on Computer Vision and Pattern Recognition*. 10.1109/CVPR.2005.188

Sivaraman, S., & Trivedi, M. M. (2013). Looking at Vehicles on the Road: A Survey of vision-based Vehicle Detection, Tracking, and Behavior Analysis. *IEEE Transactions on Intelligent Transportation Systems*, *14*(4), 1773–1795. doi:10.1109/TITS.2013.2266661

Stahlkamp, J., Schlipsing, M., Salmen, J., & Igel, C. (2012). Benchmarking Machine Learning Algorithms for Traffic Sign Recognition. *Neural Networks*, *32*, 323–332. doi:10.1016/j.neunet.2012.02.016 PMID:22394690

Sun, Z., Bebis, G., & Miller, R. (2006). On-Road Vehicle Detection: A Review. *IEEE Transactions on Pattern Analysis and Machine Intelligence*, *28*(5), 694–711. doi:10.1109/TPAMI.2006.104 PMID:16640257

Swathi, M., & Suresh, K. V. (2017). *Automatic Traffic Sign Detection and Recognition: A Review. IEEE Intern. Conf. on Algorithms, Methodology, Models and Applications in Emerging Technologies.* doi:10.1109/ICAMMAET.2017.8186650

Viola, P., & Jones, M. (2001). Robust Real-time Object Detection. *International Journal of Computer Vision*.

Wen, X., Yuan, H., Song, C., Liu, W., & Zhao, H. (2007). An Algorithm based on SVM Ensembles for Motorcycle Recognition. *In 2007 IEEE International Conference on Vehicular Electronics and Safety* (pp. 1-5). IEEE.

Wu, B., & Nevatia, R. (2005, October). Detection of multiple, partially occluded humans in a single image by bayesian combination of edgelet part detectors. In *Tenth IEEE International Conference on Computer Vision (ICCV'05), 1*(1), pp. 90-97). IEEE.

Zhu, Q., Yeh, M. C., Cheng, K. T., & Avidan, S. (2006, June). Fast human detection using a cascade of histograms of oriented gradients. In *IEEE Computer Society Conference on Computer Vision and Pattern Recognition (CVPR'06), 2,* pp. 1491-1498. IEEE.

Zofka, M. R., Kuhnt, F., Kohlhaas, R., & Zöllner, J. M. (2016, December). Simulation framework for the development of autonomous small scale vehicles. In 2016 IEEE International Conference on Simulation, Modeling, and Programming for Autonomous Robots (SIMPAR) (pp. 318-324). IEEE.

Chapter 15
A Proposed Trajectory Planning Algorithm for Mobile Robot Navigation Based on A* Algorithm

Şahin Yıldırım
Erciyes University, Turkey

Sertaç Savaş
https://orcid.org/0000-0001-8096-1140
Erciyes University, Turkey

ABSTRACT

This chapter proposes a new trajectory planning approach by improving A algorithm, which is a widely-used, path-planning algorithm. This algorithm is a heuristic method used in maps such as the occupancy grid map. As the resolution increases in these maps, obstacles can be defined more precisely. However, the cell/grid size must be larger than the size of the mobile robot to prevent the robot from crashing into the borders of the working environment or obstacles. The second constraint of the algorithm is that it does not provide continuous headings. In this study, an avoidance area is calculated on the map for the mobile robot to avoid collisions. Then curve-fitting methods, general polynomial and b-spline, are applied to the path calculated by traditional A* algorithm to obtain smooth rotations and continuous headings by staying faithful to the original path calculated. Performance of the proposed trajectory planning method is compared to others for different target points on the grid map by using a software developed in Labview Environment.*

DOI: 10.4018/978-1-7998-0137-5.ch015

INTRODUCTION

Scientific studies on mobile robots have increased considerably in recent years. These studies can be summarized in four main sections; localization, mapping, path or trajectory planning and control. The path planning topic from these four main headings is the main theme of this paper. Over the years, many methods have been proposed for path planning. Calculation time, path length and path characteristics are the focus of improvements in path planning.

Many different algorithms have been developed for the mobile robot path planning task and the improvements for these algorithms are focused on factors such as calculation time, path length and path characteristics. In addition, these algorithms are mainly divided into classical methods and heuristic methods. A survey (Mac, Copot, Tran, & De Keyser, 2016) was concentrated on heuristic algorithms based on neural network, fuzzy logic, nature-inspired algorithms and hybrid algorithms. Also Buniyamin et al. (Buniyamin, Ngah, Sariff, & Mohamad, 2011) introduced a new approach called PointBug as a local path planning algorithm and compared the proposed approach with traditional Bug algorithms, Distbug and TangenBug, in terms of path length and reachability performance to the target. Čikeš et al. (Čikeš, Đakulović, & Petrović, 2011) presented a comparative study of D*, TWD* (Two Way D*) and E* algorithms on the occupancy grid map of the environment in terms of path characteristics, searching time and the number of iterations. Subramanian et al. (Subramanian, Sudhagar, & RajaRajeswari, 2014) presented Breadth First Search (BFS) algorithm as a path planning method to search for the shortest path faster than heuristic methods. Also in another study they (Subramanian, Sudhagar, & Rajarajeswari, 2015) presented a dynamic pathfinding approach in an unknown environment using vector field histogram for obstacle avoidance and A* algorithm for a heuristic path planning.

In addition, many studies have been made on A* algorithm which is a basic but powerful path planning approach and various improvements are presented to optimize the algorithm. Guruji et al. (Guruji, Agarwal, & Parsediya, 2016) proposed a method to provide efficiency in processing time of A* algorithm. Yin and Yang (2013) proposed a multi-path algorithm based A* comparing with a K-Dijkstra-based algorithm to use in route navigation systems. Pala et al. (Pala et al., 2013) proposed a path planning algorithm called HCTNav which is focused on minimum use of computing and energy resources and they compared their method with Dijkstra's algorithm.

The A* path planning algorithm is not only used for wheeled mobile robots. The algorithm can be also used for navigation planning of water and air vehicles. For this reason, studies on the development of the method have been made for various platforms and tasks. Masudur Rahman Al-Arif et al. (Al-Arif, Ferdous, & Nijami, 2012) proposed a path planning algorithm using artificial intelligence technique considering to be used for a water vehicle during a rescue operation and presented a comparative study of Dijkstra and A* algorithms. Tseng et al. (Tseng, Liang, Lee, Chou, & Chao, 2014) proposed A* algorithm to plan flight path of a civil unmanned aerial vehicle to provide a better quality on 3G communication during flight. Cheng et al. (Cheng, Liu, & Yan, 2014) presented an improved hierarchical A* algorithm that takes into account shortest distance and time to calculate optimal parking path in a large parking guidance system.

This paper is organized as follows: In Section 2, occupancy grid map and A* algorithm are introduced. In Section 3, calculating avoidance area on grid map and curve fitting process for calculated path points are explained. After that, simulation results for the proposed approach compared with other methods are given in Section 4. Finally, discussions on the performance of improvements and suggestions for future studies are made in Section 5.

OCCUPANCY GRID MAPS AND A* ALGORITHM

Occupancy Grid Maps

In mobile robot applications, it is very important to build a map that will be an abstract representation of the real environment. This representation should be defined as empty or non-empty areas in order to provide navigation of the mobile robot. Based on this basic idea occupancy grid maps (Gonzalez-Arjona, Sanchez, López-Colino, de Castro, & Garrido, 2013; Milstein, 2008; Satapathy & Kumar, 2016) are used to represent the working environment as a 2D array of cells/grids. Since each cell only provides a square representation, occupancy grid maps can not be absolutely accurate. But this problem could be overcomed insofar by choosing a small enough cell size.

Occupancy grid mapping algorithms are based on probabilistic approaches to determine the occupancy of each cell by using the depth data which can be collected from various sensors on mobile robot with the assumption that the instant position of the robot is known. So the map is generated and updated relative to the robot's estimated position. Occupancy grid maps are represented by using intensity graphs and the intensity value of a cell is 1 or 100, respectively white if the cell is free to move or black if it is not free because of an obstacle. An illustration of an occupancy grid map and the value of each cell is shown in Figure 1.

A* Algorithm

A* is a best-first algorithm widely used for mobile robot applications to search for a least-cost path through a static map from a start point to a **goal point**. The algorithm is invented in 1968 by the researchers working on path planning problem of "Shakey the Robot" which is the first general-purpose mobile robot (Brachman & Levesque, 2004). The method is a combination of Dijkstra and Herustic algorithm. The evaluation function (Duchoň et al., 2014) used in this algorithm is given by;

$$F(v) = G(v) + H(v) \tag{1}$$

Figure 1. An illustration of an occupancy grid map and the value of each cell

1	100	100	100	100	100	100	100	100
1	1	1	1	1	100	100	100	100
100	1	1	1	1	1	100	100	100
100	100	100	1	1	1	1	1	100
100	100	100	100	100	1	1	1	1

where $v = \left(x_v, y_v\right)$ is recent expanding cell, $F\left(v\right)$ is the estimated minimum cost of the cell, $G\left(v\right)$ is the actual cost of the length from the start cell to the current cell, $H\left(v\right)$ is the herustic estimation of minimum cost of a path form the current cell to the goal cell. The algorithm uses two list arrays called open list and closed list to calculate the cost values for the neighbors of the current cell and updates the lists during calculations. So the calculations are expanded for the following cell in the sequence with the lowest value of $F\left(v\right)$. A basic example for A* algorithm implementation to reach cell B from cell A is shown in Figure 2.

MODIFICATIONS ON GRID MAP AND CURVE FITTING PROCESS

Calculating Avoidance Area for Grid Map

A square platform, in dimensions of 4.5x4.5 meter, is considered as the working environment to generate the occupancy grid map for using to implement and compare proposed path planning algorithm with other methods. The grid map, in dimensions of 150x150 cells, is generated by manuel selection of each cell value by a software developed in Labview. Each cell in the grid map represents a square area of 3x3 cm. Also some square obstacles are considered to see the heading perfomance of algorithms.

After the grid map is built, a software is developed to calculate the avoidance area to prevent the mobile robot from crashing into environment borders and obstacles. The flowchart of the developed software is given in Figure 3 for good understanding. In this flowchart, $V\left(i, j\right)$ represents the value of each cell as 1 or 100. Each cell of the $M \times N$ size grid map is examined in terms of neighboring cells and the cells found to be near the obstacle or platform boundary are marked as full, and this marking is repeated W times for the entire map. W represents the width of the avoidance area, chosen as the width of the mobile robot. As a mobile robot, The "National Instruments Robotics Starter Kit 2.0" robot is chosen in order to determine the calculation parameters of proposed path planning to be used in simula-

Figure 2. A basic diagram of A algorithm implementation.*

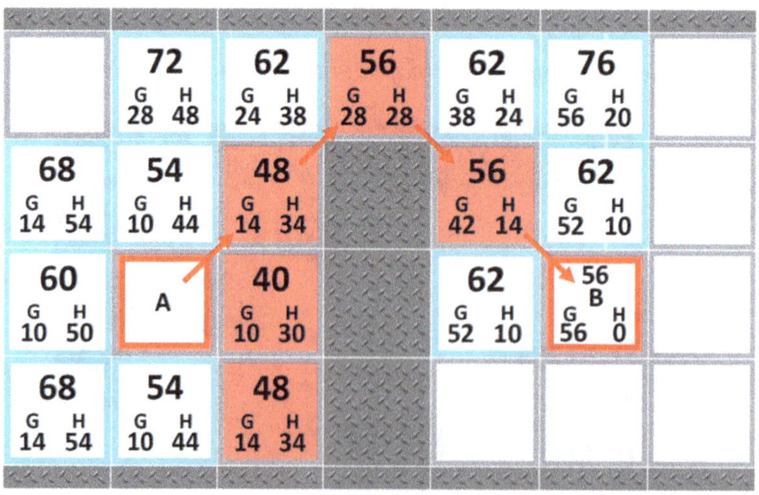

Figure 3. Flowchart of the software that calculates the avoidance area.

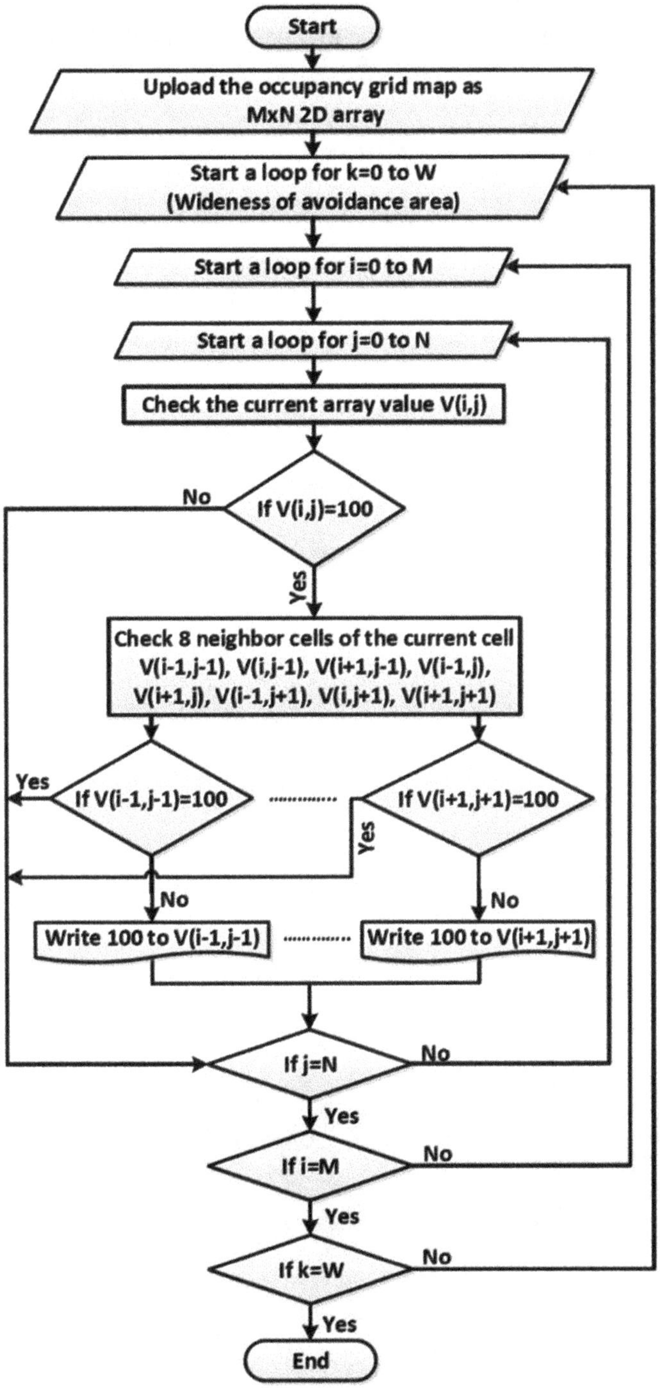

tion studies. The mobile robot is shown in Figure 4. Assuming that the width of the mobile robot (*L*) is 33 cm, the *W* value is chosen as 11 to provide an avoidance area as much as the width of the mobile robot. The occupancy grid map generated and the avoidance area calculated for this map are shown in Figure 5.

Curve Fitting for Calculated Path Points

In the study two curve fitting methods, general polynomial and b-spline, are applied to the path points *(X, Y)* calculated by A* algorithm on the occupancy grid map with avoidance area. General Polynomial Fitting process fits a data set to a polynomial function of the general form described by the following equation using the Least Square, Least Absolute Residual or Bisquare method;

$$f\left(i\right) = \sum_{j=0}^{m} a_j x_i^j \tag{2}$$

where *f* is the output array as best fit, *x* is the input array *X*, a_j is the polynomial coefficient for each portion of equation and *m* is the polynomial order ("General Polynomial Fit VI - LabVIEW 2014 Help," 2014).

Figure 4. "National Instruments Robotics Starter Kit 2.0" mobile robot.

Figure 5. Occupancy grid map with avoidance area calculated.

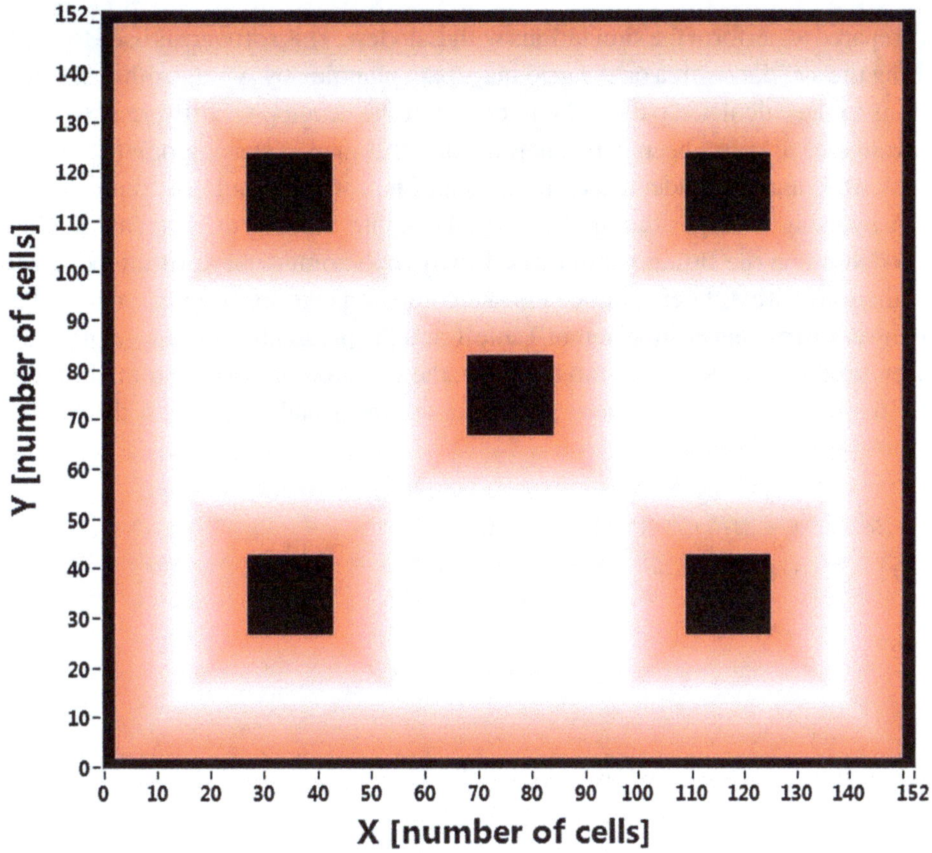

B-Spline is the other method applied to the path points calculated. The B-Spline Fitting process fits a data set by minimizing the residue according to the following equation using the equally spaced, chord length or centripetal method which computes the interim knot vector;

$$\frac{1}{N}\sum_{i=0}^{N-1} w_i \cdot \left\| (x_i, y_i) - (x_i', y_i') \right\|^2 = \frac{1}{N}\sum_{i=0}^{N-1} w_i \cdot \left[\left(x_i - x_i' \right)^2 + \left(y_i - y_i' \right)^2 \right] \quad (3)$$

where N is the length of Y, w_i is the ith element of Weight, (x_i, y_i) is the ith pair of the input sequences (X, Y), (x'_i, y'_i) is the ith pair of best fit and the norm symbols (||) on both sides of the function compute the l_2 norm of a vector ("B-Spline Fit VI - LabVIEW 2011 Help," 2011).

SIMULATION RESULTS

A software is prepared to perform the simulations with Labview. This software is used to implement three tasks; calculating the avoidance area on the grid map, path planning by A* algorithm and applying curve fitting operations to the calculated paths. The interface of the developed software is shown in Figure 6.

After the avoidance area has been calculated for the grid map, a starting point (SP) and four target points marked A, B, C and D are identified on the map. The path is calculated from start point to target point marked A by using A* algorithm on the map. Then curve fitting methods are applied to the calculated path to determine the fitting parameters. Firstly, polynomial curve fitting is performed for 30 different combinations with different parameters and compared in terms of error values. The 6 combinations with the lowest error values are given in Table 1 and the parameters in row 4 with the lowest MSE value are chosen for the comparison. In the Labview, the equation of the polynomial function obtained when applying the curve fitting with these parameters to the x and y values of the points in the path calculated with A* is given below;

$$\begin{aligned} y = & -68,06 + 28,53 \cdot x - 2,75 \cdot x^2 + 0,14 \cdot x^3 - 0 \cdot x^4 \\ & +6,23 \cdot 10^{-5} \cdot x^5 - 5,16 \cdot 10^{-7} \cdot x^6 + 1,73 \cdot 10^{-9} \cdot x^7 \end{aligned} \tag{4}$$

Figure 6. The interface view of the developed software.

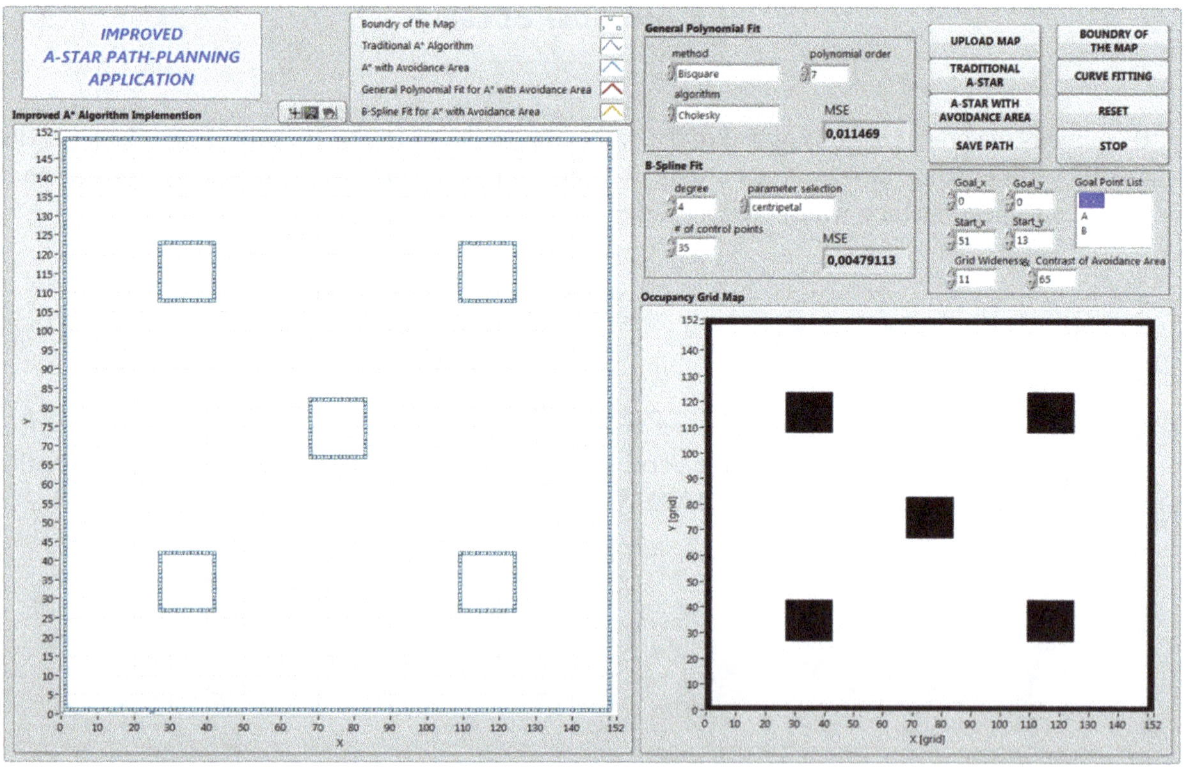

Table 1. Parameters and error values of general polynomial curve fitting for path points.

General Polynomial Fit							
	Fitting Method			Polynomial Order	MSE	Start Point Error (X-Axis)	End Point Error (X-Axis)
	Least Square	Least Absolute Residual	Bisquare				
1	X			8	0,213963	1,181257	-0,22097
2	X			9	0,150555	0,473597	0,486756
3		X		9	0,260558	0,659859	0,716724
4			**X**	**7**	**0,011469**	**1,450361**	**0,033156**
5			X	8	0,013018	1,509522	0,028266
6			X	9	0,017375	1,166607	0,189848

During the curve fitting process, it is selected 0.0001 as tolerance value, 1 as weight value for each input value and Cholesky as algorithm.

Then, B-spline curve fitting method is applied to the path planned by A* algorithm on the grid map with avoidance area. Curve fitting process is applied in 500 different combinations in terms of parameter, degree and control points, and during the curve fitting process, it is selected 1 as weight value for each input value. 6 combinations with the lowest error values are given in Table 2 and curve fitting parameters in row 5 with the lowest MSE value is chosen for comparison. The curve fitting parameters are determined for target point A and then these parameters are used for all other target points B, C, and D.

In the simulation study, four different path planning approaches have been compared with each other in terms of the paths calculated;

1. **Traditional A* Algorithm:** The path calculated by A* algorithm on the occupancy grid map without any avoidance area.
2. **A* with Avoidance Area:** The path calculated by A* algorithm on the occupancy grid map which has an avoidance area with the same width as the mobile robot.

Table 2. Parameters and error values of b-spline curve fitting for path points.

B-Spline Fit										
	Parameter Selection			Degree	Control Points	MSE	Start Point Error (X-Axis)	Start Point Error (Y-Axis)	End Point Error (X-Axis)	End Point Error (Y-Axis)
	Equally Spaced	Chord Length	Centripetal							
1	X			3	35	0,006377	0,004800	0,000000	0,063294	-0,000000
2		X		3	35	0,004740	0,001405	-0,000582	0,071522	-0,029625
3			X	3	35	0,005333	0,002642	-0,000500	0,067845	-0,012837
4	X			4	35	0,005495	0,000685	0,000000	0,021441	-0,000000
5		**X**		**4**	**35**	**0,004385**	**0,000161**	**-0,000067**	**0,025899**	**-0,010728**
6			X	4	35	0,004791	0,000338	-0,000064	0,023720	-0,004488

3. **General Polynomial Fit for A* with Avoidance Area:** The path obtained by applying general polynomial curve fitting method to the path which has calculated by A* algorithm on the occupancy grid map with avoidance area.
4. **B-Spline Fit for A* Avoidance Area:** The path obtained by applying b-spline curve fitting method to the path which has calculated by A* algorithm on the occupancy grid map with avoidance area.

The comparisons have been made for four target points on grid map. The performances of these four approaches for the target points marked A, B, C and D on the grid map are shown in Figure 7, Figure 8, Figure 9 and Figure 10, respectively.

As seen in Figure 7, if the target point is not closed too much with obstacles, the performance difference of the algorithms is not clear.However, the traditional A * algorithm could not provide a continuous trajectory by drawing a zigzag path in the last section.

Figure 7. Calculated paths for the target point marked A on the occupancy grid map

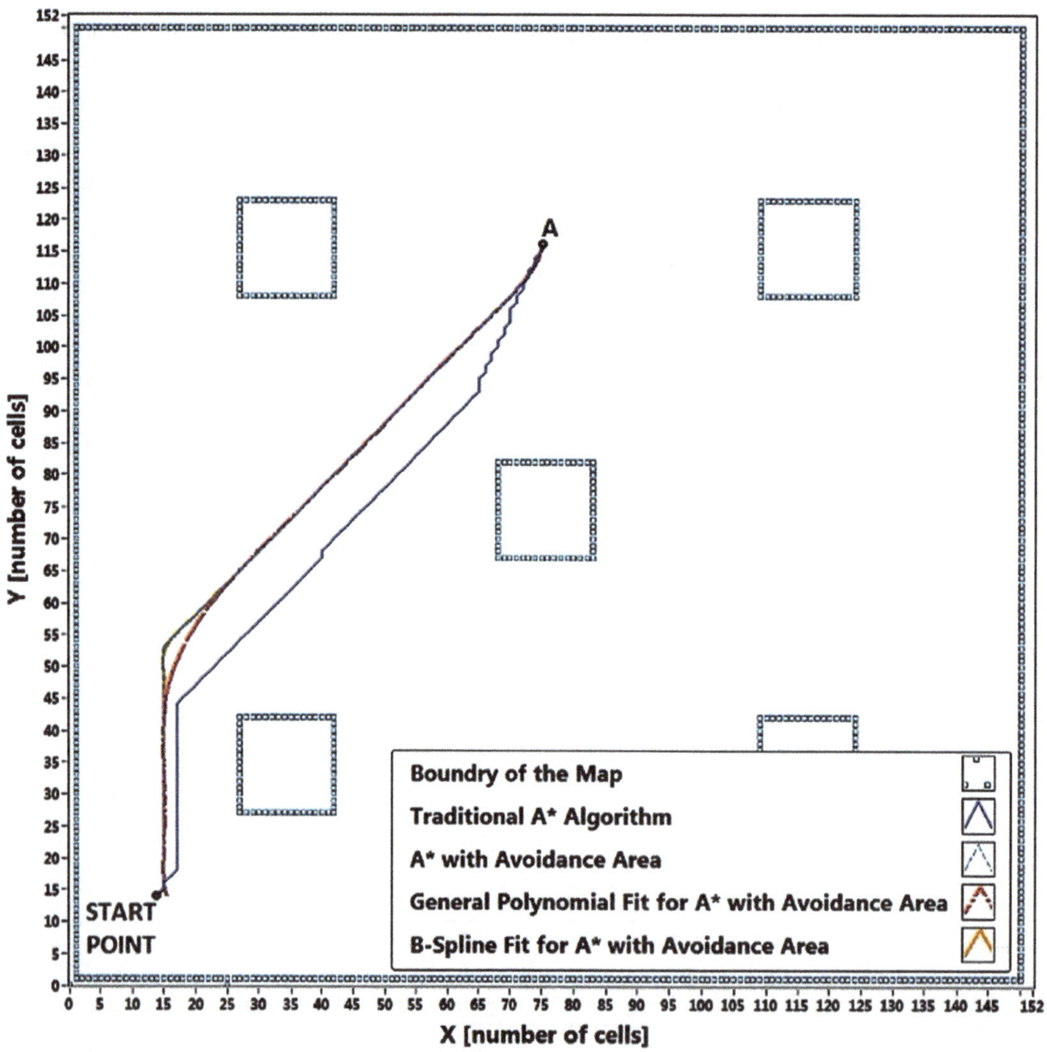

Figure 8. Calculated paths for the target point marked B on the occupancy grid map

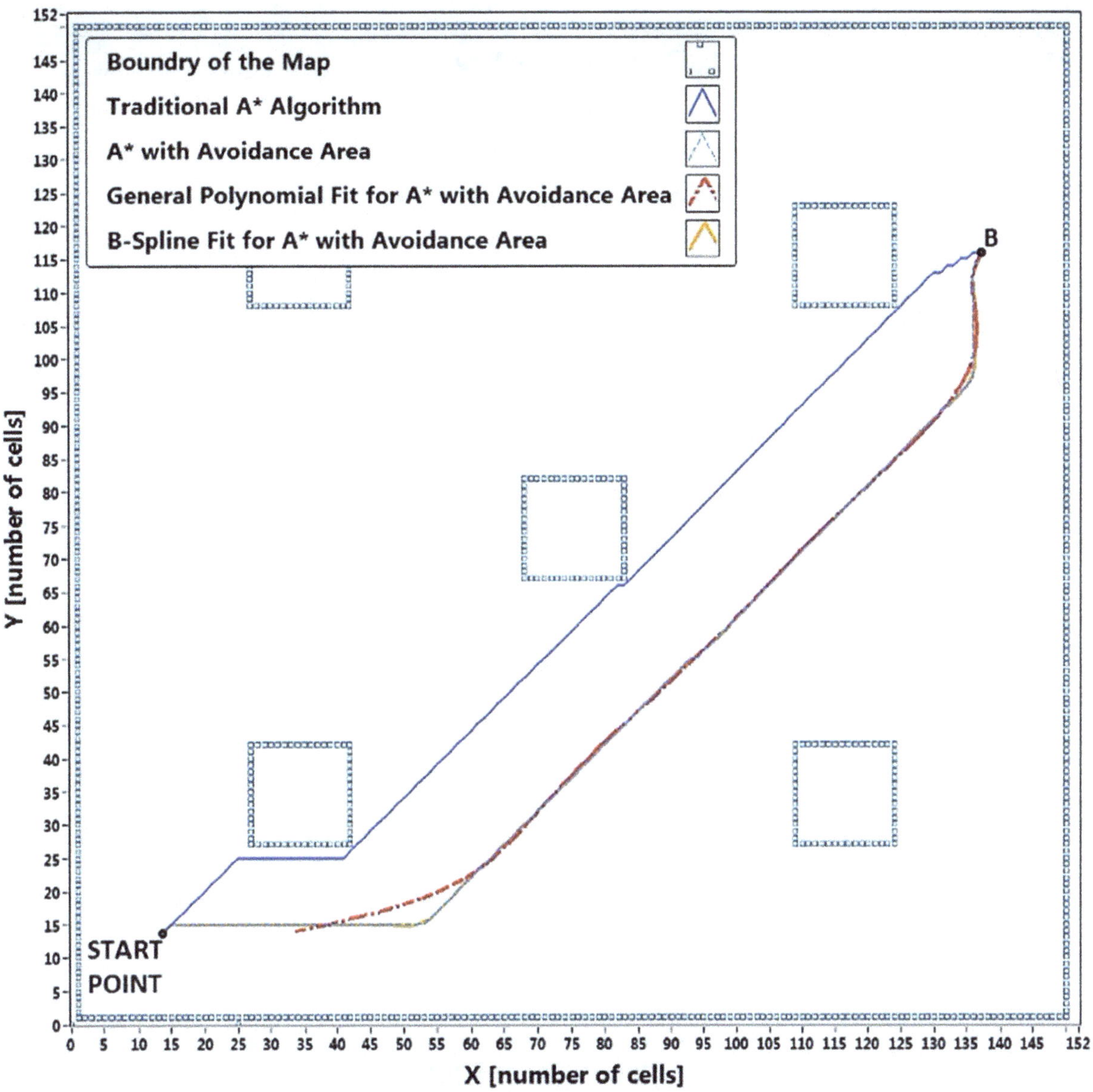

As can be seen in Figure 8 and Figure 9, the traditional A* algorithm plans a route very close to the obstacles, which can cause collisions without considering the width of the mobile robot. The trajectory obtained by applying the polynominal curve fitting performs well for the target point A in which the parameters are determined, but with these parameters a trajectory that can be used for other target points cannot be calculated.

Figure 9. Calculated paths for the target point marked C on the occupancy grid map

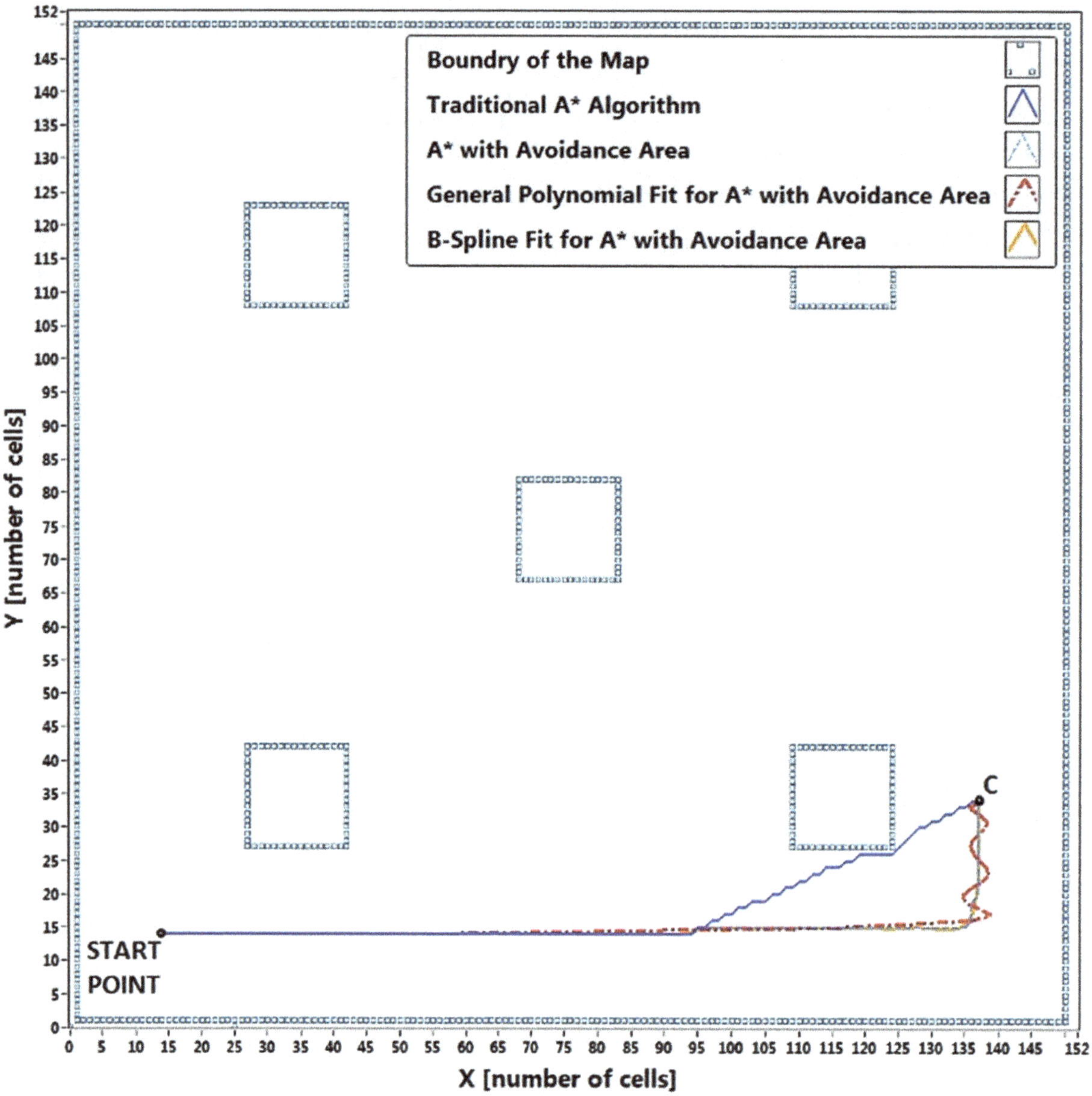

The performance of the comparative methods for rotation can be better seen in Figure 11 by referring to the Section-I in Figure 10. B-spline curve fitting applied trajectory shows the best performance. This performance can be defended by two criteria. The first criterion is that the calculated path does not hit any obstacles and map boundaries. The second criterion is that it plans a continuous trajectory with continuous headings in contrast to the traditional A* algorithm and others.

Figure 10. Calculated paths for the target point marked D on the occupancy grid map

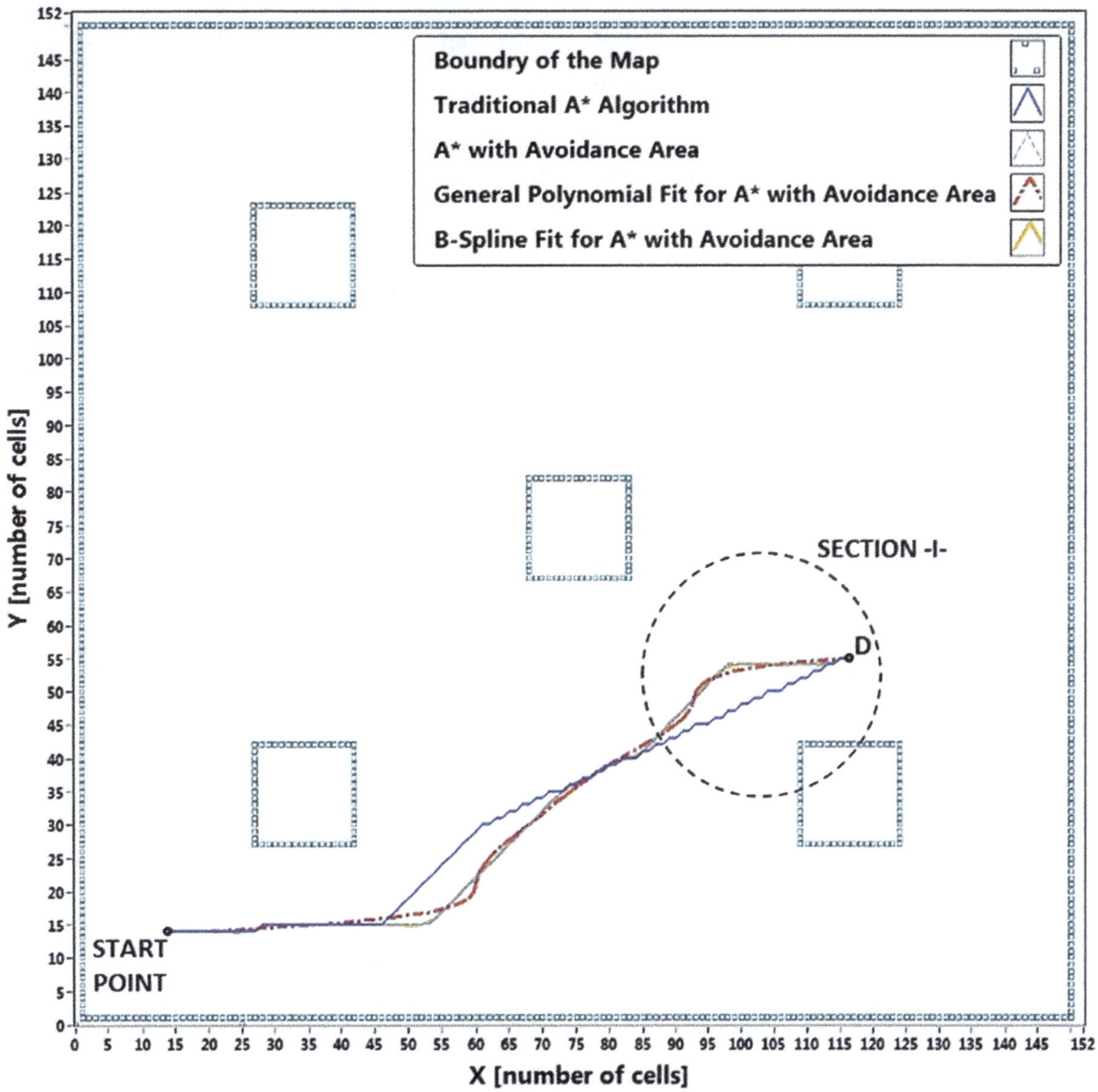

CONCLUSION

Simulation studies have shown that the path calculated by traditional A* algorithm can not avoid collisions from borders of the obstacles or working environment. For this reason, an avoidance area is calculated in the width of the experimental mobile robot. But the traditional A* algorithm calculates a zigzag path without continuous headings even for the grid map with avoidance area. So curve-fitting methods, general polynomial and b-spline, are applied to the paths calculated by A* algorithm on this grid map with avoidance area. The path obtained by general polynomial curve fitting provides a suitable path on conditions which its parameters are determined, but it does not show the same performance under dif-

Figure 11. Rotational performance of the comparative methods - Section I in Figure 10

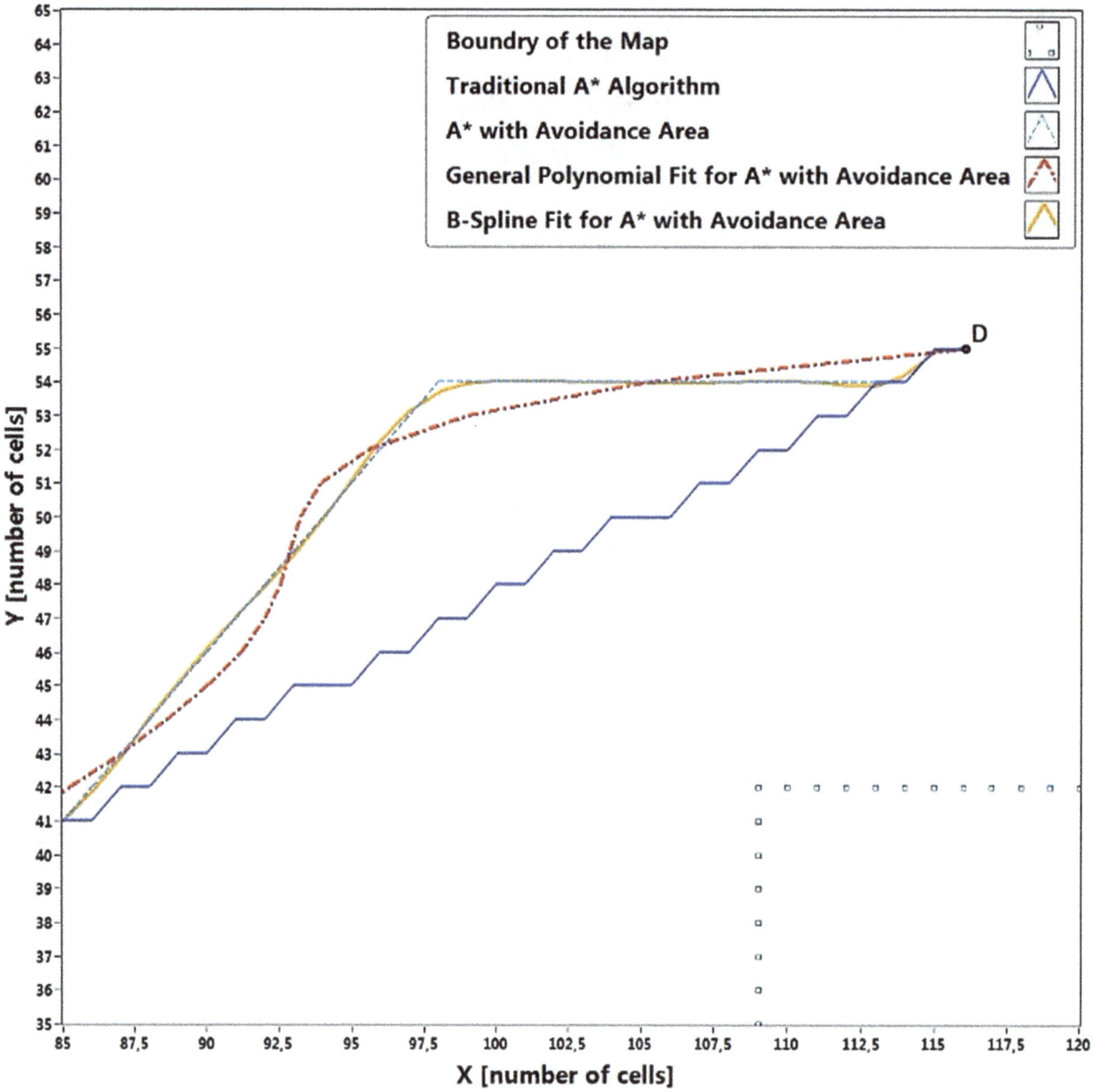

ferent conditions, especially for vertical movements. When B-spline curve fitting method is applied, it is possible to obtain smooth rotations by staying faithful to the path calculated by the A* algorithm on the grid map with avoidance area and thus continuous headings can be obtained.

Further research work will be an experimental investigation for real-time tracking performance of the mobile robot for trajectories calculated by the proposed algorithm. Also performance of the algorithm can be compared for different mobile platforms.

REFERENCES

Al-Arif, S. M. M. R., Ferdous, A. H. M. I., & Nijami, S. H. (2012). Comparative Study of Different Path Planning Algorithms: A Water based Rescue System. *International Journal of Computers and Applications*, *39*(5), 25–29. doi:10.5120/4817-7058

B-Spline Fit VI - LabVIEW 2011 Help. (2011). Retrieved from http://zone.ni.com/reference/en-XX/help/371361H-01/gmath/bspline_fit/

Brachman, R. J., & Levesque, H. J. (2004). *Planning Knowledge Representation and Reasoning* (pp. 305–325). San Francisco, CA: Morgan Kaufmann. doi:10.1016/B978-155860932-7/50100-X

Buniyamin, N., Ngah, W. A. J. W., Sariff, N., & Mohamad, Z. (2011). A Simple Local Path Planning Algorithm for Autonomous Mobile Robots. *International Journal of Systems Applications, Engineering & Development*, *5*(2), 151–159.

Cheng, L., Liu, C., & Yan, B. (2014, July). Improved Hierarchical A-star Algorithm for Optimal Parking Path Planning of the Large Parking Lot. Paper presented at the *2014 IEEE International Conference on Information and Automation (ICIA)*, Hailar, China. 10.1109/ICInfA.2014.6932742

Čikeš, M., Đakulović, M., & Petrović, I. (2011, October). *The Path Planning Algorithms for a Mobile Robot Based on the Occupancy Grid Map of the Environment - A Comparative Study.* Paper presented at the 2011 XXIII International Symposium on Information, Communication and Automation Technologies, Sarajevo, Bosnia.

Duchoň, F., Babinec, A., Kajan, M., Beňo, P., Florek, M., Fico, T., & Jurišica, L. (2014). Path Planning with Modified a Star Algorithm for a Mobile Robot. *Procedia Engineering*, *96*, 59–69. doi:10.1016/j.proeng.2014.12.098

General Polynomial Fit VI - LabVIEW 2014 Help. (2014). Retrieved from http://zone.ni.com/reference/en-XX/help/371361L-01/gmath/general_polynomial_fit/

Gonzalez-Arjona, D., Sanchez, A., López-Colino, F., de Castro, A., & Garrido, J. (2013). Simplified Occupancy Grid Indoor Mapping Optimized for Low-Cost Robots. *ISPRS International Journal of Geo-Information*, *2*(4), 959–977. doi:10.3390/ijgi2040959

Guruji, A. K., Agarwal, H., & Parsediya, D. K. (2016). Time-efficient A* Algorithm for Robot Path Planning. *Procedia Technology*, *23*, 144–149. doi:10.1016/j.protcy.2016.03.010

Mac, T. T., Copot, C., Tran, D. T., & De Keyser, R. (2016). Heuristic approaches in robot path planning: A survey. *Robotics and Autonomous Systems*, *86*, 13–28. doi:10.1016/j.robot.2016.08.001

Milstein, A. (2008). Occupancy Grid Maps for Localization and Mapping. In X.-J. Jing (Ed.), *Motion Planning*. InTech. doi:10.5772/6003

Pala, M., Osati, N., López-Colino, F., Sanchez, A., de Castro, A., & Garrido, J. (2013). HCTNav: A Path Planning Algorithm for Low-Cost Autonomous Robot Navigation in Indoor Environments. *ISPRS International Journal of Geo-Information*, *2*(3), 729–748. doi:10.3390/ijgi2030729

Satapathy, D., & Kumar, G. J. R. (2016). Path Planning Algorithm for a Mobile Robot – A Comparative Study in Different Environment. *IJCTA*, *9*(37), 453–459.

Subramanian, M. B., & Sudhagar, D. K., & RajaRajeswari, G. (2014). *Intelligent Path Planning Of Mobile Robot Agent By Using Breadth First Search Algorithm.* Paper presented at the 2014 IEEE International Conference on Innovations in Engineering and Technology (ICIET'14), Madurai, Tamil Nadu, India.

Subramanian, M. B., Sudhagar, D. K., & Rajarajeswari, G. (2015). Autonomous Mobile Robot Agent Path Planning Based on VFH Method Using Heuristic Based Optimal Algorithm. *International Journal of Applied Engineering Research*, *10*(6), 5189–5194.

Tseng, F. H., Liang, T. T., Lee, C. H., Chou, L. D., & Chao, H. C. (2014, August). *A Star Search Algorithm for Civil UAV Path Planning with 3G Communication.* Paper presented at the 2014 Tenth International Conference on Intelligent Information Hiding and Multimedia Signal Processing, Kitakyushu, Japan. 10.1109/IIH-MSP.2014.236

Yin, W., & Yang, X. (2013). A Totally Astar-based Multi-path Algorithm for the Recognition of Reasonable Route Sets in Vehicle Navigation Systems. *Procedia: Social and Behavioral Sciences*, *96*, 1069–1078. doi:10.1016/j.sbspro.2013.08.123

Chapter 16
Dynamic Modelling and Control of an Underactuated Quasi-Omnidireccional Hexapod

Edgar Alonso Martinez-Garcia
https://orcid.org/0000-0001-9163-8285
Universidad Autonoma de Ciudad Juarez, Mexico

José A. Aguilera
Universidad Autonoma de Ciudad Juarez, Mexico

ABSTRACT

This chapter presents the mechanical design, dynamic model, and walking control law of an insect-like, asymmetric hexapod robot. The proposed model is an original walking mechanism designed with three actuators to provide quasi-omnidirectionality. One of the motivational aims is to reduce the number of actuators preserving similar holonomy as compared to popular 18-servo redundant hexapods with three servos per leg. This work includes the Klann mechanism as limb, two-drive differential robot's control, one per lateral triplet of legs. The legs of a triplet are synchronized in speed with different rotary angles phase. In addition, the six limbs are synchronized with bidirectional yaw motion. The proposed mechanical design has one servo for limbs yawing, one for the right limbs triplet and one motor for the left triplet. Thus, quasi-omnidirectional mobility is achieved. Furthermore, a dynamic control law that governs the robot's mechanisms motion is deduced, with an Euler-Lagrange approach. Kinematic and dynamic results are validated through numerical simulations using a tripod gait.

INTRODUCTION

Multi-legged walker robots inherently pose great locomotion capabilities due to their hype-stability over numerous complex reliefs. Some unstructured terrains include flatter, alluvial plains, farming soils, steeper, rockier uplands, watershed boundaries, drainage characterized systems and hilly terrains (HyunGyu et al., 2016). Therefore, for human safety, multi-legged walker robots are suitable to oper-

DOI: 10.4018/978-1-7998-0137-5.ch016

ate at dangerous landscapes where landslides, downhill creep, flows, slumps, and rock falls represent hazardous environments for humans. There exist different classes of robotic hexapods developed to perform different varieties of missions and tasks that are of great utility in field applications, Gonzalez de Santos et al. (2006). Multi-legged robots pose high steerage, which is essential to carry out missions in fields such as mining, forestry, construction, planetary exploration, vulcanography, search and rescue, demining, agriculture and so forth. Unfortunately, redundant walker robots require one rotary actuator per degree of freedom (DOF) in the joint space. Furthermore, wider physical space is required to instrument joints mechanism, besides, the electric energy consumption increases as the number of electric actuators increases too (Shekhar & Khumar, 2011), which drastically reduces the robot's operating time. Traditionally, multi-legged robots may require at least two actuators (2 DOF) to control one leg, one for azimuth rotation and another for elevation. In addition, incrementing the number of actuators to control a limb, the control models become redundantly kinematic. The number of independent variablès is much greater in number than the DOF in the working space.

The work of Kazemi et al. (2013) presented the dynamics of five-link planar underactuated mechanical model of a quadruped robot with four actuated joints, its controller stabilized balance and track a desired trajectory. Kecskes et al. (2015), performed a validation study on a low-cost 18-DOF hexapod walker robot that was modeled by a 3D kinematics and dynamics model, the simulation-model realistically described the even ground contact gaits. One of our work's main contributions is the mechanical design that reduced an 18-servo model (Figure 1a) into an asymmetric three-motor underactuated version of hexapod, still preserving omnidirectional mobility (Figure 1b). In addition, a dynamic gating control law is disclosed for this type of robotic system. Haitao et al. (2016) presented a locomotion control method based on central pattern generator for hexapod walking robot to achieve gait generation with smooth transition, using a limit cycle approximation of the Van der Pol oscillator. The network governed the swing/stance phase, and locomotion control for the hexapod's leg movements during the gait cycle was developed. As a difference from our approach we control the gait speed by keep a reference torque. Knoop et al. (2017) presented a two-motor hexapod robot that walks and turn with arbitrary curvature. The walking mechanism is a modular gearbox to reciprocate sinusoidal output of continuously variable amplitude, its inputs are: a drive motor, rotating at a constant speed, and a single control input for changing the output amplitude for biomimetic robotic gaits. As a difference, our work presents two synchronization mechanisms: one for all-limb steering, and another for all-limb drive. Martinez-Garcia et al. (2014) proposed a Euler-Lagrange equation that governs the redundantly actuated legs to control

Figure 1. Asymmetric six-legged robot walkers. a) Redundant eighteen servos omnidirectional hexapod. b) Underactuated three-motor quasi-omnidirectional robot concept

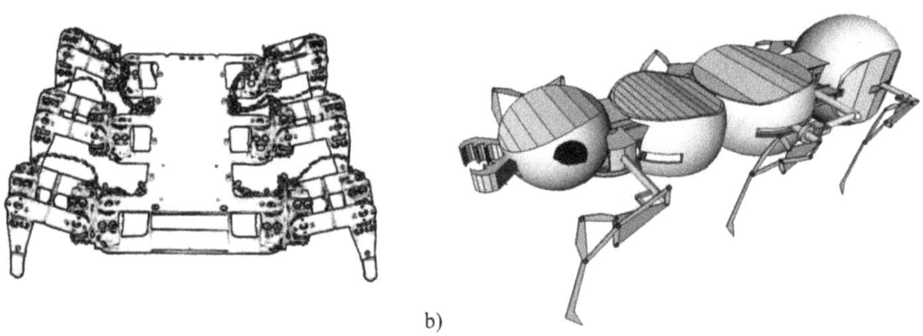

a) b)

the robot's displacements. Robot's rotation and translational velocities are feedback by motion estimated by optic flow of features of visual invariant descriptors. A general analytical solution of a derivative navigation law is proposed for hyper-static robots. This chapter presents a simulation framework and its physics model of an underactuated hexapod, and similarly to the present work, Roennau et al. (2015) developed an efficient and precise simulation system for multi-legged robots for experiments with the six-legged walking robot to evaluate accuracy and performance, analyze walking patterns and improve walking skills of designs of bio-inspired robots.

The topics of intelligent robotics tackled in this chapter combines the issues of robot design, control systems and mechanism dynamics. Such topics also fall within the mechatronics science that represents the required engineering to synergize all aspects to enhance machine intelligence. This work represents a fusion of intelligent robotics and mechatronic systems in which a novel type of underactuated multi-legged robot was designed to self-control quasi-omnidirectional mobility with a reduced number of actuators. The insight on the synergistic importance and evolution of mechatronics has been highlighted in Habib (2007, 2008).

In this work an asymmetric hexapod deploying Klann limbs is proposed (Figure 1b). Although, the Klann is a planar mechanism with seven passive joints, using one rotary actuator, the mechanism can be controlled. Thus, it yields two DOF in the walking space and naturally yielding a cycloid-type trajectory.

Kulandaidaasan et al. (2015) reported a quadruped robot using Klann mechanisms as legs and an analysis on synthesizing reconfigurable foot traces was presented. Subsequently, Kulandaidaasan et al. (2016) reported trajectory generation and its stability analysis for reconfigurable links of the Klann mechanism legs. El-Farouk, Labib et al. (2018), developed the application of a stair climbing walking robot based on reconfigurable Klann mechanisms. Similarly, Wu & Yao (2018) reported the design and analysis of a legged vehicle. Gilho & Bezerra (2012) presented the mechatronic design of a chair for disabled driven by Klann-based limbs.

As a difference from other approaches (Krishna et al., 2014), the proposed hexapod design is considered quasi-omnidirectional because it can reach any Cartesian point requiring a small latency time for the limbs to reach the targeted yaw. In order to reach quasi-omnidirectional control, this work discloses a hexapod design with three rotary actuators. One servomotor for limbs bidirectional synchronized steering, a second motor for driving the right-sided triplet of legs, and a third motor to drive the left-sided triplet. In this chapter, each lateral triplet has the limbs position with different phase to calibrate tripod gait stability. A limb's dynamic motion model includes the kinetic and potential energy interaction model, which approaches a Euler-Lagrange solution. In geometric model approaches, unlike algebraic methods, multi-joint limbs kinematics are solved by re-formulating any kinematic change. Thus, for limbs with more than three joints, the geometric-based analytic solutions are infeasible. In this chapter, the proposed gaiting control has an algebraic focus, where the walking gait configuration is setup by configuring the control vector. This modality allows to algebraically deduce the direct and inverse solutions and the input vector is augmented in case new legs are added. It is worth saying that external factors consideration such as legs slippage over the ground, dynamic frictions, terrains textures or positioning with respect to external objects are out of the scope of this manuscript. The presented approach provides simulations that concern the underactuated mobility space of tripod walking.

In section 2, the proposed robot's mechanical design is presented. In section 3, the Klann limbs kinematics model is analyzed. Section 4 deduces the dynamic control mobility model. Section 5 provides the conclusion.

UNDERACTUATED SYSTEM

Huang et al. (2012) presented a hexapod mechanism approach that differs from the present context that proposes a locomotion mechanism consisting of a differential driving system (Figures 2 and 4), an all-limb synchronous bidirectional steering mechanism (Figures 2 and 3), and the Klann-based limbs mechanism (Figure 5). The general driving system is two-drive differential (right and left sided), where lateral speed differences produce the hexapod angular velocity. The general driving structure is shown in Figure 2a. The right and left sided driving motors are in the middle of the longitudinal robot's axis. Each motor directly connects to gears e_0. The gear e_0 transmits motion to the rest of the gearing mechanism $e_1, e_2, \ldots, e_5$. The gears conduct rotary motion to the limbs' shaft of the extremes, the front-side and the backside (Figure 4b). Motion conduction is provided through gearing tracks between central limb and front/back sides. Since all limbs' direction is synchronized, then universal joints are coupled at each limb's shaft to allow simultaneous changes of orientation (Figure 4c).

The differential speed mechanisms provide two different speeds, one at each lateral triplet of limbs (Figure 4bc), interconnected front/back sides by tracks from the central drives. The gearing system (Figure 2) dynamics is subsequently deduced and analyzed in later sections. The steering mechanical system of Figure 3 shows bilateral direction angle for all-limb synchronously. All legs heads in the same direction of the robot's angle of motion (Figure 4d). The limbs yaw is zero when they are aligned with the longitudinal hexapod's axis, and can turn in yaw $-\pi/4, \ldots, \pi/4$. Thus, the limbs total yaw motion range is $\pi/2$, with forward and backward driving motion. Hence, only one rotary actuator steers all limbs synchronously.

Figure 2. Differential drive system. a) general driving structure. b) Forward and backward gearing systems

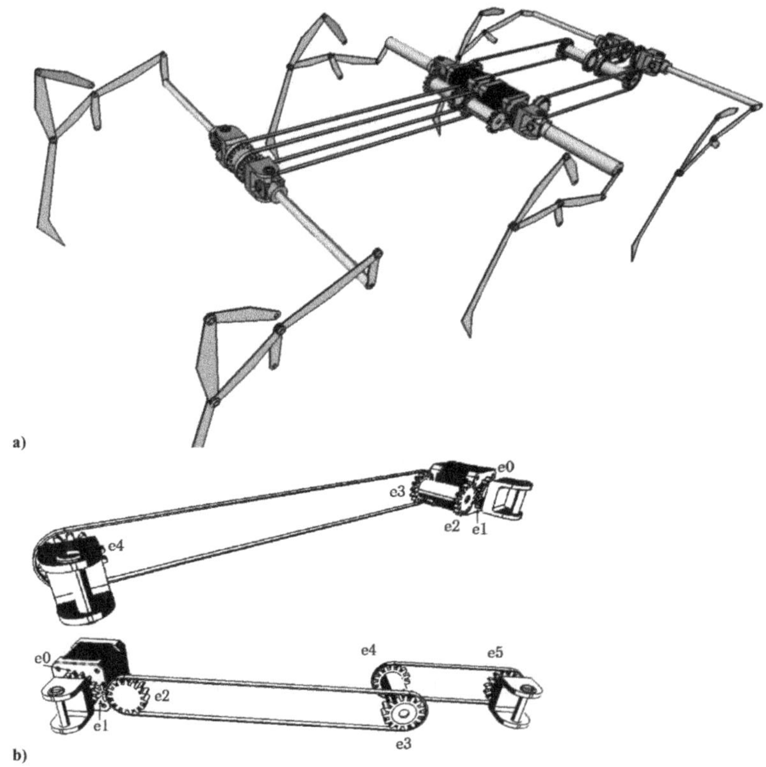

Figure 3. Hexapod steering mechanism. a) General structure. b) Steer gearing system

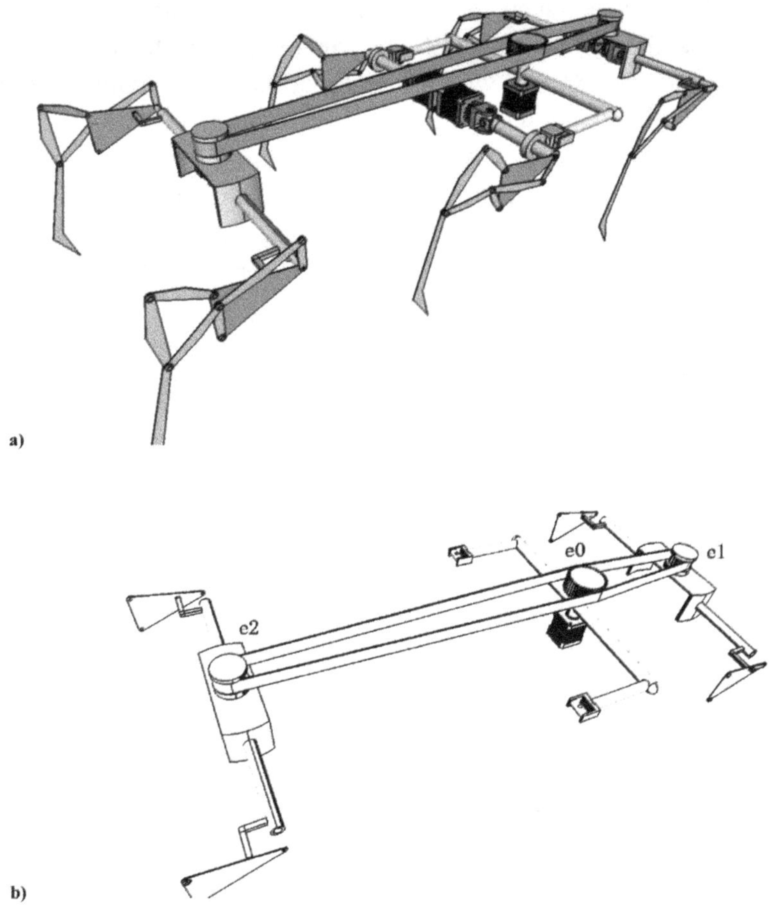

Figure 4. Quasi-omnidirectional platform kinematics and relevant mechanisms. a) Kinematic parameters (top view). b) Two-drive transmission belts. c) Steering shafts and drive belts. d) Steering actuator

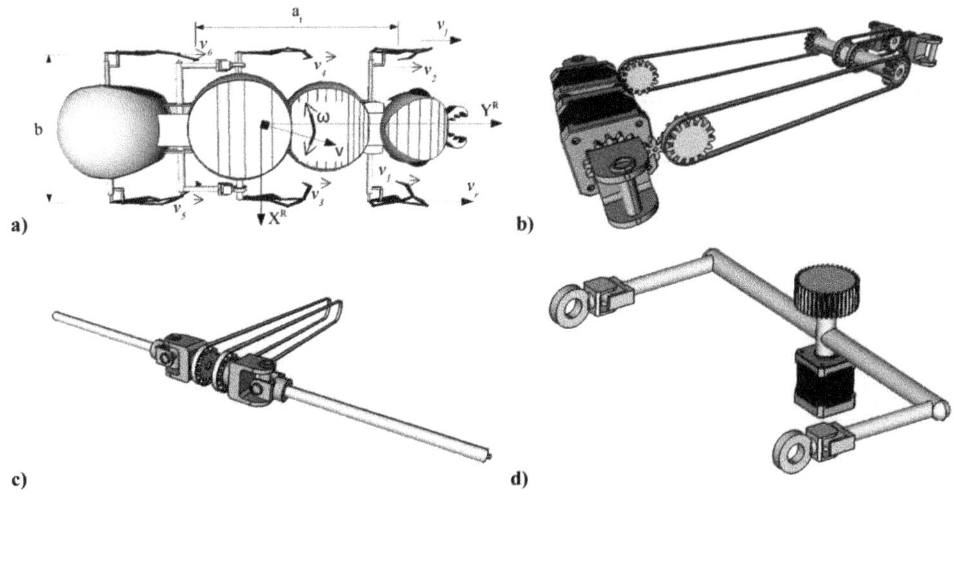

The steer mechanism is comprised of a servomotor and its vertical shaft connected to the gear e_0 above (Figure 3b and 4d), which conducts motion by means of a serrated band towards the frontal and back gears e_2 and e_1, respectively. Besides, the limbs yaw motion mechanism, the Klann shafts rotate and preserve their mechanical connection with the center of the chassis without affecting their driving performance. The universal joints allow such degrees of freedom. The steer shafts are hollow being crossed through by the rotary shafts that conduct driving motion. Figure 4abcd shows the concept.

The robotic platform design used to carry out the proposed dynamic model and simulations has the following technical specifications: a=0.612m, b=0.44m; the Klann parameters are (based on Figure 5) L1=0.0366m, L2=0.0433m, L3=0.1m, L4=0.074m, L5 0.0606m, L6=0.0883m, L7=0.1633m.

KLANN LEG KINEMATICS

The Klann mechanism (Figure 5) is an underactuated planar multi-link system, which from rotary input motion produces as output a cycloid trajectory (Figure 6). Klann (2002) presented his famous invention, the ´walking device´, which is usually deployed to emulate biological limbs motion, mostly antropods. Rooney et al. (2011) developed artificial active whiskers for underwater guidance of walking robots based on the Klann mechanism. Technologically, the Klann mechanism is widely used as walking mechanical device for robotic platforms because has several functional advantages as numerous advanced displacement systems have, such as stepping over obstacles, climb stairs, walking over all terrains, while not requiring a computing system to be controlled. The Klann mechanism is a bar-based coplanar mechanical device designed to simulate locomotion of numerous animals´ limbs, to substitute wheels particularly over unstructured terrains. The Klann links proportions are defined to optimize linear motion of the contact point at every rotary half cycle of the crank. The contact point lifts during the other rotary half cycle, before returning to the staring position. The Klann mechanism is comprised of seven rigid links (L1,…,L7, Figure 4a), a stretcher frame, a link used as crank handle L1, two as seesaw that are joined to the stretcher frame (L2 and L5), and all of them are interconnected through pivot joints (A, B,…,F). The links proportions define to optimize the link-foot linearity of motion with half rotation of the crank handle (Figure 5). The rest of the crank’s rotation allows that the link-foot move up to a defined height before motion return toward the staring position, repeating again the motion loop.

Figure 5. Klann-based limb. a) Kinematic diagram. b) Klann limb with shaft motor

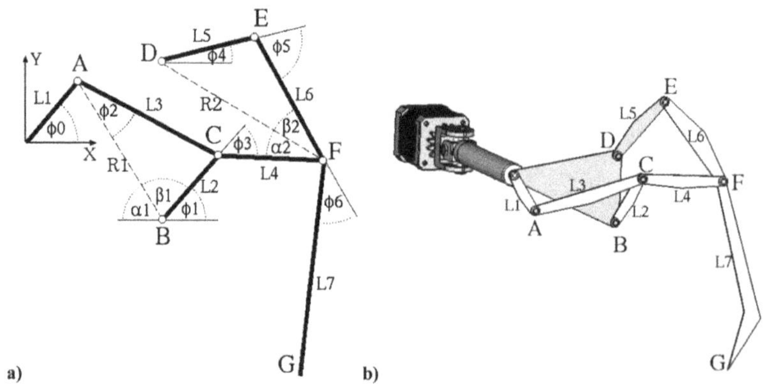

In this work, a Klann mechanism of specific proportions was designed as shown in Figure 5b for the purpose of limbs in a tripod-like walking. The kinematic model was deduced to numerically simulate and analyze the gait tracks. The link L1 in connection with L3 actuate as a rotating cam with L2 as a balancing bar. The oscillatory motion is transmitted to the coupled links L4, L5 and L6 to convert the rotary motion of L1 into linear motion in the base of the link effector L7. Let us define the crank handle Cartesian position model of link L1:

$$A_x = L_1 \cos\left(\varnothing_0\right) \tag{1a}$$

and

$$A_y = L_1 \sin\left(\varnothing_0\right), \tag{1b}$$

where ϕ_0 is the motor's rotation angle conducted to the link L1. The instantaneous distance R1 between the passive joints A and B is instantaneously calculated by

$$R_1 = \sqrt[2]{(A_x - B_x)^2 + (A_y - B_y)^2}\ ; \tag{2}$$

forming a non-stationary angle α_1 of straight-line AB with respect to (w.r.t.) the horizontal axis, such that

$$\alpha_1 = \tan^{-1}\left(\frac{A_y - B_y}{A_x - B_x}\right), \tag{3}$$

with complementary angle β_1 that comprises the triangle ABC and is obtained through the law of Cosines:

$$\beta_1 = \cos^{-1}\left(\frac{R_1^2 + L_2^2 - L_3^2}{2R_1L_2}\right). \tag{4}$$

Therefore, the angle ϕ_1 governs the motion of L2 and is inferred by

$$\phi_1 = \left(\alpha_1 - \beta_1\right) - \pi\,. \tag{5}$$

Thus, through the angle ϕ_1, the Cartesian position of the node C is deduced by the expressions

$$C_x = B_x + 2\cos\left(\phi_1\right) \tag{6a}$$

and

$$C_y = B_y + L_2 \sin(\phi_1), \tag{6b}$$

where Bx and by describe the L2 coordinates. By knowing the positions of nodes, A and C, the angle ϕ_2 can be obtained using

$$\phi_2 = \tan^{-1}\left(\frac{C_y - A_y}{C_x - A_x}\right). \tag{7}$$

Therefore, as for the angle ϕ_3 its model is:

$$\phi_3 = \phi_2 + \Delta\phi_1, \tag{8}$$

where $\Delta\phi_1 = 160.53 \cdot \pi / 180 + \pi$ being the slope angle of L3. Thus, in order to model the position of joint F, the following expression is established, where the links L1, L3 and L4 are involved in describing the ϕ_2 behavior. Thus,

$$F_x = L_1 \cos(\phi_0) + L_3 \cos(\phi_2) + L_4 \cos(\phi_3) \tag{9a}$$

and

$$F_y = L_1 \sin(\phi_0) + L_3 \sin(\phi_2) + L_4 \sin(\phi_3). \tag{9b}$$

Furthermore, to find the magnitude of R2, the position of joints D and F are used by the expression:

$$R_2 = \sqrt[2]{(F_x - D_x)^2 + (F_y - D_y)^2} \tag{10}$$

Moreover, the angle α_2 is obtained w.r.t. the absolute horizontal plane by

$$\alpha_2 = \tan^{-1}\left(\frac{F_y - D_y}{F_x - D_x}\right) \tag{11}$$

It follows that, by using the segments L5, L6 and R2 within the law of Cosines, the inner angle β_2 is obtained by

$$\beta_2 = \cos^{-1}\left(\frac{R_2^2 + L_6^2 - L_5^2}{2R_2L_6}\right). \tag{12}$$

The angle ϕ_4 determines the L6 slope w.r.t. node F,

$$\phi_4 = (\alpha_2 - \beta_2) - \pi . \tag{13}$$

By using the links L1, L3, L4 and L6 the instantaneous position of node E is obtained,

$$E_x = L_1\cos(\phi_0) + L_3\cos(\phi_2) + L_4\cos(\phi_3) + L_6\cos(\phi_4) \tag{14a}$$

and

$$E_y = L_1\sin(\phi_0) + L_3\sin(\phi_2) + L_4\sin(\phi_3) + L_6\sin(\phi_4), \tag{14b}$$

with coordinates of joints D and E known, the angle ϕ_5 is calculated with the expression:

$$\phi_5 = \tan^{-1}\left(\frac{E_y - D_y}{E_x - D_x}\right). \tag{15}$$

In addition, to determine the gait angular profile, the angle slope angle $\Delta\phi_2$ is included,

$$f_6 = f_4 + Df_2 \tag{16}$$

where $\Delta\phi_2$=90.6·π /180+π/2. Finally, the node G represents the Klann limb's end-effector

$$G_x = L_1\cos(\phi_0) + L_3\cos(\phi_2) + L_4\cos(\phi_3) + L_7\cos(\phi_6) \tag{17a}$$

and

$$G_y = L_1\sin(\phi_0) + L_3\sin(\phi_2) + L_4\sin(\phi_3) + L_7\cos(\phi_6) \tag{17b}$$

Therefore, by computing the previous kinematic formulation, all joints A,…,F and limb's contact point G are numerically simulated as shown by Figure 6.

The curves shown in Figure 6 are the Cartesian positions obtained from the components A(Ax, Ay), C(Cx, Cy), F(Fx, Fy), E(Ex, Ey) and G(Gx, Gy) provided in millimeters (mm) for this plot. In this numeric simulation, the links length produces a gaiting cycloid depicted by the contact point G, where a linear trajectory segment represents the contact of G on the ground.

Figure 6. Numerical simulation of the Klann's joints Cartesian motion

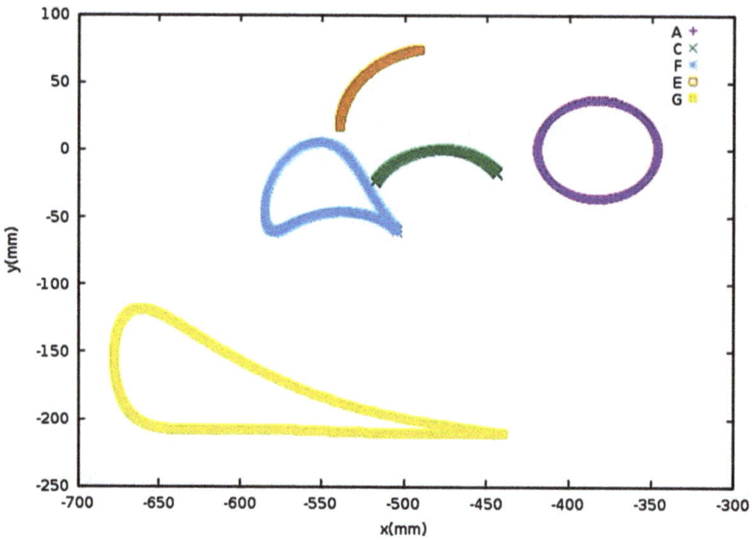

DYNAMIC MODELLING

The aim of this section is to disclose the steering and driving dynamic models based on the Euler-Lagrange approach to establish the hexapod's two main controls for the gearing mechanisms (Figures 2 and 3). Dynamic modeling allows analyzing the mechanism motion behaviors in terms of time. Based on a fundamental gears law, the tangential velocity is commonly shared by two gears that are mechanically connected while rotating. If gears radius is different, then angular speeds are different but sharing same tangential velocity. Therefore, for the kinematic gearing chain shown in Figure 2b the following expression in terms of the motor shaft angular speed $\dot{\varphi}_0$ is valid:

$$\dot{\varphi}_4 = \frac{r_0 r_3}{r_2 r_4}\dot{\varphi}_0 . \tag{18}$$

The kinetic energy of kinematic chain m_1 is deduced as merely for rotary motion $k = \frac{1}{2} I \dot{\varphi}_0^2$. The moments of inertia are assumed r^2m for cylindrical bodies (shafts). For simulation purposes, the mass of these open chains were established 0.25 kg. Thus, the total kinetic energies required by the kinematic chain m_1 are K_{1T}=K0+K1+K2+K3+K4. By factorizing these terms, the following is obtained:

$$k_{1T} = \frac{\dot{\varphi}_0^2}{2}\left(\left(m_1 + m_2 + m_3\right) r_0^2 + \left(m_3 r_3^2\right)\frac{r_0^2}{r_3^2} + m_4\left(\frac{r_0^2 r_3^2}{r_2^2}\right)\right) \tag{19}$$

Likewise, the potential energy of kinematic chain m_1 is $P = m\, g\, h$, where h is the height of the mechanism w.r.t. the ground, g is the gravity acceleration constant, and m the system mass. Thus,

$$P_{1T} = g\left(\sum_{i=0}^{4} m_i h_i\right). \tag{20}$$

From the kinetic and potential energies of chain m_1, the Lagrange operator L_{1T} is deduced as

$$L_{1T} = \frac{\dot{\varphi}_0^2}{2}\left(\left(\sum_{i=1}^{3} m_1\right) r_0^2 + \left(m_3 r_3^2\right)\frac{r_0^2}{r_3^2} + m_4\left(\frac{r_0^2 r_3^2}{r_2^2}\right)\right) - g\left(\sum_{i=0}^{4} m_i h_i\right). \tag{21}$$

Similarly, the kinematic driving chain of Figure 3b is expressed by $\dot{\varphi}_5$ as a function of the motor angular speed $\dot{\varphi}_0$, hence

$$\dot{\varphi}_5 = \frac{r_0 r_4}{r_3 r_5}\dot{\varphi}_0. \tag{22}$$

Thus, the kinetic energy model for gear e_5 is

$$k_5 = \frac{1}{2}\left(m_5 r_5^2 + \left(\frac{r_0 r_4}{r_3 r_5}\right)\dot{\varphi}_0\right)^2. \tag{23}$$

The total m_2 chain kinetic energy is K_{2T}= K0+K1+K2+K3+K4+K5. Thus, by factorizing the whole expression, the following is obtained:

$$k_{2T} = \frac{\dot{\varphi}_0^2}{2}\left(\left(m_0 + m_1 + m_2 + m_3\right) r_0^2 + \left(m_4 r_4^2\right)\frac{r_0^2}{r_3^2} + m_5\left(\frac{r_0^2 r_4^2}{r_3^2}\right)\right) \tag{24}$$

And the potential energy of kinematic chain m_2 is

$$P_{2T} = g\left(\sum_{i=0}^{5} m_i h_i\right), \tag{25}$$

then, the Lagrange operator for m_2 including the robot tripod load $w/3$ is expressed by

$$L_2 = \frac{\dot{\varphi}_0^2}{2}\left(\left(\sum_{i=0}^{3} m_i\right) r_0^2 + \left(m_4 r_4^2\right)\frac{r_0^2}{r_3^2} + m_5\left(\frac{r_0^2 r_4^2}{r_3^2}\right)\right) - g\sum_{i=0}^{5} m_i h_i - \frac{w}{3}. \tag{26}$$

Moreover, regarding the steering chain, the angular speed of gear e_1 w.r.t. the angular speed of gear e_0 is:

$$\dot{\varphi}_5 = -\frac{r_0}{r_1}\dot{\varphi}_0 . \tag{27}$$

Likewise, to calculate the speed of the gear e_2 as a function of the gear e_0:

$$\dot{\varphi}_2 = -\frac{r_0}{r_2}\dot{\varphi}_0 . \tag{28}$$

Hence, the total kinetic energy for kinematic chain m_3 is $K_{3T} = K_0 + K_1 + K_2$. Thus, by factorizing

$$K_{3T} = \frac{\dot{\varphi}_0 r_0^2}{2}\left(m_1 + m_2 + m_3\right), \tag{29}$$

then the potential energy for chain m_3 is

$$P_{3T} = g\left(m_0 h_0 + m_1 h_1 + m_2 h_2\right). \tag{30}$$

Therefore, the Lagrange operator for kinematic chain m_3 with the robot's weight w/3 supported by the tripod's contact points

$$L_2 = \frac{\dot{\varphi}_0^2}{2}\left(\left(\sum_{i=0}^{2} m_i\right) r_0^2\right) - g\sum_{i=0}^{2} m_i h_i - \frac{w}{3} . \tag{31}$$

It follows the motion equation with a Euler-Lagrange approach, also referred in this chapter as the dynamic equation that combines the geometry of motion (kinematics) and the causes of motion (inertial and gravity effects) of the robot´s mechanism. The procedure applies similarly for all Lagrange expressions presented. Thus,

$$\tau = \frac{d}{dt}\left(\frac{\partial L_2}{\partial \dot{q}}\right) - \frac{\partial L_2}{\partial q} \tag{32}$$

After solving and algebraically arranging the following dynamic expression is obtained in terms of the torque τ, since the kinematic chain is merely rotatory, and no linear forces are implicated in this model. Thus, the general motion equation of kinematic equations for driving and steering is

$$\tau_0 = \ddot{\varphi}_0 r_0^2 \sum_i m_i , \tag{33}$$

from previous expression, the dynamic equation for the actuator angular acceleration is

$$\ddot{\varphi}_0 = \tau_0 \left(r_0^2 \sum_i m_i \right)^{-1}. \tag{34}$$

WALKING CONTROL LAW

This section presents the inverse and direct solutions for the hexapod's mobility, considering the Klann's end-effectors motion equations. The presented equations are used to infer the robot's state vector and its higher order derivatives about its locomotion overtime. The independent control joint variables are the orientation angles $\dot{\varphi}_A$, which governs the Klann rotary shaft positions. The limbs' contact point position is defined by the kinematic state vector is $\mathbf{p}=(p_x,p_y,p_z)^T$. Thus, to simplify expressions, let us define hereafter the following:

$$\sin\left(\varphi_1\right) \triangleq s_1, \ldots, \cos\left(\varphi_1 + \varphi_2\right) \triangleq c_{12}$$

and so forth. Thus, defining the vector **p**:

$$p = \begin{bmatrix} s_s & 0 & 0 \\ 0 & 1 & 0 \\ 0 & 0 & c_s \end{bmatrix}^T \left(l_1 \begin{bmatrix} c_0 \\ s_0 \\ c_0 \end{bmatrix} + l_3 \begin{bmatrix} c_2 \\ s_2 \\ c_2 \end{bmatrix} + l_4 \begin{bmatrix} c_3 \\ s_3 \\ c_3 \end{bmatrix} + l_7 \begin{bmatrix} c_6 \\ s_6 \\ c_6 \end{bmatrix} \right). \tag{35}$$

Let us establish each limb's inertial frame into the robot's fixed coordinate system located ideally at its geometric center

$$q_i = R_z \left(\frac{-\pi}{2} \right) p_i + \begin{pmatrix} \Delta x \\ \Delta z \end{pmatrix}. \tag{36}$$

There is a pair of metric displacement from the robot's centroid, Δx and Δz. For all limbs they are defined by the translation coordinates [(Δx, Δz), (Δx, -Δz), (0, -Δz), (0, Δz), (-Δx, Δz), (-Δx, -Δz)] w.r.t. the center of the hexapod. The Euler rotation matrix **Rz** transforms each limb *i*, and their end-effectors position vectors $\mathbf{p}_i$. In addition, the robot's longitudinal and angular speed models described by the control vector is $\dot{\boldsymbol{u}} = \left(v, \omega\right)^T$. The absolute robot's velocity is modeled by the averaged lateral limb Cartesian speeds

$$v_t = \frac{v_r + v_l}{2} = \frac{\left\|\dot{q}_{ir} + \dot{q}_{il}\right\|}{2} = \frac{\sqrt[2]{(\dot{x}_r - \dot{x}_l)^2 + (\dot{y}_r - \dot{y}_l)^2}}{2}, \tag{37}$$

where v_t is the robot's instantaneous velocity, v_r and v_l are the limbs' averaged instantaneous lateral speeds at right and left sides respectively. The Klann limb direct solution model is defined by $\dot{p} = J \cdot \dot{\Phi}$, where **J** is Jacobian matrix and the joints vector $\dot{\Phi} = \left(\dot{\phi}_A, \dot{\phi}_C, \dot{\phi}_E, \dot{\phi}_F, \dot{\phi}_G\right)^T$ has the independent control variables. The Klann's linkage direct kinematic solution allows to determine the contact point position by in terms of the shaft crank angular motion. Thus, it is expressed:

$$\dot{p}_t = \begin{bmatrix} l_1c_0c_s + l_3c_2c_s + l_4c_3c_s + l_7c_6c_s & -l_1s_6s_s & -l_3s_2s_s & -l_4s_3s_s & -l_7s_6s_s \\ 0 & l_1c_0 & l_3c_2 & l_4c_3 & l_7c_6 \\ -l_1c_0s_s - l3c2s_s - l_4c_3s_s - l_7c_6s_s & -l_1s_0c_s & -l_3s_2c_s & -l_4s_3c_s & -l_7s_6c_s \end{bmatrix} \cdot \dot{\Phi}_t \tag{38}$$

Opposite to this, the Klann's linkage inverse kinematic solution is used with the particular interest to determine the shaft crank position as a function of the contact point position. Thus, the inverse algebraic solution is expressed by

$$\dot{\Phi}_t = \begin{bmatrix} l_1c_0c_s + l_3c_2c_s + l_4c_3c_s + l_7c_6c_s & -l_1s_6s_s & -l_3s_2s_s & -l_4s_3s_s & -l_7s_6s_s \\ 0 & l_1c_0 & l_3c_2 & l_4c_3 & l_7c_6 \\ -l_1c_0s_s - l3c2s_s - l_4c_3s_s - l_7c_6s_s & -l_1s_0c_s & -l_3s_2c_s & -l_4s_3c_s & -l_7s_6c_s \end{bmatrix}^+ \cdot \dot{p}_t, \tag{39}$$

where the pseudoinverse matrix $\mathbf{J}^+=(\mathbf{J}\ \mathbf{J}^T)^{-1}\ \mathbf{J}$ is an invertible, nonsingular and non-stationary matrix. Moreover, the robot's lateral speeds are defined in the robot's Cartesian coordinates system by

$$v_r = \frac{v_5 + v_1}{2} = \frac{\left\|\dot{q}_5 + \dot{q}_1\right\|}{2} \text{ as well as } v_l = \left\|\dot{q}_4\right\|, \tag{40}$$

and

$$v_l = \frac{v_2 + v_6}{2} = \frac{\left\|\dot{q}_2 + \dot{q}_6\right\|}{2} \text{ as well as } v_l = \left\|\dot{q}_3\right\|. \tag{41}$$

In addition, the robot's angular velocity is described by

$$\omega_t = \frac{2b\left(v_r + v_l\right)}{a_t^2 - b_t^2} \tag{42}$$

Considering that a_t and b_t are non-constant as the legs yaw ϕ_s changes, where $a_t = ||\mathbf{q}_5 - \mathbf{q}_1||$ and $b_t = |z_r - z_l|$. Moreover, $z_r = E\cos(\phi_s)$ and $z_l = E\cos(\phi_s)$, where E is the instantaneous length from the Klann's rotary point to the end-effectors contact point. When the angle $\phi_s = 0$, then b_t keeps constant and such an angle is aligned to robot's longitudinal axis, producing as a result the following specific case:

$$\omega_t = \frac{|z_r - z_l|\left(\left(\|\dot{q}_5\| + \|\dot{q}_1\|\right) - \left(\|\dot{q}_2\| + \|\dot{q}_6\|\right)\right)}{\left(\|\dot{q}_5\| - \|\dot{q}_1\|\right)^2 + |z_r - z_l|^2}. \tag{43}$$

Therefore, the control vector $\dot{u}$ is algebraically deduced with time-variant control matrix $\mathbf{K}$,

$$\begin{bmatrix} v \\ \omega \end{bmatrix} = \begin{bmatrix} v_r/2 + v_l/2 \\ \frac{2b}{a_t^2 + b_t^2}(v_l - v_l) \end{bmatrix} = \underbrace{\begin{pmatrix} 1/2 & 1/2 \\ 2ba_t^2 + b_t^2 & -2ba_t^2 + b_t^2 \end{pmatrix}}_{K} \cdot \begin{pmatrix} v_r \\ v_r \end{pmatrix}. \tag{44}$$

In addition, when v_r and v_l are at different speed rates, walking tripod gets unsynchronized because the differences yielded by the lateral contact points' speed. Therefore, the transition matrix λ forces to synchronization, varying lateral speeds to warranty synchronicity. The general expression λ·q establishes what contact points must step supporting the robot's total weight,

$$\begin{pmatrix} v_r \\ v_l \end{pmatrix} = \lambda_t \cdot \begin{pmatrix} \|\dot{q}_1\| \\ \vdots \\ \|\dot{q}_6\| \end{pmatrix}, \tag{45}$$

where λ is the commutation gait that switches either the right or the left sided tripod (Figure 7c), and considers the robot's total weight distributed into the tripod contact points approximately, defined by

$$\lambda_t = \begin{cases} \lambda 1, & P_r = g\left(\sum_i m_i h_i + \frac{w}{3}\right) \\ \lambda 2, & P_l = g\left(\sum_i m_i h_i + \frac{w}{3}\right) \end{cases}. \tag{46}$$

The tripod gait commutation is triggered between matrices λ1 and λ2 at every time a leg of either tripod completes a cycle, inferred by the coordinate (Gx, Gy) of expression (17). Other types of gait planning have been reported, Chen et al. (2013) presented a tetrapod gait planning for a leg-redundant crablike robot. Erden & Leblebicioğlu (2008) implemented the reinforcement learning algorithm for free gait generation of a six-legged robot. The approach in this work is constrained by two lateral drives, two matrices are provided as a general model solution that even may be used for up to six actuators (one drive per leg). Each row represents the right-side (row 1) and the left-side (row 2) of the robot´s limbs. Each $\lambda_{1,2}$ records the desired user gait, for any matrix element $\lambda_{1,2}(I,j)\neq 0$ establishes a contact point currently stepping ground. If a contact point does physically not match the $\lambda_{1,2}$ in turn, then the drive speed up/down to match it. Hence, the following matrices describe both commuting walking tripods:

$$\lambda_1 = \begin{bmatrix} 1/2 & 0 & 0 & 0 & 1/2 & 0 \\ 0 & 0 & 0 & 1/2 & 0 & 0 \end{bmatrix},$$

or

$$\lambda_2 = \begin{bmatrix} 0 & 0 & 1/2 & 0 & 1/2 & 0 \\ 0 & 1/2 & 0 & 0 & 0 & 1/2 \end{bmatrix} \tag{47}$$

Without loss of generality, by commuting both tripods sequentially, Figure 7ab illustrates the limbs Cartesian behaviors along the *x* component (Figure 7a), and the *y* component (Figure 7b). The plots refer to the contact points q_1 (right-front), q_4 (left-medium) and q_5 (right-back), as well as for the left-tripod contact points q_2 (left-front), q_3 (right-medium) and q_6 (left-back). The hexapod's tripods organization is depicted in Figure 7c.

Therefore, implementing an expression for feedback control, the estate equation represents the hexapod posture estimation expressed through the instantaneous state vector $\boldsymbol{\xi}= (x, y, \theta)^{\mathrm{T}}$ and the control vector $\mathbf{u}_t$. Thus, the general state equation is

$$\dot{\xi}_{t+1} = A \cdot \dot{\xi}_t + B \cdot \left[K_t \cdot \lambda \cdot \dot{Q}\right],$$

with

$$\dot{Q} = \left(\dot{q}_1, \dot{q}_2, \dot{q}_3, \dot{q}_4, \dot{q}_5, \dot{q}_6\right)^T,$$

and such an expression is disclosed with control matrix **B** explicit,

$$\dot{\xi}_{t+1} = A_t \cdot \dot{\xi}_t + \begin{pmatrix} \cos(\theta) & 0 \\ \sin(\theta) & 0 \\ 0 & 1 \end{pmatrix} \cdot \left(K_t \cdot \lambda_t \cdot \dot{q}_t\right), \tag{48}$$

where the state transition matrix $\mathbf{A}_t=\mathbf{I}$. The state equation controls the hexapod walking space it is shown in Figure 8abcd. For instance, Figure 8a shows the robots motion when $v_r>v_l$ for three different limbs' yaw: 0° (black dots), −45° (blue circles) and +45° (red crosses). The center of the plot is the robot's starting location and the trajectories are developed both: forward and backward from the Cartesian origin. Similar to Figure 8a is Figure 8b when $v_r<v_l$ for the same cases of limbs' yaw Figure 8c shows the

Figure 7. The six legs tripod walking behavior. a) Horizontal limbs position. b) Vertical limbs position. c) hexapod tripods gait

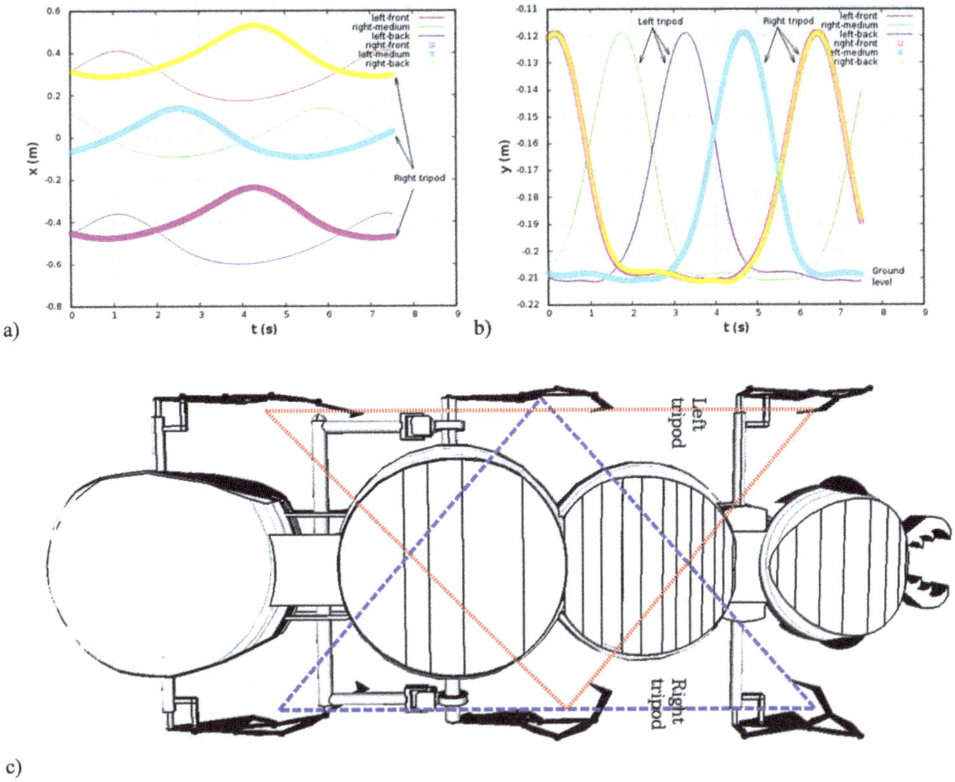

robot's forward and backward motion for $v_r=v_l$ for different limbs' yaw values. The limbs' yaw has a total range of $\left[-\frac{\pi}{4},\ldots,\frac{\pi}{4}\right]$, this range constrains the robot's limbs to turn enough as to reach the lateral sides as shown by Figure 8c, in the meantime $v_r=v_l$. By combining Figures 8a, 8b and 8c into 8d, the quasi-omnidirectional walking reachable space is obtained.

Without loss of generality, the pose estimation equation is useful to connect as an intermediate step the tracked reference trajectory by the robot with the kinematic chains of the robot's actuators. From the pose estate estimation equation, the limbs speeds are

$$\dot{Q}_t = \lambda^T(\lambda\cdot\lambda^T)^{-1}\cdot K^{-1}(B^T\cdot B)^{-1}\cdot B^T\cdot\left(\xi_{t+1} - A\cdot\dot{\xi}_t\right), \tag{49}$$

which can be used to obtain the limbs' speed from a desired robot's pose. With the first order derivative model, a second-order model is obtained to involve accelerations, torques, and the energies needed by the actuators $\tau = \left(\dot{\phi}_r, \dot{\phi}_l, \dot{\phi}_s\right)^T$. The deduction approach is the Euler-Lagrange differential equation, which will be discussed subsequently.

Figure 8. Robot's mobility space. a) when $v_R > v_L$; b) when $v_R < v_L$; c) when $v_R = v_L$ at different ϕ_s; d) nearly full holonomy

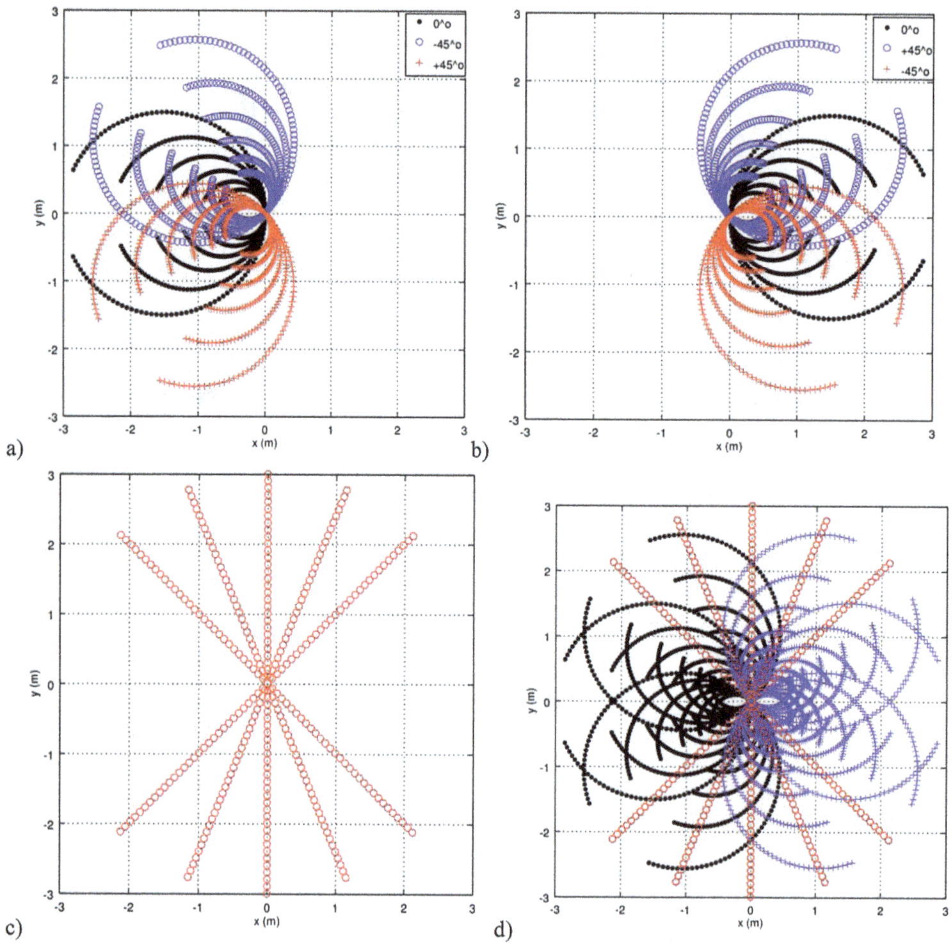

At this point, Figure 9 depicts the controlled values using the tripod gait (all trajectories develop clockwise direction). The simulations produced continuous cycloids and intentionally partial segments of the cycloids are shown, otherwise they are overlapped. Figure 9 shows the moments when a tripod contact points touch the ground, meanwhile the other tripod contact points are up. According to Figure 9, the foot trace segments develop during a period of time that starts at time t_0 and ends at t_f.

Now, assuming a trajectory reference model to be tracked by the hexapod, let us define such a reference trajectory vector with two components, both in terms of the first derivative as functions of time,

$$\dot{u}^{ref} = \begin{pmatrix} v^{ref} \\ \omega^{ref} \end{pmatrix} = \begin{pmatrix} a_0 + a_1 t + a_2 t^2 + \cdots \\ b_0 + bt + b_2 t^2 + \cdots \end{pmatrix}. \tag{50}$$

The previous reference trajectory model defines the transverse trajectory included in the inverse kinematics control that has been deduced from (43),

Figure 9. The six legs walking simultaneously synchronized with different phase angles

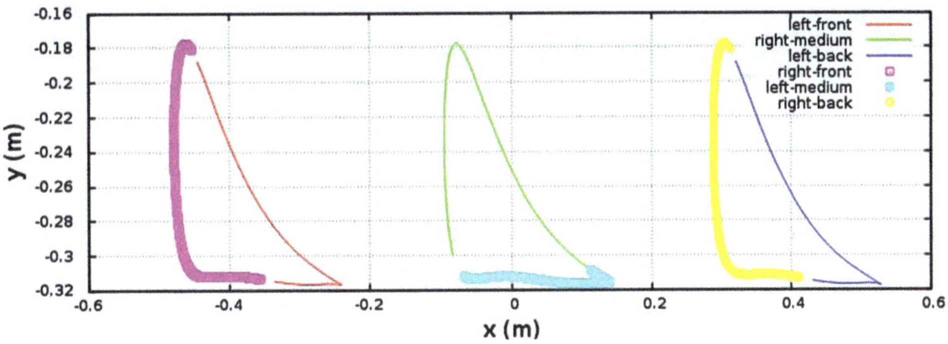

$$\dot{\mu}_{t+1} = \lambda \dot{Q}_t + K^{-1} \cdot \left(\dot{u}^{ref} - B\left(\dot{\xi}_t - \dot{\xi}_{t+1}\right)\right). \tag{51}$$

From the reference model, the vector of lateral speeds $\dot{\mu}_t = \left(v_r, v_l\right)^T$ is obtained. Thus, according to Figure 9, the right-sided tripod is constrained by two legs same speeds $\left\|\dot{p}_1\right\| = \left\|\dot{p}_5\right\| = v_r$ and $\left\|\dot{p}_4\right\| = v_l$, as well as the left-sided tripod speeds, $\left|\dot{p}_2\right| = \left|\dot{p}_6\right| = v_l$ and $\left\|\dot{p}_3\right\| = v_r$. Thus, the contact point Cartesian velocities are

$$\dot{p}_{1,5} = \frac{v_r^2}{2}, \dot{p}_4 = v_l^2, \dot{p}_{2,6} = \frac{v_l^2}{2}, and\ \dot{p}_3 = v_r^2. \tag{52}$$

By using the inverse solution of the Klann kinematics of previous expression (38), the speed contact points are known values, and the linkage vector of angular speeds ***Φ***=(ϕA, ϕC, ϕE, ϕF, ϕG)T is inferred, and of very particular interest is the Klann´s shaft/crank angular speed

$$\dot{\Phi}_{t+1} = J^{-1} \cdot \left(\dot{p}_t - \dot{J}_t \dot{\Phi}_t\right) \tag{53}$$

At this point, only the Klann crank angular speed $\dot{\phi}_A$ is of interest because represents the angular velocity of the end of the robot's mechanism chain, which is conducted by either the drive or the steer chain. There are three main kinematic chain rotary speeds $\dot{\phi}_{Ar}, \dot{\phi}_{Al}, \dot{\phi}_{As}$ for right driving, left driving and steering, respectively. For simplicity, the three kinematic rotary actuation variables are represented by the vector of independent control, $\dot{\Omega} = \left(\dot{\phi}_r, \dot{\phi}_l, \dot{\phi}_s\right)^T$. From dynamic equation (54), the following direct dynamic solution is deduced,

$$\tau = \begin{pmatrix} \tau_r \\ \tau_l \\ \tau_s \end{pmatrix} = \begin{pmatrix} r_r^2 \sum_i m_i^r & 0 & 0 \\ 0 & r_l^2 \sum_i m_i^l & 0 \\ 0 & 0 & r_s^2 \sum_i m_i^s \end{pmatrix} \cdot \dot{\Omega} \tag{54}$$

Therefore, for any chain mechanism modification, an improved general forward recursive dynamic control equation can be stated involving the inertial matrix M and the Coriolis matric C, such that in the following equation, the torque required by the entire vehicle is obtained stating as reference the desired velocity.

$$\tau_t = \frac{M}{t_2 - t_1}\left[v_{ref} - \hat{v}_t\right] + C\left[s_{t+1} - s_t\right] \tag{55}$$

With the vector of torques at hand, the inverse dynamics obtains the next desired robot's taking as a local reference the torque recursively calculated at the previous control step. Thus, by assuming $v = r\dot{\Omega}$,

$$v_{t+1} = v_t + M^{-1}\tau_t\left[t_2 - t_1\right] - M^{-1}C\left[s_{t+1} - s_t\right] \tag{56}$$

Where v_{t+1} is the prediction or next desired value at the next control loop. Since, numerical data is obtained overtime, the recursive system calculates (56) in terms of the value obtained from (55) and vice versa until reducing the numerical error s_{t+1}-s_t through successive approximations.

Figure 10 shows the torque required by a pair of geometrically opposite legs of different tripods. When the switching of tripod occurs, the term of mass of the potential energy changes by adding up the total robot's weight (4 kg) distributed equally among the three contact points. The negative/positive values of torque represent the motion direction of the Klann's crank, toward gravity/anti-gravity direction.

The robot´s total weight being supported by each contact point of the tripod is w/3, and its duration is nearly π/2 period of the rotary cycle of the Klann crank.

CONCLUSION

In this work we concluded that, as a difference from the omnidirectional walking robots that can move toward any direction at any time instant, with the proposed mechanical approach a small latency time is taken to steer the limbs yaw, and therefore the working space is considered quasi-omnidirectional. Nevertheless, as a difference from the inherently redundant hexapods with omnidirectional capability, the proposed approach is an underactuated mechanism reaching over 90% of the reachable space by a redundant walker robot. Despite that the Klann mechanism yields planar motion with inclusion of a

Figure 10. Cranks torque of two right/left legs. Experienced torques by two actuators when tripods switch and exchange the robot´s total weight (showing only two opposite legs)

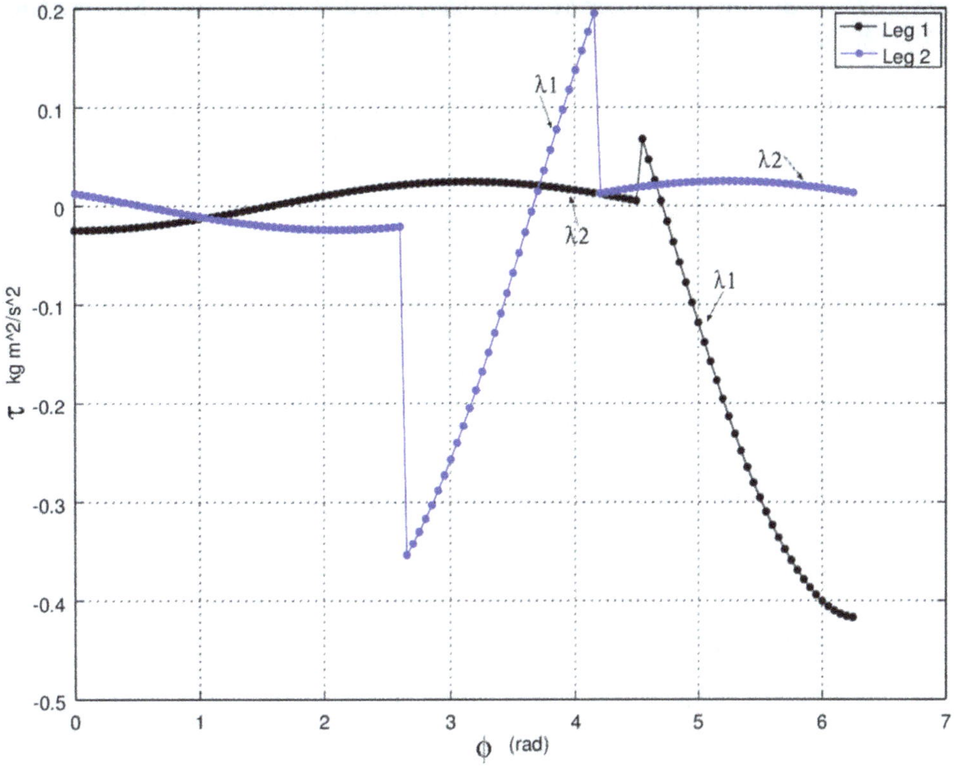

second DOF for yaw motion, one limb's working space represents a finite subset of the working space of a classical redundant limb, which exhibits quasi-omnidirectional motion over numerous terrains. In this work, only the tripod gait was considered for analysis of hyper-static walking, three contact points in the ground and three on the air synchronously distributed at the two hexapod's lateral sides. In the proposed general control law, different types of gaits can easily be implemented just by changing the walking limbs' phases and the values of matrices λ_1 and λ_2. Finally, it is concluded that with the Euler-Lagrange based dynamic control; the aim was to reach certain speeds at constant toques for missions where the hexapod's loads are different.

REFERENCES

Chen, X., Wang, L., Ye, X., Wang, G., & Wang, H. (2013). Prototype development and gait planning of biologically inspired multi-legged crablike robot. *Mechatronics*, *23*(4), 429–444. doi:10.1016/j.mechatronics.2013.03.006

El-Farouk, E., Labib, O., El-Safty, S. W., Mueller, S., Haalboom, T., & Strand, M. (2018). *Towards a Stair Climbing Robot System Based on a Re-configurable Linkage Mechanism, Intelligent Autonomous Systems 15. IAS 2018. Adv in Intelli Sys and Comp* , 867, (pp. 278–288). Cham, Switzerland: Springer; doi:10.1007/978-3-030-01370-7_22

Erden, M. S., & Leblebicioğlu, K. (2008). Free gait generation with reinforcement learning for a six-legged robot. Robotics and Autonomous Systems, 56(3), 199-212.

Gilho, B. V., & Bezerra, J. M. (2012). Mechatronic Design of a Chair for Disabled with Locomotion by Legs, *ABCM Symposium Series in Mechatronics*, 5, pp. 1142–1149.

Gonzalez de Santos, P., García, E., & Estremera, J. (2006). *Quadrupedal Locomotion: An introduction to the control of four-legged robots*. Berlin, Germany: Springer-Verlag.

Habib, M. K. (2007, Summer). Mechatronics: A Unifying Interdisciplinary and Intelligent Engineering Paradigm. *IEEE Industrial Electronics Magazine*, *1*(2), 12–24. doi:10.1109/MIE.2007.901480

Habib, M. K. (2008, August). Human Adaptive and Friendly Mechatronics (HAFM), *IEEE International Conference on Mechatronics and Automation 'ICMA2008'*, Takamatsu-Kagawa, Japan, pp. 61–65.

Huang, K., Chen, S., Tsai, M., Liang, F., Hsueh, Y., & Lin, P.-C. (2012). A Bio-inspired Hexapod Robot with Noncircular Gear Transmission System, *IEEE/ASME International Conference on Advanced Intelligent Mechatronics*, Kaohsiung, Taiwan. IEEE. 10.1109/AIM.2012.6266042

Kazemi, H., Majd, V. J., & Moghaddam, M. M. (2013). *Modeling and robust backstepping control of an underactuated quadruped robot in bounding motion, Robotica* , 31, (pp. 423–439). Cambridge, UK: Cambridge Univ Press; doi:10.1017/S0263574712000458

Kecskes, I., Burkus, E., Bazso, F., & Odry, P. (2015). *Model validation of a hexapod walker robot, Robotica* , 35, (pp. 419–462). Cambridge, UK: Cambridge Univ Press; doi:10.1017/S0263574715000673

Kim, H., Lee, D., Jeong, K., & Seo, T. (2016). Water and Ground-Running Robotic Platform by Repeated Motion of Six Spherical Footpads, IEEE/ASME, Transaction on Mechatronics, 21(1) IEEE.

Klann, J. C. (2002). *Walking Device*. Google Patents U.S. 6,478,314 B1, Nov. 12.

Knoop, E., Conn, A., & Rossiter, J. (2017). VAM: Hypocycloid Mechanism for Efficient Bionspired Robotic Gaits. In IEEE Robotics and automation letters, 2 *(2),* (pp. 1055–1061). IEEE.

Krishna, A., Krishna, N., Pradeep, S., Srihari, K., & Sivraj, P. (2014). Design a Fabrication of a Hexapod Robot, *International Conference on Embedded Systems*, India. 10.1109/EmbeddedSys.2014.6953162

Kulandaidaasan, J., Mohan, R. E., Martínez-García, E., & Tan-Phuc, L. (2015). *Synthesizing reconfigurable foot traces using a Klann mechanism, Robotica* (pp. 1–17). Cambridge, UK: Cambridge University Press; doi:10.1017/S0263574715000089

Kulandaidaasan J., Mohan R. E., Martínez-García E., & Tan-Phuc L. (2016). Trajectory Generation and Stability Analysis for Reconfigurable Klann Mechanism Based Walking Robot, *Robotics, 5*(13), pp. 1-10, MDPI, doi:. doi:10.3390/robotics5030013

Martinez-Garcia, E. A., Torres-Mendez, L. A., & Mohan, R. E. (2014). Multi-Legged Robot Dynamics Navigation Model with Optical Flow. Bingley, UK: Emerald. *International Journal of Intelligent Unmanned Systems*, *2*(2), 121–139. doi:10.1108/IJIUS-04-2014-0003

Roennau, A., Heppner, G., Klemm, S., & Dillmann, R. (2015). *RoaDS-Robot and Dynamics Simulation for Biologically-Inspired Multi-Legged Walking Robots. IEEE, Robotics and Biomimetics* (pp. 1870–1876). China: IEEE.

Rooney, T., Pearson, M. J., Welsby, J., Horsfield, I., Sewell, R., & Dogramadzi, S. (2011). Artificial Active Whiskers for Guiding Underwater Autonomous Walking Robots, CLAWAR 2011, France.

Shekhar, R. S., & Khumar, P. D. (2011). Dynamic Modeling and Energy Consumption Analysis of Crab Walking of a Six-legged Robot, IEEE Conference on Technologies for Practical Robot Applications. Woburn, MA: IEEE.

Wu, J., & Yao, Y. (2018). Design and analysis of a novel walking vehicle based on leg mechanism with variable topologies, *Mechanism and Machine Theory, 128,* pp. 663-681, Elsevier. doi:. doi:10.1016/j.mechmachtheory.2018.07.008

Yu, H., Gao, H., Ding, L., Li, M., Deng, Z., & Liu, G. (2016). Gait generation with smooth transition using CPG-based locomotion control for hexapod walking robot. IEEE Transactions on Industrial Electronics, 63(9), 5488-5500. doi:10.1109/TIE.2016.2569489

ADDITIONAL READING

Kulandaidaasan, J., & Mohan, R. E., Martinez-Garcia, E.A., & Tan-Phuc, L., (2014). Energy-Efficient Legged Robots: A Review. *International Conference on Creativity and Innovation for Business Sustainability,* Periyar, India: Maniyammai University Press.

Martinez-Garcia, E. A. (2015). *Numerical Modelling in Robotics, (2015).* Spain: OmniaScience. doi:10.3926/oss.22

Martinez-Garcia, E. A., Uribe, D. R., & Mohan, R. E. (2013). Kinematic Manipulability of Multi-Join Bio-Inspired Extremities. Amsterdam, The Netherlands: ScienceDirect, Elsevier. *Journal of Procedia Engineering*, *64*, 1533–1542. doi:10.1016/j.proeng.2013.09.235

Sheba, J. K., Elara, M. R., Martinez-Garcia, E. A., & Tan-Phuc, L. (2015). Design and Evaluation of Reconfigurable Klann Mechanism based Four Legged Walking Robot, *10th Intl. Conf. on Information, Communications and Signal Processing*, Singapore. 10.1109/ICICS.2015.7459939

Sheba, J. K., Martínez-García, E., Elara, M. R., & Tan-Phuc, L. (2014, December). Synchronization and stability analysis of quadruped based on reconfigurable Klann mechanism. In *2014 13th International Conference on Control Automation Robotics & Vision (ICARCV),* (pp. 663-668). Singapore: IEEE. 10.1109/ICARCV.2014.7064383

Vega, M., Martinez-Garcia, E.A., & Torres-Cordoba, R. (2015). *Unstructured terrain adaptive navigation of self-reconfigurable quadruped robot,* 11th IFAC Symposium on Robot Control, Salvador, Brazil: IFAC.

KEY TERMS AND DEFINITIONS

Asymmetric-Hexapod: A six-legged robotic platform where all limbs distances between their locations and the robot's geometric center are not equal.

Differential-Drive-Robot: A mobile robot that has steering control by varying two lateral drives.

Dynamic Control: A mathematical model that considers together kinematics and kinetics to form an equation with non-stationary control transition matrix.

Euler-Lagrange-Equation: A partial differential equation to find the linear generalized function of forces and angular moments of a mechanism in terms of energies and kinematic models.

Klann-Mechanism: A seven links mechanism to produce planar cycloid motion, one joint is rotary active and Conduct motion to the other six passive links.

Omnidirectional Walker: A legged robot capable to move toward any directional at any time instant.

Passive-Joint: An articulated mechanism that transmits rotary or linear motion with no explicit actuator device, but implicitly is conducted by an actuated source of motion.

Quasi-Omnidirectional Walker: A multi-legged robot that can walk toward any direction taking a small latency time to adjust such a direction of motion.

Synchronized Limbs: A multi-legged robot that steer all limbs simultaneously toward a same direction and speeds are the same, regardless of the motion phases.

Underactuated-Robot: A robot consisting of a locomotion mechanism where the number of actuators is less than the number of degrees of freedom in its working space and the inertial and gravitational forces are advantageously used.

Chapter 17
Robust and Intelligent Control Techniques for Twin Rotor System

Yener Taskin
Istanbul University – Cerrahpasa, Turkey

Yuksel Hacioglu
Istanbul University – Cerrahpasa, Turkey

ABSTRACT

During recent decades, popularity of unmanned aerial vehicles has increased throughout academics, engineers, as well as students because of their wide range of application areas such as inspection, communication, and transportation. The twin rotor system represents the nonlinear and coupled dynamics of a helicopter to a certain extent. Therefore, it provides a good test bed for engineering students' education on dynamics of aerial vehicles and control of mechatronic systems. In this study, sliding mode and fuzzy logic controllers are designed to control hovering motion of this highly nonlinear system. A robust sliding mode controller is preferred since it is insensitive to external disturbances, and intelligent fuzzy logic control is preferred since it is applicable to nonlinear systems where exact model of the system is not needed. Designed controllers, which are robust and intelligent, are then applied to the twin rotor system in real time. Then, experimental results are presented and discussed in terms of control performances of the designed controllers in detail.

INTRODUCTION

Twin rotor system which is usually called as twin rotor multiple-input multiple-output (MIMO) system (TRMS), is a laboratory experimental setup generally utilized to simulate and to control hovering motion of a reduced helicopter model. It possesses two degrees of freedom which correspond to pitch and yaw motions respectively. It is also well-known with its highly nonlinear cross coupled rotor dynamics. As a result of involved limitations of TRMS, it is a challenging system for modelling and implementation

DOI: 10.4018/978-1-7998-0137-5.ch017

of new control algorithms for academicians, researches and educators as well. In 2001, Ahmad et al. modelled a 2 DOF MIMO TRMS in hover by applying linear system identification techniques (Ahmad, Chipperfield, & Tokhi, 2001). A suitable modelling approach is concluded from their investigation for complex new-generation air vehicles. In 2004, (Aldebrez, Darus, & Tokhi, 2004) and (Darus, Aldebrez, & Tokhi, 2004) had proposed dynamic and parametric modelling of a twin rotor system using a multi-layer perceptron neuro-model and a genetic algorithm model respectively. Rahideh et al. utilized Newtonian and Lagrangian methods in order to model the TRMS in terms of vertical and horizontal 1DOF dynamics (Rahideh, Shaheed, & Huijberts, 2008). In another work, a complete dynamic model of the TRMS derived explicitly through application of the Euler–Lagrange approach that was contributed by Tastemirov et al. (Tastemirov, Lecchini-Visintini, & Morales-Viviescas, 2017). Most of above researches were done with commercially available setups. In 2015, Bayrak et al. published a paper on design and development of a TRMS in an academic setting (Bayrak, Dogan, Tatlicioglu, & Ozdemirel, 2015). This work demonstrated that the total cost of a TRMS could be lowered during the design stage compared to the commercial systems available in the market.

In general, TRMS models and experimental installations are used to evaluate and develop controller designs. There are many works published in the literature contributing to this purpose from classical to robust and intelligent techniques. Azarmi et al. proposed a simple analytical method for designing fractional order PID controller (Azarmi, Tavakoli-Kakhki, Sedigh, & Fatehi, 2015). The designed controller was implemented on a laboratory scale helicopter platform. Chemachema and Zeghlache utilized an output feedback linearization method in order to derive control laws without decoupling the TRMS into horizontal and vertical subsystems (Chemachema & Zeghlache, 2015). Choudhary proposed an H_{∞} optimal feedback controller for TRMS in the presence of cross-coupling effects with MATLAB simulation results (Choudhary, 2017). Witczak et al. presented a robust fault estimation approach for a class of nonlinear discrete-time systems with the application to a case study based on the TRMS (Witczak, Buciakowski, Puig, Rotondo, & Nejjari, 2016). As a nonlinear control strategy, Ilyas et al. considered sliding mode and backstepping controllers in order to achieve limitations of TRMS due to unstable, nonlinear and coupled dynamics with simulation results (Ilyas et al., 2016). In another work, Rashad et al. proposed a robust tracking controller for an experimental TRMS setup by combining a backstepping approach with a sliding mode disturbance observer to an integral sliding mode controller (Rashad, El-Badawy, & Aboudonia, 2017). Yang and Hsu designed an adaptive control approach based on the backstepping concept since the system parameters are assumed to be unavailable and are estimated on-line (Yang & Hsu, 2009). When the mathematical model and system parameters are not known exactly, fuzzy logic can be preferred to design an intelligent controller with expert knowledge related with the dynamics of the system. In (Taskin, 2014), (Hacioglu, 2015), and (Che Soh, Abdul Rahman, Md. Sarkan, & Yeo, 2013), numerical and experimental studies were published about fuzzy logic controller implementations to TRMS. Fuzzy logic is a powerful tool of which is applied to many fields not only as a controller approach but also as system analysis and synthesis approaches. In order to better and more understand the fuzzy systems and their applications, the following references can be taken into account (Burrascano, Callegari, Montisci, Ricci, & Versaci, 2015; Versaci et al., 2015b, 2015a, 2015c; Versaci, La Foresta, Morabito, & Angiulli, 2018; Versaci, Morabito, & Angiulli, 2017).

In this chapter, robust and intelligent control techniques that are implemented to a twin rotor system in real time are briefly described. The remainder of the chapter is arranged as follows. In next section, twin rotor model and experimental set-up are introduced. Design of sliding mode and fuzzy logic controllers are then explained respectively. Afterwards, experimental results are given and a discussion is made on responses of the implemented methods. Chapter is concluded with the assessment of the robust and intelligent control techniques in the conclusion section.

Twin Rotor Model and Experimental Set-up

In this section, the twin rotor system which is shown in Figure 1 and its mathematical model are described. The twin rotor is mounted on a fixed base having two degrees of freedom. It includes two propellers driven by two independent DC motors that are called the front and back propellers placed on perpendicular planes. If the front propeller generates a thrust force, then twin rotor will rotate around the pitch axis (horizontal axis). Similarly, a rotational movement around yaw axis (vertical axis) will occur if the back propeller generates a thrust force. The rotational displacements are measured by two encoders located at both pitch and yaw axes. The displacement resolutions are 0,0879 degrees and 0,0439 degrees for pitch and yaw encoders respectively (Quanser Inc., 2010). The range of pitch motion (θ) is limited between –40.3 and 36 degrees whereas the system can rotate freely about yaw axis and the range of yaw motion (ψ) is 360 degrees. The twin rotor reaches to the horizontal position when the pitch angle is equals to zero degrees. The control voltage commands which are calculated by the designed control algorithm, are transferred to the amplifier that powers the DC motors of the front and back propellers. DC motor control voltages are limited between ±24 V. All data is acquired by Quanser data acquisition terminal and input/output boards.

The equations of motion of the twin rotor system were derived via Lagrange's equations and given below. The parameters of the mathematical model and their values are listed in Table 1.

$$m_{hel} l^2 \ddot{\theta} + I_{pitch} \ddot{\theta} + m_{hel} l^2 \dot{\psi}^2 \cos(\theta) \sin(\theta) + B_{pitch} \dot{\theta} + m_{hel} g l \cos(\theta) = \tau_\theta + \bar{d}_\theta \quad (1)$$

$$m_{hel} l^2 \ddot{\psi} \cos^2(\theta) - 2 m_{hel} l^2 \dot{\theta} \dot{\psi} \cos(\theta) \sin(\theta) + I_{yaw} \ddot{\psi} + B_{yaw} \dot{\psi} = \tau_\psi + \bar{d}_\psi \quad (2)$$

Figure 1. Twin rotor MIMO system experimental set-up

Figure 2. Physical model of Twin rotor system

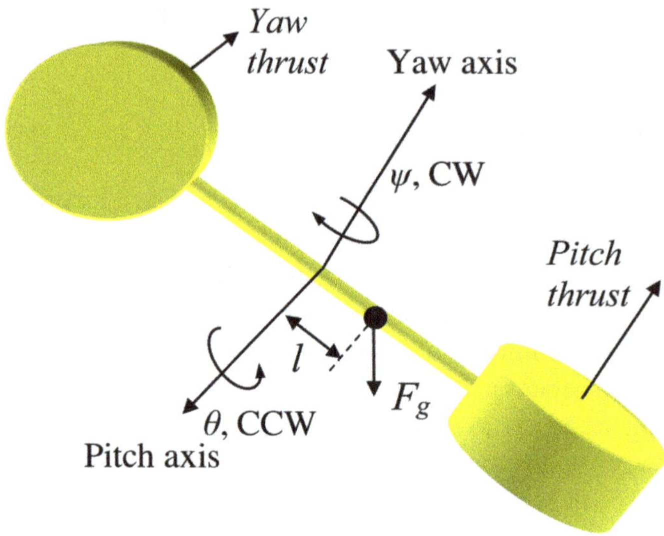

Table 1. Parameters of the twin rotor system

Parameter	Value	Unit
m_{hel}	1 .3872	kg
l	0.186	m
B_{pitch}	0.8	N/V
B_{yaw}	0.318	N/V
I_{pitch}	0.0384	kgm²
I_{yaw}	0.0432	kgm²

Here m_{hel} denotes total moving mass of the helicopter, l denotes the center of mass length along helicopter body from pitch axis, B_{pitch} and B_{yaw} denote equivalent viscous damping about pitch and yaw axis, respectively. Similarly, I_{pitch} and I_{yaw} denote the total mass moment of inertia about pitch and yaw axis, respectively. Finally, $\overline{d}_\theta$ and $\overline{d}_\psi$ are the external disturbances acting to the system.

Assigning the state variables as

$$\begin{bmatrix} x_1 & x_2 & x_3 & x_4 \end{bmatrix}^T = \begin{bmatrix} \theta & \dot{\theta} & \psi & \dot{\psi} \end{bmatrix}^T$$

the system equations may be presented in a more convenient form for the controller design as

$$\dot{x}_1 = x_2$$

$$\dot{x}_2 = \frac{-m_{hel}l^2\left(x_4\right)^2 \cos\left(x_1\right)\sin\left(x_1\right) - B_{pitch}\left(x_2\right) - m_{hel}gl\cos\left(x_1\right)}{\left(m_{hel}l^2 + I_{pitch}\right)} + \frac{\tau_\theta}{\left(m_{hel}l^2 + I_{pitch}\right)} + d_\theta$$

$$\dot{x}_3 = x_4$$

$$\dot{x}_4 = \frac{2m_{hel}l^2\left(x_2\right)\left(x_4\right)\cos\left(x_1\right)\sin\left(x_1\right) - B_{yaw}\left(x_4\right)}{m_{hel}l^2\cos^2\left(x_1\right) + I_{yaw}} + \frac{\tau_\psi}{m_{hel}l^2\cos^2\left(x_1\right) + I_{yaw}} + d_\psi$$

Sliding Mode Controller Design

In this section as a robust control method, a sliding mode controller will be designed for the control of the twin rotor system. For the sliding mode controller design, it is convenient to present the state equations of the dynamic system as:

$$\dot{x}_1 = x_2 \tag{8}$$

$$\dot{x}_1 = f\left(x_1, x_2\right) + g\left(x_1, x_2\right)u + d \tag{9}$$

where x_1 and x_2 are the states, u is the control signal and d is a bounded external disturbance such that, $\left|d\right| \leq \Delta, \Delta > 0$. Sliding surface is chosen as:

$$\sigma = \alpha\left(x_{1r} - x_1\right) + \left(\dot{x}_{1r} - \dot{x}_1\right) \tag{10}$$

where $\alpha > 0$ for stability and subscript r stand for reference value, $e = x_{1r} - x_1$ is the error. A Lyapunov function candidate which must be positive definite and its derivative must be at least negative semi-definite is defined as

$$V = \frac{1}{2}\sigma^2 \tag{11}$$

$$\dot{V} = \sigma\dot{\sigma} \tag{12}$$

Consider now that the limit condition is applied to eq. (12) then

$$\dot{\sigma} = \alpha\left(\dot{x}_{1r} - \dot{x}_1\right) + \left(\ddot{x}_{1r} - \ddot{x}_1\right) = 0 \tag{13}$$

Using eqs. (8-9) and (13), the control law for the limit case, which is called as equivalent control u_{eq}, is obtained as below

$$u_{eq} = g^{-1}\left(x_1, x_2\right)\left\{\alpha\left(x_1 - x_2\right) + \dot{x}_{2r} - f\left(x_1, x_2\right)\right\} \tag{14}$$

Let the total control signal is chosen to be;

$$u = u_{eq} + kg^{-1}\left(x_1, x_2\right) sign(\sigma) \tag{15}$$

Where $sign(\cdot)$ is the signum function. Then the derivative of the Lyapunov function becomes:

$$\begin{aligned} \dot{V} &= \sigma\dot{\sigma} \\ &= \sigma\left\{\alpha\left(x_{2r} - x_2\right) + \left(\dot{x}_{2r} - f\left(x_1, x_2\right) - g\left(x_1, x_2\right)u - d\right)\right\} \\ &= -k\sigma sign(\sigma) - \sigma d \\ &\leq -\left|\sigma\right|\left(k - \Delta\right) \end{aligned} \tag{16}$$

From this result it is seen that by choosing $k > \Delta$, then derivative of the Lyapunov function is negative definite $\dot{V} < 0$, thus the system is enforced to reach the sliding surface. During real time applications there may be a difference between the calculated and actual equivalent control inputs since there may be some uncertainty in some of the model parameters in $f\left(x_1, x_2\right)$. To overcome this problem in this study, an estimation for equivalent control $\hat{u}_{eq}$ will be used which is obtained via passing the total control signal through a low-pass filter. The estimation rule is:

$$\hat{u}_{eq} = \frac{\varepsilon}{s + \varepsilon} u \tag{17}$$

Here the main idea is that low frequencies determine the characteristics of the signal and high frequencies come from unmodeled dynamics. If the cut-off frequency ε of the low-pass filter is tuned properly then the output of the filter tends to equivalent control (Utkin, Guldner, & Shi, 2009). Thus, the stability of the system is maintained. Eventually, the total control input is

$$u = \hat{u}_{eq} + kg^{-1}\left(x_1, x_2\right) sign(\sigma) \tag{18}$$

By using the system equations - for the pitch and yaw motions the control signals of SMC for twin rotor system are derived as

$$\tau_{\theta} = \hat{\tau}_{eq\theta} + k_{\theta}\left(m_{hel}l^{2} + I_{pitch}\right) sign\left\{\left(\alpha_{\theta}\left(x_{1r} - x_{1}\right) + \dot{x}_{1r} - x_{2}\right)\right\} \quad (19)$$

$$\dot{\hat{\tau}}_{eq\theta} = \varepsilon\left(\tau_{\theta} - \hat{\tau}_{eq\theta}\right) \tag{20}$$

$$\tau_{\psi} = \hat{\tau}_{eq\psi} + k_{\psi}\left(m_{hel}l^{2}\cos^{2}\left(x_{1}\right) + I_{yaw}\right) sign\left\{\left(\alpha_{\psi}\left(x_{3r} - x_{3}\right) + \dot{x}_{3r} - x_{4}\right)\right\} \tag{21}$$

$$\dot{\hat{\tau}}_{eq\psi} = \varepsilon\left(\tau_{\psi} - \hat{\tau}_{eq\psi}\right) \tag{22}$$

For the experimental system the following control voltage equations are used

$$V_{mp} = \frac{K_{py}\tau_{\psi} - K_{yy}\tau_{\theta}}{K_{yp}K_{py} - K_{yy}K_{pp}} \tag{23}$$

$$V_{my} = \frac{K_{pp}\tau_{\psi} - K_{yp}\tau_{\theta}}{K_{pp}K_{yy} - K_{yp}K_{py}} \tag{24}$$

Here K_{pp} is the thrust force constant of yaw motor/propeller, K_{py} is the thrust torque constant acting on pitch axis from yaw motor/propeller, K_{yp} is the thrust torque constant acting on yaw axis from pitch motor/propeller and K_{yy} is the thrust torque constant acting on yaw axis from yaw motor/propeller. Numerical values of these parameters are presented in Table 2.

A nonlinear feedforward controller was also designed for the pitch motion of the twin rotor system which is designed to compensate for the gravitational torque as presented in following equation

Table 2. Parameters of the electrical motors of the propellers

K_{pp}	0.204	Nm/V
K_{py}	0.0068	Nm/V
K_{yp}	0.0219	Nm/V
K_{yy}	0.072	Nm/V

$$(\tau_\theta)_{ff} = \left(K_{ff}/K_{pp}\right) mgl\cos(\theta_r) \qquad (25)$$

Finally, the block diagram of the sliding mode controller for pitch motion is presented in Figure 3. In this figure the SMC block uses the equations (19)-(20) to calculate the control signal for SMC part and the block FF uses the equation (25) to calculate the feedforward part of the controller. Finally, the total control signal is obtained by adding those signals and it is applied to the twin rotor system by using equations (23)-(24).

Fuzzy Logic Controller Design

In this section as an intelligent control method a fuzzy logic controller will be designed for the control of the twin rotor system. Thus, a fuzzy logic controller (FLC) along with fuzzy feedforward compensator (FFC) is designed. Fuzzy logic has several attractive properties. The first one is that precise mathematical model of the system is not needed and the second one is that it may be applied to both linear and nonlinear systems.

Fuzzy logic gives the ability to use the information coming from experts, which is in linguistic form, by using fuzzy rules. Thus, experience of the experts can be incorporated in controller design. It is based on fuzzy set theory that was first presented by Zadeh in 1965 (Zadeh, 1965). Contrary to classical set theory where grade of membership is equal to zero or one, in fuzzy sets the grade of membership may be any number between zero and one. Fuzzy logic gives the ability to define and use the linguistic variables such as small, big, short, tall, cold, warm, and hot by defining corresponding fuzzy membership functions. Those membership functions may have triangular, trapezoidal, Gaussian and many other shapes. There are three main design steps in fuzzy-logic control: fuzzification, inference and defuzzification. In the fuzzification step the crisp/certain variables are converted to fuzzy variables. In other words, the types, numbers and ranges of the membership functions are described for the input and output variables.

Figure 3. Block diagram of the sliding mode controller with feedforward compensator for pitch motion

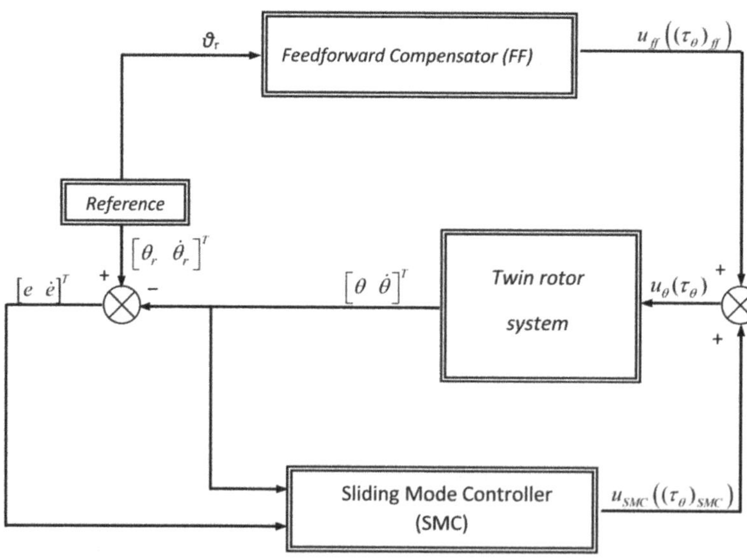

In the inference step the linguistic expressions are used to form the rules that constitute the rule base. The output of the fuzzy inference step is a fuzzy quantity and since the output values cannot be used directly, in the defuzzification step the fuzzy variables are converted to crisp values with an appropriate defuzzification method. In this study the centroid method, which is widely used in the literature, is used for the defuzzification of the fuzzy variables.

In this study, the FLC controller is composed of PD and PI type FLCs. In the PD part of the FLC the inputs are error (e) and its time derivative ($\dot{e}$) and control signal (u) is the output. In the PI part of the FLC the inputs are error (e) and its derivative ($\dot{e}$) while output is the incremental change in control signal (Δu). Here the PI part is included to remove the possible steady state errors. Furthermore, a fuzzy feedforward compensator is also designed for pitch motion to compensate for the gravitational torque.

The block diagram of the intelligent controller for pitch motion is shown in Fig. 4. The resultant control signal for the pitch motion u_θ is

$$u_\theta = u_{FLC} + u_{FFC} = SF_{\Delta u}\Delta u_n + u_{PI}(t-\delta t) + SF_u u_n + SF_{ffc} u_{ffcn} \tag{26}$$

Figure 4. Block diagram of the fuzzy logic controller with fuzzy feedforward compensator for pitch motion

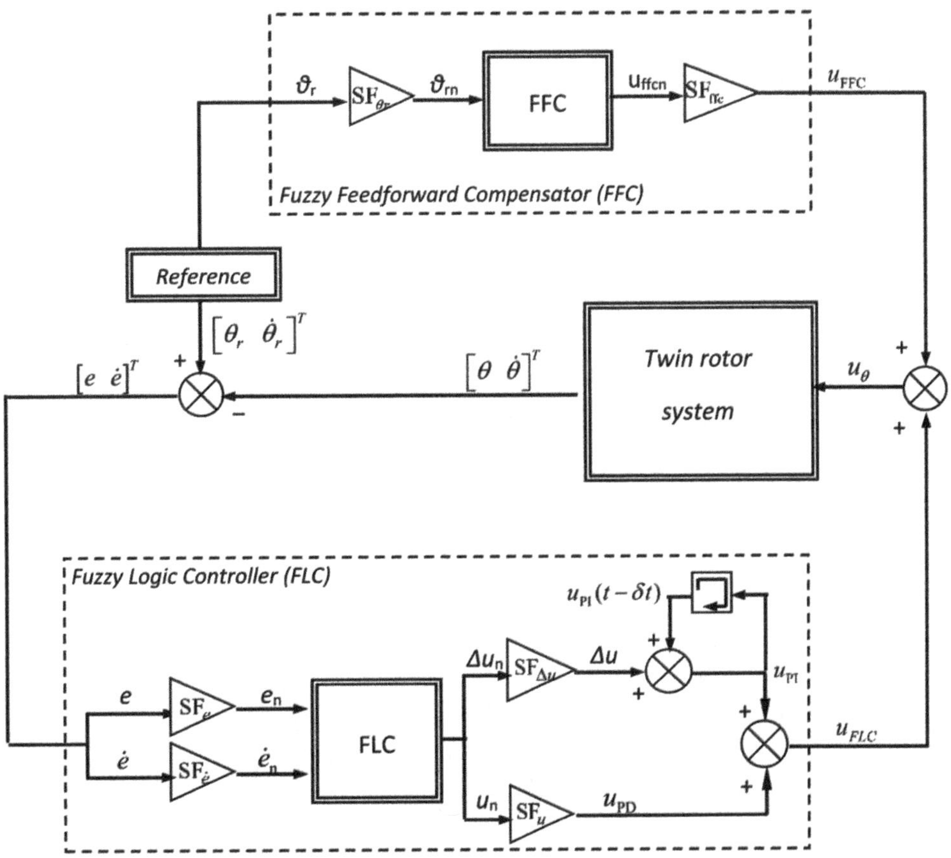

The structure of the intelligent controller for the yaw motion is similar to that of pitch motion controller but there is no feedforward compensator for yaw motion. Finally the control signal for the yaw motion u_ψ is

$$u_\psi = u_{FLC} = SF_{\Delta u}\Delta u_n + u_{PI}(t-\delta t) + SF_u u_n \tag{27}$$

To represent the input and output variables in fuzzy universe of discourse, triangular shaped membership functions are utilized as shown in Figure 5. For the membership functions on the figure the PVB, PB, PM, PS, Z, NS, NM, NB, NVB stand for positive very big, positive big, positive medium, positive small, zero, negative small, negative medium, negative big and negative very big, respectively. The output membership functions of the feedforward compensator are defined on [0 1] interval, and all other membership functions are defined on [-1 1] interval. In order to map the crisp input and output variables to their fuzzy values, scaling factors (*SF*) are used and subscript n denotes the normalized variable. For the designed intelligent controller, the inference rules are in the following form:

$$IF\ (e_n\ is\ A_1)\ and\ (\dot{e}_n\ is\ A_2)\ THEN\ (\Delta u_n\ is\ B_1)\ and\ (u_n\ is\ B_2) \tag{28}$$

where A_1, A_2 and B_1, B_2 are input and output membership functions, respectively. It is seen that during the construction of the fuzzy rules the exact mathematical model of the system is not needed and this is one of the main advantage of the fuzzy logic controller.

The rule base of the fuzzy logic controller shown in Table 3 was constructed by using following manner. Consider the case that the system's output is at a great distance from the desired value i.e. error e_n is PB and its time derivative $\dot{e}_n$ is Z then u_n and Δu_n are chosen to be PM to reduce the error. Consider another situation that e_n is Z but it tends to increase because $\dot{e}_n$ is NB, then u_n and Δu_n should not be zero and they are chosen to be NM. Consider a different situation where both e_n and $\dot{e}_n$ are Z, that is the desired case, thus u_n and Δu_n are chosen to be Z. Thus, it is seen that by using this

Figure 5. Membership functions of the a) input variables b) output variable of the fuzzy feedforward compensator c) output variables of the fuzzy logic controller

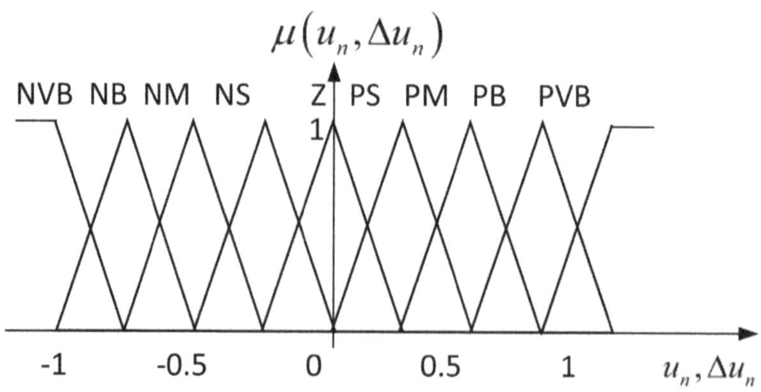

Table 3. Rule table of fuzzy logic controller for computing u_n and Δu_n

		$\dot{e}_n$				
		NB	**NS**	**Z**	**PS**	**PB**
e_n	**NB**	NVB	NB	NM	NS	Z
	NS	NB	NM	NS	Z	PS
	Z	NM	NS	Z	PS	PM
	PS	NS	Z	PS	PM	PB
	PB	Z	PS	PM	PB	PVB

manner the fuzzy rules are constructed to regulate the system error to zero that is the main aim of the closed-loop controller.

For the rules of the pitch motion fuzzy feedforward compensator in Table 4 the following manner is used. The gravitational torque due to the weight of the twin rotor is acting about the pitch axis and it attains biggest values when pitch angle is approximately zero. Therefore, the feedforward control u_{ffcn} is chosen to be PB when θ_{rn} is Z. Since gravitational torque decreases with increasing pitch angle then feedforward control part u_{ffcn} is also decreased.

Results and Discussions

In this section experimental results obtained via applications of SMC and FLC are discussed. Twin rotor system is under closed-loop control in order to follow the reference trajectories determined for pitch and yaw axes. At the beginning of each experiment, the twin rotor began its motion from the initial position where the initial pitch and yaw angles were $\theta(0) = -40{,}3$ and $\psi(0) = 0$ degrees, respectively. Each designed controller was examined under the same trajectory with and without disturbance moment acting on pitch motion demonstrated in Figure 8. This disturbance was added to the system to test the robustness of the controller. Though it is applied to the pitch axis, due to coupling effect it will also affect the yaw motion of the system to a certain extent. The types of the reference trajectories for the pitch and yaw motions are both square waves where the amplitude is 20 degrees and frequency is 0,05 Hz. The initial angle values of reference trajectory for pitch and yaw motions are -30 and -10 degrees respectively. The experiments are carried out with 1 ms sampling time.

Figure 6 presents the time responses for the pitch and yaw motions of the twin rotor system with SMC. It is seen that the system is following the reference trajectories for the pitch and yaw motions,

Table 4. Rules for computing u_{ffcn} for the fuzzy feedforward compensator

θ_{rn}	NB	NS	Z	PS	PB
u_{ffcn}	PS	PM	PB	PM	PS

Figure 6. Time responses for a) pitch motion b) yaw motion under SMC without disturbance

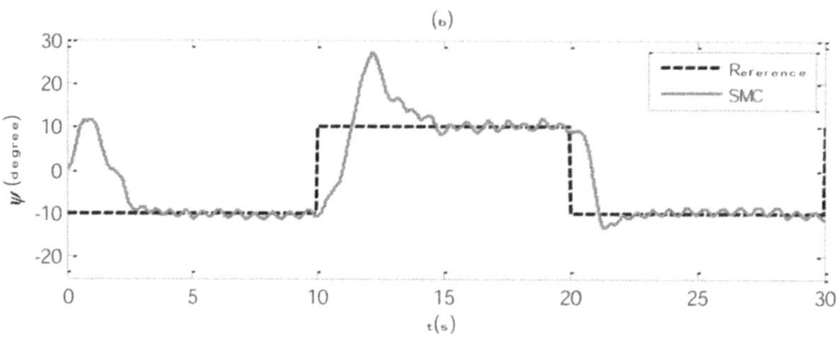

successfully. Figure 9 presents the time responses for the SMC but now in the case of a disturbance applied to the pitch propeller. It is seen that the SMC recovers in reasonable time. This shows the robust behavior of the SMC.

Figure 11 presents the time responses of the twin rotor system with FLC. It is seen that the system is tracking the reference trajectories for pitch and yaw motions successfully. Figure 13 shows the time responses for the FLC in the case of a disturbance applied to the pitch propeller. From this figure it is observed that after application of the disturbance the FLC recovers very fast. From these results it can be stated that the designed FLC also has a robust character.

Coupling effects are easily observable at all time response diagrams (Figures 6, 9,11 and 13) with different amounts according to the success of the designed controllers. Especially, when the pitch angle reaches to -30 degrees starting from -40.3 degrees, yaw angle first increases to a positive magnitude as a result of coupling effect, then follows back to reference trajectory with the control action. Consequently, pitch thrust forces generates more cross-coupling loads when compared to yaw trust forces.

Figure 7, Figure 10, Figure 12 and 14 present the control voltages applied to the DC motors of both axes which were calculated by the designed controllers with and without disturbance application. It is observed that in SMC application, chattering is present in control signals. This may reduce the life of the electric motors. On the other hand, the control signals of the FLC are more smooth, therefore they may be classified as more applicable control signals from the viewpoint of life of electric motors.

To compare the results more quantitatively, step response characteristics for both SMC and FLC are obtained and normalized from the square waveform parts of the time responses for 10 s duration and given in Figure 15-16 and in Table 5. As it is observed from the time responses, the fastest rise time is

Figure 7. Control voltage signals for SMC without disturbance

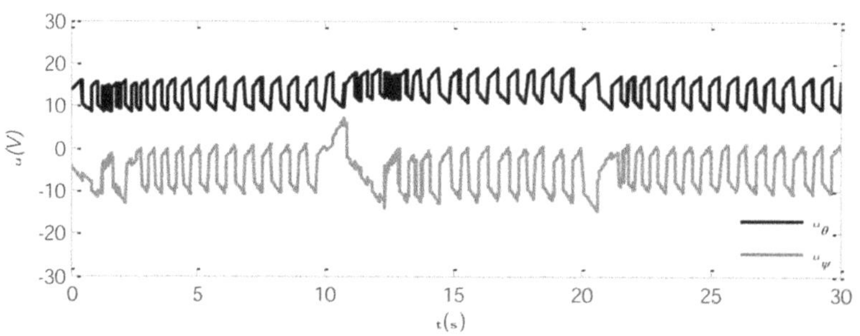

Figure 8. Disturbance moment acting on pitch motion

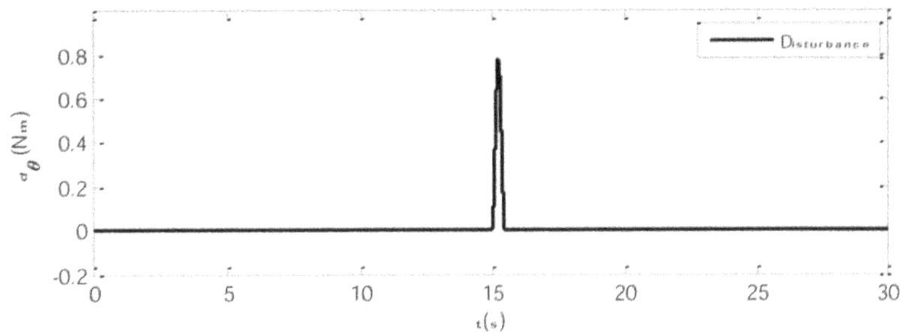

Figure 9. Time responses for a) pitch motion b) yaw motion under SMC with disturbance

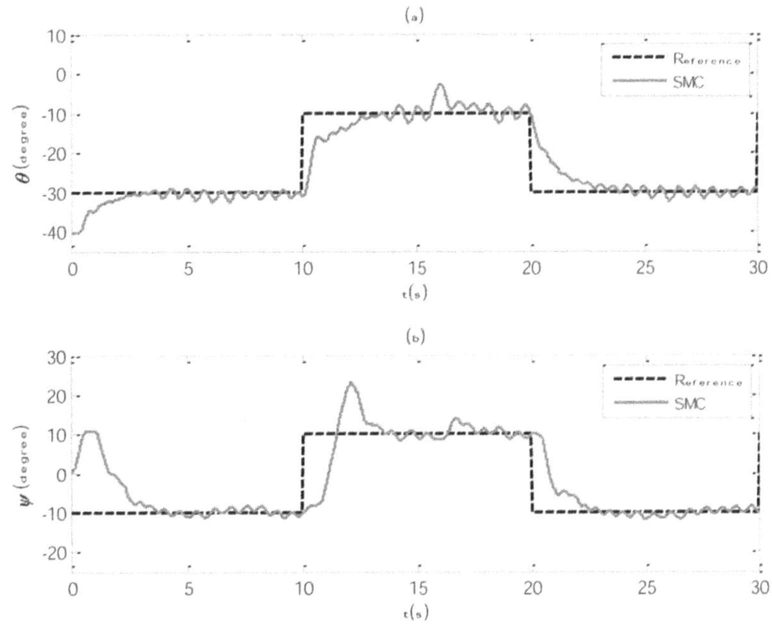

Figure 10. Control voltage signals for SMC with disturbance

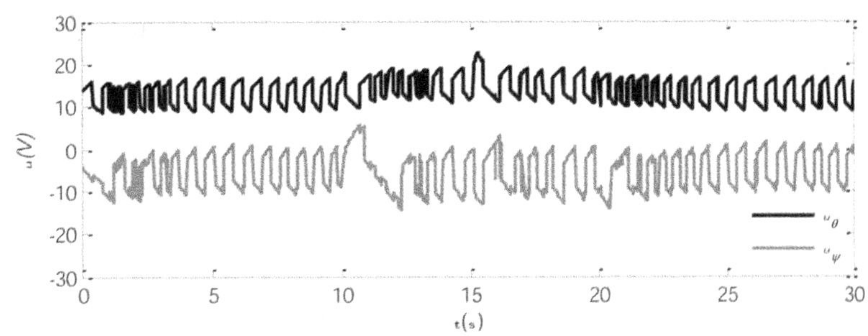

Figure 11. Time responses for a) pitch motion b) yaw motion under FLC without disturbance

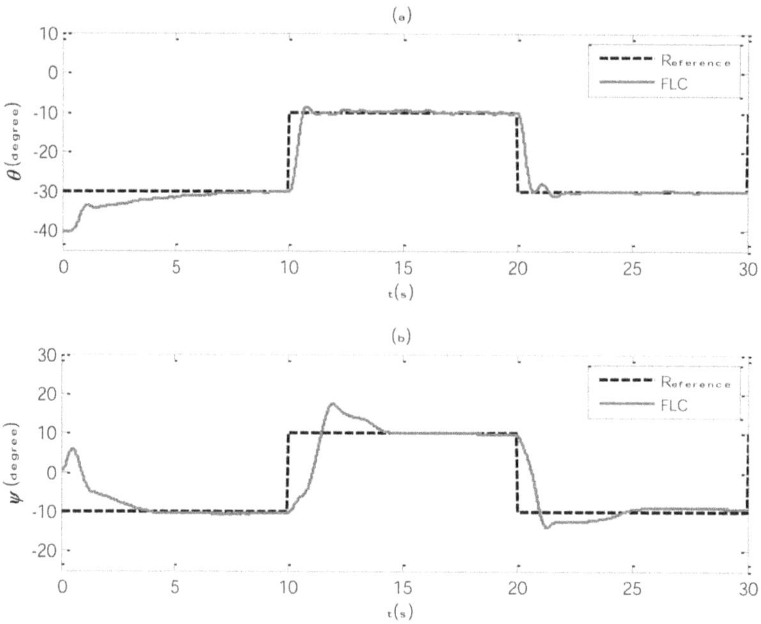

Figure 12. Control voltage signals for FLC without disturbance

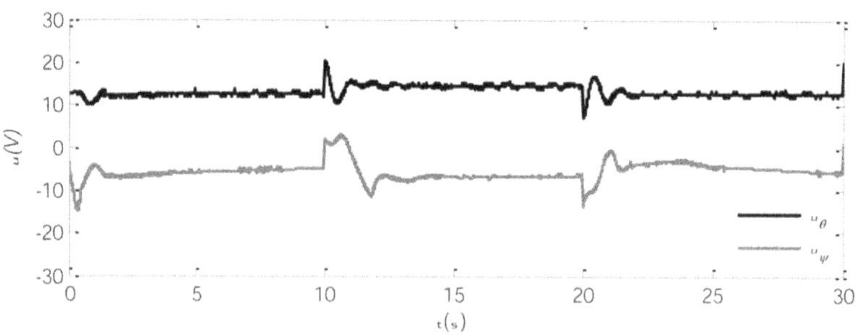

obtained from pitch motion of FLC. Also it has the lowest settling time and very little overshoot comparing the other cases. Mean squared error values of FLC case is lower than SMC case when the results compared for each pitch and yaw motions.

CONCLUSION

In this study, the hovering motion of a twin rotor system has been investigated. In order to cope with this highly nonlinear system, robust and intelligent controllers, namely, sliding mode controller and fuzzy logic controllers have been designed. Sliding mode controller was preferred since it is well known with its robust behavior and the fuzzy logic controller was preferred since it can be applied to nonlinear systems especially when the exact mathematical model is not known. Those controllers were then applied to the experimental twin rotor system which is to track specified trajectories for pitch and yaw motions.

Figure 13. Time responses under disturbance for a) pitch motion b) yaw motion under FLC with disturbance

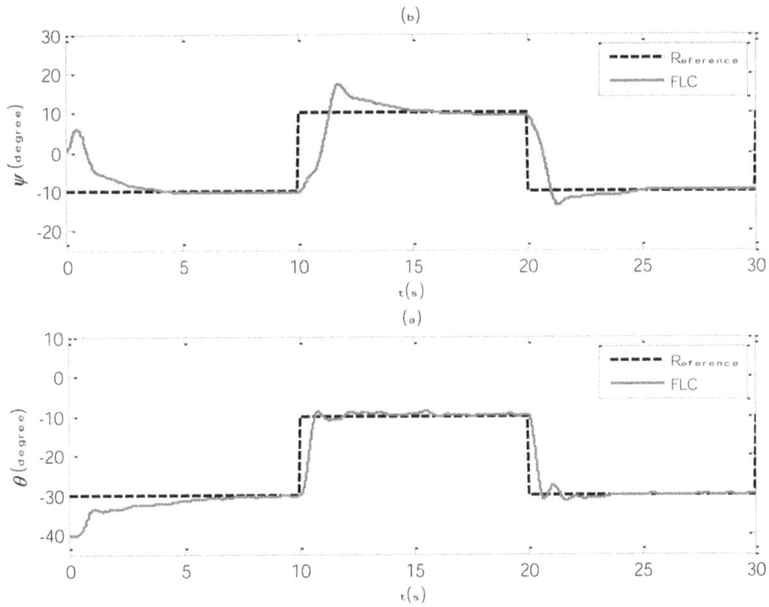

Figure 14. Control voltage signals for FLC with disturbance

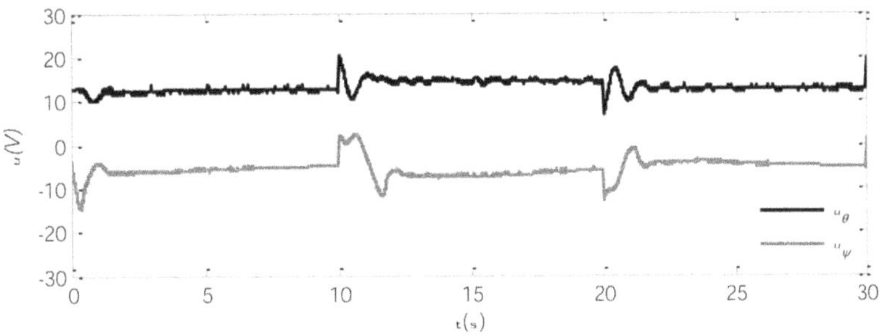

Figure 15. Normalized step response for pitch and yaw motions under SMC

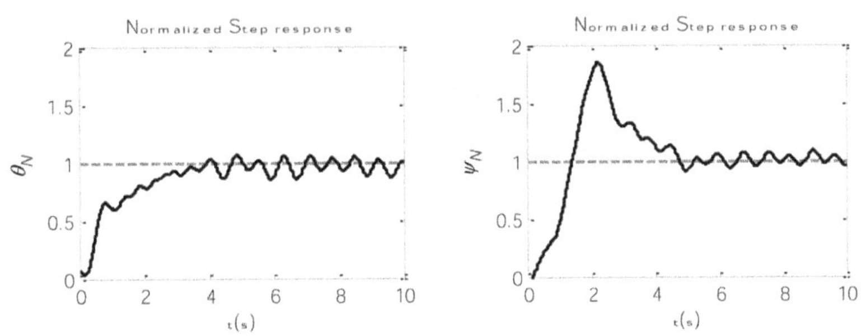

Figure 16. Normalized step response for pitch and yaw motions under FLC

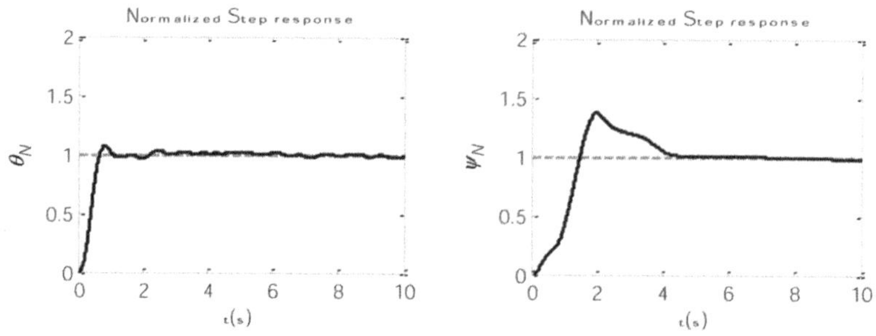

Table 5. Comparison of the step response characteristics

Pitch (P) and Yaw (Y) Motion Characteristics		Rise Time (s)	Overshoot (%)	Settling Time (s)	Mean-Squared Error
SMC	P	2.637	6.453	9.798	0.063
	Y	1.302	85.131	9.087	0.133
FLC	P	0.585	6.893	0.932	0.031
	Y	1.416	37.67	4.054	0.094

The results have demonstrated that designed controllers performed successfully even in the presence of external disturbances and high coupling effect between pitch and yaw motions. Especially the fuzzy logic controller performed better than the sliding mode controller in terms of the time response performance criteria. Furthermore, from the viewpoint of the control voltages, the fuzzy logic controller produced more smooth signals than the sliding mode controller, and this may protect the actuators and prolong their lifetime. Therefore, the results demonstrated that fuzzy logic controller may be suggested for future works in unmanned aerial vehicles.

REFERENCES

Ahmad, S. M., Chipperfield, A. J., & Tokhi, M. O. (2001). Parametric modelling and dynamic characterization of a two-degree-of-freedom twin-rotor multi-input multi-output system. *Proceedings of the Institution of Mechanical Engineers, Part G: Journal of Aerospace Engineering*. 10.1243/0954410011531772

Aldebrez, F. M., Darus, I. Z. M., & Tokhi, M. O. (2004). Dynamic modelling of a twin rotor system in hovering position. *First International Symposium on Control, Communications and Signal Processing*, 2004, 823–826. 10.1109/ISCCSP.2004.1296572

Azarmi, R., Tavakoli-Kakhki, M., Sedigh, A. K., & Fatehi, A. (2015). Analytical design of fractional order PID controllers based on the fractional set-point weighted structure: Case study in twin rotor helicopter. *Mechatronics*, *31*, 222–233. doi:10.1016/j.mechatronics.2015.08.008

Bayrak, A., Dogan, F., Tatlicioglu, E., & Ozdemirel, B. (2015). Design of an experimental twin-rotor multi-input multi-output system. *Computer Applications in Engineering Education*, *23*(4), 578–586. doi:10.1002/cae.21628

Burrascano, P., Callegari, S., Montisci, A., Ricci, M., & Versaci, M. (Eds.). (2015). *Ultrasonic Nondestructive Evaluation Systems*. doi:10.1007/978-3-319-10566-6

Che Soh, A., Abdul Rahman, R. Z., Md. Sarkan, H., & Yeo, L. T. (2013). Intelligent control of twin-rotor MIMO system using fuzzy inference techniques. *International Review of Automatic Control*, *6*(1), 35–46.

Chemachema, M., & Zeghlache, S. (2015). Output Feedback Linearization Based Controller for a Helicopter-like Twin Rotor MIMO System. *Journal of Intelligent and Robotic Systems: Theory and Applications*, *80*(1), 181–190. doi:10.100710846-014-0129-3

Choudhary, S. K. (2017). Optimal feedback control of a twin rotor MIMO system. *International Journal of Modelling and Simulation*, *37*(1), 46–53. doi:10.1080/02286203.2016.1233008

Darus, I. Z. M., Aldebrez, F. M., & Tokhi, M. O. (2004). Parametric modelling of a twin rotor system using genetic algorithms. *First International Symposium on Control, Communications and Signal Processing*, 115–118. 10.1109/ISCCSP.2004.1296232

Hacioglu, Y. (2015). Multiple-input multiple-output proportional-integral-proportional-derivative type fuzzy logic controller design for a twin rotor system. *Journal of Vibroengineering*, *17*(6), 3104–3110.

Ilyas, M., Abbas, N., UbaidUllah, M., Imtiaz, W. A., Shah, M. A. Q., & Mahmood, K. (2016). Control Law Design for Twin Rotor MIMO System with Nonlinear Control Strategy. *Discrete Dynamics in Nature and Society*, 1–10. doi:10.1155/2016/2952738

Quanser Inc. (2010). *Quanser 2 DOF Helicopter User and Laboratory Manuals, Revision 2.3.*

Rahideh, A., Shaheed, M. H., & Huijberts, H. J. C. (2008). Dynamic modelling of a TRMS using analytical and empirical approaches. *Control Engineering Practice*, *16*(3), 241–259. doi:10.1016/j.conengprac.2007.04.008

Rashad, R., El-Badawy, A., & Aboudonia, A. (2017). Sliding mode disturbance observer-based control of a twin rotor MIMO system. *ISA Transactions*, *69*, 166–174. doi:10.1016/j.isatra.2017.04.013 PMID:28449822

Taskin, Y. (2014). Improving pitch and yaw motion control of twin rotor MIMO system. *Journal of Vibroengineering*, *16*(4), 1650–1660.

Tastemirov, A., Lecchini-Visintini, A., & Morales-Viviescas, R. M. (2017). Complete dynamic model of the Twin Rotor MIMO System (TRMS) with experimental validation. *Control Engineering Practice*, *66*(July), 89–98. doi:10.1016/j.conengprac.2017.06.009

Utkin, V., Guldner, J., & Shi, J. (2009). Sliding Mode Control in Electro-Mechanical Systems, Second Edition. In Automation and Control Engineering: Vol. 31. doi:10.1201/9781420065619

Versaci, M., Calcagno, S., Cacciola, M., Morabito, F. C., Palamara, I., & Pellicanò, D. (2015a). Fuzzy Geometrical Techniques for Characterizing Defects in Ultrasonic Non-destructive Evaluation. In Ultrasonic Nondestructive Evaluation Systems (pp. 259–269). doi:10.1007/978-3-319-10566-6_10

Versaci, M., Calcagno, S., Cacciola, M., Morabito, F. C., Palamara, I., & Pellicanò, D. (2015b). Innovative Fuzzy Techniques for Characterizing Defects in Ultrasonic Nondestructive Evaluation. In Ultrasonic Nondestructive Evaluation Systems (pp. 201–232). doi:10.1007/978-3-319-10566-6_7

Versaci, M., Calcagno, S., Cacciola, M., Morabito, F. C., Palamara, I., & Pellicanò, D. (2015c). Standard Soft Computing Techniques for Characterization of Defects in Nondestructive Evaluation. In Ultrasonic Nondestructive Evaluation Systems (pp. 175–199). doi:10.1007/978-3-319-10566-6_6

Versaci, M., La Foresta, F., Morabito, F. C., & Angiulli, G. (2018). A fuzzy divergence approach for solving electrostatic identification problems for NDT applications. *International Journal of Applied Electromagnetics and Mechanics*, *57*(2), 133–146. doi:10.3233/JAE-170043

Versaci, M., Morabito, F. C., & Angiulli, G. (2017). Adaptive Image Contrast Enhancement by Computing Distances into a 4-Dimensional Fuzzy Unit Hypercube. *IEEE Access: Practical Innovations, Open Solutions*, *5*, 26922–26931. doi:10.1109/ACCESS.2017.2776349

Witczak, M., Buciakowski, M., Puig, V., Rotondo, D., & Nejjari, F. (2016). An LMI approach to robust fault estimation for a class of nonlinear systems. *International Journal of Robust and Nonlinear Control*, *26*(7), 1530–1548. doi:10.1002/rnc.3365

Yang, J. H., & Hsu, W. C. (2009). Adaptive backstepping control for electrically driven unmanned helicopter. *Control Engineering Practice*, *17*(8), 903–913. doi:10.1016/j.conengprac.2009.02.012

Zadeh, L. A. (1965). Fuzzy Sets- [*Information and Control.*]. *Information and Control*, 1965.

APPENDIX

Table 6. Acronyms and Symbols

Acronym / Symbol	Definition
B_{pitch}	Equivalent viscous damping about pitch axis
B_{yaw}	Equivalent viscous damping about yaw axis
CW	Clockwise
CCW	Counter clockwise
DC	Direct current
DOF	Degrees of freedom
$\bar{d}_{\psi}$	External disturbance
$\bar{d}_{\theta}$	External disturbance
F_g	Weight of twin rotor system
FFC	Fuzzy feedforward compensator
FLC	Fuzzy logic controller
I_{pitch}	Total moment of inertia about pitch axis
I_{yaw}	Total moment of inertia about yaw axis
K_{pp}	Thrust force constant of yaw motor/propeller
K_{py}	Thrust torque constant acting on pitch axis from yaw motor/propeller
K_{yp}	Thrust torque constant acting on yaw axis from pitch motor/propeller
K_{yy}	Thrust torque constant acting on yaw axis from yaw motor/propeller
l	Center of mass length along helicopter body from pitch axis
m_{hel}	Total moving mass of the helicopter
MIMO	Multiple-input multiple-output
NB	Negative big
MN	Negative medium
NS	Negative small
NVB	Negative very big

continued on the following page

Table 6. Continued

Acronym / Symbol	Definition
ψ	The range of yaw motion
$\psi(0)$	ψ initial value
PB	Positive big
PD	Proportional derivative
PI	Proportional integral
PM	Positive medium
PS	Positive small
PVB	Positive very big
SF	Scaling factor
SMC	Sliding mode controller
θ	The range of pitch motion
$\theta(0)$	θ initial value
τ_θ	Torque of pitch motor/propeller
τ_ψ	Torque of yaw motor/propeller
TRMS	Twin rotor MIMO system
V_{mp}	Pitch motor limit voltage
V_{my}	Yaw motor limit voltage
Z	Zero

Compilation of References

"Myomo," [Online]. (2016). Available at http://www.myomo.com/myopro-upper-limb-orthosis-technologymyomo-

Abdel-Aal, H. A. (2016, November 3). Functional surfaces for tribological applications: Inspiration and design. *Surface Topography: Metrology and Properties.*, *4*(4), 043001.

Abdel-Aal, H. A., & El Mansori, M. (2014, October 23). Characterization of load bearing metrological parameters in reptilian exuviae in comparison to precision finished cylinder liner surfaces. *Surface Topography: Metrology and Properties.*, *2*(4), 045002.

Abdel-Aal, H. A., El Mansori, M., & Mezghani, S. (2010, March 1). Multi-scale investigation of surface topography of ball python (Python regius) shed skin in comparison to human skin. *Tribology Letters*, *37*(3), 517–527. doi:10.100711249-009-9547-y

Abdi, J., Al-Hindawi, A., Ng, T., & Vizcaychipi, M. P. (2018). Scoping review on the use of socially assistive robot technology in elderly care. *BMJ Open*, *8*(2). doi:10.1136/bmjopen-2017-018815 PMID:29440212

Abouaf, J. (1998). Trial by Fire: Teleoperated Robot Targets Chernobyl. *IEEE Computer Graphics and Applications*, *18*(4), 10–14. doi:10.1109/38.689654

Abràmoff, M. D., Magallhas, P. J., & Ram, S. J. (2004). Image Processing with ImageJ. *Biophotonics International*, *11*(4), 36–42.

Aghaei, M., Quater, P. B. F., Grimaccia, F., Leva, S., & Mussetta, M. (2014). Unmanned aerial vehicles in photovoltaic systems monitoring applications. *Proceedings of 29th European Photovoltaic Solar Energy Conference and Exhibition*, *1*(1), 2734-2739.

Ahmad, S. M., Chipperfield, A. J., & Tokhi, M. O. (2001). Parametric modelling and dynamic characterization of a two-degree-of-freedom twin-rotor multi-input multi-output system. *Proceedings of the Institution of Mechanical Engineers, Part G: Journal of Aerospace Engineering*. 10.1243/0954410011531772

Ahmed, E., Harwood, D., & Davis, H. (2000). Non-parametric model for background subtraction. In Computer Vision—ECCV, pp. 751-767. Berlin, Germany: Springer.

Akbar, M. F., Ramadhani, N. A. Y., & Putri, R. A. (2018). Assistive and wearable technology for elderly. *Bulletin of Social Informatics Theory and Application*, *2*(1), 8–14. doi:10.31763/businta.v2i1.101

Akula, A., Ghosh, R., & Sardana, H. K. (2011). Thermal imaging and its application in defence systems. *Optics: Phenomena, Materials, Devices, and Characterization*. In *AIP Conference Proceedings*, *1391*(1), 333–335.

Al-Arif, S. M. M. R., Ferdous, A. H. M. I., & Nijami, S. H. (2012). Comparative Study of Different Path Planning Algorithms: A Water based Rescue System. *International Journal of Computers and Applications*, *39*(5), 25–29. doi:10.5120/4817-7058

Aldebrez, F. M., Darus, I. Z. M., & Tokhi, M. O. (2004). Dynamic modelling of a twin rotor system in hovering position. *First International Symposium on Control, Communications and Signal Processing*, 2004, 823–826. 10.1109/ISCCSP.2004.1296572

Alexander, E., Chow, Y., Cohen, J., & Jeske, S. (2010). WINROBO Window-Washing Robot. *Docstoc.com*. Retrieved from http://www.docstoc.com/docs/ 23539888/Winrobo -Window-Washing-Robot

Alexander, R. M. N., Bennett, M. B., & Ker, R. F. (1986). Mechanical properties and function of the paw pads of some mammals. *J. Zool., 209*, 405-419. .1986.tb03601.x doi:10.1111/j.1469-7998

Alexander, R., Bennett, M. B., & Ker, R. F. (1986, July 1). Mechanical properties and function of the paw pads of some mammals. *Journal of Zoology*, *209*(3), 405–419. doi:10.1111/j.1469-7998.1986.tb03601.x

Alfred, D., Kolody, B., & Wood, P. (1990). Effects of social support from various sources on depression in elderly persons. *Journal of Health and Social Behavior*, 148-161. PMID:2102494

Altuzarra, O., Pinto, C., Avilés, R., & Hernández, A. (2004). A practical procedure to analyze singular configurations in closed kinematic chains. *IEEE Transactions on Robotics*, *20*(6), 929–940. doi:10.1109/TRO.2004.832798

Anderson, J. D. (2001). *Fundamentals of aerodynamics* (3rd ed.).

Andreasen, M. M., & McAloone, T. C. (2001). Joining Three Heads - Experiences From Mechatronic Projects. In *12th Design For X Symposium,* Neukirchen, Germany. pp 151-156.

Andrikopoulos, G., Nikolakopoulos, G., & Manesis, S. (2011). A Survey on Applications of Pneumatic Artificial Muscles, *19th Mediterranean Conference on Control and Automation*, Corfu, Greece. 10.1109/MED.2011.5982983

Angiulli, G., Jannelli, A., Morabito, F. C., & Versaci, M. (2018). Reconstructing the Membrane Detection of a 1D Electrostatic-Driven MEMS Device by the Shooting Method: Convergence Analysis and Ghost Solutions Identification. *Computational & Applied Mathematics*, *37*(September), 4484–4498. doi:10.100740314-017-0564-4

Antuvan, C. W., Bisio, F., Marini, F., Yen, S.-C., Cambria, E., & Masia, L. (2016). Role of Muscle Synergies in Real-Time Classification of Upper Limb Motions using Extreme Learning Machines. *Journal of Neuroengineering and Rehabilitation*, *13*(76). PMID:27527511

Aoyagi, D., Ichinose, W. E., Reinkensmeyer, D. J., & Bobrow, J. (2004). Human Step Rehabilitation Using a Robot Attached To the Pelvis. In *2004 ASME International Mechanical Engineering Congress and Exposition* (p. 7).

Apple Inc. (2019). panoramaPhoto – INPhotoAttributeOptions. Retrieved from https://developer.apple.com/documentation/sirikit/inphotoattributeoptions/1902458-panoramaphoto

Armanini, S. F., de Visser, C. C., de Croon, G. C. H. E., & Mulder, M. (2016). Time-Varying Model Identification of Flapping-Wing Vehicle Dynamics Using Flight Data. *Journal of Guidance, Control, and Dynamics*, *39*(3), 526–541. doi:10.2514/1.G001470

Ates, S., Sluiter, V. I., Lammertse, P., & Stienen, A. H. A. (2014). ServoSEA Concept: Cheap, Miniature Series-Elastic Actuators for Orthotic, Prosthetic and Robotic Hands, *2014 5th IEEE RAS & EMBS International Conference on Biomedical Robotics and Biomechatronics (BioRob),* Sao Paulo, Brazil. 10.1109/BIOROB.2014.6913868

Aung, Y. M., & Al-Jumaily, A. (2013). Estimation of Upper Limb Joint Angle Using Surface EMG Signal. *International Journal of Advanced Robotic Systems*, 1–8.

Azarmi, R., Tavakoli-Kakhki, M., Sedigh, A. K., & Fatehi, A. (2015). Analytical design of fractional order PID controllers based on the fractional set-point weighted structure: Case study in twin rotor helicopter. *Mechatronics*, *31*, 222–233. doi:10.1016/j.mechatronics.2015.08.008

Babaiasl, M., Mahdioun, S. H., Jaryani, P., & Yazdani, M. (2015). A review of technological and clinical aspects of robot-aided rehabilitation of upper-extremity after stroke. *Disability and Rehabilitation. Assistive Technology*, 1–18. do i:10.3109/17483107.2014.1002539 PMID:25600057

Baek, S. S., Garcia Bermudez, F. L., & Fearing, R. S. (2011). Flight control for target seeking by 13 gram ornithopter. In *2011 IEEE/RSJ International Conference on Intelligent Robots and Systems* (pp. 2674–2681). IEEE. 10.1109/IROS.2011.6094581

Bailey, P. B., Shampine, L. F., & Waltman, P. E. (1968). *Nonlinear Two Point Boundary Value Problems*. New York, NY: Academic Press.

Bandera, J. P., Marfil, R., Romero-Garcés, A., & Voilmy, D. (2019). A new paradigm for autonomous human motion description and evaluation: Application to the Get Up & Go test use case. *Pattern Recognition Letters*, *118*, 51–60. doi:10.1016/j.patrec.2018.02.003

Banerjee, K., Krishnan, P., & Mridha, N. (2018). Application of thermal imaging of wheat crop canopy to estimate leaf area index under different moisture stress conditions. *Biosystems Engineering*, *166*(1), 13–27.

Bannwart, M., Rauter, G., Curt, A., & Bolliger, M. (2015). The FLOAT: Interactive BWS training of everyday walking challenges in 4 dimensions. Retrieved from https://www.research-collection.ethz.ch/handle/20.500.11850/105182

Barbosa, A. M. (2013). *Desenvolvimento de uma estrutura robótica paralela atuada por cabos para reabilitação dos movimentos do ombro*. Universidade Federal de Uberlândia.

Baum, M. J., Kovalev, A. E., Michels, J., & Gorb, S. N. (2014, May 1). Anisotropic friction of the ventral scales in the snake Lampropeltis getula californiae. *Tribology Letters*, *54*(2), 139–150. doi:10.100711249-014-0319-y

Bayrak, A., Dogan, F., Tatlicioglu, E., & Ozdemirel, B. (2015). Design of an experimental twin-rotor multi-input multi-output system. *Computer Applications in Engineering Education*, *23*(4), 578–586. doi:10.1002/cae.21628

Beer, F. P., Johnston, E. R., Jr. ., Mazurek, D. F., León Cárdenas, J., Ríos Sánchez, M. A., & Zamora Gallardo, E. (2013). *Mecánica vectorial para ingenieros : estática*. McGraw-Hill Interamericana.

Bengtson, V. (2018). *Global aging and challenges to families*. Routledge. doi:10.4324/9781351328166

Bennett, M. B., & Ker, R. F. (1990, August). The mechanical properties of the human subcalcaneal fat pad in compression. *Journal of Anatomy*, *171*, 131. PMID:2081699

Bensrhair, A., Bertozzi, M., Broggi, A., Mich'e, P., Mousset, S., & Toulminet, G. (2001). A Cooperative Approach to Vision-based Vehicle Detection, *IEEE Intelligent Transportation Systems Conference*. 10.1109/ITSC.2001.948657

Berns, K., & von Puttkamer, E. (2009). Autonomous land vehicles: Steps towards service robots, Springer-Vieweg, Wiesbaden doi:10.1007/978-3-8348-9334-5

Biewener, A. A. (1983, July 1). Allometry of quadrupedal locomotion: The scaling of duty factor, bone curvature and limb orientation to body size. *The Journal of Experimental Biology*, *105*(1), 147–171. PMID:6619724

Biewener, A. A. (1990). Biomechanics of mammalian terrestrial locomotion. *Science, 250*(4984), 1097–1103. doi:10.1126cience.2251499 PMID:2251499

Bischoff, A. & Bruhm, H. (2016). *GIS-basierte Cloud-Services für die Geländeexploration mit kooperierenden autonomen Systemen*, GIS Talk, Munich, Oct. 25, 2016 (German Presentation)

Björk, E., Palm, B., & Nordenberg, J. (2010). A thermographic study of the on-off behavior of an all-refrigerator. *Applied Thermal Engineering, 30*(14-15), 1974–1984.

Bluman, G. W., & Anco, S. C. (2002). *Symmetry and Integration Methods for Differential Equations* (Vol. 154). New York, NY: Springer-Verlag.

Bocca, M., Kaltiokallio, O., Patwari, N., & Venkatasubramanian, S. (2014). Multiple target tracking with RF sensor networks. *IEEE Transactions on Mobile Computing, 13*(18), 1787–1800. doi:10.1109/TMC.2013.92

Bogado Martínez, C. F., Dutra, M. S., & Raptopoulos, L. S. C. (2018). Proposed control for wing movement, type flat plate, for ornithopter autonomous robot. *Journal of the Brazilian Society of Mechanical Sciences and Engineering, 40*(9), 456. doi:10.100740430-018-1362-7

Boian, R. F., Bouzit, M., Burdea, G. C., Lewis, J., & Deutsch, J. E. (2005). Dual stewart platform mobility simulator. In *Proceedings of the 2005 IEEE 9th International Conference on Rehabilitation Robotics* (Vol. 2005, pp. 550–555). https://doi.org/10.1109/ICORR.2005.1502023

Bolton, W. (2003). *Mechatronics - Electronic Control Systems in Mechanical and Electrical Engineering.* Edinburg Gate Harlow Essex CM20 2JE England: Pearson Prentice Hall .

Borràs, J., Thomas, F., Ottaviano, E., & Ceccarelli, M. (2009). A reconfigurable 5-DoF 5-SPU parallel platform" Proceedings of the 2009 ASME/IFToMM International Conference on Reconfigurable Mechanisms and Robots, London, UK, 5173892, pp. 617-623.

Brachman, R. J., & Levesque, H. J. (2004). *Planning Knowledge Representation and Reasoning* (pp. 305–325). San Francisco, CA: Morgan Kaufmann. doi:10.1016/B978-155860932-7/50100-X

Brasil. (2013). *Diretrizes de Atenção à Reabilitação da Pessoa com Acidente Vascular Cerebral.*

Brasil. (2017). Acidente Vascular Cerebral - AVC. Ministério da Saúde. Retrieved from http://portalms.saude.gov.br/saude-de-a-z/acidente-vascular-cerebral-avc

Breazeal, C., & Scassellati, B. (2002). Robots that imitate humans. *Trends in Cognitive Sciences, 6*(11), 481-487. doi:10.1016/S1364-6613(02)02016-8 PMID:12457900

Bretl, T. (2006). Motion planning of multi-limbed robots subject to equilibrium constraints: The free-climbing robot problem. *The International Journal of Robotics Research, 25*(4), 317–342. doi:10.1177/0278364906063979

British Geriatric Society. (2019). Comprehensive Geriatric Assessment Toolkit for Primary Care Practitioners. British Geriatric Society. Retrieved from https://www.bgs.org.uk/sites/default/files/content/resources/files/2019-02-08/BGS%20Toolkit%20-%20FINAL%20FOR%20WEB_0.pdf

Broadbent, E., Stafford, R., & MacDonald, B. (2009). Acceptance of healthcare robots for the older population: Review and future directions. *International Journal of Social Robotics, 1*(4), 319-330. doi:10.100712369-009-0030-6

Brooks, R. (1986). A robust layered control system for a mobile robot. *IEEE Journal on Robotics and Automation, 2*(1), 14-23. doi:10.1109/JRA.1986.1087032

Brown, J. C., & Yalden, D. W. (1973, December). The description ofmammals– 2 Limbs and locomotion of terrestial mammals. *Mammal Review*, *3*(4), 107–134. doi:10.1111/j.1365-2907.1973.tb00178.x

B-Spline Fit VI - LabVIEW 2011 Help. (2011). Retrieved from http://zone.ni.com/reference/en-XX/help/371361H-01/gmath/bspline_fit/

Budaker, B. (2014). Biomechatronic System Engineering, *5th International Conference on Biomedical Robotics and Biomechatronics (BioRob)*, Sao Paulo, Brazil, pp. 768-773.

Budsberg, S.C., Verstraete, M.C., & Soutas-Little, R.W. (1987). Force plate analysis of the walking gait in healthy dogs. *American Journal of Veterinary Research, 48*(6).

Budsberg, S. C., Verstraete, M. C., Soutas-Little, R. W., Flo, G. L., & Probst, C. W. (1988, September). Force plate analyses before and after stabilization of canine stifles for cruciate injury. *American Journal of Veterinary Research, 49*(9), 1522–1524. PMID:3223659

Buniyamin, N., Ngah, W. A. J. W., Sariff, N., & Mohamad, Z. (2011). A Simple Local Path Planning Algorithm for Autonomous Mobile Robots. *International Journal of Systems Applications, Engineering & Development*, *5*(2), 151–159.

Burghart, C., Mikut, R., Stiefelhagen, R., Asfour, T., Holzapfel, H., Steinhaus, P., & Dillmann, R. (2005). A cognitive architecture for a humanoid robot: A first approach. In *5th IEEE-RAS International Conference on Humanoid Robots.* 10.1109/ICHR.2005.1573593

Burlayenko, V. N., & Sadowski, T. (2010). Effective elastic properties of foam-filled honeycomb cores of sandwich panels. *Composite Structures*, *92*(12), 2890–2900. doi:10.1016/j.compstruct.2010.04.015

Burrascano, P., Callegari, S., Montisci, A., Ricci, M., & Versaci, M. (Eds.). (2015). *Ultrasonic Nondestructive Evaluation Systems.* doi:10.1007/978-3-319-10566-6

Bustos, P., Manso, L. J., Bandera, A., Bandera, J. P., García-Varea, I., & Martínez-Gómez, J. (2019). The CORTEX cognitive robotics architecture: Use cases. *Cognitive Systems Research*, *55*, 107–123. doi:10.1016/j.cogsys.2019.01.003

Buur, J. (1989). *Mechatronics design in Japan.* Institute for Engineering Design, Technical University of Denmark.

Cabrera, A. A., Foeken, M. J., Tekin, O. A., Woestenenk, K., Erden, M. S., De Schutter, B., ... Tomiyama, T. (2010). Towards Automation of Control Software: A Review of Challenges in Mechatronic Design. *Mechatronics*, *20*(8), 876–886. doi:10.1016/j.mechatronics.2010.05.003

Calcagno, S., La Foresta, F., & Versaci, M. (2014). Independent component analysis and discrete wavelet transform for artifact removal in biomedical signal processing. *American Journal of Applied Sciences*, *11*(1), 57–68. doi:10.3844/ajassp.2014.57.68

Camargos, A. C. R., Lacerda, T. T. B., Taise, V. B., Silva, G. C., Parreiras, J. T., & Vidal, T. H. J. (2012). Relação entre independência funcional e qualidade de vida na paralisia cerebral Relationship between functional independence and quality of life in cerebral palsy, *25*(1), 83–92.

Cao, Z., Hidalgo, G., Simon, T., Wei, S.-E., & Sheikh, Y. (2018). OpenPose: Realtime Multi-Person 2D Pose Estimation using Part Affinity Fields. In *arXiv preprint arXiv:1812.08008.*

Cardinaels, R., Maes, J., Deseure, J., Saldien, J., Aelterman, K., Verstockt, S., & Wyffels, F. (2017). GECO: the development of a wall-climbing robot. *20th International Conference on Climbing and Walking Robots and the Support Technologies for Mobile Machines (CLAWAR 2017).*

Castelli, G., & Ottaviano, E. (2010). Modelling, Simulation and Testing of a Reconfigurable Cable-Based Parallel Manipulator as Motion Aiding System. *Applied Bionics and Biomechanics*, *7*(4), 253–268. doi:10.1155/2010/548197

Castelli, G., Ottaviano, E., & Rea, P. (2014). A Cartesian Cable-Suspended Robot for improving end-users' mobility in an urban environment. *Robotics and Computer-integrated Manufacturing*, *30*(3), 335–343. doi:10.1016/j.rcim.2013.11.001

Cavagna, G. A., Heglund, N. C., & Taylor, C. R. (1977, November 1). Mechanical work in terrestrial locomotion: Two basic mechanisms for minimizing energy expenditure. *American Journal of Physiology. Regulatory, Integrative and Comparative Physiology*, *233*(5), R243–R261. doi:10.1152/ajpregu.1977.233.5.R243 PMID:411381

Che Soh, A., Abdul Rahman, R. Z., Md. Sarkan, H., & Yeo, L. T. (2013). Intelligent control of twin-rotor MIMO system using fuzzy inference techniques. *International Review of Automatic Control*, *6*(1), 35–46.

Chemachema, M., & Zeghlache, S. (2015). Output Feedback Linearization Based Controller for a Helicopter-like Twin Rotor MIMO System. *Journal of Intelligent and Robotic Systems: Theory and Applications*, *80*(1), 181–190. doi:10.100710846-014-0129-3

Chen, D., Magee, R., & Yakubov, N. (2005). Static and Kinetic Friction. *Revitalizing Achievement by using Instrumentation in Science Education*. Retrieved from http://raise.poly.edu/Information/RAISE_Workshop_PowerPointFiles/Static_&_ _Friction.ppt

Cheng, L., Liu, C., & Yan, B. (2014, July). Improved Hierarchical A-star Algorithm for Optimal Parking Path Planning of the Large Parking Lot. Paper presented at the *2014 IEEE International Conference on Information and Automation (ICIA)*, Hailar, China. 10.1109/ICInfA.2014.6932742

Cheng, M., & Xiang, D. (2019). The Design and Application of a Track-type Autonomous Inspection Robot for Electrical Distribution Room. *Robotica*, *1*(22).

Chenkun, Q., Feng, G., Qiao, S., Xianbao, C., Yilin, X., & Xianchao, Z. (2015). A foot force sensing approach for a legged walking robot using the motor current. In *Robotics and Biomimetics (ROBIO), 2015 IEEE International Conference on 2015 Dec 6* (pp. 1078-1083). IEEE. 10.1109/ROBIO.2015.7418915

Chen, X., Wang, L., Ye, X., Wang, G., & Wang, H. (2013). Prototype development and gait planning of biologically inspired multi-legged crablike robot. *Mechatronics*, *23*(4), 429–444. doi:10.1016/j.mechatronics.2013.03.006

Chen, Y., Zhao, X., & Han, J. (2013). Hierarchical projection regression for online estimation of elbow joint angle using EMG signals. *Neural Computing & Applications*, *23*(3-4), 1129–1138. doi:10.100700521-012-1045-8

Cheung, J. T.-M., Zhang, M., Leung, A. K.-L., & Fan, Y.-B. (2005). Three-dimensional finite element analysis of the foot during standing—A material sensitivity study. *Journal of Biomechanics*, *38*(5), 1045–1054. doi:10.1016/j.jbiomech.2004.05.035 PMID:15797586

Chi, K. J. (2005). *Functional morphology and biomechanics of mammalian footpads* (PhD thesis). Duke University, Durham, NC.

Chirarattananon, P., Ma, K. Y., & Wood, R. J. (2014). Adaptive control of a millimeter-scale flapping-wing robot. *Bioinspiration & Biomimetics*, *9*(2). doi:10.1088/1748-3182/9/2/025004 PMID:24855052

Choudhary, S. K. (2017). Optimal feedback control of a twin rotor MIMO system. *International Journal of Modelling and Simulation*, *37*(1), 46–53. doi:10.1080/02286203.2016.1233008

Chung-Cheng, C., Min-Yu, K., & Hung-Tsung, C. (2007). Motorcycle Detection and Tracking System with Occlusion Segmentation, *Eighth International Workshop on Image Analysis for Multimedia Interactive Services*, pp. 23–32.

Chung, J. H., & Yi, B. J. (2015). Analysis of Internal Loading at Multiple Robotic Systems. *Journal of Mechanical Science and Technology. KSME International Journal, 19*(8), 14.

Ciampa, F., Mahomoodi, P., Pinto, F., & Meo, M. (2018). Recent analysis in active infrared thermography for non-destructive testing of aerospace components. *Sensors (Basel), 18*(609), 1–37.

Cid-Ruzafa, J., & Damián-Moreno, J. (1997). Valoración de la discapacidad física: El índice de Barthel. *Revista Espanola de Salud Publica, 71*(2), 127–137. doi:10.1590/S1135-57271997000200004 PMID:9546856

Čikeš, M., Đakulović, M., & Petrović, I. (2011, October). *The Path Planning Algorithms for a Mobile Robot Based on the Occupancy Grid Map of the Environment - A Comparative Study.* Paper presented at the 2011 XXIII International Symposium on Information, Communication and Automation Technologies, Sarajevo, Bosnia.

Coeckelbergh, M. (2011). Humans, animals, and robots: A phenomenological approach to human-robot relations. *International Journal of Social Robotics, 3*(2), 197-204. doi:10.100712369-010-0075-6

Colasanto, L., Van der Noot, N., & Ijspeert, A. J. (2015). Bio-inspired walking for humanoid robots using feet with human-like compliance and neuromuscular control. In Humanoids 2015 Nov 3 (pp. 26-32). Academic Press. doi:10.1109/HUMANOIDS.2015.7363518

Colorado, J. (2012). BaTboT : A biologically inspired flapping and morphing bat robot actuated by SMA-based artificial muscles., 223. Retrieved from http://digital.csic.es/handle/10261/109918

Crişan, L.-M., & Borza, I. (2019). Management of Maintenance for Photovoltaic Systems. *Nonconventional Technologies Review, 23*(1), 40–45.

Cunningham, C. B., Schilling, N., Anders, C., & Carrier, D. R. (2010). The influence of foot posture on the cost of transport in humans. *JEB, 213*(5), 790–797. doi:10.1242/jeb.038984 PMID:20154195

Dahiya, R. S., Metta, G., Valle, M., & Sandini, G. (2010, February). Tactile sensing—From humans to humanoids. *IEEE Transactions on Robotics, 26*(1), 1–20. doi:10.1109/TRO.2009.2033627

Dalal, N., & Triggs, B. (2005). Histograms of Oriented Gradients for Human Detection, *IEEE Computer Society Conference on Computer Vision and Pattern Recognition.* 10.1109/CVPR.2005.177

Daltorio, K. A., Gorb, S., Peressadko, A., Horchler, A. D., Ritzmann, R. E., & Quinn, R. D. (2006). A robot that climbs walls using micro-structured polymer feet. In *Climbing and walking robots* (pp. 131–138). Berlin: Springer. doi:10.1007/3-540-26415-9_15

Dario, P., & Buttazzo, G. (1987, September). An anthropomorphic robot finger for investigating artificial tactile perception. *The International Journal of Robotics Research, 6*(3), 25–48. doi:10.1177/027836498700600302

Dario, P., Carrozza, M. C., Marcacci, M., D'Attanasio, S., Magnani, B., Tonet, O., & Megali, G. (2000). A Novel Mechatronic Tool for Computer-Assisted Arthroscopy. *IEEE Transactions on Information Technology in Biomedicine, 4*(1), 15–29. doi:10.1109/4233.826855 PMID:10761770

Darrow, C. W., & Gullickson, G. R. (1970). The Peripheral mechanism of the galvanic skin response. *Psychophysiology, 6*(5), 597–600. doi:10.1111/j.1469-8986.1970.tb02248.x PMID:5485543

Darus, I. Z. M., Aldebrez, F. M., & Tokhi, M. O. (2004). Parametric modelling of a twin rotor system using genetic algorithms. *First International Symposium on Control, Communications and Signal Processing*, 115–118. 10.1109/ISCCSP.2004.1296232

De Croon, E., Groen, M. A., De Wagter, C., Remes, B., Ruijsink, R., & Van Oudheusden, B. W. (2012). Design, aerodynamics and autonomy of the DelFly Recent citations Mimicking Insect Wings: The Roadmap to Bioinspiration Jafar Hasan et al Design, aerodynamics and autonomy of the DelFly. *Bioinspiration & Biomimetics*, *7*, 16. doi:10.1088/1748-3182/7/2/025003

de Hollander, G., Forstmann, B. U., & Brown, S. D. (2016). Different ways of linking behavioral and neural data via computational cognitive models. *Biological Psychiatry: Cognitive Neuroscience and Neuroimaging*, *1*(2), 101-109. PMID:29560872

De Wagter, C., Tijmons, S., Remes, B. D. W., & de Croon, G. C. H. E. (2014). Autonomous flight of a 20-gram Flapping Wing MAV with a 4-gram onboard stereo vision system. In *2014 IEEE International Conference on Robotics and Automation (ICRA)* (pp. 4982–4987). IEEE. 10.1109/ICRA.2014.6907589

Debor, S., Borgeest, K. & Imhof, Chr. (2015). Evolutionäre Algorithmen bei der Optimierung und beim Test mobiler Roboter (in German language). *12. AALE-Konferenz*, Jena

Delaurier, J. D. (1988). The Development and Testing of a Full-Scale Piloted Ornithopter. *Canadian Aeronautics and Space Journal*. Retrieved from http://www.ornithopter.net/Publications/TheDevelopmentandTestingofaFull-ScalePilotedOrnithopter.pdf

DeLaurier, J. D., & Harris, J. (1993). *A study of mechanical flapping-wing flight. Aeronautical Journal*. Cambridge University Press; doi:10.10170001924000026452

Desplenter, T., Lobo-Prat, J., Stienen, A. H. A., & Trejor, A. L. (2016). Extension of the WearME Framework for EMG-driven Control of a Wearable Assistive Exoskeleton, *IEEE International Conference on Advanced Intelligent Mechatronics (AIM)*, Alberta, Canada. 10.1109/AIM.2016.7576781

Di Barba, P., Fattorusso, L., & Versaci, M. (2017). Electrostatic Field in Terms of Geometric Curvature in Membrane MEMS Devices. *Communications in Applied and Industrial Mathematics*, *8*(1), 165–184. doi:10.1515/caim-2017-0009

Dietl, J. M., & Garcia, E. (2008). Stability in Ornithopter Longitudinal Flight Dynamics. *Journal of Guidance, Control, and Dynamics*, *31*(4), 1157–1163. doi:10.2514/1.33561

Ding, Q. C., Xiong, A. B., Zhao, X. G., & Han, J. D. (2011). A Novel EMG-driven State Space Model for the Estimation of Continuous Joint Movements. In *IEEE conference on Systems*. Alaska, USA: Man, and Cybernetics. doi:10.1109/ICSMC.2011.6084104

do Carmo, M. P. (2016). *Differential Geometry of Curves and Surfaces*. Englewood, NJ: Prentice-Hall.

Dolan, R. J. (2002). Emotion, cognition, and behavior. science 298(5596), pp. 1191-1194. doi:10.1126cience.1076358

Dometios, A. C., Papageorgiou, X. S., Tzafestas, C. S., & Vartholomeos, P. (2016, June). Towards ICT-Supported Bath Robots: Control Architecture Description and Localized Perception of User for Robot Motion Planning, *24th Mediterranean Conference on Control and Automation (MED)*, Athens, Greece. 10.1109/MED.2016.7535954

Dorfmeister, M., Koessl, B., Schneider, M., & Schmid, U. (2019). A Novel Bi-Stable MEMS Membrane Concept Based on a Piezoelectric Thin Film Actuator for Integrated Switching. 1-4. doi:10.3390/proceedings2130912

Duchoň, F., Babinec, A., Kajan, M., Beňo, P., Florek, M., Fico, T., & Jurišica, L. (2014). Path Planning with Modified a Star Algorithm for a Mobile Robot. *Procedia Engineering*, *96*, 59–69. doi:10.1016/j.proeng.2014.12.098

Duffy, J. (1996). Statics and Kinematics with application to Robotics. Cambridge University Press, Amazon.com

Dupeyroux, J., Viollet, S., & Serres, J. (2017). Providing an autonomous hexapod walking robot with the ability to reorientate: application of a novel ant-inspired celestial compass. *Journée des Jeunes Chercheurs en Robotique.*

Durandau, G., & Suleiman, W. (2016). User-safe Orthosis Based on Compliant Actuators: Mechanical Design and Control Framework. In *Proceedings of the SICE Annual Conference.* 10.1109/SICE.2016.7749223

Eckert, P., & Ijspeert, A. (2016). *Adaptive foot design for small quadruped robots.* Academic Press.

Ehrampoosh, A., Yousefi-koma, A., & Ayati, M. (2015). Estimation of elbow joint angle by NARX model using EMG data, *RSI International Conference on Robotics and Mechatronics*, Tehran, Iran.

El Rafei, M., Badr, S., Lefebvre, G., Machuron, F., Capon, B., Flipo, R.-M., & Cotten, A. (2018). Sacroiliac joints: Anatomical variations on MR images. *European Radiology, 28*(12), 5328–5337. doi:10.100700330-018-5540-x PMID:29876707

El-Farouk, E., Labib, O., El-Safty, S. W., Mueller, S., Haalboom, T., & Strand, M. (2018). *Towards a Stair Climbing Robot System Based on a Re-configurable Linkage Mechanism, Intelligent Autonomous Systems 15. IAS 2018. Adv in Intelli Sys and Comp* , 867, (pp. 278–288). Cham, Switzerland: Springer; doi:10.1007/978-3-030-01370-7_22

El-Kalay, A. K., Akyurt, M., Aljawi, A. A. N., & Dehlawi, F. M. (1995). On Gripping Mechanism for Industrial Robots. *King Abdulaziz University Official Website.* Retrieved from http://www.kau.edu.sa/Files/320/Researches/46479_17.pdf

Ellis, G., Gardner, M., Tsiachristas, A., Langhorne, P., Burke, O., Harwood, R. H., ... Wald, H. (2017). Comprehensive geriatric assessment for older adults admitted to hospital. *Cochrane Database of Systematic Reviews*, 9. PMID:28898390

Entrop, A. G., & Vasenev, A. (2017). Infrared drones in the construction industry: Designing a protocol for building thermography procedures. *Energy Procedia, 137*(1), 63–68.

Erden, M. S., & Leblebicioğlu, K. (2008). Free gait generation with reinforcement learning for a six-legged robot. Robotics and Autonomous Systems, 56(3), 199-212.

Esposito, P., Ghoussoub, N., & Guo, Y. (2010). Mathematical Analysis of Partial Differential Equations Modeling Electrostatic MEMS (Vol. 20). American Mathematica Society.

Esposito, R., Fiorini, L., Limosani, R., Bonaccorsi, M., Manzi, A., Cavallo, F., & Dario, P. (2016). "Supporting active and healthy aging with advanced robotics integrated in smart environment." Optimizing assistive technologies for aging populations (pp. 46–77). Hershey, PA: IGI Global.

Fattorusso, L., & Versaci, M. (2018). A new one-dimensional electrostatic model for membrane MEMS devices. Lecture Notes in Engineering and Computer Science, 2235. London, UK.

Fazio, R., & Jannelli, A. (2014). Finite Difference Schemes on Quasi-Uniform Grids for BVPs on Infinite Intervals. *Journal of Computational and Applied Mathematics, 269,* 14–23. doi:10.1016/j.cam.2014.02.036

Fazio, R., & Jannelli, A. (2017). BVPs on Infinite Intervals: A Test Problem, a Nonstandard Finite Difference Scheme and a Posteriori Error Estimator. *Mathematical Methods in the Applied Sciences, 40*(18), 6285–6294. doi:10.1002/mma.4456

Figliolini, G., & Rea, P. (2012). Effects of the design parameters on the synthesis of Geneva mechanisms". Proceedings of the Institution of Mechanical Engineers. *PART C, Journal of Mechanical Engineering Science, vol. 2012,* pp. 1-10. doi:10.1177/0954406212466194

Figliolini, G., Rea, P., & Angeles, J. (2016). The synthesis of the axodes of RCCC linkages, *Journal of Mechanisms and Robotics, 8*(2). doi:. doi:10.1115/1.4031950

Figliolini, G., Rea, P., & Conte, M. (2010). Mechanical Design of a Novel Biped Climbing and Walking Robot, ROMANSY 2010, 18th CISM-IFToMM Symposium on Robot Design, Dynamics, and Control, pp.199-206.

Fliess, M., Levine, J., Martin, P., Rouchon, P., Lévine, J., Martin, P., & Rouchon, P. (1995). Flatness and defect of non-linear systems: Introductory theory and examples. *International Journal of Control*, *61*(6), 1327–1361. doi:10.1080/00207179508921959

Frascati Manual. (2015). *The Measurement of Scientific, Technological and Innovation Activities- Guidelines for Collecting and Reporting Data on Research and Experimental Development*. OECD.

Fry, S. N., Sayaman, R., & Dickinson, M. H. (2005). The aerodynamics of hovering flight in Drosophila. *The Journal of Experimental Biology*, *208*(Pt 12), 2303–2318. doi:10.1242/jeb.01612 PMID:15939772

Fujimori, B., & Yokota, T. (1962). Central and peripheral mechanisms of galvanic skin reflex. *Acta Neurovegetativa*, *24*(1-4), 241–249. doi:10.1007/BF01233000

Fukuda, T., Nakajima, M., Ahmad, M., Shen, Y., & Kojima, M. (2010). Micro- and Nanomechatronics. *IEEE Industrial Electronics Magazine*, *4*(4), 13–22. doi:10.1109/MIE.2010.938721

Fukuoka, Y., & Akama, J. (2014, October). Dynamic bipedal walking of a dinosaur-like robot with an extant vertebrate's nervous system. *Robotica*, *32*(6), 851–865.

Fu, M.-Y., & Huang, Y.-S. (2010). *A Survey of Traffic Sign Recognition. IEEE Intern. Conf. on Wavelet Analysis and Pattern Recognition*. doi:10.1109/ICWAPR.2010.5576425

Gad-el Hak, M. (2006). *MEMS: Design and Fabrication*. Boca Raton, FL: Chapman & Hall, CRC Taylor & Francis.

Galiano, A., Massaro, A., Barbuzzi, D., Pellicani, L., & Birardi, V. (2016). Heat Monitoring Plan in Building for Energy Efficiency using Thermal Imaging Analysis. [JNIT]. *Journal of Next Generation Information Technology*, *7*(4), 8–18.

Galli, M., Cimolin, V., De Pandis, M. F., Le Pera, D., Sova, I., Albertini, G., ... Franceschini, M. (2016). Robot-assisted gait training versus treadmill training in patients with Parkinson's disease: A kinematic evaluation with gait profile score. *Functional Neurology*, *31*(3), 163–170. doi:10.11138/FNeur/2016.31.3.163 PMID:27678210

Gálvez Cano, M., Varela Pinedo, L. F., Helver Chávez, J., Cieza Zevallos, J., & Méndez Silva, F. (2010). Correlación del Test" Get Up and Go. *Acta Médica Peruana*, *27*(1).

Gao, X. S., Shao, J., Dai, F. Q., Zong, C. G., & Guo, W. Z. (2014). A Gecko-Inspired Robot for Wind Power Tower Inspection. Applied Mechanics and Materials, 461, 831-837.

Gausemeier, J., Frank, U., Donoth, J., & Kahl, S. (2009). Specification Technique for the Description of Self-optimizing Mechatronic Systems. *Research in Engineering Design*, *20*(4), 201–223. doi:10.100700163-008-0058-x

General Polynomial Fit VI - LabVIEW 2014 Help. (2014). Retrieved from http://zone.ni.com/reference/en-XX/help/371361L-01/gmath/general_polynomial_fit/

Ghidoni, S., Antonello, M., Nanni, L., & Menegatti, E. (2015). A thermographic visual inspection system for crack detection in metal parts exploiting a robotic workcell. *Robotics and Autonomous Systems 74*(B), 351-359.

Gilho, B. V., & Bezerra, J. M. (2012). Mechatronic Design of a Chair for Disabled with Locomotion by Legs, *ABCM Symposium Series in Mechatronics*, 5, pp. 1142–1149.

Gill, H. S., & O'Connor, J. J. (2003b). Heelstrike and the pathomechanics of osteoarthrosis: A simulation study. *Journal of Biomechanics*, *36*(11), 1617–1624. doi:10.1016/S0021-9290(03)00190-8 PMID:14522202

GitHub, Inc. (2019). Retrieved from http://github.com/.

Giusti, G. (2000). *Analisi Matematica*. Turin, Italy: Bollati Boringhieri.

Goffer, A. (2006). *US7153242 B2.*

Golubitsky, M., Stewart, I., Buono, P. L., & Collins, J. J. (1999, October). Symmetry in locomotor central pattern generators and animal gaits. *Nature*, *401*(6754), 693–695. doi:10.1038/44416 PMID:10537106

Gonçalves, R. S., Hamilton, T., & Krebs, H. I. (2017). MIT-Skywalker: On the use of a markerless system. In *IEEE International Conference on Rehabilitation Robotics* (pp. 205–210). https://doi.org/10.1109/ICORR.2017.8009247

Gonçalves, R. S., Hamilton, T., Daher, A. R., Hirai, H., & Hermano, I. (2017). MIT-Skywalker : Evaluating Comfort of Bicycle / Saddle Seat *. In *2017 International Conference on Rehabilitation Robotics (ICORR)* (pp. 516–520). London, UK.

Gonçalves, B. J., de Oliveira Giarola, T. M., Pereira, D. F., de Barros Vilas Boas, E. V., & de Resende, J. V. (2016). Using infrared thermography to evaluate the injuries of cold-stored guava. *Journal of Food Science and Technology*, *53*(2), 1063–1070.

Gonçalves, R. S. (2010). Desenvolvimento de uma estrutura robótica paralela atuada por cabos para reabilitação dos movimentos do ombro. In *VI Congresso Nacional de Engenharia Mecânica*. Campina Grande.

Gonzalez de Santos, P., García, E., & Estremera, J. (2006). *Quadrupedal Locomotion: An introduction to the control of four-legged robots*. Berlin, Germany: Springer-Verlag.

Gonzalez-Arjona, D., Sanchez, A., López-Colino, F., de Castro, A., & Garrido, J. (2013). Simplified Occupancy Grid Indoor Mapping Optimized for Low-Cost Robots. *ISPRS International Journal of Geo-Information*, *2*(4), 959–977. doi:10.3390/ijgi2040959

Gosselin, C. M., Masouleh, M. T., Duchaine, V., Richard, P.-L., Foucault, S., & Kong, X. (2007). Parallel Mechanisms of the Multipteron Family: Kinematic Architectures and Benchmarking. In *Proceedings 2007 IEEE International Conference on Robotics and Automation* (pp. 555–560). IEEE. https://doi.org/10.1109/ROBOT.2007.363045

Gray, J. (2013). *How animals move*. Cambridge University Press.

Greiner, C., & Schäfer, M. (2015, June 30). Bio-inspired scale-like surface textures and their tribological properties. *Bioinspiration & Biomimetics*, *10*(4), 044001. doi:10.1088/1748-3190/10/4/044001 PMID:26125522

Grys, S. (2018). Determining the dimension of subsurface defects by active infrared thermography- experimental research. *Journal of Sensors and Sensor System*, *7*(1), 153–160.

Guruji, A. K., Agarwal, H., & Parsediya, D. K. (2016). Time-efficient A* Algorithm for Robot Path Planning. *Procedia Technology*, *23*, 144–149. doi:10.1016/j.protcy.2016.03.010

Habib, M. K. (2007). Mechatronics: A Unifying Interdisciplinary and Intelligent Engineering Paradigm, IEEE Industrial Electronics Magazine, 1(2), Summer, pp. 12-24.

Habib, M. K. (2008). Human adaptive and friendly mechatronics (HAFM). In *2008 IEEE International Conference on Mechatronics and Automation* (pp. 61–65). IEEE. https://doi.org/10.1109/ICMA.2008.4798726

Habib, M. K. (2008, August). Human Adaptive and Friendly Mechatronics (HAFM), *IEEE International Conference on Mechatronics and Automation 'ICMA2008'*, Takamatsu-Kagawa, Japan, pp. 61–65.

Habib, M. K. (2008, August). Human Adaptive and Friendly Mechatronics (HAFM), *IEEE International Conference on Mechatronics and Automation*, Takamatsu-Kagawa, Japan, pp. 61-65.

Habib, M. K. (2008, August). Human Adaptive and Friendly Mechatronics (HAFM). In *IEEE International Conference on Mechatronics and Automation 'ICMA'*, Takamatsu-Kagawa, Japan, pp. 61–65.

Habib, M. (2007). Mechatronics - A unifying interdisciplinary and intelligent engineering science paradigm. *IEEE Industrial Electronics Magazine*, *1*(2), 12–24. doi:10.1109/MIE.2007.901480

Habib, M. K. (2007). Mechatronics: A Unifying Interdisciplinary and Intelligent Engineering Paradigm. *IEEE Industrial Electronics Magazine*, *1*(2), 12–24.

Habib, M. K. (2008, August) Human Adaptive and Friendly Mechatronics (HAFM). *IEEE International Conference on Mechatronics and Automation 'ICMA2008'*, Takamatsu-Kagawa, Japan, 61-65.

Hacioglu, Y. (2015). Multiple-input multiple-output proportional-integral-proportional-derivative type fuzzy logic controller design for a twin rotor system. *Journal of Vibroengineering*, *17*(6), 3104–3110.

Haney, P. (2004). Rubber Friction. *Inside Racing Technology*. Retrieved from http://insideracingtechnology.com/tirebkexerpt1.htm

Harada, K., Kajita, S., Kanehiro, F., Fujiwara, K., Kaneko, K., Yokoi, K., & Hirukawa, H. (2007, February). Realtime planning of humanoid robot's gait for force-controlled manipulation. *IEEE/ASME Transactions on Mechatronics*, *12*(1), 53–62. doi:10.1109/TMECH.2006.886254

Hari Krishnan, R., & Pugazhenthi, S. (2014). Mobility assistive devices and self-transfer robotic systems for elderly, a review. *Intelligent Service Robotics*, *7*(1), 37-49. doi:10.100711370-013-0142-6

Hasegawa, N., & Takahashi, Y. (2017). Preliminary Research into Control System using Quantified Human Emotions. In *Proceedings of International Conference on Control, Automation and Systems*, pp. 1519-1524. 10.23919/ICCAS.2017.8204230

Hasegawa, N., & Takahashi, Y. (2019). How recognition of human facial expression can be incorporated in robot control. *International Conference on Research and Education in Mechatronics*.

Hawkes, E. W., Christensen, D. L., & Okamura, A. M. (2016). Design and Implementation of a 300% Strain Soft Artificial Muscle, *IEEE International Conference on Robotics and Automation (ICRA)*, Stockholm, Sweden, pp. 1-10. 10.1109/ICRA.2016.7487592

Hayashi, R., & Kato, S. (2015). Chobonyan: soft-stuffed robot in palmtop size and light weight for therapy. In *Consumer Electronics (GCCE), 2015 IEEE 4th Global Conference on 2015 Oct 27* (pp. 112-113). IEEE. 10.1109/GCCE.2015.7398537

He, Z., Meng, F., Liu, H., Wang, L., Zhu, X., Fan, X., . . . Huang, Q. (2019). Method of Design Optimization and Trajectory Implementation on a Small Cat-Like Robot. In ROMANSY 22–Robot Design, Dynamics and Control 2019 (pp. 323-330). Springer. doi:10.1007/978-3-319-78963-7_41

Hekman, E. E. G. , Prange, G., Aalsma, A. M., van der Helm, F. C., & van der Kooij, H. (2009). Dampace: Design of an Exoskeleton for Force-Coordination Training in Upper-Extremity Rehabilitation. *Journal of Medical Devices*, 1–10.

Herrnstadt, G., & Menon, C. (2016). Voluntary-Driven elbow Orthosis with speed-controlled Tremor suppression. *Frontiers in Bioengineering and Biotechnology*, *4*(29). PMID:27066477

Hesse, S., Schmidt, H., Werner, C., & Bardeleben, A. (2003). *Upper and lower extremity robotic devices for rehabilitation and for studying motor control. Current Opinion in Neurology*. Lippincott Williams & Wilkins. doi:10.1097/00019052-200312000-00010

Hesse, S., & Uhlenbrock, D. (2000). A mechanized gait trainer for restoration of gait. *Journal of Rehabilitation Research and Development*, *37*(6), 701–708. Retrieved from https://www.rehab.research.va.gov/jour/00/37/6/pdf/Hesse.pdf PMID:11321006

Hicks, J. H. (1954, January). The mechanics of the foot: II. The plantar aponeurosis and the arch. *Journal of Anatomy, 88*(Pt 1), 25. PMID:13129168

Hirai, K., Hirose, M., Haikawa, Y., & Takenaka, T. (1998). The development of Honda humanoid robot. In *Robotics and Automation, 1998. Proceedings. 1998 IEEE International Conference on 1998 May 16* (Vol. 2, pp. 1321-1326). IEEE. 10.1109/ROBOT.1998.677288

Hogan, N., & Sternad, D. (2013). Dynamic primitives in the control of locomotion. *Frontiers in Computational Neuroscience, 7.* doi:10.3389/fncom.2013.00071 PMID:23801959

How to make Robot. (2014). Retrieved from http://hi-robotics.blogspot.in

Howes, A., & Young, R. M. (1997). The role of cognitive architecture in modeling the user: Soar's learning mechanism. *Human-Computer Interaction, 12*(4), 311-343. doi:10.1207153270511hci1204_2

Hu, C., Wang, X., Mandal, M. K., Meng, M., & Li, D. (2003, May). Efficient face and gesture recognition techniques for robot control. In *CCECE 2003-Canadian Conference on Electrical and Computer Engineering. Toward a Caring and Humane Technology, 3,* (pp. 1757-1762). IEEE.

Hu, W., Tan, T., Wang, T., & Maybank, S. (2004). A survey on visual surveillance of object motion and behaviors. *Systems, Man, and Cybernetics, Part C: Applications and Reviews, IEEE Transactions on 34,* 3, pp. 334-352.

Huang, K., Chen, S., Tsai, M., Liang, F., Hsueh, Y., & Lin, P.-C. (2012). A Bio-inspired Hexapod Robot with Noncircular Gear Transmission System, *IEEE/ASME International Conference on Advanced Intelligent Mechatronics*, Kaohsiung, Taiwan. IEEE. 10.1109/AIM.2012.6266042

Hunt, A., Szczecinski, N., & Quinn, R. (2017, April 4). Development and training of a neural controller for hind leg walking in a dog robot. *Frontiers in Neurorobotics, 11*, 18. doi:10.3389/fnbot.2017.00018 PMID:28420977

IBGE - Instituto Brasileiro de Geografia e Estatística. (2010). *Antropometria e Estado Nutricional no Brasil 2008-2009.* Retrieved from https://doi.org/978-85-240-4131-0

Ilyas, M., Abbas, N., UbaidUllah, M., Imtiaz, W. A., Shah, M. A. Q., & Mahmood, K. (2016). Control Law Design for Twin Rotor MIMO System with Nonlinear Control Strategy. *Discrete Dynamics in Nature and Society*, 1–10. doi:10.1155/2016/2952738

Infitechm Co. Ltd. (2017). Robot-type watching system. Available at http://infitechm.com/en/index.html

Inoshita, H., Ogata, M., Tokunaga, A., Bando, M., Yamada, Y., & Marumoto, K. (1993). Studies on Galvanic skin response (GSR) in relation to psychological condition test for the students. *Bull. Sch. Health Sci. Okayama Univ., 4,* 99–103, (in Japanese).

Inoue, T. & Hirai, S. (2009). *Mechanics and Control of Soft-fingered Manipulation.* Springer.

Isermann, R. (2005). Mechatronic systems - Innovative products with embedded control. In *IFAC Proceedings Volumes (IFAC-PapersOnline).*

Ishida, H., Wada, Y., & Matsukura, H. (2012). Chemical Sensing in Robotic Applications: A Review. *IEEE Sensors Journal, 12*(11), 3163–3173. doi:10.1109/JSEN.2012.2208740

Ishimwe, R., Abutaleb, K., & Ahmed, F. (2014). Applications of thermal imaging in agriculture – a review. *Advances in Remote Sensing, 3*(3), 128–140.

Iwama, K., & Abe, M. (1953). Conditioned galvanic skin reflex and electroencephalogram. *The Tohoku Journal of Experimental Medicine, 57*(4), 327–335. doi:10.1620/tjem.57.327 PMID:13090218

Jain, P., Firke, P. N., Kapadnis, K. N., Patil, T. S., & Rode, S. S. (2014). RF based Spy robot. *International Journal of Engineering Research and Applications, 4*(4), version 2, pp. 06-09.

Jatsun, S. F., & Yatsun, A. S. (2018). Criterion of the rehabilitation process effectiveness on the basis of biomehatronic system ExoLite Rehab. In *2018 IEEE Workshop on Advanced Robotics and its Social Impacts (ARSO)* (pp. 105-110). IEEE. 10.1109/ARSO.2018.8625840

Jazar, R. N. (2010). Theory of applied robotics: kinematics, dynamics, and control. Berlin, Germany: Springer Science & Business Media. doi:10.1007/978-1-4419-1750-8

Jezernik, S., Colombo, G., & Morari, M. (2004). Automatic gait-pattern adaptation algorithms for rehabilitation with a 4-DOF robotic orthosis. *IEEE Transactions on Robotics and Automation, 20*(3), 574–582. doi:10.1109/TRA.2004.825515

Jha, A., Singh, A., Turna, R., & Chauhan, S. (2015). War Field Spying Robot with Night Vision Camera. *Journal of Network Communications and Emerging Technologies (JNCET), 1*(2).

Jiang, H., Zhou, C., & Xie, P. (2016). Design and kinematic analysis of seagull inspired flapping wing robot. In *2016 IEEE International Conference on Information and Automation (ICIA)* (pp. 1382–1386). IEEE. 10.1109/ICInfA.2016.7832035

Johansson, R. S., & Vallbo, A. B. (1979, January 1). Tactile sensibility in the human hand: Relative and absolute densities of four types of mechanoreceptive units in glabrous skin. *The Journal of Physiology, 286*(1), 283–300. doi:10.1113/jphysiol.1979.sp012619 PMID:439026

Jouaneh, M. (2012). *Fundamentals of Mechatronics*. CI-Engineering.

Kai, N. (1962). An Experimental Study of Emotional Stability in Galvanic Skin Reflection (GSR), Memoirs of the Faculty of Education, Kumamoto University, *Humanistic science, 10,* pp.109-121 (in Japanese).

Kalamdani, A., Messom, C., & Siegel, M. (2007, October). Robots with sensitive feet. *IEEE Instrumentation & Measurement Magazine, 10*(5), 46–53. doi:10.1109/MIM.2007.4343567

Kamaruzzaman, M., Haque, R., & … . (2017). *Enhancement of Railway Safety Systems using Anti Collision Device, ICSMART*. Durgapur, India: CMERI.

Kaneko, K., Harada, K., Kanehiro, F., Miyamori, G., & Akachi, K. (2008). Humanoid robot HRP-3. In *Intelligent Robots and Systems, 2008. IROS 2008. IEEE/RSJ International Conference on 2008 Sep 22* (pp. 2471-2478). IEEE.

Kass, M., Witkin, A., & Terzopoulos, D. (1987). Snakes: Active contour models. *International Journal of Computer Vision, 1*(4), 321–331.

Kato, N. (2013). Incipient Scientific Thinking in Five-year-old Children's Activities: Through Sandbox Play, Soap Bubble Play, Playing in the Water Color and Mud Ball Play, *Cross-cultural business and cultural studies, 18*(1), pp. 17-28 (in Japanese).

Kato, T. A., Shinfuku, N., Sartorius, N., & Kanba, S. (2017). *Loneliness and single-person households: issues of Kodoku-Shi and Hikikomori in Japan* (pp. 1-15). Mental Health and Illness in the City.

Kaura, H. K., Honrao, V., Patil, S., & Shetty, P. (2013). Gesture controlled robot using image processing. *International Journal of Advanced Research in Artificial Intelligence, 2*(5), 69–76.

Kazemi, H., Majd, V. J., & Moghaddam, M. M. (2013). *Modeling and robust backstepping control of an underactuated quadruped robot in bounding motion, Robotica*, 31, (pp. 423–439). Cambridge, UK: Cambridge Univ Press; doi:10.1017/S0263574712000458

Kecskes, I., Burkus, E., Bazso, F., & Odry, P. (2015). *Model validation of a hexapod walker robot, Robotica* , 35, (pp. 419–462). Cambridge, UK: Cambridge Univ Press; doi:10.1017/S0263574715000673

Ker, R. F. (1996, July 1). The time-dependent mechanical properties of the human heel pad in the context of locomotion. *The Journal of Experimental Biology*, *199*(7), 1501–1508. PMID:8699155

Ker, R. F. (1999, December 1). The design of soft collagenous load-bearing tissues. *The Journal of Experimental Biology*, *202*(23), 3315–3324. PMID:10562514

Khamassi, M., Lallée, S., Enel, P., Procyk, E., & Dominey, P. F. (2011). Robot cognitive control with a neurophysiologically inspired reinforcement learning model. *Frontiers in Neurorobotics*, *5*, 1. doi:10.3389/fnbot.2011.00001 PMID:21808619

Khan, M. J., Brodie, G., Gupta, D., & Foletta, S. (2018). Microwave soil treatment improves weed management in Australian dryland wheat. *Transactions of the ASABE*, *61*(2), 671–680.

Kharb, A., Saini, V., Jain, Y. K., & Dhiman, S. (2011). *A REVIEW OF GAIT CYCLE AND ITS PARAMETERS. IJCEM International Journal of Computational Engineering & Management* (Vol. 13). Retrieved from www.IJCEM.orgIJCEM-www.ijcem.org

Khoshnoud, F., & de Silva, C. W. (2012). *Recent Advances in MEMS Sensor Technology - Biomedical Applications.* IEEE Instrumentation & Measurement Magazine; doi:10.1109/MIM.2012.6145254

Kiguchi, K., & Hayashi, Y. (2013). Motion Estimation based on EMG and EEG Signals to Control Wearable Robots, *IEEE International Conference on Systems, Man, and Cybernetics*, Washington, USA. 10.1109/SMC.2013.718

Kim, H., Lee, D., Jeong, K., & Seo, T. (2016). Water and Ground-Running Robotic Platform by Repeated Motion of Six Spherical Footpads, IEEE/ASME, Transaction on Mechatronics, 21(1) IEEE.

Kim, S., Asbeck, A., Cutkosky, M., & Provancher, W. (2008). Spinybot II: Climbing hard walls with compliant microspines. In Proc. of the Kim, S. et al. Smooth vertical surface climbing with directional adhesion. IEEE Trans. Robot., 24, 65–74.

Kimura, M. (2009). Micro Temperature Sensors and Their Applications to MEMS Thermal Sensors. In *S. C. Mukhopadhyay, G. S. Gupta, & R. Y. Huang (Eds.), Recent Advances in Sensing Technology. Lecure Notes in Electrical Engineering* (Vol. 49). Berlin, Germany: Springer; doi:10.1007/978-3-642-00578-7_19

Kirkpatrick, S. (1990). The Moment of Inertia of Bird Wings. *The Journal of Experimental Biology*, (January): 489–494. Retrieved from http://d.wanfangdata.com.cn/NSTLQK_NSTL_QK9557394.aspx

Kitamura, K., Takemura, N., Iwai, Y., & Sato, K. (2014). Detection of Unconscious Facial Reactions to Uncomfortable Illumination, IEEJ Transactions on Electronics. *Information Systems*, *134*(2), 218–224 (in Japanese).

Klann, J. C. (2002). *Walking Device*. Google Patents U.S. 6,478,314 B1, Nov. 12.

Knoop, E., Conn, A., & Rossiter, J. (2017). VAM: Hypocycloid Mechanism for Efficient Bionspired Robotic Gaits. In IEEE Robotics and automation letters, 2 *(2),* (pp. 1055–1061). IEEE.

Knuth, D. (1992). Two Notes on Notation. *The American Mathematical Monthly*, *99*(5), 403–422. doi:10.1080/00029890.1992.11995869

Koh, J. S., Yang, E., Jung, G. P., Jung, S. P., Son, J. H., Lee, S. I., ... Cho, K. J. (2015, July 31). Jumping on water: Surface tension–dominated jumping of water striders and robotic insects. *Science*, *349*(6247), 517–521. doi:10.1126cience.aab1637 PMID:26228144

Kojima, K., Karasawa, T., Kozuki, T., Kuroiwa, E., Yukizaki, S., Iwaishi, S., . . . Nozawa, S. (2015). Development of life-sized high-power humanoid robot jaxon for real-world use. In *Humanoid Robots (Humanoids), 2015 IEEE-RAS 15th International Conference on 2015 Nov 3* (pp. 838-843). IEEE. 10.1109/HUMANOIDS.2015.7363459

Korchut, A., Szklener, S., Abdelnour, C., Tantinya, N., Hernández-Farigola, J., Ribes, J. C., ... Rejdak, K. (2017). Challenges for service robots—Requirements of elderly adults with cognitive impairments. *Frontiers in Neurology, 8*, 228. doi:10.3389/fneur.2017.00228 PMID:28620342

Kress, R. (1998). *Numerical Analysis* (Vol. 181). New York, NY: Springer-Verlag; doi:10.1007/978-1-4612-0599-9

Krishna, A., Krishna, N., Pradeep, S., Srihari, K., & Sivraj, P. (2014). Design a Fabrication of a Hexapod Robot, *International Conference on Embedded Systems*, India. 10.1109/EmbeddedSys.2014.6953162

Kuhnt, F., Pfeier, M., Zimmer, P., & Zöllner, J. M. (2016). *Robust Environment Perception for the Audi Autonomous Driving Cup. IEEE Intern. Conf. on Intelligent Transportation Systems.* doi:10.1109/ITSC.2016.7795744

Kulandaidaasan J., Mohan R. E., Martínez-García E., & Tan-Phuc L. (2016). Trajectory Generation and Stability Analysis for Reconfigurable Klann Mechanism Based Walking Robot, *Robotics, 5*(13), pp. 1-10, MDPI, doi:. doi:10.3390/robotics5030013

Kulandaidaasan, J., Mohan, R. E., Martínez-García, E., & Tan-Phuc, L. (2015). *Synthesizing reconfigurable foot traces using a Klann mechanism, Robotica* (pp. 1–17). Cambridge, UK: Cambridge University Press; doi:10.1017/S0263574715000089

Kumar, K. S., Kavitha, G., Subramanian, R., & Ramesh, G. (2011). Visual and Thermal Image Fusion for Based Target Tracking, U. A. V. MATLAB - A Ubiquitous Tool for the Practical Engineer, Prof. Clara Ionescu (Ed.): ch. 16.

Kummer, F., & Michelson, J. D. (1992). Investigations into the fat pads of the sole of the foot: heel pressure studies. *Foot Ankle Int., 13*, 227-232. doi:10.1177/107110079201300501

Kwon, J. Y., Lee, B. H., Mun, K. H., & Jung, J. S. (2015, September 1). A study on the effects of using an Eco-friendly cat robot to treat children with autism spectrum disorder. *Indian Journal of Science and Technology, 8*(23), 1.

La Rosa, G., & Risitano, A. (2000). Thermographic methodology for rapid determination of the fatigue limit of materials and mechanical components. *International Journal of Fatigue, 22*(1), 65–73.

Lackland, D. (2017). *Heart Disease and Stroke Statistics — 2017 Update A Report From the American Heart Association.* https://doi.org/ doi:10.1161/CIR.0000000000000485

Lamers, K., Tijmons, S., De Wagter, C., & de Croon, G. (2016). Self-supervised monocular distance learning on a lightweight micro air vehicle. In *2016 IEEE/RSJ International Conference on Intelligent Robots and Systems (IROS)* (pp. 1779–1784). IEEE. 10.1109/IROS.2016.7759284

Langley, P., Laird, J. E., & Rogers, S. (2009). Cognitive architectures: Research issues and challenges. *Cognitive Systems Research, 10*(2), 141-160. doi:10.1016/j.cogsys.2006.07.004

Lascelles, B. D., Roe, S. C., Smith, E., Reynolds, L., Markham, J., Marcellin-Little, D., ... Budsberg, S. C. (2006, February). Evaluation of a pressure walkway system for measurement of vertical limb forces in clinically normal dogs. *American Journal of Veterinary Research, 67*(2), 277–282. doi:10.2460/ajvr.67.2.277 PMID:16454633

LeCun, Y., Bottou, L., Bengio, Y., & Haffner, P. (1998). Gradient-Based Learning Applied to Document Recognition. *Proceedings of the IEEE, 86*(11), 2278–2324. doi:10.1109/5.726791

Ledoux, W. R., & Blevins, J. J. (2007). The compressive material properties of the plantar soft tissue. *Journal of Biomechanics, 40*(13), 2975–2981. doi:10.1016/j.jbiomech.2007.02.009 PMID:17433335

Lee, M. D., & Vanpaemel, W. (2018). Determining informative priors for cognitive models. *Psychonomic Bulletin & Review, 25*(1), 114-127. doi:10.375813423-017-1238-3 PMID:28194721

Lemaignan, S., Warnier, M., Sisbot, E. A., Clodic, A., & Alami, R. (2017). Artificial cognition for social human–robot interaction: An implementation. *Artificial Intelligence, 247*, 45-69. doi:10.1016/j.artint.2016.07.002

Leong, T. W., & Johnston, B. (2016). Co-design and robots: a case study of a robot dog for aging people. In *International Conference on Social Robotics 2016 Nov 1* (pp. 702-711). Springer. 10.1007/978-3-319-47437-3_69

Levine, J. (2009). Analysis and control of nonlinear systems: A flatness-based approach. Berlin, Germany: Springer Science & Business Media. doi:10.1007/978-3-642-00839-9

Lieberman, D. E., Polk, J. D., & Demes, B. (2004, February). Predicting long bone loading from cross-sectional geometry. *American Journal of Physical Anthropology: The Official Publication of the American Association of Physical Anthropologists., 123*(2), 156–171. doi:10.1002/ajpa.10316 PMID:14730649

Linner, T., Georgoulas, C., & Bock, T. (n.d.). Advanced building engineering: Deploying mechatronics and robotics in architecture, *Gerontechnology, 11*(2), doi:. doi:10.4017/gt.2012.11.02.158.002012

Liu, G., Liu, Y., Wang, X., Wu, X., & Mei, T. (2016). Design and experiment of a bioinspired wall climbing robot using spiny grippers. In *Mechatronics and Automation (ICMA), 2016 IEEE International Conference on 2016 Aug 7* (pp. 665-670). IEEE. 10.1109/ICMA.2016.7558642

Liu, Q., Sa, M., Lee, H., Lee, M., Kim, S.W., & Eune, J. (2017). Peekabot: Robot that Helps Children's Cognitive/Physical Development. In *Proceedings of the 2017 CHI Conference Extended Abstracts on Human Factors in Computing Systems 2017 May 6* (pp. 468-468). ACM.

Liu, T. (2006). Comparative Scaling of Flapping- and Fixed-Wing Flyers. *AIAA Journal, 44*(1), 24–33. doi:10.2514/1.4035

Li, Y., Qian, C., Zhu, B., & Fang, Y. (2017). *Kinematic, Static and Dynamic Analyses of Flapping Wing Mechanism Based on ANSYS Workbench* (pp. 316–323). Cham, Switzerland: Springer; doi:10.1007/978-3-319-70136-3_34

Ljungberg, S. A. (1994). *Infrared techniques in buildings and structures: operation and maintenance. Infrared Methodology and Technology, 7, 211.* Gordon and Breach Science Publishers.

Lloyd-Jones, D., Adams, R. J., Brown, T. M., Carnethon, M., Dai, S., & De Simone, G. … Wylie-Rosett, J. (2010). Executive summary: Heart disease and stroke statistics-2010 update: A report from the American Heart Association. *Circulation*. doi:10.1161/CIRCULATIONAHA.109.192667

Lobato, F. S., Steffen, V. Jr, & Oliveira-Lopes, L. C. (2007). *AN EVOLUTIONARY APPROACH BASED ON SEARCH CHAOTIC PATTERN ASSOCIATED TO DIFFERENTIAL EVOLUTION ALGORITHM. In 17º* (p. 9). Uberlandia: Simpósio do Programa de Pós-graduação em Engenharia Mecânica.

Looned, R., Webb, J., Zheng, G. X., & Menon, C. (2014). Assisting drinking with an affordable BCI-controlled wearable robot and electrical stimulation: A preliminary investigation. *Journal of Neuroengineering and Rehabilitation*, 1–13. PMID:24708603

Lorens, S. A. Jr, & Darrow, C. W. (1962). Eye movements, EEG, GSR and EKG during mental multiplication. *Electroencephalography and Clinical Neurophysiology, 14*(5), 739–746. doi:10.1016/0013-4694(62)90088-3

Louie, D. R., & Eng, J. J. (2016). Powered robotic exoskeletons in post-stroke rehabilitation of gait: A scoping review. *Journal of Neuroengineering and Rehabilitation, 13*(1), 53. doi:10.118612984-016-0162-5 PMID:27278136

Lu, Z., Aoyama, T., Sekiyama, K., Hasegawa, Y., & Fukuda, T. (n.d.). Motion transfer control from walking to brachiation through vertical ladder climbing for a multi-locomotion robot. IEEE/ASME Transactions on Mechatronics, 19(5), 1681–1693.

Luhmann, N. (1993). *Communication and Social Order: Risk: A Sociological Theory*. Piscataway, NJ: Transaction Publishers.

Lum, P., Reinkensmeyer, D., Mahoney, R., Rymer, W. Z., & Burgar, C. (2002). Robotic Devices for Movement Therapy After Stroke: Current Status and Challenges to Clinical Acceptance. *Topics in Stroke Rehabilitation*, *8*(4), 40–53. doi:10.1310/9KFM-KF81-P9A4-5WW0 PMID:14523729

Luo, Y., Liu, J., Gao, Y., & Lu, Z. (2014). Smartphone-controlled robot snake for urban search and rescue. *In International Conference on Intelligent Robotics and Applications* 2014 Dec 17 (pp. 352-363). Springer. 10.1007/978-3-319-13966-1_35

Mac, T. T., Copot, C., Tran, D. T., & De Keyser, R. (2016). Heuristic approaches in robot path planning: A survey. *Robotics and Autonomous Systems*, *86*, 13–28. doi:10.1016/j.robot.2016.08.001

Magrab, E. B. (2012). *Vibrations of Elastic Systems with Applications to MEMS and NEMS* (Vol. 184). Springer Netherlands; doi:10.1007/978-94-007-2672-7

Ma, K. Y., Chirarattananon, P., Fuller, S. B., & Wood, R. J. (2013). Controlled flight of a biologically inspired, insect-scale robot. *Science*, *340*(6132), 603–607. doi:10.1126cience.1231806 PMID:23641114

Ma, K. Y., Felton, S. M., & Wood, R. J. (2012). Design, fabrication, and modeling of the split actuator microrobotic bee. In *2012 IEEE/RSJ International Conference on Intelligent Robots and Systems* (pp. 1133–1140). IEEE. 10.1109/IROS.2012.6386192

Manama, M., Arrayo, A., Ortiz, A., Renedo, C. J., Perez, S., & Delgado, F. (2011). Field winding fault diagnosis in DC motors during manufacturing using thermal monitoring. *Applied Thermal Engineering*, *31*(5), 978–983.

Mann, E., Koller, M., Mann, C., van der Cammen, T., & Steurer, J. (2004). Comprehensive Geriatric Assessment (CGA) in general practice: Results from a pilot study in Vorarlberg, Austria. *BMC Geriatrics*, *4*(1), 4. doi:10.1186/1471-2318-4-4 PMID:15151704

Markus, E. D., Agee, J. T., & Jimoh, A. A. (2017). Flat control of industrial robotic manipulators. *Robotics and Autonomous Systems*, *87*, 226–236. doi:10.1016/j.robot.2016.10.009

Marlene, J. A., & Cooper, J. (1995). *Biomechanics of Human Movement*. Brown & Benchmark.

Martin, P., Murray, R. M., & Rouchon, P. (1997). Flat Systems. In *European Control Conference* (p. 54). Brussels. Retrieved from http://pierre.rouchon.chez-alice.fr/publications/PR1997/eccflat.pdf

Martin, Philippe, & Paden, B. (1994). A different look at output tracking : control of a VTOL aircraft * Searching for a flat output, (December), 2376–2381.

Martinez-Garcia, E. A., Torres-Mendez, L. A., & Mohan, R. E. (2014). Multi-Legged Robot Dynamics Navigation Model with Optical Flow. Bingley, UK: Emerald. *International Journal of Intelligent Unmanned Systems*, *2*(2), 121–139. doi:10.1108/IJIUS-04-2014-0003

Massaro, A., Ekuakille, A. L., Caratelli, D., Palamara, I., & Morabito, F. C. (2012). Optical performance evaluation of oil spill detection methods: Thickness and extent. *IEEE Transactions on Instrumentation and Measurement*, *61*(12), 3332–3339.

Massaro, A., & Galiano, A., Meuli, G., & Massari, S. F. (2018). Overview and application of enabling technologies oriented on energy routing monitoring, on network installation and on predictive maintenance. *International Journal of Artificial Intelligence and Applications*, *9*(2), 1–20.

Matarić, M. J. (2017). Socially assistive robotics: Human augmentation versus automation. *Science Robotics, 2*(4).

Matsumura, K., & Shiono, M. (2016). Interference Experiments of Soap-Bubble Films Using a Flashlight as a Light Source and Its Applications to Teaching Material of Science Educations, Studies in human sciences, Seinan gakuin Univ, 12(1), pp. 83-100 (in Japanese).

Mekki, M., Delgado, A. D., Fry, A., Putrino, D., & Huang, V. (2018). Robotic Rehabilitation and Spinal Cord Injury: A Narrative Review. *Neurotherapeutics; the Journal of the American Society for Experimental NeuroTherapeutics*, *15*(3), 604–617. doi:10.100713311-018-0642-3 PMID:29987763

Messelodi, S., Modena, C. M., & Cattoni, G. (2007). Vision-based Bicycle/Motorcycle Classification. *Pattern Recognition Letters*, *28*(13), 1719–1726. doi:10.1016/j.patrec.2007.04.014

Meyer, W., Bartels, T., Tsukise, A., & Neurand, K. (1990). Histochemical aspects of stratum corneum function in the feline foot pad. *Archives of Dermatological Research*, *281*(8), 541–543. doi:10.1007/BF00412742 PMID:2322014

Miller-Young, J. E., Duncan, N. A., & Baroud, G. (2002, December 1). Material properties of the human calcaneal fat pad in compression: Experiment and theory. *Journal of Biomechanics*, *35*(12), 1523–1531. doi:10.1016/S0021-9290(02)00090-8 PMID:12445605

Milstein, A. (2008). Occupancy Grid Maps for Localization and Mapping. In X.-J. Jing (Ed.), *Motion Planning*. InTech. doi:10.5772/6003

Minh, V. T., & Hashim, M. (2010). Adaptive teleoperation system with neural network-based multiple model control, *Mathematical Problems in Engineering*, Open Access, 592054.

Minh, V.T., & Rani A. M. (2009). Modeling and control of distillation column in a petroleum process, *Mathematical Problems in Engineering*, Open Access, 404702.

Mistry, K. K., & Nahapatra, A. (2012). Design and Simulation of a hermo Transfer Type MEMS Based Micro Flow Sensor for Arterial Blood Flow Measurement. *Microsystem Technologies*, *18*(6), 683–692. doi:10.100700542-012-1515-z

Mogelmose, A., Trivedi, M. M., & Moeslung, T. B. (2012). Traffic Sign Detection and Analysis: Recent Studies and Emerging Trends. Anchorage, AK: In 15th International IEEE Conference on Intelligent Transportation Systems.

Mohammadi, E., Zohoor, H., & Khadem, S. M. (2014). Control System Design of an Active Assistive Exoskeletal Robot for Rehabilitation of Elbow and Wrist. In *Proceedings of the 2nd RSI/ISM International Conference on Robotics and Mechatronics*, Tehran, Iran. 10.1109/ICRoM.2014.6991008

Mon, T. T., Mohd Romlay, F. R., & Tamin, M. N. (2012). Role of Finite Element Analysis in Designing Multi-axes Positioning for Robotic Manipulators. In E. Hall (Ed.), Advances in Robot Manipulators, InTech.

Moreira, N. F., Luz, V. G., Moreira, C. C., Pereira, R. A., Sichieri, R., Ferreira, M. G., ... Rodrigues, P. R. M. (2018). Self-reported weight and height are valid measures to determine weight status: Results from the Brazilian National Health Survey (PNS 2013). *Cadernos de Saude Publica*, *34*(5). doi:10.1590/0102-311x00063917 PMID:29768583

Motor selection. (n.d.). *Panasonic Global*. Retrieved from http://industrial.panasonic.com/ww/i_e/25000/fa_pro_sgeard_shing1_e/fa_pro_sgeard_shing1_e/ctlg_geared_e_9.pdf

Motor Torque Calculation. (n.d.). *Selection Guides-Leadshine Technology Co., Ltd.* Retrieved from http://www.leadshine.com/UploadFile/documents_software/ Applications/Motor_Torque_Calculation.pdf

Mountjoy, K. (2008). Use of A Hill-based Muscle Model in the Fast Orthogonal Search Method to Estimate Wrist Force and Upper Arm Physiological Parameters, IEEE conference on Medicine and Biology, Vancouver, Canada.

Mühlberger, M., Rohn, M., Danzberger, J., Sonntag, E., Rank, A., Schumm, L., ... Trappl, E. (2015, June 15). UV-NIL fabricated bio-inspired inlays for injection molding to influence the friction behavior of ceramic surfaces. *Microelectronic Engineering*, *141*, 140–144. doi:10.1016/j.mee.2015.02.051

Muijres, F. T., Bowlin, M. S., Johansson, L. C., & Hedenström, A. (2012). Vortex wake, downwash distribution, aerodynamic performance and wingbeat kinematics in slow-flying pied flycatchers. *Journal of the Royal Society, Interface*, *9*(67), 293–303. doi:10.1098/rsif.2011.0238 PMID:21676971

Mukhtar, A., & Tang, T. B. (2015, October). Vision based motorcycle detection using HOG features. In *2015 IEEE International Conference on Signal and Image Processing Applications (ICSIPA),* (pp. 452-456). IEEE.

Müller-Schneiders, S. (2019). Vision-based Detection of Road Users and Infrastructure Elements for Automated Driving, *IEEE Intern. Conference on Research and Education in Mechatronics (REM 2019).* 10.1109/REM.2019.8744089

Müller-Schneiders, S., Nunn, C., & Meuter, M. (2008). Performance Evaluation of a Real Time Traffic Sign Recognition System, *IEEE Intelligent Vehicle Symposium.* 10.1109/IVS.2008.4621164

Mun, K. R., Yu, H., Zhu, C., & Cruz, M. S. (2014). Design of a novel robotic over-ground walking device for gait rehabilitation. In *IEEE 13th international workshop - Advanced Motion Control(AMC).* https://doi.org/10.1109/AMC.2014.6823325

Mun, K. R., & Lim, S. Bin, Guo, Z., & Yu, H. (2017). Biomechanical effects of body weight support with a novel robotic walker for over-ground gait rehabilitation. *Medical & Biological Engineering & Computing*, *55*(2), 315–326. doi:10.100711517-016-1515-8 PMID:27193227

Murphy, I. P., Kurdila, A. J., Bayandor, J., & Burdisso, R. A. (2013). *Modeling and Control of Flapping Wing Robots.* Retrieved from https://vtechworks.lib.vt.edu/bitstream/handle/10919/19275/Murphy_IP_T_2013.pdf?sequence=1

Murray, R. M., Sluis, W., Rathinam, M., & Sluis, W. (1995). Differential flatness of mechanical control systems: A catalog of prototype systems. *Proc. of the ASME Mechanical Engineering Congress and Expo*, 1-9. doi:10.1109/INFCOM.2011.5934932

Muscolo, G. G., Caldwell, D., & Cannella, F. (2017). Biomechanics of human locomotion with constraints to design flexible-wheeled biped robots. In *Advanced Intelligent Mechatronics (AIM), 2017 IEEE International Conference on 2017 Jul 3* (pp. 1273-1278). IEEE. 10.1109/AIM.2017.8014193

Nabulsi, S., Sarria, J. F., Montes, H., & Armada, M. A. (2009, October). High-resolution indirect feet–ground interaction measurement for hydraulic-legged robots. *IEEE Transactions on Instrumentation and Measurement*, *58*(10), 3396–3404. doi:10.1109/TIM.2009.2017650

Nadeau, S. E., Dobkin, B., Wu, S. S., Pei, Q., & Duncan, P. W., & Team, for the L. I. (2016). The Effects of Stroke Type, Locus, and Extent on Long-Term Outcome of Gait Rehabilitation. *Neurorehabilitation and Neural Repair*, *30*(7), 615–625. doi:10.1177/1545968315613851 PMID:26498434

Nakamura, M., Inaba, G., Tamaoki, J., Hoshino, J., & Shiratori, K. (2006). Mounting and Application of Soap Bubble Display Method. *The Virtual Reality Society of Japan*, *11*(2), 339–346 (in Japanese).

Narioka, K., Tsugawa, S., & Hosoda, K. (2009). 3D limit cycle walking of musculoskeletal humanoid robot with flat feet. In *Intelligent Robots and Systems, 2009. IROS 2009. IEEE/RSJ International Conference on 2009 Oct 10* (pp. 4676-4681). IEEE.

Nasr, S., Mekki, H., & Bouallegue, K. (2019). A multi-scroll chaotic system for a higher coverage path planning of a mobile robot using flatness controller. *Chaos, Solitons, and Fractals, 118*, 366–375. doi:10.1016/j.chaos.2018.12.002

Natali, A. N., Fontanella, C. G., & Carniel, E. L. (2010). Constitutive formulation and analysis of heel pad tissues mechanics. *Medical Engineering & Physics, 32*(5), 516–522. doi:10.1016/j.medengphy.2010.02.018 PMID:20304698

Nguyen, Q.-V., & Chan, W. L. (2018). Development and flight performance of a biologically-inspired tailless flapping-wing micro air vehicle with wing stroke plane modulation. *Bioinspiration & Biomimetics, 14*(1). doi:10.1088/1748-3190/aaefa0 PMID:30523879

Nicholas, A., Iphigenia, K., & Kiranoudis, C. T. (2011). A background subtraction algorithm for detecting and tracking vehicles. *Elsevier Expert Systems with Applications, 38*(3), 1619–1631. doi:10.1016/j.eswa.2010.07.083

Nielsen, J. (1994). Usability engineering. Amsterdam, The Netherlands: Elsevier.

Ninomiya, H., Akiyama, E., Simazaki, K., Oguri, A., Jitsumoto, M., & Fukuyama, T. (2011, December). Functional anatomy of the footpad vasculature of dogs: Scanning electron microscopy of vascular corrosion casts. *Veterinary Dermatology, 22*(6), 475–481. doi:10.1111/j.1365-3164.2011.00976.x PMID:21438930

Nunes, W. M. (2012). *Desenvolvimento de uma estrutura robótica atuada por cabos para reabilitação/recuperação dos movimentos do ombro humano.* Universidade Federal De Uberlândia. Retrieved from https://repositorio.ufu.br/bitstream/123456789/14900/1/d.pdf

Odashima, K., Hata, A., & Goto, Y. (2014). How to satisfactorily enjoy an amusement park, Kokyuroku, Research Institute for Mathematical Sciences. *Kyoto University, 1887*, 28–35 (in Japanese).

Ogata, K. (2010). Modern control engineering. Upper Saddle River, NJ: Prentice-Hall.

Open Source Robotics Foundation. (2019). *ROS documentation.* Retrieved from http://wiki.ros.org/Documentation

Oshiro, T. (2018). *Soft Robotics in Radiation Environments: A Prospective Study of an Emerging Automated Technology for Existing Nuclear Applications*, (Master thesis, Oregon State University, Oregon).

Otsu, P. (1975). Threshold selection method from gray-level histogram. automatic, 11, page 285-296.

Ottaviano, E., Ceccarelli, M., & De Ciantis, M. (2007). A 4-4 Cable-Based Parallel Manipulator for an Application in Hospital Environment, 15th Mediterranean Conference on Control and Automation – MED07, Athens, Greece. 10.1109/MED.2007.4433839

Ottaviano, E., Ceccarelli, M., & Castelli, G. (2004). *Experimental Results of a 3-DOF Parallel Manipulator a an Earthquake Motion Simulator, ASME Design Engineering Technical Conference and Computers in Engineering Conference* (pp. C2004–C57075). Salt Lake City, Utah: Paper DET.

Ottaviano, E., Rea, P., & Grandinetti, A. (2019). Mechatronic Design of a Low-Cost Control System for Assisting Devices. In J. Machado, F. Soares, & G. Veiga (Eds.), *Innovation, Engineering and Entrepreneurship. HELIX 2018. Lecture Notes in Electrical Engineering* (Vol. 505). Cham, Switzerland: Springer. doi:10.1007/978-3-319-91334-6_21

Pala, M., Osati, N., López-Colino, F., Sanchez, A., de Castro, A., & Garrido, J. (2013). HCTNav: A Path Planning Algorithm for Low-Cost Autonomous Robot Navigation in Indoor Environments. *ISPRS International Journal of Geo-Information, 2*(3), 729–748. doi:10.3390/ijgi2030729

Palmer, L. R., Diller, E. D., & Quinn, R. D. (2010). Toward a rapid and robust attachment strategy for vertical climbing. *Proc. of the IEEE Int. Conf. on Robotics and Automation*, 2810-2815. 10.1109/ROBOT.2010.5509680

Papageorgiou, X. S., Tzafestas, C. S., Vartholomeos, P., Laschi, C., & Lopez, R. (2015). ICT-Supported Bath Robots: Design Concepts, 17th International Conference on Social Robotics (ICSR 2015). Retrieved from http://www.i-support-project.eu/web/wp-content/uploads/2018/08/Papageorgiou_et-al_ICSR2015.v.2.pdf

Parivash, F., & Bamdad, M. (2015). Independent Position-Stiffness Control for Elbow Rehabilitation Robot with Cable-Based Series Elastic Actuator, *22nd Iranian Conference on Biomedical Engineering(ICBME 2015)*, Tehran, Iran. 10.1109/ICBME.2015.7404168

Park, I. W., Kim, J. Y., Lee, J., & Oh, J. H. (2005). Mechanical design of humanoid robot platform KHR-3 (KAIST humanoid robot 3: HUBO). In *Humanoid Robots, 2005 5th IEEE-RAS International Conference on 2005 Dec 5* (pp. 321-326). IEEE.

Park, J. H., & Yoon, K.-J. (2008). Designing a Biomimetic Ornithopter Capable of Sustained and Controlled Flight. *Journal of Bionics Engineering*, *5*(1), 39–47. doi:10.1016/S1672-6529(08)60005-0

Patra, A. K., & Ray, S. J. (2007). Guiding robots using mobile phone International Symposium on automation and robotics in construction. In *ISARC,* Canada.

Patwari, N., & Wilson, A. (2010). RF sensor networks for device-free localization, measurement, models and algorithms. IEEE 98(11), pp. 1961-1973. doi:10.1109/JPROC.2010.2052010

Pedrocchi, A., Ferrante, S., Ambrosini, E., Gandolla, M., Casellato, C., Schauer, T. ... Reichenfelser, W. (2013). MUNDUS project: MUltimodal Neuroprosthesis for daily Upper limb Support. Journal of neuroengineering and rehabilitation, 10(1).

Peko, D., Basu, S., Kraft, S., Mizokami, S., & Rempe, J. (2018). Working together to enhance nuclear reactor safety; Information from examinations at the affected reactors at Fukushima Daiichi is being used to enhance reactor safety in U.S. operating plants, plants under construction, and future nuclear power plants, Nuclear News, American Nuclear Society, 36-42.

Pelesko, J. A. (2004). Electrostatics in MEMS and NEMS. In *V. M. Harik, & L. S. Luo (Eds.), Miromechanics and Nanoscale Effects* (Vol. 10). Dordrecht, Germany: Springer.

Pelesko, J. A., & Bernstein, D. H. (2003). *Modeling MEMS and NEMS.* Boca Raton, FL: Chapman & Hall, CRC Press Company.

Pennestri, E. (1991). On The Kinematic Analysis Of Geared Robotic Wrists. Springer. *Meccanica*, 1991.

Petkov, N., & Azzopardi, G. (2015). *Lecture Notes in Computer Science*, *9256*, Springer International Publishing.

Petrie, H., & Bevan, N. (2009). The Evaluation of Accessibility, Usability, and User Experience. *The universal access handbook, 1,* 1-16.

Pettersson, N., Petersson, L., & Andersson, L. (2008). The histogram feature - a resource-efficient Weak Classifier, *IEEE Intelligent Vehicles Symposium.* 10.1109/IVS.2008.4621174

Peurala, S. H., Airaksinen, O., Huuskonen, P., Jäkälä, P., Juhakoski, M., Sandell, K., ... Sivenius, J. (2009). Effects of intensive therapy using gait trainer or floor walking exercises early after stroke. *Journal of Rehabilitation Medicine*, *41*(3), 166–173. doi:10.2340/16501977-0304 PMID:19229450

Pfeifer, R., Iida, F., & Gomez, G. (2006). Morphological computation for adaptive behavior and cognition. *International Congress Series*, *1291*, 22–29. doi:10.1016/j.ics.2005.12.080

Pfeifer, R., Lungarella, M., & Iida, F. (2007). Self-organization, embodiment, and biologically inspired robotics. *Science*, *318*(5853), 1088–1093. doi:10.1126cience.1145803 PMID:18006736

Pfeiffer, A. T., Lee, J.-S., Han, J.-H., & Baier, H. (2010). Ornithopter flight simulation based on flexible multi-body dynamics. *Journal of Bionics Engineering*, *7*(1), 102–111. doi:10.1016/S1672-6529(09)60189-X

Phan, H. P., Nguyen, T. K., Dinh, T., Iacopi, A., Hold, L., Shiddiky, M. J., & Nguyen, N. T. (2018). Robust Free-Standing Nano-Thin SiC Membranes Enable Direct Photolithography for MEMS Sensing Applications. In *Advanced Engineering Materials, 20(1)*. Wiley Online Library; doi:10.1002/adem.201700858

Phan, H. V., Kang, T., & Park, H. C. (2017). Design and stable flight of a 21 g insect-like tailless flapping wing micro air vehicle with angular rates feedback control. *Bioinspiration & Biomimetics*, *12*(3). doi:10.1088/1748-3190/aa65db PMID:28281468

Piaget, J. (1964). Part I: Cognitive development in children: Piaget development and learning. *Journal of Research in Science Teaching*, *2*(3), 176-186. doi:10.1002/tea.3660020306

Pietrusinski, M., Cajigas, I., Mizikacioglu, Y., Goldsmith, M., Bonato, P., & Mavroidis, C. (2010). Gait rehabilitation therapy using robot generated force fields applied at the pelvis. *2010 IEEE Haptics Symposium, HAPTICS 2010*, 401–407. https://doi.org/10.1109/HAPTIC.2010.5444624

Pissulla, D., & Borgeest, K. (2016). *Energieeffizienter Antrieb eines Mobilroboters mit einem kombinierten DC/BLDC-Motor (in German). 13*. Lübeck, Germany: AALE-Konferenz.

Plander, I., & Stepanovsky, M. (2017). MEMS Technology in Optical Switching. *IEEE 14th International Scientific Conference on Informatics*. Poprad, Slovakia.

Pons, J. L. (2008). *Wearable Robots: Biomechatronic Exoskeletons*. Hoboken, NJ: John Wiley & Sons. doi:10.1002/9780470987667

Porikli, F. (2005). Integral histogram: a fast way to extract histograms in Cartesian spaces, *IEEE Computer Society Conference on Computer Vision and Pattern Recognition*. 10.1109/CVPR.2005.188

Potemkin, E., Astafurov, P., Osipov, A., Malenkov, M., Mishkinyuk, V., & Sologub, P. (1992). Remote-controlled robots for repair and recovery in the zones of high radiation levels, *Proceedings of the 1992 IEEE International Conference on Robotics and Automation*, Nice, France 10.1109/ROBOT.1992.220331

Pulido, J. C., González, J. C., Suárez-Mejías, C., Bandera, A., Bustos, P., & Fernandez, F. (2017). Evaluating the Child–Robot Interaction of the NAOTherapist Platform in Pediatric Rehabilitation, *Int J of Soc Robotics*, 9, 343. doi:10.100712369-017-0402-2

Purushotham, A., & Anjeneyulu, M. J. (2013). Kane's Method for Robotic Arm Dynamics: a Novel Approach. *IOSR Journal of Mechanical and Civil Engineering (IOSR-JMCE)* *6*(4): 07-13.

Qian, Z., Miao, H., Shang, Z., & Ren, L. (2014). The foot-ground contact analysis of german shepherd dog based on normal walking, trotting and jumping gait. *J. Jilin Univ.*, *44*, 1692–1697.

Quan, W., Toda, Y., Woo, J., & Kubota, N. (2018). Integrated Robotic Control System for Public Nursing. World Automation Congress. *16th International Symposium on Robotics and Applications (ISORA 2018)*, Washington, DC. 10.23919/WAC.2018.8430398

Quanser Inc. (2010). *Quanser 2 DOF Helicopter User and Laboratory Manuals, Revision 2.3*.

Quan, W., Woo, J., Toda, Y., & Kubota, N. (2019). Human Posture Recognition for Estimation of Human Body Condition. [JACIII]. *Journal of Advanced Computational Intelligence and Intelligent Informatics*, *23*(3), 519-527. doi:10.20965/jaciii.2019.p0519

Radford, N. A., Strawser, P., Hambuchen, K., Mehling, J. S., Verdeyen, W. K., Donnan, A. S., ... Berka, R. (2015, May). Valkyrie: Nasa's first bipedal humanoid robot. *Journal of Field Robotics*, *32*(3), 397–419. doi:10.1002/rob.21560

Raghatate, R. P., Rajurkar, S. S., Waghmare, M. P. & Ambakar, P. V. (2013). *International Journal of Modern Engineering research*, *3*(2), 816-820.

Ragulskis, M., Jiang, H., Quan, Q., Fedaravicius, A., & Xie, G. (2014). Mechatronics and Mechanical Engineering in Cyber-Physical Systems. *Hindawi Publishing Corporation Advances in Mechanical Engineering*. doi:10.1155/2014/591629

Rahideh, A., Shaheed, M. H., & Huijberts, H. J. C. (2008). Dynamic modelling of a TRMS using analytical and empirical approaches. *Control Engineering Practice*, *16*(3), 241–259. doi:10.1016/j.conengprac.2007.04.008

Rahn, C. D. (2002). Mechatronic Control of Distributed Noise and Vibration: A Lyapunov Approach. *Measurement Science & Technology*, *13*(4), 643–644. doi:10.1088/0957-0233/13/4/704

Raibert, M., Blankespoor, K., Nelson, G., & Playter, R. (2008). Bigdog, the rough-terrain quadruped robot. *Proc. 17th World Congress, The Int. Federation of Automatic Control*, 10822–5.

Rakrueangdet, K., Nunak, N., Suesut, T., & Sritham, E. (2016). Emissivity measurements of reflective materials using infrared thermography. *Proceedings of the International MultiConference of Engineers and Computer Scientists 2016* Vol I.

Rampa, S., Savazzi, M. N., & Amico, M. D. (2015). Physical modeling and performance bounds for device-free localization systems. IEEE Signal Process, 22(11), pp. 1864-1868.

Rashad, R., El-Badawy, A., & Aboudonia, A. (2017). Sliding mode disturbance observer-based control of a twin rotor MIMO system. *ISA Transactions*, *69*, 166–174. doi:10.1016/j.isatra.2017.04.013 PMID:28449822

Razali, Z. B., & Lee, G. B. (2012, February) *High-end technology for Effective Operating Environment in a Glass Manufacturing Industry.* Paper presented at the International Conference on Application and Design in Mechanical Engineering (ICADME2012), Penang, Malaysia.

Razali, Z. B., & Lee, G. B. (2013). High-End Technology for Effective Glass Manufacturing Industry: Innovative Operating Environment. Germany: LAP Lambert Academic Publishing.

Razali, Z. B., & Madasamy, G. (2013). Automated Mechanical Sorting Device for Mixed Household Waste: Integration Sorting Conveyor with Neural Network. Germany: LAP Lambert Academic Publishing, Germany.

Razali, Z. B., & Madasamy, S. (2012, February). *An Application of Image Processing for Automated Mixed Household Waste Sorting System.* Paper presented at the International Conference on Man-Machine System (ICoMMS2012), Penang, Malaysia.

Razali, Z. B., & Mat Nawi, N. A. (2013). Automated Segregating Device for Reuse and Recycle Waste: Application on size based, friendly recycling design. Germany: LAP Lambert Academic Publishing, Germany.

Rea, P., Ottaviano, E., & Castelli, G. A. (2013). Procedure for the Design of Novel Assisting Devices for the Sit-to-Stand, *Journal of Bionic Engineering, 10*(4), 488-496.

Rea, P., & Ottaviano, E. (2018). Functional Design for Customizing Sit-To-Stand Assisting Devices. *Journal of Bionics Engineering*, *15*(1), 83–93. doi:10.100742235-017-0006-4

Rea, P., Ottaviano, E., Conte, M., D'Aguanno, A., & De Carolis, D. (2013). The design of a novel tilt seat for inversion therapy. *Int. Journal of Imaging and Robotics*, *11*(3), 1–10.

Renteria-Bilbao, A. (2019). Ethical Concerns when Robots Assist Eldery People. *Dilemata*, *30*, 43-49.

Ripel, T., Krejsa, J., Hrbacek, J., & Cizmar, I. (2014). Active Elbow Orthosis. *International Journal of Advanced Robotic Systems*, 1–10.

Roberts, M. (2004). A robot for gait rehabilitation. Retrieved from http://dspace.mit.edu/handle/1721.1/34562

ROBOTIS. (2017). DYNAMIXEL. Retrieved from http://www.robotis.us/dynamixel/

Roccia, B. A., Preidikman, S., & Massa, J. (2007). Desarrollo de un código computacional para simular y analizar la cinemática de alas batientes Diseño estructural mecánico View project. *Mecanica Computacional*, *XXVI*, 3224–3245. Retrieved from http://www.ing.unrc.edu.ar

Rodrigues, L. A. O. (2017). *DESENVOLVIMENTO DE DISPOSITIVOS NÃO MOTORIZADOS PARA REABILITAÇÃO DO MEMBRO INFERIOR HUMANO*. Universidade Federal de Uberlândia.

Roebroeck, M. E., Doorenbosch, C. A. M., Harlaar, J., Jacobs, R., & Lankhorst, G. J. (1994). Biomechanics and muscular activity during sit-to-stand transfer. *Clin. Biom.*, *9*(4), 235–244. doi:10.1016/0268-0033(94)90004-3 PMID:23916233

Roennau, A., Heppner, G., Klemm, S., & Dillmann, R. (2015). *RoaDS-Robot and Dynamics Simulation for Biologically-Inspired Multi-Legged Walking Robots. IEEE, Robotics and Biomimetics* (pp. 1870–1876). China: IEEE.

Rogalski, A., & Chrzanowski, K. (2002). Infrared devices and techniques. *Opto-Electronics Review*, *10*(2), 111–136.

Rooney, T., Pearson, M. J., Welsby, J., Horsfield, I., Sewell, R., & Dogramadzi, S. (2011). Artificial Active Whiskers for Guiding Underwater Autonomous Walking Robots, CLAWAR 2011, France.

Rubenstein, D. A., Yin, W., & Frame, M. D. (2012). Fundamentals of Fluid Mechanics. In Biofluid Mechanics (pp. 11–48). doi:10.1016/B978-0-12-381383-1.00002-3

Rubenstein, L. Z. (2004). Comprehensive geriatric assessment: From miracle to reality (JT Freeman Award Lecture). *J Gerontol A Biol. Medical Science*, *59*, M473–M477.

Ryan, J. V. (2010). Energy Solutions from Glass Road Surfaces. *Glass Manufacturing Industry Council*. Retrieved from http://www.gmic.org/News/ Joe_Ryan_ Energy_Solutions_ from_Glass_Road_ Surfaces.pdf

Ryu, J.-C., & Agrawal, S. K. (2011). Differential flatness-based robust control of mobile robots in the presence of slip. *The International Journal of Robotics Research*, *30*(4), 463–475. doi:10.1177/0278364910385586

Sagan, V., Maimaitijiang, M., Sidike, P., Eblimit, K., Peterson, K. T., Hartling, S., ... Mockler, T. (2019). UAV-Based High Resolution Thermal Imaging for Vegetation Monitoring, and Plant Phenotyping Using ICI 8640 P, FLIR Vue Pro R 640, and thermoMap Cameras. *Remote Sensing*, *11*(1), 330.

Salguero-Beltran, A., Yamhure, G., Manrique, M., Jimenez, L. C., Hernandez, A. M., & Cotrino, C. (2012). On the design of an ischiatic body weight support system (IBWS) for gait rehabilitation. *Proceedings of the IEEE RAS and EMBS International Conference on Biomedical Robotics and Biomechatronics*, (5), 1434–1439. https://doi.org/10.1109/BioRob.2012.6290940

Salim, V. V. (2018). *DESENVOLVIMENTO DE UM SISTEMA SEM MARCADORES PARA CONTROLE DE UM DISPOSITIVO ROBÓTICO PARA REABILITAÇÃO DA MARCHA HUMANA*. Universidade Federal de Uberlândia.

Salter, T., Dautenhahn, K., & te Boekhorst, R. (2006). Learning about natural human-robot interaction styles. *Robotics and Autonomous Systems*, *54*(2), 127-134. doi:10.1016/j.robot.2005.09.022

Samanian, N., & Mohebbi, M. (2016). Thermography, a new approach in food science studies: A review. *MOJ Food Processing & Technology*, *2*(3), 110–119.

Sane, S. P., & Dickinson, M. H. (2001). Control of Flight force by a flapping wing. *The Journal of Experimental Biology*, *2626*(15), 2607–2626. PMID:11533111

Sasaki, T., Saraiji, M. H. D., Minamizawa, K., & Inami, M. (2018). MetaArmS: Body remapping using feet-controlled artificial arms. In *The 31st Annual ACM Symposium on User Interface Software and Technology Adjunct Proceedings* (pp. 140-142). ACM.

Satapathy, D., & Kumar, G. J. R. (2016). Path Planning Algorithm for a Mobile Robot – A Comparative Study in Different Environment. *IJCTA*, *9*(37), 453–459.

Sedlaček, M., Podgornik, B., & Vižintin, J. (2012, April 1). Correlation between standard roughness parameters skewness and kurtosis and tribological behavior of contact surfaces. *Tribology International*, *48*, 102–112. doi:10.1016/j.triboint.2011.11.008

Seidl, D. (2004). *Luhmann's theory of autopoietic social systems* (pp. 36-37). Ludwig-Maximilians-Universität München-Munich School of Management.

Selig, M., Guglielmo, J. J., Broeren, A. P., & Giguere, P. (2012). *Summary of Low-Speed Airfoil Data - Vol 5* (Vol. 5). doi:10.1115/1.1793208

Selig, J. M. (2000). *Geometrical Foundations of Robotics*. WORLD SCIENTIFIC. doi:10.1142/4257

Sellaouti, R., Stasse, O., Kajita, S., Yokoi, K., & Kheddar, A. (2006). Faster and smoother walking of humanoid HRP-2 with passive toe joints. In *Intelligent Robots and Systems, 2006 IEEE/RSJ International Conference on 2006 Oct 9* (pp. 4909-4914). IEEE.

Shahid, M. K. (2015). Fletcher, Robati S and Pemmaraju G, "The Biomechanical Forces that Act on the Elbow Joint. *EC Orthopaedics*, *1*(1), 1–11.

Sheba, J. K., Salman, A. A., Kumar, S., Elara, M. R., & Martínez-García, E. (2018, January 1). Development of Rehabilitative Multimodal Interactive Pet Robot for Elderly Residents. *Procedia Computer Science*, *133*, 401–408. doi:10.1016/j.procs.2018.07.049

Shekhar, R. S., & Khumar, P. D. (2011). Dynamic Modeling and Energy Consumption Analysis of Crab Walking of a Six-legged Robot, IEEE Conference on Technologies for Practical Robot Applications. Woburn, MA: IEEE.

Sheridan, T. B. (2016). Human–robot interaction: Status and challenges. *Human Factors*, *58*(4), 525-532. doi:10.1177/0018720816644364 PMID:27098262

Shirai, S., & Takahashi, Y. (2018). Personal Mobility Vehicle User's Psychological Reaction to Surrounding People. In *Proceedings of International Conference on Control, Automation and Systems*, pp. 1730-1735.

Singh, P., Kumar, A., & Vashisth, M. (2013). Design of a Robotic Arm with Gripper & End Effector for Spot Welding. *Universal Journal of Mechanical Engineering*, *1*(3), 6. doi:10.13189/ujme.2013.010303

Sivaraman, S., & Trivedi, M. M. (2013). Looking at Vehicles on the Road: A Survey of vision-based Vehicle Detection, Tracking, and Behavior Analysis. *IEEE Transactions on Intelligent Transportation Systems*, *14*(4), 1773–1795. doi:10.1109/TITS.2013.2266661

Smania, N., Berto, G., Melotti, C., & Miditi, A. (2012). Rehabilitation of brachial plexus injuries in adults and children. *European Journal of Physical and Rehabilitation Medicine*, *43*(3), 483–506. PMID:23075907

So, W.-C., Wong, M. K.-Y., Lam, C. K.-Y., Lam, W.-Y., Chui, A. T.-F., Lee, T.-L., ... Fok, D. C.-W. (2018). Using a social robot to teach gestural recognition and production in children with autism spectrum disorders. *Disability and Rehabilitation. Assistive Technology*, *13*(6), 527-539. doi:10.1080/17483107.2017.1344886 PMID:28673117

Spenko, M. J., Haynes, G. C., Saunders, J. A., Cutkosky, M. R., Rizzi, A. A., Full, R. J., & Koditschek, D. E. (2008, April). Biologically inspired climbing with a hexapedal robot. *Journal of Field Robotics*, *25*(4-5), 223–242. doi:10.1002/rob.20238

Spentzas, K. N., & Kanarachos, S. A. (2002). An incremental finite element analysis of mechanisms and robots.

Stahlkamp, J., Schlipsing, M., Salmen, J., & Igel, C. (2012). Benchmarking Machine Learning Algorithms for Traffic Sign Recognition. *Neural Networks*, *32*, 323–332. doi:10.1016/j.neunet.2012.02.016 PMID:22394690

Stout, K. J., Sullivan, P. J., Dong, W. P., Mainsah, E., Luo, N., Mathia, T., & Zahouani, H. (1994), The development of methods for the characterization of roughness on three dimensions. Publication no. EUR 15178 EN of the Commission of the European Communities, Luxembourg.

Stryker-InTouch. (2019). Retrieved from https://patientcare.stryker.com/en/products/beds/critical-care-beds/intouch

Stuck, A. E., & Iliffe, S. (2011). Comprehensive geriatric assessment for older adults.

Subramanian, M. B., & Sudhagar, D. K., & RajaRajeswari, G. (2014). *Intelligent Path Planning Of Mobile Robot Agent By Using Breadth First Search Algorithm.* Paper presented at the 2014 IEEE International Conference on Innovations in Engineering and Technology (ICIET'14), Madurai, Tamil Nadu, India.

Subramanian, M. B., Sudhagar, D. K., & Rajarajeswari, G. (2015). Autonomous Mobile Robot Agent Path Planning Based on VFH Method Using Heuristic Based Optimal Algorithm. *International Journal of Applied Engineering Research*, *10*(6), 5189–5194.

Sui, P., Yao, L., Lin, Z., Yan, H., & Dai, J. S. (2009). Analysis and synthesis of ankle motion and rehabilitation robots. In *2009 IEEE International Conference on Robotics and Biomimetics, ROBIO 2009* (pp. 2533–2538). IEEE. https://doi.org/10.1109/ROBIO.2009.5420487

Sun, Z., Bebis, G., & Miller, R. (2006). On-Road Vehicle Detection: A Review. *IEEE Transactions on Pattern Analysis and Machine Intelligence*, *28*(5), 694–711. doi:10.1109/TPAMI.2006.104 PMID:16640257

Susko, T. G. (2015). *MIT Skywalker: A novel robot for gait rehabilitation of stroke and cerebral palsy patients.*

Susko, T., Swaminathan, K., & Krebs, H. I. (2016). MIT-Skywalker: A Novel Gait Neurorehabilitation Robot for Stroke and Cerebral Palsy. *IEEE Transactions on Neural Systems and Rehabilitation Engineering*, *24*(10), 1089–1099. doi:10.1109/TNSRE.2016.2533492 PMID:26929056

Swathi, M., & Suresh, K. V. (2017). *Automatic Traffic Sign Detection and Recognition: A Review. IEEE Intern. Conf. on Algorithms, Methodology, Models and Applications in Emerging Technologies.* doi:10.1109/ICAMMAET.2017.8186650

Swinnen, E., Baeyens, J. P., Hens, G., & Knaepen, K., Beckwee, D., Michielsen, M., … Kerckhofs, E. (2015). Body weight support during robot-assisted walking: Influence on the trunk and pelvis kinematics. *NeuroRehabilitation.* doi:10.3233/NRE-141195

Swinnen, E., Baeyens, J. P., Knaepen, K., Michielsen, M., & Clijsen, R., Beckwee, D., & Kerckhofs, E. (2015). Robot-assisted walking with the Lokomat: The influence of different levels of guidance force on thorax and pelvis kinematics. *Clinical Biomechanics (Bristol, Avon)*, *30*(3), 254–259. doi:10.1016/j.clinbiomech.2015.01.006 PMID:25662678

Swinnen, E., Baeyens, J.-P., Knaepen, K., Michielsen, M., Hens, G., Clijsen, R., ... Kerckhofs, E. (2015). Walking with robot assistance: The influence of body weight support on the trunk and pelvis kinematics. *Disability and Rehabilitation. Assistive Technology*, *10*(3), 252–257. doi:10.3109/17483107.2014.888487 PMID:24512196

Szczecinski, N. S., Chrzanowski, D. M., Cofer, D. W., Terrasi, A. S., Moore, D. R., Martin, J. P., . . . Quinn, R. D. (2015). Introducing mantisbot: hexapod robot controlled by a highfidelity, real-time neural simulation. In *Intelligent Robots and Systems (IROS), 2015 IEEE/RSJ International Conference on 2015 Sep 28* (pp. 3875-3881). IEEE. 10.1109/IROS.2015.7353922

Szeles, J., Kubota, N., & Woo, J. (2017). Weather Forecast Support System Implemented into Robot Partner for Supporting Elderly People Using Fuzzy Logic. *Joint 17th World Congress of International Fuzzy Systems Association and 9th International Conference on Soft Computing and Intelligent Systems (IFSA-SCIS 2017), Otsu, Japan.* 10.1109/IFSA-SCIS.2017.8023309

Takahashi, Y., Manabe, G., Takahashi, K., & Hatakeyama, T. (2002). Experiments on Burden Measurement of Human and Transfer Aid System for Developing Self-Transfer Aid System. *The transactions of the Institute of Electrical Engineers of Japan. C, 68*(674), pp. 2977-2983 (in Japanese).

Takahashi, Y., Goto, H., & Shigemoto, H. (2006). Facial expressions of simplified robot face for health care system. *Transactions of the Institute of Measurement and Control*, *28*(2), 163–176. doi:10.1191/0142331206tm144oa

Takahashi, Y., Hasegawa, H., Takahashi, K., & Hatakeyama, T. (2001). Human Interface Using PC Display with Head pointing Device for Eating Assist Robot and Emotional Evaluation by GSR Sensor. In *Proceedings of International Conference on Robotics and Automation*, pp. 3674-3679. 10.1109/ROBOT.2001.933189

Takahashi, Y., Hasegawa, H., Takahashi, K., & Hatakeyama, T. (2002). Experimental Evaluations of Human Interface Using Head Space Pointer for Eating Assist Robot and Human's Emotional Response to Robot Movement, *Journal of Japan Society for Design Engineering*, *37*(3), 122–128 (in Japanese).

Tan, Y.G., Li, Z., Chen, Y.H., & Wang, H. (n.d.). *Bionic mechanism and kinetic characteristic for quadruped robot dog*. Academic Press.

Tanaka, M., & Tanaka, K. (2015, April). Control of a snake robot for ascending and descending steps. *IEEE Transactions on Robotics*, *31*(2), 511–520. doi:10.1109/TRO.2015.2400655

Tanaka, R., Woo, J., & Kubota, N. (2019). Nonverbal Communication based on Instructed Learning for Socially Embedded Robot Partners. [JACIII]. *Journal of Advanced Computational Intelligence and Intelligent Informatics*, *23*(3), 584-591. doi:10.20965/jaciii.2019.p0584

Tang, D., Yusuf, B., Botzheim, J., Kubota, N., & Chan, C. S. (2015). A novel multimodal communication framework using robot partner for aging population. *Expert Systems with Applications*, *42*(9), 4540–4555. doi:10.1016/j.eswa.2015.01.016

Taskin, Y. (2014). Improving pitch and yaw motion control of twin rotor MIMO system. *Journal of Vibroengineering*, *16*(4), 1650–1660.

Tastemirov, A., Lecchini-Visintini, A., & Morales-Viviescas, R. M. (2017). Complete dynamic model of the Twin Rotor MIMO System (TRMS) with experimental validation. *Control Engineering Practice*, *66*(July), 89–98. doi:10.1016/j.conengprac.2017.06.009

Tenenbaum, R. A. (2013). *Fundamentals of applied dynamics. Choice Reviews Online* (Vol. 41). USA: Springer; doi:10.5860/choice.41-6560

Tenreiro, J. A., & Silva, M. F. (2006). An Overview of Legged Robots. *Proceedings of the International Symposium on Mathematical Methods in Engineering*, 27-29.

Ting, K. L. H., Voilmy, D., Iglesias, A., Pulido, J. C., García, J., Romero-Garcés, A., . . . Dueñas, Á. (2017, August). Integrating the users in the design of a robot for making Comprehensive Geriatric Assessments (CGA) to elderly people in care centers. In *2017 26th IEEE International Symposium on Robot and Human Interactive Communication (RO-MAN)* (pp. 483-488). IEEE.

Tobalske, B., & Dial, K. (1996). Flight kinematics of black-billed magpies and pigeons over a wide range of speeds. *The Journal of Experimental Biology*, *199*(2). PMID:9317775

Tomiyama, T., Amelio, V. D., Urbanic, J., & ElMaraghy, W. (2007). Complexity of Multi-Disciplinary Design. *Annals of the CIRP*, *56*(1), 185–188. doi:10.1016/j.cirp.2007.05.044

Trieu Minh, Vu., Hashim, M., & Awang, F. (2012). Development of a real-time clutch transition strategy for a parallel hybrid electric vehicle. *Journal of Systems and Control Engineering*, *226*(2), 188–203.

Tse, D., & Viswanath, P. (2005). *Fundamentals of Wireless Communication.* Cambridge, UK: Cambridge University Press. doi:10.1017/CBO9780511807213

Tseng, F. H., Liang, T. T., Lee, C. H., Chou, L. D., & Chao, H. C. (2014, August). *A Star Search Algorithm for Civil UAV Path Planning with 3G Communication.* Paper presented at the 2014 Tenth International Conference on Intelligent Information Hiding and Multimedia Signal Processing, Kitakyushu, Japan. 10.1109/IIH-MSP.2014.236

Tsitsimpelis, I., Taylor, C. J., Lennox, B., & Joyce, M. J. (2019). A review of ground-based robotic systems for the characterization of nuclear environments. *Progress in Nuclear Energy*, *111*, 109–124. doi:10.1016/j.pnucene.2018.10.023

Tully, E. A., Fotoohabadi, M. R., & Galea, M. P. (2005). Sagittal spine and lower limb movement during sit-to-stand in healthy young subjects. *Gait & Posture*, *22*(4), 338–345. doi:10.1016/j.gaitpost.2004.11.007 PMID:16274916

Tummala, R. L., Mukherjee, R., Xi, N., Aslam, D., Dulimarta, H., Xiao, J., ... Dang, G. (2002, December). Climbing the walls. *IEEE Robotics & Automation Magazine*, *9*(4), 10–19. doi:10.1109/MRA.2002.1160067

Uicker, J. J., Pennock, G. R., Shigley, J. E., & McCarthy, J. M. (2003). Theory of Machines and Mechanisms. *Journal of Mechanical Design.* Oxford Press. https://doi.org/ doi:10.1115/1.1605769

Ullah, I., Yang, F., Khan, R., Liu, L., Yang, H., Gao, B., & Sun, K. (2017). Predictive mainyenance of power substation equipment by infrared thermography using a machine-learning approach. *Energies*, *10*(12), 1–13.

Ullah, K., & Kim, J.-H. (2009). A Mathematical Model for Mapping EMG Signal to Joint Torque for the Human Elbow Joint using Nonlinear Regression, *Proceedings of the 4th International Conference on Autonomus Robots and Agents*, pp. 103-108, Wellington, New Zealand.

Umehara, H., & Sekido, T. (1976). Measurement of Human Emotion by GSR. *The SICE Journal of Control, Measurement, and System Integration*, *15*(9), 753–757 (in Japanese).

Umetsu, K., Woo, J., & Kubota, N. (2019). Audience Attention in Robot Interactive Theater. In *Proceedings of the 2019 JSME Conference on Robotics and Mechatronics (ROBOMECH2019)*, Hiroshima, Japan.

Usamentiaga, R., Venegas, P., Guerediaga, J., Vega, L., Molleda, J., & Bulnes, F. G. (2014). Infrared thermography for temperature measurement and non-destructive testing. *Sensors (Basel)*, *14*(7), 12305–12348.

Utkin, V., Guldner, J., & Shi, J. (2009). Sliding Mode Control in Electro-Mechanical Systems, Second Edition. In Automation and Control Engineering: Vol. 31. doi:10.1201/9781420065619

Van Nieuwstadt, M. J. (1996). *Trajectory Generation for Nonlinear Control Systems*. Pasadena, CA: California Institute of Technology.

Van Nieuwstadt, M. J., & Murray, R. M. (1997). Real Time Trajectory Generation for Differentially Flat Systems. *International Journal of Robust and Nonlinear Control*, 28.

Vazquez-Santacruz, E., Morales-Cruz, C., Portillo-Velez, R., & Gamboa-Zuniga, M. (2015). Mechatronic Approach to an Intelligent Machine: The Case of an Assistive Device. *Research in Computing Science*, *100*, 115–127.

Veneman, J. F., Kruidhof, R., Hekman, E. E. G., Ekkelenkamp, R., Van Asseldonk, E. H. F., & Van Der Kooij, H. (2007). Design and evaluation of the LOPES exoskeleton robot for interactive gait rehabilitation. *IEEE Transactions on Neural Systems and Rehabilitation Engineering*, *15*(1), 379–386. doi:10.1109/TNSRE.2007.903919 PMID:17894270

Veneman, J. F., Menger, J., van Asseldonk, E. H. F., van der Helm, F. C. T., & van der Kooij, H. (2008). Fixating the pelvis in the horizontal plane affects gait characteristics. *Gait & Posture*, *28*(1), 157–163. doi:10.1016/j.gaitpost.2007.11.008 PMID:18207406

Venkatesh, V., Morris, M. G., Davis, G. B., & Davis, F. D. (2003). User acceptance of information technology: Toward a unified view. *Management Information Systems Quarterly*, *27*(3), 425–478. doi:10.2307/30036540

Verboom, J. L., Tijmons, S., De Wagter, C., Remes, B., Babuska, R., & de Croon, G. C. H. E. (2015). Attitude and altitude estimation and control on board a Flapping Wing Micro Air Vehicle. In *2015 IEEE International Conference on Robotics and Automation (ICRA)* (pp. 5846–5851). IEEE. 10.1109/ICRA.2015.7140017

Vercelli, A., Rainero, I., Ciferri, L., Boido, M., & Pirri, F. (2018). Robots in Elderly Care. *DigitCult-Scientific Journal on Digital Cultures*, *2*(2), 37–50.

Versaci, M., Calcagno, S., Cacciola, M., Morabito, F. C., Palamara, I., & Pellicanò, D. (2015a). Fuzzy Geometrical Techniques for Characterizing Defects in Ultrasonic Non-destructive Evaluation. In Ultrasonic Nondestructive Evaluation Systems (pp. 259–269). doi:10.1007/978-3-319-10566-6_10

Versaci, M., Calcagno, S., Cacciola, M., Morabito, F. C., Palamara, I., & Pellicanò, D. (2015b). Innovative Fuzzy Techniques for Characterizing Defects in Ultrasonic Nondestructive Evaluation. In Ultrasonic Nondestructive Evaluation Systems (pp. 201–232). doi:10.1007/978-3-319-10566-6_7

Versaci, M., Calcagno, S., Cacciola, M., Morabito, F. C., Palamara, I., & Pellicanò, D. (2015c). Standard Soft Computing Techniques for Characterization of Defects in Nondestructive Evaluation. In Ultrasonic Nondestructive Evaluation Systems (pp. 175–199). doi:10.1007/978-3-319-10566-6_6

Versaci, M., Morabito, F. C., & Angiulli, G. (2017). Adaptive Image Contrast Enhancement by Computing Distances into a 4-Dimensional Fuzzy Unit Hypercube. *IEEE Access: Practical Innovations, Open Solutions*, *5*(1), 26922–26931.

Versaci, M., Angiulli, G., Fattorusso, L., & Jannelli, A. (2019). On the Uniqueness of the Solution for a Semi-Linear Elliptic Boundary Value Problem of the Membrane MEMS Device for Reconstructing the Membrane Profile in Absence of Ghost Solutions. *International Journal of Non-linear Mechanics*, *109*, 24–31. doi:10.1016/j.ijnonlinmec.2018.10.014

Versaci, M., La Foresta, F., Morabito, F. C., & Angiulli, G. (2018). A fuzzy divergence approach for solving electrostatic identification problems for NDT applications. *International Journal of Applied Electromagnetics and Mechanics*, *57*(2), 133–146. doi:10.3233/JAE-170043

Veslin Diaz, E., Slama, J., Dutra, M., Lengerke, O., & Morales Tavera, M. (2011). Trajectory tracking for robot manipulators using differential flatness. *Ingenieria e Investigacion*, *31*(2), 84–90.

Vespignani, M., Melo, K., Mutlu, M., & Ijspeert, A. J. (2015). Compliant snake robot locomotion on horizontal pipes. In *Safety, Security, and Rescue Robotics (SSRR), 2015 IEE International Symposium on 2015 Oct 18* (pp. 1-8). IEEE. 10.1109/SSRR.2015.7442941

Viola, P., & Jones, M. J. (2004). Robust real-time face detection. International Journal of Computer Vision, 57(2), 137-154.

Viola, P., & Jones, M. (2001). Robust Real-time Object Detection. *International Journal of Computer Vision*.

Voigt, D., Karguth, A., & Gorb, S. (2012). Shoe soles for the gripping robot: Searching for polymer based materials maximizing friction. *Robotics and Autonomous Systems*, *60*(8), 1046–1055. doi:10.1016/j.robot.2012.05.012

Volavka, J., Matousek, M., & Roubicek, J. (1967). Mental arithmetic and eye opening. An EEG frequency analysis and GSR study. *Electroencephalography and Clinical Neurophysiology*, *22*(2), 174–176. doi:10.1016/0013-4694(67)90158-7 PMID:4163688

Vollmer, M. (2010). *Mollmann. K.-P.* Infrared Thermal Imaging. WILEY-VCH Verlag GmbH & Co.

Voltmer, D. (2007). *Foundament of Electromagnetics*. Synthesis Lectures on Computational Electromagnetics.

Wang, W., & Soper, S. A. (Eds.). (2006). Bio-MEMS Technologies and Applications. Boca Raton, FL: CRC Press.

Wang, L. (2015). Collaborative robot monitoring and control for enhanced sustainability. *International Journal of Advanced Manufacturing Technology*, *81*(9), 1433–1445. doi:10.100700170-013-4864-6

Ward, K. T., & Reuben, D. B. (2016). Comprehensive geriatric assessment. *UpToDate, Waltham, MA. Accessed, 4,* 03-19. Retrieved from http://www. uptodate. com/contents/comprehensive-geriatric-assessment

Warren, M. E. (2008). Micro-Opto-Mechanical Systems. doi:10.1036/1097-8542.757727

Watanabe, K., & Amano, A. (1994). A Study on Vital Reactions of School Children during the Class - Fluctuations of GSR, Heart Rate, and Subcutaneous Temperature -, *Japanese journal of school health, 36*(7), pp. 470-478 (in Japanese).

Waterman, B. R., Owens, B. D., Davey, S., Zacchilli, M. A., & Belmont, P. J. (2010). The epidemiology of ankle sprains in the United States. *Journal of Bone and Joint Surgery - Series A*, *92*(13), 2279–2284. doi:10.2106/JBJS.I.01537 PMID:20926721

Weissengruber, G. E., Egger, G. F., Hutchinson, J. R., Groenewald, H. B., Elsässer, L., Famini, D., & Forstenpointner, G. (2006). The structure of the cushions in the feet of African elephants (Loxodonta africana). *Journal of Anatomy*, *209*(6), 781–792. doi:10.1111/j.1469-7580.2006.00648.x PMID:17118065

Wen, X., Yuan, H., Song, C., Liu, W., & Zhao, H. (2007). An Algorithm based on SVM Ensembles for Motorcycle Recognition. *In 2007 IEEE International Conference on Vehicular Electronics and Safety* (pp. 1-5). IEEE.

Wisse, M., Hobbelen, D. G., Rotteveel, R. J., Anderson, S. O., & Zeglin, G. J. (2006). Ankle springs instead of arc-shaped feet for passive dynamic walkers. In *Humanoid Robots, 2006 6th IEEE-RAS International Conference on 2006 Dec 4* (pp. 110-116). IEEE. 10.1109/ICHR.2006.321371

Witczak, M., Buciakowski, M., Puig, V., Rotondo, D., & Nejjari, F. (2016). An LMI approach to robust fault estimation for a class of nonlinear systems. *International Journal of Robust and Nonlinear Control*, *26*(7), 1530–1548. doi:10.1002/rnc.3365

Woo, J., & Kubota, N. (2017). Interaction Content Design for Information Support based on Robot Partner, The 10th International Conference on Human System Interactions (HSI 2017), Ulsan, Republic of Korea, 35-40.

Woo, J., & Kubota, N. (2017). Interaction Content Design for Information Support based on Robot Partner, *The 10th International Conference on Human System Interactions (HSI 2017), Ulsan, Republic of Korea,* pp. 35-40. 10.1109/HSI.2017.8005019

Woo, J., Botzheim, J., & Kubota, N. (2014). Facial and Gestural Expression Generation for Robot Partners. In *25th 2014 International Symposium on Micro-NanoMechatronics and Human Science (MHS2014),* Nagoya, Japan, 218-223.

Woo, J., Botzheim, J., & Kubota, N. (2015). Verbal conversation system for a socially embedded robot partner using emotional model. In *Proceedings of the 24th IEEE International Symposium on Robot and Human Interactive Communication (RO-MAN2015),* Kobe, Japan, pp. 37-42. 10.1109/ROMAN.2015.7333685

Woo, J., János, B., & Kubota, N. (2017). A modular cognitive model of socially embedded robot partners for information support. *ROBOMECH Journal 4*(1), 10.

Woo, J., Shimazaki, J., Matsuo, Y., Masuta, H., Hun-ok, L., & Kubota, N. (2013). Communication based on Frankl's Psychology for Humanoid Robot Partners using Emotional Model. In *2013 IEEE International Conference on Fuzzy Systems (FUZZ-IEEE 2013),* Hyderabad, India.

Woo, J., Tanaka, R., & Kubota, N. (2017). *A Robot Partner Interaction System based on Persona with Modular Cognitive Model.* The 18th International Symposium on Advanced Intelligent Systems, Daegu, Republic of Korea, pp. 51-58.

Woo, J., Wada, K., & Kubota, N. (2012). Robot Partner System for elderly people care by using sensor network. In *4th IEEE RAS & EMBS International Conference on Biomedical Robotics and Biomechatronics (BioRob2012),* Rome, Italy, 1329-1334.

Woo, J., Wada, K., & Kubota, N. (2012). Robot Partner System for elderly people care by using sensor network. In *Proceedings of 4th IEEE RAS & EMBS International Conference on Biomedical Robotics and Biomechatronics (BioRob2012),* Rome, Italy, pp. 1329-1334. 10.1109/BioRob.2012.6290933

Woo, J., Yamamoto, K., & Kubota, N. (2018). Interaction Design Using Multi-Robot System for Information Support. *2018 International Conference on Fuzzy Theory and Its Applications (iFuzzy 2018),* Daegu, Korea.

Woo, J., Botzheim, J., & Kubota, N. (2017). Emotional empathy model for robot partners using recurrent spiking neural network model with Hebbian-LMS learning. *Malaysian Journal of Computer Science, 30*(4), 258-285. doi:10.22452/mjcs.vol30no4.1

Woo, J., Botzheim, J., & Kubota, N. (2018). A Socially Interactive Robot Partner Using Content-based Conversation System for Information Support. [JACIII]. *Journal of Advanced Computational Intelligence and Intelligent Informatics, 22*(6), 989-997. doi:10.20965/jaciii.2018.p0989

Woo, J., Botzheim, J., & Kubota, N. (2018). System Integration for Cognitive Model of a Robot Partner. [AUTOSOFT]. *Intelligent Automation & Soft Computing, 24*(4), 829-842.

Wu, B., & Nevatia, R. (2005, October). Detection of multiple, partially occluded humans in a single image by bayesian combination of edgelet part detectors. In *Tenth IEEE International Conference on Computer Vision (ICCV'05), 1*(1), pp. 90-97). IEEE.

Wu, J., & Yao, Y. (2018). Design and analysis of a novel walking vehicle based on leg mechanism with variable topologies, *Mechanism and Machine Theory, 128,* pp. 663-681, Elsevier. doi:. doi:10.1016/j.mechmachtheory.2018.07.008

Wu, Y., Hwang, M., Ren, Y., Gaebler-spira, D., & Zhang, L. (2011). Combined Passive Stretching and Active Movement Rehabilitation of Lower-Limb Impairments in Children With Cerebral Palsy Using a Portable Robot. https://doi.org/ doi:10.1177/1545968310388666

Yabe, T. (2012). Development Process of Musical Expression under Varying Environmental Configurations: Practical study of bubble play in four-year-old preschool class. *Journal of the Study of School Music Education Practice, 16,* 49–56 (in Japanese).

Yamada, S. (2017). A Study on Environmental Setting to Expression Activity (1): The Case of Two Soap Bubble Arts, Shukutoku University Junior College bulletin, 57, pp. 73-87 (in Japanese).

Yang, H., Yao, X.-F., Wang, S., Ke, Y.-C., & Liu, Y.-H. (2019). Simultaneous determination of gas leakage location and leakage rate based on local temperature gradient. *Measurement, 133*(1), 233–240.

Yang, J. H., & Hsu, W. C. (2009). Adaptive backstepping control for electrically driven unmanned helicopter. *Control Engineering Practice, 17*(8), 903–913. doi:10.1016/j.conengprac.2009.02.012

Yang, S., Shen, Y., Li, B., Li, Y., & Zhang, J. (2018). Modeling and Simulation of a Flapping-Wing Robot with Active Tails for Balancing Control during Wheeled Running. In *2018 IEEE International Conference on Mechatronics and Automation (ICMA)* (pp. 1806–1811). IEEE. 10.1109/ICMA.2018.8484330

Yin, W., & Yang, X. (2013). A Totally Astar-based Multi-path Algorithm for the Recognition of Reasonable Route Sets in Vehicle Navigation Systems. *Procedia: Social and Behavioral Sciences, 96,* 1069–1078. doi:10.1016/j.sbspro.2013.08.123

Yokota, T., Sato, A., & Fujimori, B. (1963). Analysis of inhibitory influence of bulbar reticular formation upon sudomotor activity. *Japanese Journal of Physiology, 15*(13), 145–154. doi:10.2170/jjphysiol.13.145 PMID:14002614

Yokota, T., Takahashi, T., Kondo, M., & Fujimori, B. (1959). Studies on the diphasic wave form of the galvanic skin reflex. *Electroencephalography and Clinical Neurophysiology, 11*(4), 687–696. doi:10.1016/0013-4694(59)90109-9 PMID:13846558

Youn, W., & Kim, J. (2010). Estimation of elbow flexion force during isometric muscle contraction from mechanomyography and electromyography. *Medical & Biological Engineering, 48*(11), 1149–1157. PMID:20524072

Yu, H., Gao, H., Ding, L., Li, M., Deng, Z., & Liu, G. (2016). Gait generation with smooth transition using CPG-based locomotion control for hexapod walking robot. IEEE Transactions on Industrial Electronics, 63(9), 5488-5500. doi:10.1109/TIE.2016.2569489

Yuan, B. S., & Book, W. J. (1993). Dynamics of Flexible Manipulator Arms: Alternative Derivation, Verification, and Characteristics for Control. ASME 115.

Yu, Z., Shi, Y., Xie, J., Yang, S. X., & Dai, Z. (2018). Design and analysis of a bionic adhesive foot for gecko robot climbing the ceiling. *International Journal of Robotics and Automation, 33*(4). doi:10.2316/Journal.206.2018.4.206-5412

Zadeh, L. A. (1965). Fuzzy Sets- [*Information and Control.*]. *Information and Control*, 1965.

Zhang, D., Dubey, V. N., Yu, W., & Low, K. H. (2018). Editorial: Biomechatronics: Harmonizing Mechatronic Systems With Human Beings. *Frontiers in Neuroscience.* doi:10.3389/fnins.2018.00768 PMID:30459543

Zhang, Q., Hosoda, R., & Venture, G. (2013). Human Joint Motion Estimation for Electromyography (EMG)-Based Dynamic Motion Control, *35th IEEE conference on EMBS*, Osaka, Japan.

Zhao, Y., Mehenen, J., Xu, W., Alrashed, M., Abineri, S., & Roy, R. (2015). Degradation assessment of industrial composite using thermography. *Procedia CIRP, 38*(1), 147-152.

Zhu, Q., Yeh, M. C., Cheng, K. T., & Avidan, S. (2006, June). Fast human detection using a cascade of histograms of oriented gradients. In *IEEE Computer Society Conference on Computer Vision and Pattern Recognition (CVPR'06), 2,* pp. 1491-1498. IEEE.

Zielinska, T. (2004). Development of walking machines; historical perspective. In *International Symposium on History of Machines and Mechanisms 2004* (pp. 357-370). Springer.

Zimny, G. H., & Weidenfeller, E. W. (1963). Effects of Music upon GSR and Heart-Rate. *The American Journal of Psychology, 76*(2), 311–314. doi:10.2307/1419170 PMID:14003728

Zofka, M. R., Kuhnt, F., Kohlhaas, R., & Zöllner, J. M. (2016, December). Simulation framework for the development of autonomous small scale vehicles. In 2016 IEEE International Conference on Simulation, Modeling, and Programming for Autonomous Robots (SIMPAR) (pp. 318-324). IEEE.

About the Contributors

Maki Habib obtained his Doctor of Engineering Sciences in Intelligent and Autonomous Robot from the University of Tsukuba, Japan. He was a senior research scientist with GMD, Japan, leading the Telecooperation group, Associate Professor with Monash University and lead of the Mechatronics Engineering Programme. He was appointed as a full Professor of Robotics and Mechatronics at Swinburne University. Then, he was an invited Professor at KAIST, Korea, Visiting Professor at Saga University, Japan, and now a full Professor at AUC. His main areas of research are focusing on human adaptive and friendly Mechatronics, autonomous navigation, service robots and humanitarian demining, telecooperation, Intelligent Control, distributed teleoperation and collaborative control, wireless sensor networks and ambient intelligence, deep Learning, biomemtic, bioinspiration and biomedical robots.

* * *

Hisham A. Abdel-Aal obtained his first degree in Mechanical Engineering from Alexandria University, Alexandria, Egypt. He received his Doctoral degree in Precision Engineering from the University of North Carolina in Charlotte in 1998. Since his graduation, he has been active in research and teaching in the US and abroad. He has occupied several invited positions in Europe, South America and Asia. Including an invited Research Professorship at the Laboratoire de Materiux et Procedes at Arts et Métiers ParisTech in France where he founded the first research group for Bio-inspired sustainable surfaces for manufacturing. Dr. Abdel-Aal has lectured extensively on the subjects of tribology, Bio-inspired surfaces, and thermodynamics of wear.

José A. Aguilera graduated in Mechatronics Engineering in 2017 from Universidad Autónoma de Ciudad Juarez, Mexico. He carries out R&D in the industrial sector. His academic interests are mechanisms design, mobile robotics, modelling, and simulation.

Antonio Bandera-Rubio received a Bachelor Degree in Telecommunications Eng. and a PhD. degree, both from the UMA, in 1995 and 2000, respectively. He belongs to the ISIS research group. His research in the fields of robotics and computer vision has resulted in more than 150 refereed publications in international journals and conferences. He has coordinated several national and international projects along with a number of contracts in the field of robotics.

César F. Bogado-Martínez received his Bachelor degree in Electronical Engineering from the Ingeneering Faculty of the Universidad Nacional de Asunción, Paraguay, in 2012; and his Master degree in Mechanical Engineering from the Universidade Federal do Rio de Janeiro, Brasil, in 2016. Currently, he is doing the doctoral course in Computer Science at the Politechnic Faculty in Universidad Nacional de Asunción.

Kai Borgeest received his PhD in 1998 from Technical University of Hamburg-Harburg. From 1998 to 2003, he worked in different functions from system engineer to project manager at Bosch Diesel Systems in Stuttgart. Since 2003, he has been a professor of automotive electronics and combustion engines at Aschaffenburg University of Applied Sciences, where he is involved in industrial and academic projects mainly about control units and engines. He also works in the research unit ZeWiS in Obernburg on projects about mobile robotics.

Mohd Hisam Daud is a Master student at Universiti Malaysia Perlis.

Alvaro Dueñas-Ruiz studied Physiotherapy at the University of Seville, Spain (2009-2013). Since 2014, he has been working for FISEVI (Andalusian Public Foundation for the Management of Health Research in Seville), which is located in the Virgen del Rocío University Hospital. He specialized in the clinical adaptation of new technologies in rehabilitation and robotics, and has participated in several European and national projects in this area: REWIRE, CLARK, Therapist, etc.

Max Dutra graduated as Mechanical Engineer from Universidade Federal Fluminense (1987), earned a Mechanical Engineering Master Degree from Universidade Federal do Rio de Janeiro (1990), and earned a doctoral degree in Mechanical Engineering from Gerhard Mercator Universitat Duisburg – Germany (1995). He is Titular Professor in the Mechanical Engineering Program of COPPE/UFRJ. Dutra had experience in the Mechanical Engineering area with an emphasis in Robotics. He is also working in mechatronics project, security instrumental systems, accidental analysis, risk analysis, robotics, instrumentation, industrial automation, machine design, biomechanics, non-linear dynamics, multi-body systems, and domotics.

Angelo Galiano is director of Dyrecta Lab srl.

Rogério S. Gonçalves was born in Barretos, Brazil, in 1981. He received the mechanical engineering degree from the Federal University of Uberlândia in 2004, Master degree in 2006 and Ph.D. at the same University in 2009. Goncalves is Professor at Federal University of Uberlândia. He is member of RoboCup Brazil and ABCM (Brazilian Society of Engineering and Mechanical Sciences). He authored or co-authored about 100 papers, which have been presented in national and international conferences or published in national and international journals. His research interest includes kinematics and dynamics of serial and parallel structure, stiffness, cable-driven parallel structure, mobile robots, biorobotics, and humanoid robots.

Yuksel Hacioglu received the B.S., M.S., and Ph.D. degrees from the Department of Mechanical Engineering, Faculty of Engineering, Istanbul University, Istanbul, Turkey, in 2002, 2004, and 2009, respectively. He is an Associate Professor with the Department of Mechanical Engineering, Faculty of Engineering, Istanbul University-Cerrahpasa. His research areas include modeling and control of robotic manipulators and vehicle suspension systems, nonlinear control theory, and fuzzy logic.

Rafiqul Haque is working as an Assistant Professor in the Department of Mechanical Engineering of Aliah University, Kolkata. He was awarded B.Tech degree from Jalpaiguri Government Engineering College and M.E. from Bengal Engineering and Science University (known as Indian Institute of Engineering Science and Technology), Shibpur. He is pursuing Ph.D degree in the domain of Advance Manufacturing. He has worked as an Assistant Professor in the Department of Mechanical Engineering at Birbhum Institute of Engineering and Technology, Suri. He also worked as an Assistant Engineer in the Dept. of Water Resources Investigation & Development, Govt. of West Bengal. He published two International and National Journal papers of repute and three research papers in reputed national and international conferences proceedings. His research interests are in Machine Design, Advance Manufacturing, Composite Materials, Materials Synthesis, Nano Materials and Foundry Technology. He is the council member of The Institute of Indian Foundrymen and life member of International Association of Engineers.

Naoya Hasegawa graduated from master course at Kanagawa Institute of Technology in 2000 and then worked at Robotics-related companies. He is a PhD student while running his venture company. He is interested in the researches on amusement robots for which human emotions are considered.

Ana Iglesias received the M.Sc. in Computer Science from Carlos III University of Madrid in 1999, and her Ph.D. in Computer Science from the same university in 2004. Since 1999, she has worked at the Advanced Databases Group in the Computer Science Department at Carlos III University of Madrid. Her research interests include Accessibility, Adaptive Intelligent Educational Systems, Natural Language Processing, Information Retrieval, Database Design and Advanced Database Technologies, among others. Home page: http://www.inf.uc3m.es/profile/aiglesia.

Md. Kamaruzzaman is working as a faculty member in the Department of Mechanical Engineering at Aliah University, Kolkata. He was awarded with a B.Tech degree from B. I. E. T., Suri, and M.Tech. from N. I. T. T. T. R., Kolkata. He is pursuing a Ph.D degree from Jadavpur University, Kolkata. He has published two International and National Journal papers of repute and about three research papers in reputed national and international conferences proceedings. His research interest is in Low power CMOS technology and Wireless sensor Network, intelligent Robotics, Mechatronics. He is the life member of International Association of Engineers.

Daniel Kern earned a bachelor's degree from Aschaffenburg University of Applied Science in March 2015 and a master's degree at the same university in September 2016. Since October 2016, he has worked at Opel Automobile GmbH.

Pavel Kovalenko received his Specialist Diploma in Mechatronics and PhD in System Analysis, Control, and Information Processing from ITMO University, Saint Petersburg, Russia, in 2008 and 2011, respectively. Since 2012, he is associate professor in the Faculty of Control Systems and Robotics. His research interests are robotics, mechatronics, automation systems, hydraulic walking robots, gate modelling, and biomechatronics.

Naoyuki Kubota graduated from Osaka Kyoiku University in 1992, received the M.E. degree from Hokkaido University in 1994, and the D.E. degree from Nagoya University, Japan, in 1997. He joined Osaka Institute of Technology in 1997, and the Department of Human and Artificial Intelligence Systems, Fukui University, as an associate professor in 2000. Kubota joined the Department of Mechanical Engineering, Tokyo Metropolitan University, in 2004. He was an associate professor from 2005 to 2012, and has been a professor since 2012 at the Department of System Design in Tokyo Metropolitan University, Japan.

Mohamed Mydin M. Abdul Kader is a Lecturer at Universiti Malaysia Perlis.

Rebeca Marfil received her title of Telecommunication Engineering and Ph.D. degree from the University of Malaga, Spain, in 2001 and 2006, respectively. Her research is focused on artificial vision, image processing, and social robotics.

Edgar Alonso Martínez García is an associate professor at the Institute of Engineering and Technology, Universidad Autónoma de Ciudad Juárez, Mexico. He has been the founder and head of the Laboratory of Robotics since 2007. Dr. Martínez received his PhD Eng degree from the University of Tsukuba, Japan, in 2005. From mid-2005 to 2007, he worked as a Research Fellow at the School of Electrical and Electronic Engineering, Nanyang Technological University, Singapore. Dr. Martínez leads the academic body of Mechatronics in the same institution. He poses numerous scientific publications and patents in the robotics field. His research interests are in modeling and applied control of mobile robots, planning, computational mechanics, design and control of mechanisms, scientific computing in robotics, and analysis of non-linear sensing and actuating systems.

Alessandro Massaro received the Laurea degree in electronic engineering and the Ph.D. degree in telecommunication engineering from the Università Politecnica delle Marche, Ancona, Italy, in 2001 and 2004, respectively. From 2004 to 2006, he worked as Research Scientist (post-doc) in the Department of Electromagnetism and Bioengineering at Università Politecnica delle Marche. In 2006, he spent one year in Research and Development at medical and industrial optics industry (endoscope design and optical systems). He worked for two years with National Nanotechnology Laboratory of CNR-INFM, Università del Salento, as principal investigator. He was team leader in Robotics Lab. and in Smart Materials platforms of the Center for Bio-Molecular Nanotechnology of Italian Institute Technology (IIT), Arnesano, Lecce, Italy. His research interests are in the design and modeling of photonic band gap circuits, in the development of computer aided design (CAD) tools in the area of integrated optics, MEMS technology and systems, and smart material implementation. He worked with the National Council of Research (CNR) – Istituto di Metodologie Inorganiche e dei Plasmi (IMIP now Nanotech) of Bari - Italy, in diamond and nanodiamond technology for aerospace applications. Actually, he is responsible for the Research and Development laboratory of DyrectaLab s.r.l.

Vu Trieu Minh received the B.Eng in Mechanical Engineering from Hanoi University of Technology, Vietnam, in 1983; M.E and Ph.D in Mechatronics from Asian Institute of Technology (AIT), Thailand, in 1998 and 2004, respectively. Since 2012, he has worked as a professor of mechatronics in Tallinn University of Technology. His research interests are model predictive control, advanced control techniques, automotive, microcontrollers, and artificial intelligence.

Iurii Monakhov is an associate professor in Department of Mechatronics, Faculty of Control Systems and Robotics. He earned a PhD in methods and means of measurements in 2013 from ITMO University, Russia. His research interests are robotics, mechatronics and automation systems, biomechatronics, and rehabilitation system.

Francesco Carlo Morabito is a Full Professor of Electrical Engineering with the Mediterranea University of Reggio Calabria, Italy, and the Former Dean of the Faculty of Engineering. He is also serving as the Vice-Rector for International and Institutional Relations. He has authored or co-authored over 150 papers in international journals/conference proceedings in various fields on engineering (radar data processing, nuclear fusion, biomedical signal processing, nondestructive testing and evaluation, and computational intelligence). He is co-author of 10 books and has held three international patents. He is a Foreign Member of the Royal Academy of Doctors, Spain. He served as the Governor of the International Neural Society for 12 years and as the President of the Italian Network Society from 2008 to 2014. He is a member of the Editorial Boards of international journals, including the International Journal of Information Acquisition, Neural Networks, Renewable Energy, and International Journal of Neural Systems.

Stefan Müller-Schneiders received the Dipl.-Ing. and the Dr.-Ing. degrees from Gerhard-Mercator University, Duisburg, Germany, in 1996 and 2001, respectively. From 1996 to 2002, he was a Research Assistant with the Fraunhofer Institute and the University of Duisburg. From 2002 to 2006, he was a Researcher with the Surround Sensing Competence Center of Robert Bosch. From 2006 to 2014, he has been a Research Engineer and Project Manager with Delphi Electronics & Safety, Wuppertal, Germany. Since 2017, he is Professor for Computer Science at the Bochum University of Applied Sciences. He is the author or coauthor of about 40 papers in the field of pattern recognition and image processing.

Victor Musalimov received his PhD degree in 1972 and Dsc degree in 1986 from Tomsk Polytechnic University, Russia. From 2012 to 2015, he was a head of the Department of Mechatronics, ITMO University. His main fields of study are mechanics of deformable solids, biomechatronics, control systems, tribology, image processing and pattern recognition, human gait modelling, and modelling of nonlinear dynamics of mechatronics.

Ana Belén Naranjo-Saucedo graduated from the University of Seville with a Master in Psychology of Organizations and Work. Throughout her career in investigation, she has worked in projects of national and international character. They were based on the creation and validation of prototypes specialized in social robotics applied to therapeutic treatments in patients and the elderly, as well as in the application of telemedicine (mHealth) for the cognitive treatment and of adult patients with acquired brain damage and neurodegenerative diseases. Likewise, she has participated in some projects whose lines of research have focused on the application of virtual reality for surgical planning of various clinical specialties, and even in the development and validation of tools to support the clinical decision at the foot of the bed.

Erika Ottaviano received the Master Degree summa cum laude in 1998 and Ph.D. in Industrial Engineering on Feb. 2002 at the University of Cassino. Since 2002, she has worked as Assistant Prof. for Mechanics of Machines and Mechanical Systems. Since 2013, she has served as a member of the Editorial Board for Mechanical Engineering of the Scientific World Journals, Hindawi Publishing Corp. Since 2012, she is Editor for the International Journals of Imaging and Robotics. Since 2011, she is a member of the Editorial Board of the International Journal of Advanced Robotic Systems, INTECH Open Access Pub. Since 2004, she is member of the IFToMM Technical Committee for Computational Kinematics, and has served as secretary since 2009. Her research interests are Mechanics, Robotics, and Mechanical Design. She is author and co-author of more than 150 papers published in Books, Proc. of Nat. and International Conferences, or in Journals.

Luciano Raptopoulos graduated from the Federal University of Rio de Janeiro in 1997. He earned a Master degree in Mechanical Engineering from the Federal University of Rio de Janeiro in 1999 and a Ph.D. in Mechanical Engineering from the same university in 2003. He works as professor level Associate III of the Federal Center of Technological Education Celso Suckow da Fonseca, teaching the courses of Control and Automation Engineering, and Mechanical Engineering, of the Campus of Nova Iguaçu. His areas of interest are machine design, robotics, dynamic systems, biomechanics, and fault analysis.

Zol Bahri Razali is Associate Professor at University Malaysia Perlis, Malaysia. He has over 28 years of experience in higher education as a lecturer, researcher, and administrator, especially in Mechanical Engineering and Engineering Education. He holds BE in Mechanical/Aeronautical Engineering and a Ph.D. in Higher Education from The University of Western Australia. His research focus is on understanding students—laboratory interaction issues as it relates to measuring students' Practical Intelligence after performing 'hands-on' laboratory exercises in Electrical Engineering Technology Laboratory. His other research interests are mainly in designing Robotics Systems and Automated Sorting Devices for green technology. His research findings have been published in a variety of Engineering Education conferences, journals, book chapters, and books. He serves as Head of Department for Robotics and Automation Technology Program at Faculty of Engineering Technology, University Malaysia Perlis.

Pierluigi Rea earned the Master Degree in Mechanical Engineering at the Faculty of Engineering of the University of Cassino and Southern Lazio in 2002, and obtained the Ph.D. Degree in Civil and Mechanical Engineering there in 2007. In 2014, he received the Italian National Scientific Qualification to become an Associate Professor in the field of Mechanics of Machines. He authored or co-authored more than 80 scientific papers which appeared on Proceedings of National and International Conferences or National and International Journals. From 2007 to 2010, he was Assistant Researcher at the Unicas for the development of research activity regarding the Design and Experimentation of electro-pneumatic servo-systems for the actuation of robotic devices. Since 2002, he was Mechanical Engineer Consultant for several public and private research projects, such as those with FCA Group, Rockwell Automation, Thyssen Krupp, KMI, WAIRCOM-MBS, INAIL, and SKF. From 2016 to 2017. he was Mechatronic Engineer Consultant for several activities on solving problems for Industrial Automation. From 2010 to 2015, he was Assistant Professor at the University of Cassino for the Mechanical design disciplines. I was Assistant Researcher for the courses of Mechanics of Machines, Design of Mechanisms, Fluid Automation, and Fluid Servo-systems. In 2015, he was Visiting Professor at the University "UCLM" in Toledo, Spain, for the course of Mechatronics and Mechanical Design. He is proposal Engineer at Icap-Group, working directly with customers to understand their processes, recognize and understand their needs, and quickly develop the best solutions for their requirements. Typical applications he develops concern robotic assembly, material handling, machine tending, palletizing, packaging, and machine vision, including vision inspection, measurement, and robot guidance. My main activity interests regard topics on Automatic Machinery and Robotics Design/Simulation, Mechatronics, and Fluid Automation.

Lucas Antônio O. Rodrigues received the B.S. degree in mechatronics engineering in 2014 and the M.S. degree in mechanical engineering in 2017, both from Federal University of Uberlândia, Minas Gerais, Brazil. He is pursuing the Ph.D. degree in mechanical engineering at Federal University of Uberlândia, Minas Gerais, Brazil. From 2010 to 2014, he was a Research Assistant with the Robotics and Automation laboratory at the mechanical engineering department of Federal University of Uberlândia. His research interests include the development of Parallel rehabilitation robotic structures, nonmotorized rehabilitation systems, active bodyweight supports, and treadmill gait rehabilitation devices. Rodrigues' awards and honors include the Best Research Assistant Paper Prize of the 2011 National Conference of Biomechanics Engineering (ENEBI 2011), for presenting a novel cable-driven rehabilitation structure for human upper limb, its computational simulations, and experimental results.

Aleksei Safonov received his Master degree in Mechatronics in 2017 from the dual mechatronics master program of Tallinn University of Technology and ITMO University. He was working on the research topic of smart wearable elbow orthosis. His research interests include bio-mechatronics, human gait modeling and simulation, microelectronics, and control systems.

Sertaç Savaş is Research Assistant at Department of Mechatronics Engineering, Faculty of Engineering, Erciyes University. He is studying his Ph.D education at the same university on the topics of mobile robots, system dynamics and control. He has many articles and conference proceedings about mobile robots.

Cristina Suárez Mejías is a Telecommunication Engineer, PhD in Telecomunication Engineer, PhD in Molecular Biology and Biomedicine, and MSC in Electronic and Signal and Communications Theory. She is researching responsible treatment of medical images. Concretely, she participated in 38 R&D national and international projects. She is coordinator of European project CLARC from ECHORD, which funded the European Commission. Also, she is IP of SocketSense (H2020) and HERO (Human Evaluation and Rehabilitation after Oncology, Fundación Inocente Inocente) projects. She has six patents and two property registers, 51 publications in journals of impact and more than 65 international and national congresses, and six research awards. She is member of DICOM (Digital Imaging and Communication in Medicine) Standards Committee, Member of IBIS, the Committee for Standardization NEMA DICOM image and voice in the Technical Standards Committee AEN / CTN 139 "Information technologies and communications to health". Also, she is Honorary Lecturer at University of Seville and lecturer of the Biomedical Image subject of Health Engineering Grade at University of Seville. She is also founder of Digitalica Health enterprise.

Yoshihiko Takahashi graduated from master and doctor courses at Tokyo University of Agriculture and Technology in Japan in 1980 and 1995. He joined Toshiba Corporation in 1985 as a researcher of Mechatronics. Since 1996, he has been a faculty member of Kanagawa Institute of Technology in Japan. He is a professor of Department of Robotics and Mechatronics and is interested in human assist robots, agricultural engineering, micro vehicles, etc.

Mart Tamre received Dipl.Eng in Fine Mechanics from Tallinn University of Technology, and a Ph.D from the same university in 1995. He is professor of mechatronics and head of mechatronics, Department of Electrical Power Engineering and Mechatronics, Tallinn University of Technology. His research interests include design of mechatronics and robotics systems, smart production and service systems, and machine vision.

Yener Taskin received BS and MS degrees in 1996 and 1999 from the Department of Mechanical Engineering of Istanbul University, and the Ph.D. degree in 2008 from the Department of Mechanical Engineering of Yildiz Technical University, Istanbul. He is an Associate Professor in the Department of Mechanical Engineering of Istanbul University-Cerrahpasa. His research interests include modeling and control of mechatronic systems and mechanical vibrations.

Mario Versaci received a degree in civil engineering in 1994 from the "Mediterranea" University, Italy, where he also received a Ph.D. in electronic engineering in 1999. In the same university, he serves as both Associate Professor of electrical engineering and scientific head of the NDT/NDE Lab. In 2013, he received a degree in mathematics from the University of Messina and in 2018, the higher degree in mathematics from the same University. His research focuses on 1) soft computing techniques for NDT/NDE; 2) image processing; 3) MEMS for mechatronics; 4) Magnetorheological Fluids; and 5) Microstrip antennas. He is a Member of the Italian Society for Industrial and Applied Mathematics and has been a Senior Member of IEEE since 2017. Mario Versaci is also a member of Editorial Board of the International Journal of Computer Intelligence and Neuroscience (Hindawi Coorporation), and co-author of more than 150 papers in international journals/conference proceedings.

Elkin Veslin Díaz is a Mechatronics Engineer with an MSC in Mechanical Engineering and DSC in Mechanical Engineering. He is a Researcher doing Post-doctoral work at the Federal University of Rio de Janeiro COPPE/UFRJ. Diaz works at the Robotics Laboratory in the areas of biomechatronics, robotics, and assistive devices, as well as integrating biological signal processing and non-linear control.

Stephen Khor Wen Hwooi is a Bachelor of Mechatronics Engineering student at Universiti Malaysia Perlis.

Rajen Wiemers holds a Bachelor's degree in Computer Science (2018, Bochum University of Applied Sciences). He is pursuing his M.Sc in Computer Science while working as a Research Assistant.

Jinseok Woo graduated from the Kumoh National Insitute of Technology, Republic of Korea, in 2009, and received a Master of Engineering degree from Tokyo Metropolitan University, Japan, in 2011. He received his Ph.D. degree from Tokyo Metropolitan University, Japan, in 2017. He joined the Tokyo Metropolitan University, Tokyo, Japan, in 2017. He joined the Tokyo University of Technology, Tokyo, Japan, in 2019. His interests are in the fields of mechatronics, computational intelligence, and cognitive robotics for human-friendly robot partner.

Şahin Yıldırım is Professor at Department of Mechatronics Engineering, Faculty of Engineering, Erciyes University. He received his Ph.D. degree from the Systems Engineering of University of Wales, Cardiff, England. He has implemented 11 research and development projects, and two projects are in progress. He has more than 70 articles in journals, science citation indexed. His scientific interests focus on artificial neural networks, robotics, system dynamics and control.

Index

H

I

K

L

M

N

O

P

Q

R

S

T

U

W

www.ingramcontent.com/pod-product-compliance
Lightning Source LLC
LaVergne TN
LVHW060256011026
849161LV00011B/2727
* 9 7 8 1 7 9 9 8 0 1 3 7 5 *